Methods in Comparative Plant Population Ecology

Methods in Comparative Plant Population Ecology

Second Edition

David J. Gibson
Department of Plant Biology
Center for Ecology
Southern Illinois University
Carbondale, Illinois

OXFORD
UNIVERSITY PRESS

Great Clarendon Street, Oxford, OX2 6DP,
United Kingdom

Oxford University Press is a department of the University of Oxford.
It furthers the University's objective of excellence in research, scholarship,
and education by publishing worldwide. Oxford is a registered trade mark of
Oxford University Press in the UK and in certain other countries

First Edition published in 2002
Second Edition published in 2015

Impression: 1

Published in the United States of America by Oxford University Press
198 Madison Avenue, New York, NY 10016, United States of America

British Library Cataloguing in Publication Data
Data available

Library of Congress Control Number: 2014939140

ISBN 978–0–19–967146–5 (hbk.)
ISBN 978–0–19–967147–2 (pbk.)

Printed and bound by
CPI Group (UK) Ltd, Croydon, CR0 4YY

Preface to the Second Edition

There have been substantial and significant advances in the methodology used by plant population ecologists in the 12 years that have elapsed since the first edition of this book was published. While some things such as the need for and approach to estimating population density through the use of sample plots have not changed, the advances in molecular and physiological methods and new and improved statistical techniques necessitated, I felt, the need for a second edition. Fortunately, Southern Illinois University Carbondale granted me a sabbatical leave in spring 2013 that, along with a private office in the Morris Library that I told no one about, allowed time for most of the work to be done. A grateful thanks to my colleagues and friends, students, staff, and faculty for not bothering me too much during that time.

The format and outline of the first edition has not changed, with four major sections: What is Plant Population Ecology? (Chapter 1), Planning the Study (Chapters 2 and 3), Doing the Study (Chapters 4, 5, and 6), and Analysis (Chapters 7 and 8). Libby John's 'Planning, choosing, and using statistics' (Chapter 4 in the first edition) has been revised and moved into the Analysis section to become Chapter 7. The other chapters have been renumbered accordingly. I moved this chapter to provide a clearer section on analysis rather than have statistical methods separated at different ends of the book. Chapter 8 is now prefaced as 'Advanced statistical techniques'. The text on Bayesian analysis has been updated and moved out of the old Chapter 2 into Chapter 7 where I feel it better belongs.

Other major changes include the replacement of Sarakhán's classic 1960s buttercup study as Case Study 1 with a recent study by Peiter Zuidema and colleagues on tree recruitment in tropical forests. This and the other three case studies are referred to repeatedly throughout the book. I hesitated to drop Sarakhán's magnificent work, especially since it was conducted under the mentorship of the late John Harper (someone who had a huge influence on my development as an ecologist), but I wanted to bring in a more recent and tropical-based study. I expect Zuidema et al.'s work to stand the test of time as well as Sarakhán's has. I also added a 'Roadmap for using this book' with a new accompanying figure at the end of Chapter 1 that I hope will help users navigate the text. I added a new section on 'Some things that might cause panic, but shouldn't' to Chapter 2. The suggestions here are drawn heavily from comments made by Glenn Matlack in his extraordinarily comprehensive review of the proposal for this second edition. There's also a new section and figure in Chapter 2 describing the 'hierarchy of hypotheses approach'. Other new sections include 'Generalized linear mixed models' (Chapter 7), and 'Life table response experiments', 'Integral projection models', and 'Population viability analysis' in Chapter 8. The section on 'Molecular methods' has been brought up to date with the inclusion of subsections on single nucleotide polymorphisms, gene expression profiling, epigenetics, and functional metagenomics. The scientific computing environment R has now become the statistical package of choice for many ecologists because of its flexibility and wide range of available packages, and because it's free. Suitable R packages are mentioned throughout the text and summarized in the Appendix. The inclusion of R script is beyond the scope of this book, although a short example is provided in the worked example in Box 8.1 in Chapter 8.

Throughout, citations and examples have been updated, keeping reference to older studies that are still relevant but adding the most recent examples of methodology where appropriate. Over the years I had kept a list of minor typographical errors, which along with lists provided by Marcel Rejmanik and Lauren Schwartz have been used to make corrections. Thanks to both of them for passing their lists on to me. There are 12 new figures, 7 new tables, and countless new citations in this second edition.

A number of colleagues have provided comments and feedback on drafts of the revised text, and my grateful thanks go out to Paul Barnes, Gregg Cheplick, Stephen Ebbs, Matt Geisler, Libby John, Mike Hutchings, and Eric Menges. I also thank colleagues including Rob Freckleton and Rob Salguero-Gómez who responded to my twitter request for suggestions of suitable R packages to include in the Appendix. It was Rob Freckleton who made the percipient observation that R packages are 'a bit of a moving feast'.

Thanks to the editorial and production staff at Oxford University Press for their help. Ian Sherman encouraged me to take on the second edition of a 'never again' book, and Helen Eaton and Lucy Nash were terrific, acting as my OUP editorial contacts and helping to streamline everything. Bridget Johnson and Suganiya Karumbayeeram did a terrific job in copy editing and production. I also have to thank the British Ecological Society, which over the last several years has provided me with the opportunity to read a lot of cutting edge population ecology papers as part of my duties as an editor for the *Journal of Ecology*. I got to read the good, the bad, and the ugly, but all helped shaped my appreciation and understanding of the methods used by plant population ecologists. Similarly, mentoring and teaching my graduate students, and the students who have taken my plant population ecology course, has helped me better appreciate the issues involved in designing and conducting good experiments.

Finally, a thanks to my wife Lisa, and Lacey and Dylan, the two young adults who are our children, for their support and encouragement. I promise not to write another book; at least until I start the next one.

Carbondale, Illinois, April 2014
@DavidJohnGibson

Preface to the First Edition

My purpose in writing this book is to fill a need that I perceive exists for beginning researchers in plant population ecology. The book is intended for senior undergraduates with a background in plant ecology, and, in particular, for graduate students planning research in the topic. In addition, I hope that more experienced researchers will find the material covered in the book to be of use too.

The need for this book became apparent the first time that I taught a graduate course in plant population ecology. While Silvertown and Lovett-Doust's (1993; with a new edition due out as I write) book provides an excellent introduction to the basic theory and concepts important to the topic, I felt that there was little available to assist the student actually conduct research in the laboratory/field portion of the course. A number of books discuss basic ecological methodology (e.g. Krebs 1999) and the approach to research in ecology (Ford 2000), but these are not specific enough. I use the inquiry approach in my course and so was also dissatisfied with the 'cook book' approach to learning new methods provided by some other textbooks. I want students to think for themselves, critically evaluate the problem that they are addressing, and determine the most suitable approach. Indeed, the variation and vagaries of the environment preclude the use of many standard methods in field ecology. Hence, this book.

If this book has a message it is this: think carefully about the problem that you are about to tackle. Develop a research plan, outline hypotheses, objectives and the experimental approach to address them. And, most importantly, consider the statistical tests that you will use before starting field or greenhouse work.

This book will provide you with an assessment of the methodology used in current plant population ecology. Many of the methods used by other researchers will have to be adapted to your own situation. I provide the sources for standard methods, but many of these too may have to be adapted. Always be aware of the limitations of the methods that you do use.

Time to get started on the two-year journey that writing this book represents was provided by a sabbatical leave from Southern Illinois University. Cathy Kennedy. Libby John, and Mike Hutchings helped immensely in getting the initial outline and direction for the book off the ground. A grateful thanks to Libby John who wrote Chapter 4, taking on a task that would daunt and put off most of us number-challenged ecologists. Thanks to the many colleagues who critically read the following chapters, in whole or part, including Eric Adams (Chapter 1), Sara Baer (Chapter 5), Mark Basinger (Chapter 3), Greg Cheplick (Chapters 1 and 3), John Connolly (Chapter 5), Stephen Ebbs (Chapter 6), Joe Ely (Chapters 1 and 6), Danny Gustafson (Chapters 1 and 5), Rod Hunt (Chapter 6), Libby John (Chapters 1, 2, 3, and 5), Alan Knapp (Chapter 6), Marilyn Mathis (Chapters 2 and 3), Glenn Matlack (Chapter 5), Eric Menges (Chapter 8), Paige Mettler (Chapters 1 and 2), Chris Peterson (Chapter 2), Malcolm Press (Chapter 6), Mark Rees (Chapters 1, 2, 3, 5, 7, and 8), Ken Thompson (Chapter 6), Jake Weiner (Chapter 1), and Andrew Wood (Chapter 6). The text is much better for the time and care that they put into correcting my mistakes and misconceptions. Tim Seastedt and Brian Beckage provided access to manuscripts in advance of publication, and Miguel Franco, Mike Hutchings, Brian McCarthy, Rodney Mauricio, and Dan Nickrent kindly provided some of the photographs. Thanks also to Cathy Kennedy, Esther Browning, Jonathan Crowe, and John Grandidge at Oxford University Press for their help in moving the book along through to publication. The inspiration to tackle problems in plant population ecology derives from my interactions with many friends and colleagues through the years, not the least of all the group of students, post-docs, and researchers that debated the contrasting virtues of population versus community ecology in the 'Bangor School' of John Harper and Peter Greig-Smith during the late 1980s.

Finally, thanks to my wife Lisa, and our two children Lacey and Dylan for their patience, support, and love.

Carbondale, Illinois, September 2001

Contents

PART 1

What is Plant Population Ecology?

CHAPTER 1

The scope of plant population ecology

On fine days when the grass was dry, I used to lie down on it and draw the blades as they grew, with the ground herbage of buttercup or hawkweed mixed among them, until every square foot of meadow, or mossy bank, became an infinite picture and possession to me

(John Ruskin 1887, Præterita II. X)

- **Populations and plant population ecology defined**
- **A brief history of plant population ecology**
- **The goals of the field**
- **Four case studies illustrating the scope of the topic and the range of methodology and analyses**

Preamble

An early example of an ecological study is Dürer's watercolour *Das große Rasenstück* [Sod of Grass, Great Piece of Turf] painted in 1503 (Fig. 1.1 and Plate 1). Notice how the painting shows a ground-level view of the meadow from the plant's perspective. Flowering stalks of *Poa pratensis* (smooth meadow-grass) are abundant (about 22 can be counted), but noticeably they are not all of the same height. Which of these flower stalks will produce seeds that ultimately germinate to produce seedlings for the next generation? Where might there be a suitable and safe site for seedling growth? Note too the high density of grass tillers. How might this have affected the success of an individual plant in flowering? Finally, notice the low-growing but large leaves of *Plantago major* (greater plantain). What mechanism of interaction with other species allows it to succeed in this habitat? Competition for nutrients, light, or moisture? Or perhaps its shading leaves help keep the soil moist, facilitating the establishment of seedlings. Are the meadow grass or plantain populations stable, expanding, or declining?

Figure 1.1 *Das große Rasenstück*. Watercolour by Dürer, 1503. Albertina, Vienna. (See also Plate 1.)

These are some of the typical questions asked by plant population ecologists, and illustrate a small part of the broad scope of the discipline. In this first chapter, plant population ecology is defined and

Methods in Comparative Plant Population Ecology. Second Edition. David J. Gibson.

the breadth of topics addressed by its practitioners is outlined.

1.1 Plant populations

A *population* of plants is a collection of individuals belonging to the same species, living in the same area (Silvertown and Charlesworth 2001). Populations can range in size from a few small bryophytes colonizing a rock surface, through the thousands of stems of a clonal grass such as big bluestem (*Andropogon gerardii*) spread across several hectares of tallgrass prairie, to the majestic groves of giant redwood (*Sequoia sempervirens*) in California. These examples illustrate the variation in size of the different species that can comprise a population. What makes plant population ecology fascinating and important, though, is the structural variation within and among populations. Structural variation takes four forms, all of which change through time (this change is referred to as *temporal variation*)—the spatial, size, age, and genetic structure of a population (Silvertown and Charlesworth 2001):

- *Spatial structure* is the placement of individuals in space.
- *Size structure* is the relative numbers of large and small individuals.
- *Age structure* is the relative numbers of young and old individuals.
- *Genetic structure* is the variation in gene frequency and genotypes among individuals.

Methods of assessing these different types of variation and how they change through time are described in this book.

Given this definition of a plant population, and considering the four forms of variation within populations, *plant population ecology* is the study of plant populations in their habitat. The Greek root for the word ecology '*oikos*', meaning 'house', suggests that we are interested in understanding the origin, influences, and temporal and spatial dynamics of plant populations in the environment in which they are growing. We are concerned with determining how the biotic and abiotic environments influence plant populations. The biotic environment consists of all the other organisms that affect a plant population, whereas the abiotic environment comprises the non-living influences, such as the soil, bedrock, and climate. The broader term *plant population biology* is used by some authors and includes genetic and evolutionary factors (Hastings 1997). Indeed, the authors of a popular text about plant populations changed its title from *Introduction to plant population ecology* to *Introduction to plant population biology* between the second and third editions, as they expanded their coverage (Silvertown 1987; Silvertown and Lovett Doust 1993).

1.2 History of plant population ecology

Perhaps the first experiment in plant population ecology was Darwin's study of seedling recruitment in charlock (*Sinapis arvensis*) from the seed bank of an arable field (Darwin 1855a) (see Chapter 3, Box 3.1). However, with only a few exceptions (e.g. Tansley 1917; Clements et al. 1929; Salisbury 1942), population ecology had been the exclusive remit of zoologists until the late 1950s. At that time, a number of botanists such as John Harper (1925–2009; see Turkington 2009) in the United Kingdom realized that many of the population-based questions being addressed by zoologists such as Elton (1966) applied to plants too. Reflecting the influence of the agricultural scientists G. E. Blackman and J. L. Harley, with whom he worked at Oxford University, the early work of Harper and his students emphasized the population biology of a number of weed plants such as buttercups, ragwort, plantain, sorrel, and poppy (see the list of Harper's publications in White 1985b). Harper realized that the dynamics of weed plants applied to all plant populations and that populations of plants were regulated by a reaction to density (Sheail 1987). Strongly influenced by reading Charles Darwin, Harper synthesized these ideas in two, now classic, reviews: 'The individual in the population' (Harper 1964) and his Presidential Address to the British Ecological Society, 'A Darwinian approach to plant ecology' (Harper 1967). In his Presidential Address, Harper showed that ecological thinking in the 1960s had its origin in Darwin's ideas on evolution in *The origin of species* (Darwin 1859). Perhaps more importantly, he called for the application of well-developed concepts from animal ecology to the study of plant populations, especially demography. His insight was that plasticity and vegetative reproduction, properties of plants that set them aside from most animals, make plants ideal organisms for questions such as those concerning the allocation of resources—the central idea of life-history strategies (Kingsolver and Paine 1991). He argued that an 'ultimate explanation' of the behaviour and distribution of plants depends upon a reductionist approach—'a concentration of effort on the lives and deaths of individual plants' (Harper 1982). Plant population ecology is generally agreed to have been codified as a discipline in 1977 with the publication of Harper's *Population biology of plants*. This 892-page book attempted to provide the first real synthesis and a focus for this area of ecological study. Harper's ideas subsequently provided a research programme for a generation of plant population ecologists.

As the millennium drew to a close, plant population ecology had ceased to be a radical new idea, and it is now an integral component of plant ecology. Plant population ecology provides a conceptual springboard for understanding patterns and processes in vegetation. The incorporation of molecular methods in particular (see Chapter 5) has allowed detailed evolutionary questions to be addressed. The field is now so large that most attempts at a synthesis deal with either a specific group of plants or a subdisciplinary topic such as plant life histories or plant resource allocation. A list of some important books and review articles is provided in Table 1.1.

1.3 The goals of plant population ecology

The primary goal of plant population ecology studies, although not the focus of all, is to understand changes in plant populations through time and across space (Harper 1977; Hutchings 1986). At its simplest, the number of any particular organism in an area is related to its number a year earlier. Hence, the number of plants (or *modules* or *genets*; see below) per unit area N_t at time t is related to the number N_{t-1} the previous year

$$N_t = N_{t-1} + B - D + I - E,$$

where B is the number of births, D the number of deaths, and I and E are the number of immigrants and emigrants into and out of the population, respectively. A *genet* is an individual plant that has arisen from a seed, whereas a *module* is a repeating subunit of growth such as a tiller or branch. Change, or flux, in populations is summarized as:

$$N_t / N_{t-1}.$$

This ratio is the *annual* or *finite rate of increase*, and is denoted by lambda (λ). When $\lambda = 1$ the population is unchanging or in stasis, when $\lambda > 1$ the population is increasing in number per unit time, and when $\lambda < 1$ the population is decreasing. Hence, a fundamental goal of plant population ecology is to determine the biotic and abiotic factors affecting B, D, I, and E, if not λ itself.

More formally, λ is the antilogarithm of the intrinsic (infinitesimal) rate of increase of a population (r) (Birch 1948) and r is the constant in the differential equation for population growth in an unlimited environment

$$dN/dt = rN,$$

which is equivalent to the integrated form $N_t = N_0 e^{rt}$, where N_0 is the number of plants at time zero, N_t is the number of plants at time t, and $r = B - D$.

Important issues and questions

There are many theoretical issues that concern plant population ecologists. It is hoped that by solving these we will be able to improve the integration of population issues into the disciplines of ecology and biology as a whole and allow a better use of plant population ecology in applied contexts. The aim is that processes in vegetation can be understood mechanistically in terms of the behaviour of individuals and populations. Some of these theoretical issues include furthering our understanding of the importance of the following topics (relevant discussion is provided in the referenced chapters):

Table 1.1 Reading list for plant population ecologists: important and influential books and review articles. Most of the books are specifically about plants, although some of them are broad introductory texts on population ecology.

Title	Author(s), date
Books	
Plant resource allocation	Bazzaz and Grace (1997)
Diseases and plant population biology	Burdon (1987)
Population biology of grasses	Cheplick (1998)
Plant population ecology	Davy et al. (1988)
The ecology and evolution of clonal plants	de Kroon and van Groenendael (1997)
Perspectives in plant population ecology	Dirzo and Sarukhán (1984)
Plant and animal populations: methods in demography	Ebert (1999)
Perspectives on plant competition	Grace and Tilman (1990)
Population biology of plants	Harper (1977)
Plant reproductive ecology: patterns and strategies	Lovett Doust and Lovett Doust (1988)
Introduction to population biology	Neal (2004)
Reproductive allocation in plants	Reekie and Bazzaz (2005)
Introduction to population ecology	Rockwood (2006)
Introduction to plant population biology	Silvertown and Charlesworth (2001)
Plant life histories: ecology, phylogeny and evolution	Silvertown et al. (1997)
Topics in plant population biology	Solbrig et al. (1979)
Demography and evolution in plant populations	Solbrig (1980a)
Population ecology: first principles	Vandermeer and Goldberg (2003)
The population structure of vegetation	White (1985b)
Studies on plant demography: a festschrift for John L. Harper	White (1985c)
Some 'must read' review articles (some classic, some modern)	
The ecological and genetic consequences of density-dependent regulation in plants	Antonovics and Levin (1980)
Sibling competition in plants	Cheplick (1992)
Morphological plasticity in clonal plants: the foraging concept revisited	de Kroon and Hutchings (1995)
The individual in the population	Harper (1964)
A Darwinian approach to plant ecology	Harper (1967)
A metapopulation perspective in plant population biology	Husband and Barrett (1996)
Mast seeding in perennial plants: why, how, where?	Kelly and Sork (2002)
The allometry of reproduction within plant populations	Weiner et al. (2009)
Constant final yield	Weiner and Freckleton (2010)
Correlated changes in plant size and number in plant populations	White and Harper (1970)
The plant as a metapopulation	White (1979a)

1. Phylogenetic considerations for comparing plants to account for shared evolutionary history (Chapter 3).
2. Genetic variation and plant demography (Chapter 5).
3. Life-cycle components and the finite rate of increase of populations (Chapter 8).
4. Separating and understanding the role of competitive effects and responses (Chapter 4).
5. Plant life histories and metapopulation dynamics (Chapter 3).
6. The application of plant population ecology for understanding invasion ecology and rare species (e.g. population viability analysis; see Case Study 2 and Chapter 8) and testing concepts in an agricultural context (e.g. the evolution of herbicide tolerance; see Section 4.3).

A number of important fundamental questions in population ecology were identified by Sutherland et al. (2013) and are listed in Table 1.2. A more narrowly focused set of questions related to plant senescence are provided in Salgueo-Goméz and Shefferson (2013). Overall, the methodologies discussed in this book provide the background knowledge and tools needed to design investigations to address these and other questions.

1.4 Four case studies

In this section, four selected case studies are described to illustrate the scope of plant population ecology. These studies were chosen to illustrate both the diversity of methodological approaches and the application of concepts used by plant population ecologists. In describing the case studies, reference is made to the chapters in this book in which full details of the procedures used by the investigators are discussed. While reading the case studies think about the methods that are used. Are the methods appropriate for addressing the researchers' objectives? Are there other things that should have been measured? Are there obvious sources of error in their methods?

Apart from the very general goal of furthering our understanding of the influences of the biotic and abiotic environments on plant populations,

Table 1.2 Fourteen fundamental questions in population ecology—from a larger list of fundamental ecological questions by Sutherland et al. (2013). Questions unrelated to plant population ecology are omitted.

1. What are the evolutionary and ecological mechanisms that govern species' range margins?
2. How can we upscale detailed processes at the level of individuals into patterns at the population scale?
3. How do species and population traits and landscape configuration interact to determine realized dispersal distances?
4. What is the heritability/genetic basis of dispersal and movement behaviour?
5. Do individuals in the tails of dispersal or dormancy distributions have distinctive genotypes or phenotypes?
6. Do different demographic rates vary predictably over different spatial scales, and how do they then combine to influence spatio-temporal population dynamics?
7. How does demographic and spatial structure modify the effects of environmental stochasticity on population dynamics?
8. How does environmental stochasticity and environmental change interact with density dependence to generate population dynamics and species distributions?
9. To what degree do trans-generational effects on life histories, such as maternal effects, impact on population dynamics?
10. What are the magnitudes and durations of carry-over effects of previous environmental experiences on an individual's subsequent life history and consequent population dynamics?
11. How does covariance among life-history traits affect their contributions to population dynamics?
12. What is the relative importance of direct (consumption, competition) versus indirect (induced behavioural change) interactions in determining the effect of one species on others?
13. How important is individual variation to population, community, and ecosystem dynamics?
14. What demographic traits determine the resilience of natural populations to disturbance and perturbation?

ecologists today make use of the concepts which emerge from plant population ecology to address a number of pressing societal and environmental issues (e.g. the effects of global environmental change). The first case study illustrates how subpopulations of a single species growing in close proximity can affect each other's population growth rate. In the second case study, demographic parameters are used to evaluate the influences of management and other environmental factors upon the projected persistence of a rare species. In contrast to these field observations, the third and fourth case studies are experimental, in the field and greenhouse, respectively. The third case study considers insect

Case Study 1 Recruitment of trees into non-preferred tropical forest habitats (Zuidema et al. 2010)

In this case study, Zuidema et al. (2010) studied the demography of a tropical tree, *Scaphium borneense* (Sterculiaceae), in habitats where it is abundant and in non-preferred habitats where it occurs at a low density. *Scaphium borneense* is a monoecious canopy tree showing strong habitat preferences for ridges compared with slopes and valleys in the Lambir Hills equatorial rainforest region of the east Malaysian island of Borneo where the study took place. By studying subpopulations of the same species in different habitats, the environmental constraints on population growth can be contrasted.

1. Aims of the study. The authors addressed the question of 'how subpopulations in non-preferred habitats are able to persist'. The most likely suggestion that they offered was that populations in non-preferred habitats (slopes and valleys) could be considered 'sinks', maintained by 'recruitment subsidies' from source populations on ridge tops. Specifically, the authors stated that they were testing the hypothesis that 'subpopulations of a Bornean tree species in non-preferred habitats are strongly supported by recruitment subsidies from high-density habitat'.

2. Methodology. All trees > 1 cm d.b.h. (diameter at breast height) in 1300 20 m × 20 m subplots within a 52-ha study plot were tagged, mapped, identified, and their d.b.h. measured in 1992 (Fig. 1.2). The trees were remeasured in 1997 and 2002, adding new recruits since previous censuses. These data allowed the spatial distribution, growth, mortality, and recruitment of all *S. borneense* adults and recruits to be calculated. The subplots were categorized as ridge, slope, or valley based on topographic position.

3. Analytical methods. Zuidema and colleagues used two approaches to their data analysis. First, they used a first-order method of spatial pattern analysis (Chapter 8) to determine if recruits were distributed at random with respect to adult individuals in each habitat using data from two censuses. Secondly, they used what could be considered to be fairly standard actuarial (in this case demographic) matrix analyses of their data (Chapter 8). These methods are derived from procedures used in the life insurance industry for the analysis of human mortality data. They constructed a multi-habitat stage-based Lefkovitch transition matrix model allowing population dynamics within habitats and seed exchange between habitats to be simulated. Submatrices described dynamics within habitats. From these models asymptotic (λ_a) and projected population growth rates over 100 years (λ_{100}) were calculated for each habitat, and elasticities for λ_{100} to allow the evaluation of subpopulation recruitment subsidies (i.e. to assess how important recruitment from one subpopulation to another was).

4. Findings and relevance. Tree recruits occurred close to adults: 80–95% were less than 40 m from nearest adult. Most (83–91%) of the recruits in valleys had likely mother trees in other habitats, while only 13–24% of recruits on the ridge had likely mothers in the valley or on the slopes. Projected λ_{100} was slightly greater than one for each growth period, indicating a growing population overall. Elasticity analysis showed that ridge subpopulations contributed seven to nine times more to λ_a than valley subpopulations and two times more to λ_a than slope subpopulations (Fig. 1.3). These findings support the hypothesis of ridge subpopulations being recruitment sources, and slope and valley subpopulations as sinks.

The relevance of this study for plant population ecology is that it shows how data that might appear straightforward to collect without special equipment (repeated mapping of individuals, although certainly not routine!) can be used to generate a tremendous amount of information about the demography of populations. In this case the demographic comparison of different populations of the same species growing in different habitats allowed a useful comparison of source–sink dynamics. The careful observations carried out by Zuidema et al. (2010) provide an excellent procedural example that can be utilized by similar investigators.

Case Study 1 *Continued*

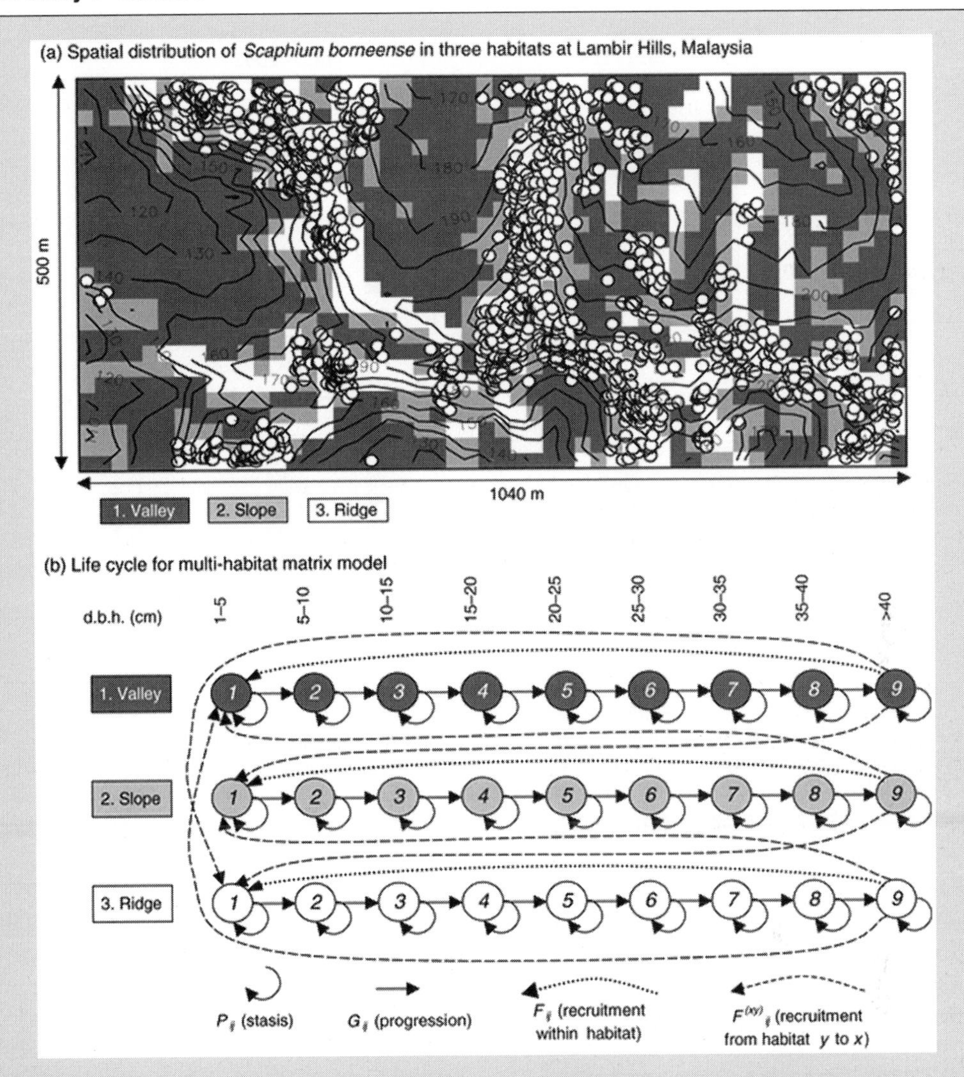

Figure 1.2 (a) Spatial distribution of trees > 1 cm d.b.h. (diameter at breast height) Isolines are metres above sea level. (b) Life-cycle graph of *Scaphium borneense*, recruitment arrows for categories 5–8 are omitted for clarity. From Zuidema et al. (2010). Reproduced with permission from John Wiley & Sons.

continued

Case Study 1 *Continued*

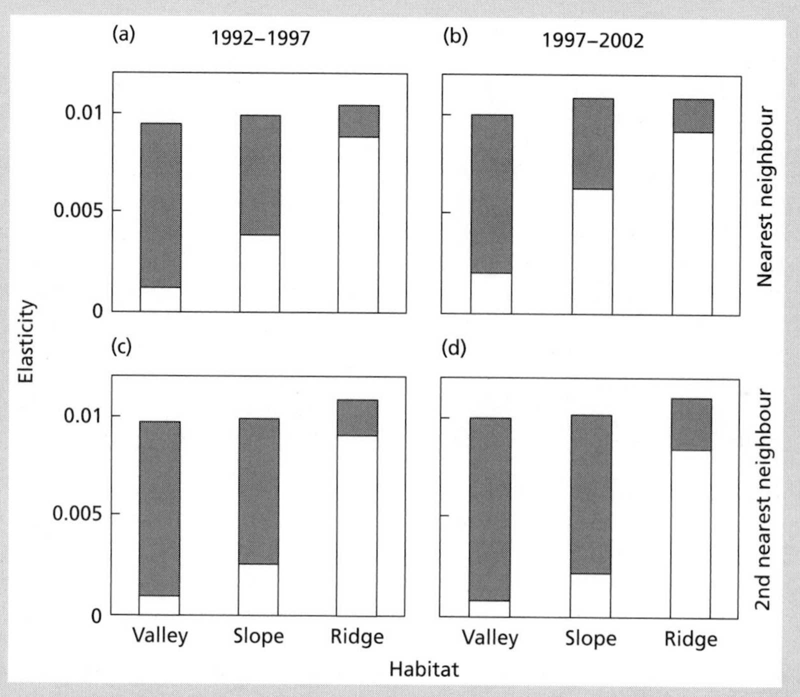

Figure 1.3 Elasticity of the projected 100-year growth rate, $\lambda 100$, of *Scaphium borneense* subpopulations for recruitment exchange (grey bars) and recruitment within own habitat (open bars) for two time periods (a, c, and b, d) based upon nearest neighbours (a, b) and 2nd nearest neighbours (c, d). From Zuidema et al. (2010) Reproduced with permission from John Wiley & Sons.

Case Study 2 Demographic viability of populations of a rare perennial herb in relation to management, location, isolation, and genetic variation: an application of population viability analysis (Menges and Dolan 1998)

There is widespread concern about the worldwide loss of biodiversity. Species are being lost (going extinct) at a faster rate than ever before in the Earth's history, principally due to loss of habitat. Conservation of rare plants requires a thorough knowledge of their ecology, especially their population ecology. Indeed, some investigators claim that demographic criteria provide a better indication of the biological status of a rare species than does knowledge of the distribution of genetic variation (Schemske et al. 1994). Information on the population ecology of rare plants is required in many species recovery plans in the United States and species action plans in the United Kingdom. For example, when considering the need for conservation of rare broomrape (*Orobanche* spp.) species in the United Kingdom, Headley et al. (1998) called for the following: (a) monitoring of population sizes, (b) determination of the life span of each species, (c) ecology and population dynamics of host plants, and (d) molecular studies of the breeding system to evaluate the prevalence of inbreeding depression.

Menges and Dolan's (1998) study provides a good example of how knowledge of the demographic viability of a rare species can be useful in making recommendations for management based on the application of population viability analysis.

Case Study 2 *Continued*

1. Aims of the study. To determine how the demographic viability of a long-lived, rare perennial, *Silene regia* (Caryophyllaceae), could be influenced by management regimes, population size, location, isolation, and genetic variation.

2. Methodology. Sixteen populations of *S. regia* from across the Midwestern region of the United States were monitored for 8 years. All non-seedlings (over 2500 individuals) were marked and mapped annually within permanent study plots. For each plant, the number of stems, height, stem basal diameter, flowering status, and number of flowering whorls was recorded. Seedling recruitment was estimated from counts within 1-m^2 quadrats. The following environmental data were collected at the location of each plant: percentage vegetation cover, litter depth, and substrate class (undisturbed, disturbed, gravel, moss). Data on genetic variation using isozymes (Chapter 5) were obtained from an earlier study (Dolan 1994).

3. Analytical methods. The authors used population viability analysis (PVA; see Chapter 8) to extrapolate from the field-collected demographic data to extinction probabilities for each population. PVA consists of a suite of techniques that use demographic data and models to estimate the likelihood that a population will persist for a given period of time. Six life-history stages (seedlings, vegetative plants, alive–undefined, and three height classes) were used in the construction of 98 stage-based Lefkovitch transition matrices (Chapter 8) for individual populations and years. Finite rates of increase were based upon projections of 1000 replicate runs of 100 or 1000 years with (λ_R) and without (λ_N) recruitment. Initial population sizes in the simulations were based upon the last field census for the population. The life-history components of λ were quantified by calculating elasticities representing growth, survival, and fecundity. Elasticity is a measure that allows a comparison of the sensitivity of λ to each component of the organism's life history (Chapter 8). Univariate (analysis of variance) and multivariate (principal components analysis) (Chapter 7) analyses were used to relate λ and elasticity to the management regime, population size, geographic region, degree of isolation, genetic variation, and habitat characteristics.

4. Findings and relevance. *Silene regia* was found to be a long-lived plant with low rates of adult mortality (5–17% per year). Over 75% of adult plants flowered each year. However, seedling recruitment was episodic and more frequent in sites managed by fire. Analysis revealed three groups of populations: one group of four populations whose extinction appeared assured within 100 years, three populations that were likely to persist for 100 years, and eight populations that had no risk of extinction in 1000 years. λ varied among years and populations, and was related to the management regime and whether or not recruitment occurred (Fig. 1.4). Sites managed using fire had a higher λ with higher elasticities for growth and fecundity and lower elasticity for survival than unburned sites. The multivariate analysis showed that fire and geographic region had the strongest relationship to λ, although genetic variation, population size, and isolation were also important. λ was highest for large populations with a high genetic diversity.

This study is a good example of the application of plant population ecology to a better understanding of the biology of rare species. Habitat for *S. regia* can be better managed with the knowledge gained from this study. It is worth noting that in this case study demographically defined population viability was only weakly related to genetic variation. The authors speculated that the current genetic structure may reflect the original regional levels in place before modern-day habitat loss. PVA is an increasingly important use of demographic population data for the study of rare species—other examples include Nield et al. (2009), Yates and Ladd (2010), and Bucharová et al. (2010). Guidelines for using PVA for species reintroductions are provided by Knight (2012), with more information provided here in Chapter 8.

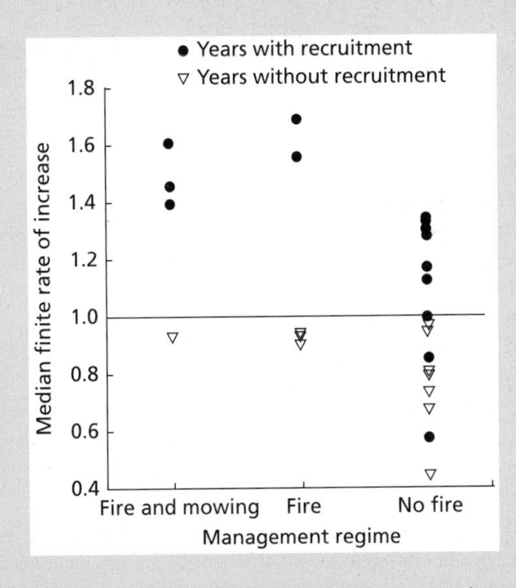

Figure 1.4 Finite rates of increase (λ) of *Silene regia* populations as a function of management regime in years with or without recruitment. From Menges and Dolan (1998). Reproduced with permission from John Wiley & Sons.

Case Study 3 Disturbance, competition, and the effects of herbivory on an introduced species (McEvoy et al. 1993)

Introduced, non-native, alien, or exotic species (the same thing in common ecological parlance) are a major problem in the maintenance of biodiversity in native habitats. Plant population ecology allows us to learn about the life-history characteristics of these species with a view to extirpating or at least controlling them (Mack 1985). At the same time, study of the often rapid growth and sometimes high fecundity of introduced species illuminates population dynamic processes of interest to the plant population ecologist. Examples of comprehensive population ecology studies of introduced plants include Luken and Mattimiro's (1991) study of *Lonicera maackii* in the eastern United States, Tripathi's (1985) study of *Eupatorium* spp. and *Galinsoga* spp. in India, and Mack and Pyke's (1983, 1984) and Pyke's (1986, 1987) studies of *Bromus tectorum* in the western United States.

1. Aims of the study. The purpose of the study by McEvoy et al. (1993) was to investigate whether biological control of an introduced weed, *Senecio jacobaea* (ragwort, Asteraceae), by introduced insects was related to local disturbance and plant competition. *Senecio jacobaea* is a short-lived perennial native to Europe and introduced to North America (the site of this study), Australia, and New Zealand. Two natural herbivores of *S. jacobaea*, the cinnabar moth (*Tyria jacobaeae*) and the flea beetle (*Longitarsus jacobaeae*), were introduced as biological control agents from France and Italy, respectively. Three specific hypotheses were tested: H_1 (the activation hypothesis)—localized disturbance and buried seed combine to create incipient weed outbreaks; H_2 (the inhibition hypothesis)—insect herbivory and interspecific plant competition combine to oppose the increase and spread of incipient weed outbreaks; and H_3 (the stability hypothesis)—the balance between short-range activation and long-range inhibition leads to local instability and stable average spatial concentration of the pest. A field experiment was used to address these hypotheses.

2. Methodology. The hypotheses were tested in a 0.9-ha meadow in Oregon, United States, where *S. jacobaea* was ranked 16th in abundance with a large buried seed bank. A randomized complete block design (Chapters 3 and 7) was used to assign 0.25-m^2 treatment plots to combinations of four treatments: two disturbance times (autumn, spring), three levels of background vegetation (removed, clipped, unaltered), two levels of exposure to cinnabar

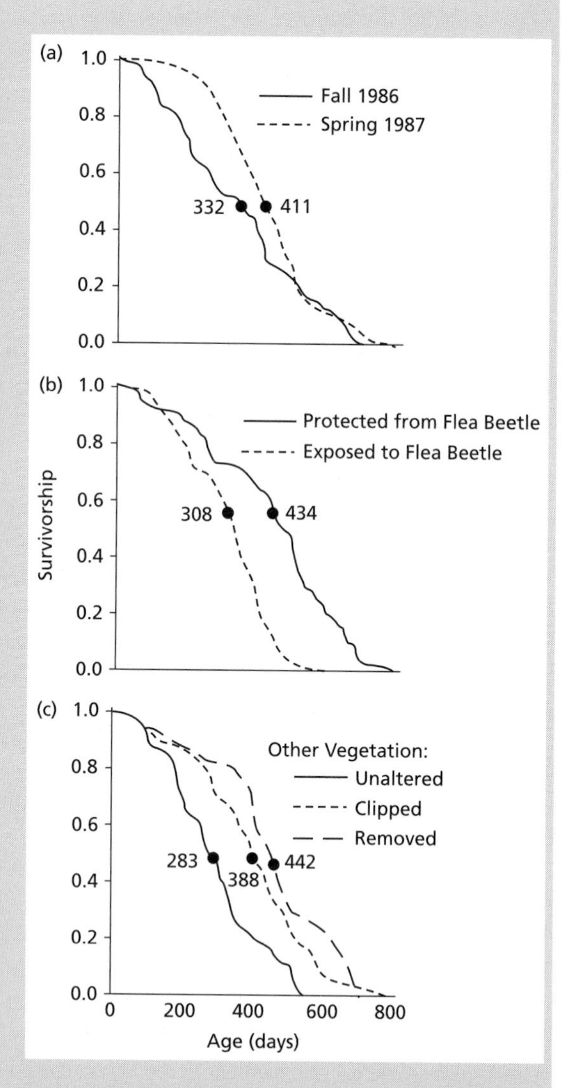

Figure 1.5 Survivorship curves for *Senecio jacobaea* comparing (a) time of disturbance, (b) protection from herbivory by the flea beetle, and (c) competition. Solid symbols and adjacent numbers represent the mean length of life for each level of a given treatment, averaged over all levels of the other treatments. From McEvoy et al. (1993). Reproduced with permission from the Ecological Society of America.

continued

Case Study 3 *Continued*

moth (exposed, protected), and two levels of exposure to flea beetle (exposed, protected). Experimental disturbances were achieved by digging to a depth of up to 30 cm with a shovel, breaking the sod, and sieving out roots and rhizomes. The insect treatments were applied by allowing exposure of the target plots to natural attack at the appropriate times of the year with protection to other plots using screening. Following establishment of the experiment, the location of all *S. jacobaea* seedlings in a 0.06-m^2 area in the centre of each study plot was recorded on a regular basis. At each sampling time, the position of each seedling was circled on a sheet of acetate placed over the plot. In this way, the fate of each individual was followed beginning with emergence, through growth, reproduction, and death. The basal and rosette diameters of each plant, reproductive status, and length of the longest leaf were recorded. These data allowed plant dry mass to be estimated through the use of linear regression (Chapter 7).

3. Analytical methods. The experiment was analysed using analysis of variance (Chapter 7) on *S. jacobaea* characters (colonization, survival, reproduction, biomass). Orthogonal contrasts (Chapter 7) were used to test whether the response variables differed depending on whether background vegetation was present or absent, or clipped or unaltered. The frequent occurrence of zero values for reproductive responses meant that analysis of these data was carried out on only subsets of the treatment combinations.

4. Findings and relevance. Outbreaks of this introduced species were related to rapid germination and establishment from abundant buried seed initiated from localized disturbances. This is consistent with H_1, the activation hypothesis. Time of disturbance was the only treatment that affected the number of seedlings, with 32% more seedlings established in the autumn compared with in the spring. In contrast to establishment, spring seedlings survived for 24% longer (mean length of life) than autumn seedlings. Survivorship was also increased by protection from the flea beetle and in the absence of competition from other plants (Fig. 1.5). This observation is consistent with H_2, the inhibition hypothesis. The natural enemies readily colonized the experimental *S. jacobaea* outbreaks and over the course of the experiment reduced biomass to zero except where plants were protected (removed competition treatment and protected from moth and beetle treatments). The cinnabar moth, by reducing the number of seeds produced, reduced the reproduction of surviving adult plants. The flea beetle reduced reproduction by reducing the number of adult plants. These interactions between the agents of biological control (the two insects) and disturbance and competition are consistent with H_3, the stability hypothesis. At the spatial scale studied in this experiment, the populations of *S. jacobaea* are locally unstable and depend upon short-term activation from the seed bank through disturbance and inhibition through biological control. This interaction allows large-scale maintenance of pest populations because of their ability to seek out the plant.

This study is important because it illustrates the complexity of biological factors acting upon the life cycle and population dynamics of an introduced plant. The introduction of herbivores is a classic form of biological weed control that can be extremely successful, as in this case, or a complete failure when there is not a full understanding of the relationships between the organisms. An important part of this is understanding the population ecology of the plant in its new habitat. Unfortunately, very few of the studies on the biocontrol of weeds are effectively replicated or randomized to allow adequate statistical rigour (Crawley 1997a).

This case study is also a good example of the use of a field experiment to test hypotheses. As noted in Chapter 3, field experiments allow precise and rigorous control of experimental treatments along with a high degree of realism, precision, and repeatability.

Case Study 4 The effects of herbivory on fitness (Mauricio et al. 1993)

All plants are attacked by one or more herbivores (Crawley 1997b). As shown in Case Study 3, herbivores can dramatically affect the population biology of a plant, potentially affecting all aspects of its life cycle. This case study illustrates the use of a greenhouse experiment to determine the effects of leaf damage on the *fitness* of an annual plant. Fitness can be defined as the 'chance of leaving descendants' (Harper 1977, p. 410; see also Chapter 5) and has obvious and important evolutionary consequences (e.g. a higher fitness means an increased contribution to the gene pool).

1. Aims of the study. This study was designed to assess the effects of the spatial pattern of herbivore damage (leaf damage) upon plant fitness. This assessment is important because some herbivores will forage on a few individual leaves while others, such as the larva of the cabbage butterfly (*Pieris rapae*), produce a pattern of damage that is dispersed across the host plant. There can also be differences in the age of the leaves that are damaged, as new leaves may have higher nitrogen concentrations and levels of secondary compounds than older leaves on the same plant. A new leaf with a high rate of photosynthesis may be more important to a plant than an older leaf as a source of carbohydrates. In this situation, damage to new leaves may reduce fitness.

2. Methodology. In a greenhouse setting, 20 6-week-old four-leaf seedlings of *Raphanus sativus* (wild radish, Brassicaceae) were assigned to each of six treatments of leaf removal: (1) control, no leaf damage; (2) one entire mature leaf; (3) one entire new leaf; (4) 50% of two mature leaves; (5) 50% of two new leaves; (6) 25% of all four leaves removed, respectively. These treatments ensured that all damaged plants had equal amounts of their total leaf area removed (25%). Leaf damage was achieved using a hole punch. Six weeks after imposing the treatments the plants were harvested and the dry mass of the following components measured: stem, rosette leaves, stem leaves, flowers and buds, and roots. The following additional parameters were recorded: number of flowers, living and senesced rosette leaves, and stem leaves. Flower number, reproductive biomass, and total plant biomass were considered as measures of fitness.

3. Analytical methods. After testing for normality and homogeneity of variances and taking appropriate transformations, the variables were analysed using separate one-way analyses of variance (Chapter 7). A priori (single degree of freedom contrasts) and a posteriori (Ryan–Einot–Gabriel–Welsch) tests were conducted to determine differences among treatment means (Chapter 7). Preliminary analysis showed that there were no significant effects of leaf age (new versus mature leaves) on any measured variable, so these treatments (numbers 2 and 3, and 4 and 5) were pooled into treatments referred to as concentrated (removal of an entire leaf) or intermediate (removal of 50% of a leaf) for further analysis.

4. Findings and relevance. The experiment showed that the pattern of leaf removal influenced the fitness of *R. sativus*. Intermediate damage (where 50% of two mature or two new leaves was removed) caused a significant reduction in fitness, whereas the effects of concentrated and dispersed damage were no different from the control (Fig. 1.6). The means for the concentrated damage treatment were consistently lower than those for the control plants, although they were not significantly different. These findings are important because they demonstrate that the pattern of herbivory, not just the imposition of herbivory per se, can have important fitness consequences for a plant. Furthermore, a herbivore that disperses damage over many leaves may have less of a negative effect on fitness than a herbivore that concentrates its damage on a single or a few leaves.

As a case study in methodology for plant population ecology this work is important because it illustrates how a carefully controlled greenhouse experiment can be used to test a very specific effect—in this case the spatial pattern of herbivory. A number of potentially *confounding effects* were held constant to avoid their influence upon the results; that is, the amount of leaf tissue removed per plant (always 25%), plant age and phenology, growing conditions (temperature, moisture, soil, photoperiod), herbivore, and method of damage (hole punch). Confounding effects are non-experimental (unplanned) variations in the target plant (e.g. differences in plant size due to slight differences in the germination rate) or its environment (growing conditions) that can influence the outcome of the experiment. The investigators would probably have had to contend with high levels of variation in their data set if the experiment had been conducted in the field, especially with a self-sown natural population. A major limitation of the experiment is that it was carried out in the greenhouse and extrapolation of the findings to the field has to be made with caution.

Case Study 4 *Continued*

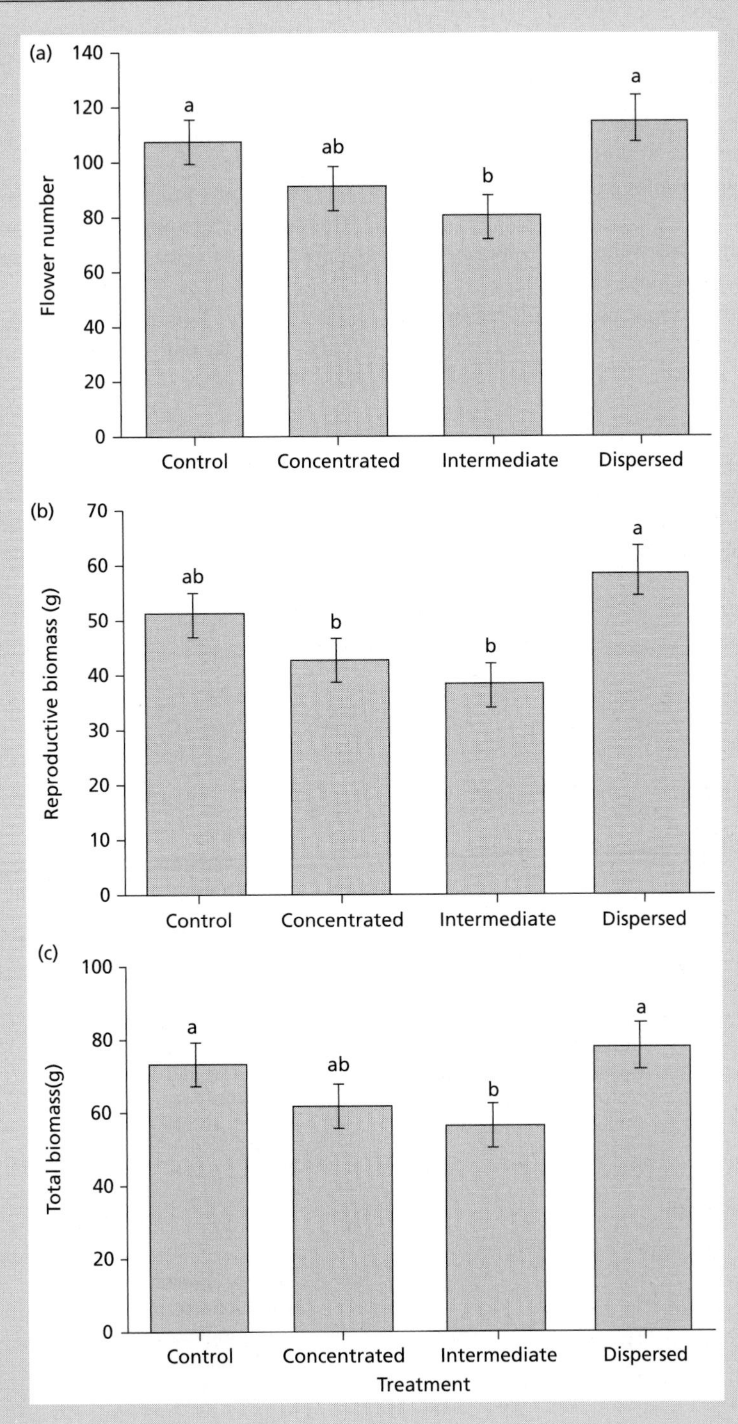

Figure 1.6 The effect of leaf damage on measures of fitness of wild radish, *Raphanus sativus*. Means with the same letter are not significantly different (Ryan–Einot–Gabriel–Welsch test; $\alpha = 0.05$, d.f. = 116). From Mauricio et al. (1993). Reproduced with permission from the Ecological Society of America.

control of an introduced species, while the fourth case study investigates the effects of herbivory on the fitness of an annual plant. The following aspects are emphasized in each case study:

1. The questions being asked by the investigators, including any specific hypotheses.
2. The methodology employed, and any problems that arise.
3. The analytical methods (refer to Chapters 7–8 for details about specific methods).
4. The relevance of their findings for addressing the broader issues.

1.5 A roadmap for using this book

The purpose of writing this book is to provide guidance for conducting research in plant population ecology. As a result, the book is organized to facilitate the process in a sequence that follows the usual order in which research is conducted. After Part 1, which briefly outlines the scope of plant population ecology in this chapter, there are three main sections in the book: Part 2, Planning the Study (Chapters 2 and 3); Part 3, Doing the Study (Chapters 4–6); and Part 4, Analysis (Chapters 7–8). In reality this is a somewhat false dichotomy as the procedures that are described in

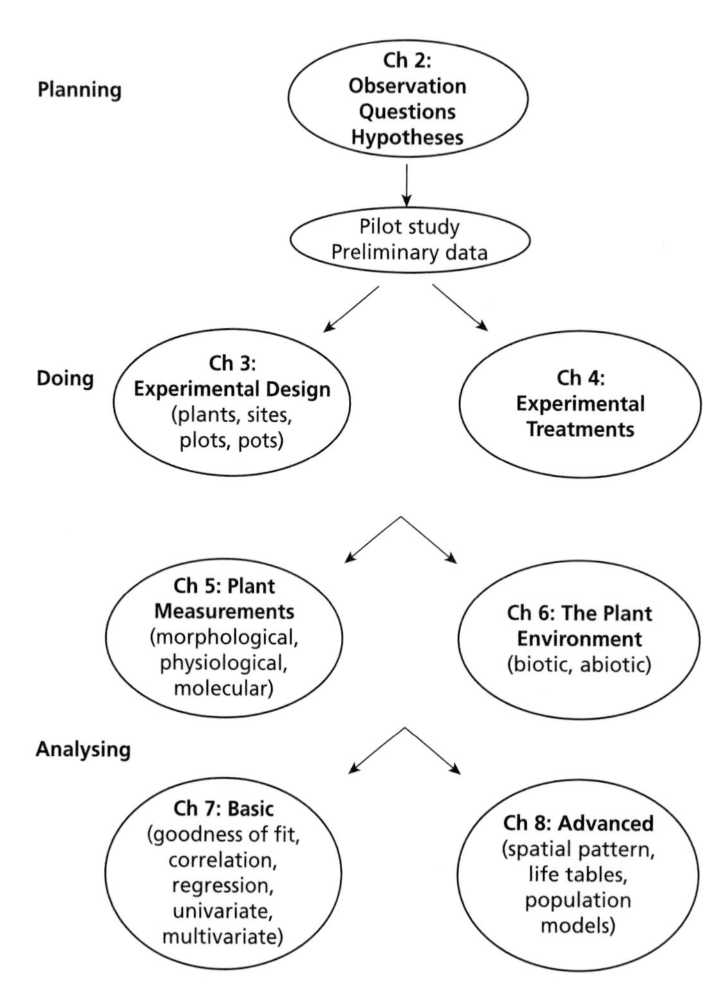

Figure 1.7 Roadmap for this book showing the chapters in which the main topics are covered.

Part 3 have to be conceived in the planning stage. Nevertheless, the sequence of these topics is illustrated in Fig. 1.7 along with the chapters in which they are covered.

1.6 Follow-up exercises

1. How important are the papers listed in Table 1.1? Use a citation index to see how often these papers have been cited since they were published. Why are they cited? What sort of studies might find these papers useful? Summarize the work of five papers that cite one of these studies.
2. Section 1.3 lists six areas of plant population ecology that are the topics of current theoretical interest. Find and report on a recent study that addresses one of these topics.
3. What sort of experimental work, in the field or greenhouse, could be carried out to follow up the findings for Case Studies 1 and 2?
4. What sort of observational field work could be carried out to follow up the findings for Case Studies 3 and 4?

PART 2

Planning the study

CHAPTER 2

The question and the approach

The most important result of a rational inquiry into nature is, therefore, to establish the unity and harmony of this stupendous mass of force and matter, to determine with impartial justice what is due to the discoveries of the past and to those of the present, and to analyze the individual part of natural phenomena without succumbing beneath the weight of the whole.

(Humboldt 1849)

As a rule, I begin my lectures on Scientific Method by telling my students that scientific method does not exist. I add that I ought to know, having been, for a time at least, the one and only professor of this non-existent subject within the British Commonwealth.

(Karl Popper 1956)

- **How do you frame an important question?**
- **Ways of addressing the question**
- **What should you measure? Ecological variables**
- **The importance of pilot studies**

Preamble

Planning the study before beginning to collect any data is of paramount importance. In this and the next two chapters the important factors that should be considered at this stage are described. In particular the planning should include the following:

- A statement of problem to be solved (questions, hypotheses) (this chapter).
- Specification of the response (dependent) variables, explanatory (independent) variables, and treatments (this chapter and Chapter 5).
- Description of the objects for study—experimental or sampling units (Chapter 3).
- The design of (random) procedures (Chapter 3).
- Determination of the desired accuracy and precision of the observational methods (Chapter 3).
- Establishment of the desired duration and size (number of study objects) of the research (Chapter 3).
- Specification of the statistical methods (Chapters 7 and 8).
- Re-evaluation of the objectives in light of the above.

In this chapter the first and part of the second (dependent and independent variables) points above are discussed. The other topics are discussed in later chapters, as specified. An outline for a research programme is given in Fig. 2.1. Note how this scheme is centred very clearly on the testing of hypotheses. In this chapter the discussion is based upon this approach. In Chapter 3 the use of natural and mensurative experiments where a less rigorous and investigative approach can be used is discussed. Hypotheses can still be tested, but experimental treatments are chosen from existing (unmanipulated) conditions in nature (e.g. plant fitness may be measured in naturally flooded versus unflooded sites).

Progress in ecology is centred on three actions: (a) making observations about nature, (b) attempting to explain the observations by proposing theories or models, and (c) distinguishing among competing theories or models by testing specific hypotheses (Underwood 1997). As noted below, a project

Methods in Comparative Plant Population Ecology. Second Edition. David J. Gibson.

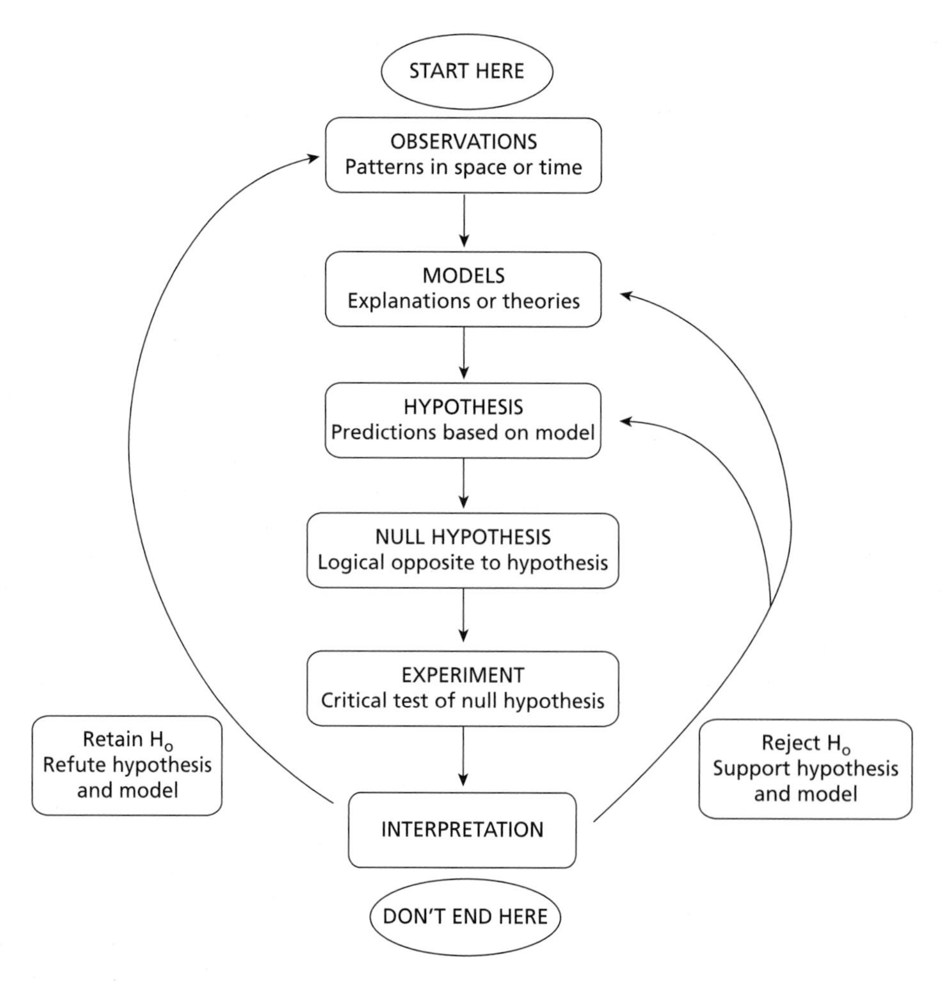

Figure 2.1 Generalized scheme of the components in a hypothesis-based research programme. From Underwood (1997). Reprinted with permission of Cambridge University Press.

may start with one's own observations (point a) or at point (b) before moving to point (c). Moreover, most research questions do not begin de novo but are based upon what has been stated earlier or proposed by others. One doesn't begin in a vacuum.

Some things that might cause panic, but shouldn't

- *I know what I want to do; do I really have to read all this old stuff?* Yes, it is important to read the earlier literature. The question that you start with might have been asked before, so there's no point in reinventing the wheel. Confirmatory studies that ask a question about an interesting species are all very well, and do help to reinforce the importance of old ideas, but they are hard to publish in a top journal. A project should be driven by relevant ecological questions, which may be refined as the project proceeds—but they must be grounded in the literature, including the old stuff.
- *My idea seems important but it's not amenable to experiments. Why can't I just go out and measure what I'm seeing?* There are lots of things that can be measured (Chapters 5 and 6), and there is great value in taking extra measurements along the way. But even descriptive studies need to be framed in terms of falsifiable a priori questions

(Section 2.1). If they are not, then one is just doing natural history in the most traditional sense, and again it may be difficult to publish the findings.

- *So, what should I measure?* One needs to identify measureable variables that address the hypothesis being tested—see Section 2.3. But this should not always be a limiting factor. The question-driven approach is important, but one should remain observant when in the field (or greenhouse). Opportunistically collected data may suggest a path forward when the initial hypothesis fails, as it often does. In the study of *Ceratiola ericoides* to be described in Chapter 8 (Gibson and Menges 1994) it was an unexpected observation that some old plants were reproducing vegetatively through layering at one of the field sites that completely changed the view of how this shrub persists in storm-disturbed coastal habitats.
- *I'm not sure which of several variables is important—do I need a massive factorial design?* A very carefully designed experiment may be needed to tease out some effects. But designing complex studies in ignorance of the basic natural history of the system and its plants can lead to a lot of wasted time in the end. Before executing a scientifically rigorous study, there is great value in running a few simple, perhaps unreplicated, heuristic experiments (see Section 2.4). These experiments or observations are not rigorous and can be misleading, but they may quickly clear away competing hypotheses, allowing one to focus on the important stuff. For a contrary and provocative view of the value of simplicity see Grubb's 1992 Presidential Address to the British Ecological Society.
- *Why isn't this proposed statement acceptable as a hypothesis?* An unacceptable hypothesis is one that cannot be tested. It is also useful to make the distinction between a question (an apparent inconsistency in nature), a hypothesis (an abstract model or 'best guess' accounting for that inconsistency), and a prediction/expectation (the observable, testable behaviour which the model predicts) (see Table 2.2).
- *How can spatial variability be captured when I can only afford a single instrument?* There's always a trade-off between how many measurements can be made with a limited number of items of expensive equipment and fully accounting for spatial variability. Sometimes, precision has to be sacrificed (the benefit of expense) to gain replication. Sometimes, cheap and easy-to-make sampling devices can be manufactured. For example, a simple integrating meter based upon photochemical paper can provide good documentation of relative light levels across a site for little cost (see Matlack 1993).
- *How can long-term processes be measured?* Important plant demographic processes may occur very slowly, taking longer than the time available for a graduate programme or a funded grant. Most people don't have the fortitude or forethought to record long-term data like the 32-year orchid dataset reported by Hutchings (2010). Space-for-time substitutions, in which sites of different time since (say) disturbance, hosting populations of a species of different ages, may be available, especially across the mosaic of agriculturally dominated areas that characterize many modern landscapes. There are important assumptions inherent in using this approach (e.g. similarity in the post-disturbance history of sites of different ages), but otherwise all one has to do after identifying appropriate sites is to go out and measure the plants. Alternatively, one may be fortunate and have access to a rich data set containing demographic data recorded by someone else over decades; one can then just do a modern analysis and write it all up. For example, data from more than 50 permanent quadrats mapped in a Kansas grassland from 1932 to 1972 allowed the demography and population dynamics of prairie forbs and grasses to be assessed, providing novel information on life expectancy and life span (Adler and HilleRisLambers 2008; Laurenroth and Adler 2008).
- *How should I record and store my data?* This is not a trivial matter, and is something that requires careful thought before the first data are collected. A notebook should be kept in which something is written down every time lab or field work is carried out. The notes may be as simple as '1 June 2014, scouted around, found new field sites . . .', or may be pages of raw data. Ideally, a field notebook should be of waterproof ('write-in-rain') paper or an electronic device (e.g. a tablet).

Actual data should be recorded on datasheets (electronic or paper) that are set up with variable names as column headings. The important thing to bear in mind is that these data will soon be uploaded or entered into a spreadsheet for analysis. Most programs that are used for analysis will be set up to read columns of numbers. One of Krebs' (1999) rules for ecological data states, 'Code all your ecological data and enter it on a computer in some machine-readable format'. Useful advice and guidelines for effective data recording and management are provided by Borer et al. (2009). Additionally, mandatory data archiving is increasingly being adopted by funding agencies and journals, which means that data and metadata (instructions on how to make sense of the data) have to be stored digitally in an approved archive (e.g. Dryad, <http://datadryad.org/>) so that others can use it. Some best practice guidelines for data archiving are provided by Whitlock (2011).

- *These data don't fit any statistical test I know of. What can I do?* This problem may be especially relevant for non-normal, heavily skewed data, or data with lots of zero values (e.g. many plants that did not reproduce; Brophy et al. 2007) or missing cases (e.g. treatments that failed so there are no data). There are a large range of *non-parametric* statistical tests that do not make traditional assumptions about data distributions because they work on ranks (see Chapter 7). Alternatively, Monte Carlo methods offer a powerful and flexible approach that is generally under-used. For these methods one must define a null condition and, using a random number generator, generate a large number of random permutations of the data for comparison with the actual data. If one's data lie outside a 95% confidence window of the randomly permutated data then they are significantly different from the null condition. Full details about testing null models are available in Gotelli and Graves (1996).

2.1 How can an important question be framed?

The last paragraph or two of the introduction of just about any article in an ecological journal should mention the goal that guided the study. The 'goal' means the aim or purpose of the study. The goal of a study is not synonymous with its objective ('point or thing aimed at'; Sykes 1976), as a goal is more general than an objective. The goal and its subservient operational actions (Table 2.1) need to be es-

Table 2.1 Operational terms and definitions for plant population ecology studies with an example of the sequence and actions involved in a study on plant–fungal relationships. Definitions according to Sykes (1976).

Operational term	Definition	Example of action
Goal/aim	Object of effort or ambition/purpose	Study plant–fungal relationships
Objective(s)	Point or thing aimed at	To determine the importance of VAM fungi on belowground processes in plant populations
Hypothesis	Supposition made as a basis for reasoning, without assumption of its truth, or as a starting point for further investigation from known facts	The presence or absence of VAM fungi has no effect on the competitive balance between plant species or their relative abundance
Question	Problem requiring solution	Does the presence/absence of VAM fungal infection significantly influence the outcome of plant competition between pairs of tallgrass prairie plants grown in mixture?
Plan (experiments)	Formulated or organized method by which thing is to be done; way of proceeding	Undertake greenhouse and field competition experiments on plant species of known mycorrhizal dependency in the presence/absence of VAM fungi
Outcome	Result, visible effect	The hypothesis was rejected (Hartnett 1993; Hetrick et al. 1994; Wilson and Hartnett 1997)

VAM, vesicular–arbuscular mycorrhizal.

tablished right at the beginning before any field or greenhouse work is even contemplated (apart, perhaps, from some semi-serious scouting of potential field sites). Defining the goal of the study right at the beginning is of paramount importance. As noted by Peters (1991), 'A good choice [of goal] directs the entire research programme by providing a tangible endpoint so that any part of the study may be judged by its relevance to the goal'. There may be one or more objectives, and these describe very specific outcomes of a project (see Table 2.1). Goals and aims are more general than objectives. For example, in education an aim may be to teach students a lot about ecology while an objective may include getting the students to read and comment on 10 important papers.

Knowing the goals and specific objectives of a study will save time, it will make one work more efficiently, it will ensure that the questions answered and hypotheses tested are the ones that were intended, and it will ensure that no time is wasted taking unnecessary or inappropriate measurements. Appropriate goals and objectives will ensure the choice of an appropriate study organism (species, population) and the correct life-history stage (seeds, seedlings, immature adults, reproductive adults, etc.) and measure the necessary features of the plant (e.g. number of vegetative shoots, height of flowering stems). For example, each seed on a plant does not need to be weighed separately if total seed mass is all that is required to estimate reproductive fitness.

Important, relevant, and interesting goals are necessary if the work is to be published. What makes a goal important? It is difficult to say. Everyone thinks that his or her own study is important, and hopefully to them at least it is interesting. However, an important study is one that addresses a conceptual or theoretical issue in the current literature or is of practical/applied value. Journals with their differing emphases vary in their definition of what constitutes an important study.

The genesis of a project may derive from one's personal knowledge of a particular species or some, perhaps paradoxical, field observations (Underwood 1990). For example, Hutchings and Barkham (1976) observed high field densities of the shoots of *Mercurialis perennis*, a clonal herb, but upon investigation found that these did not conform to an expected pattern of self-thinning. This led the authors to a more general suggestion, namely that the –1.5 power equation does not fit the development of parts of rhizomatous perennial herbs. The Nobel prize winner, Peter Medawar (1969) noted that 'Observation is the generative act in scientific discovery'. But it is important to bear in mind that science (and this is especially true for plant population ecology) progresses by challenging theories and ideas that have been proposed to explain observations, and not by simply studying natural history phenomena.

Two particular goals of ecology are *generality* and *predictability*. Theories developed from scientific enterprise need to be generalizable from the realm of the study to the wider, real world. Ideally, the findings from a study must be set into the context of the discipline of plant population ecology as a whole, i.e. generalized ('to bring into general use'; Sykes 1976). For example, it may be found that beetle herbivory on the petals of the study plant reduces seed set and ultimately fitness. This finding cannot be generalized to all other plant species with a lot of confidence. However, if it were found to be true for five different species of plant subject to herbivory by five different species of beetle, for example, then the finding that beetle herbivory of flowers reduces fitness would be generalizable with a fair degree of confidence.

Ecology must be predictive (Peters 1991), thereby allowing us to foretell events. Using the beetle example above, we would be able to predict a loss of fitness in plants that are in flower when beetles are present. Ideally, we should be able to predict the magnitude of the loss in fitness too. This would be a prediction, and should be a goal of plant population ecology studies. A good example is Case Study 2 in Chapter 1 where a prediction is made about which of the populations of *Silene regia* will go extinct within 1000 years (five of sixteen populations that are managed without fire or mowing). Of course, only time will tell if the authors are correct, but their predictions were made according to the conventional level of statistical confidence (Chapter 3) and are accepted with that caveat in mind.

It is also important to distinguish between phenomenological and mechanistic questions.

Phenomenological studies demonstrate the occurrence of events, such as competition, seed recruitment, or mortality. These studies can be very worthy and useful, especially if the phenomena were previously unknown; for example, the rare case of insect pollination of a grass that was demonstrated by Adams et al. (1981). Of far greater value are mechanistic studies that can explain why a phenomenon is observed (e.g. resource competition, allelopathy). An understanding of mechanisms can provide a short cut to prediction. In the case of the insect-pollinated grass, larger than normal pollen grains were suggested by the authors of the study as a possible reason behind the need for insect pollination.

Krebs (1999) proposed four rules regarding the collection of ecological data that are pertinent for plant population ecologists. These are:

Rule 1: not everything that can be measured should be.

Rule 2: find a problem and ask a question.

Rule 3: collect data that will answer the question being asked and make a statistician happy.

Rule 4: some ecological questions are impossible to answer at the present time.

A corollary to Rule 4 is that some ecological questions are impossible to answer within the time frame of a post-graduate research programme! Attempt the impossible, but within reason!

So, to answer the problem posed at the beginning of this chapter—how should an important question be framed? Follow the points below:

- Consider the explanation for any paradoxical field observations that have been made.
- Try to challenge an established idea, theory, or concept.
- Consider the hierarchical order of operational terms in Table 2.1. Look at Fig. 2.1 again.
- Start with an overall goal and objective(s); these will then guide the hypotheses, question(s), and finally the plan or experimental design. As noted below (Section 2.2), explicitly stated hypotheses are not always required, although they certainly do help.
- Address questions/hypotheses in a simple context to avoid unnecessary complexity (the principle of Ockham's razor; Peters 1991). An example of the hierarchical process is shown in Table 2.1, and some examples of paired hypotheses and questions from some recent studies are provided in Table 2.2.
- Familiarity with the literature on a topic at an early stage enables you to better frame your own questions and avoid addressing a question that's been answered before. Modern online search engines have a remarkable utility for researching the published literature on a specific chosen topic, but also for finding and developing a suitable research topic.

2.2 Ways of addressing the question

As described in Section 2.1 and shown in Fig. 2.1, the hypothesis forms a central part of the hierarchical order of operational terms used in framing the question(s) critical for a study. In this section, the use of hypotheses in what is known as the *hypothetico-deductive* approach (i.e. classical hypothesis testing) is described. The use of two alternatives to this approach has been advocated, namely the likelihood approach and Bayesian analysis (see Chapter 7).

The philosophical basis for classical hypothesis testing is the existence of an objective universe independent of human observations and human beings (Suppe 1977). Assuming this to be true, models can be built and theories erected about the functioning of this universe. These are assumed to be part of some grand truth that we are seeking to find, and our theories and hypotheses are derived through induction and deduction. Inductive reasoning (induction being the 'inferring of general law from particular instances'; Sykes 1976) allows us to derive patterns from non-experimental observations that we make. From these patterns we can induce generalizations that we call theories. Returning to the beetle example from Section 2.1, from observations of poor seed set in plants infested with beetles we might induce that beetle herbivory reduces plant fitness. However, we have to be careful. Induction cannot be used to prove a theory to be true, because no matter how much supporting evidence we have we cannot be sure that contrary evidence

Table 2.2 Paired hypotheses and research questions from five plant population ecology studies. In all cases the hypotheses and questions appear in the introduction to the study.

Study	Hypothesis	Questions
Fischer and Matthies (1998)	Plant fitness is reduced in small populations because of genetic factors	(1) Are population growth rate and plant performance in *Gentianella germanica* associated with plant size? (2) Is this correlation due to habitat parameters? (3) Do progeny from plants in large populations have a higher relative fitness than progeny from plants in small populations?
Christianini and Oliveira (2010)	The fruit crop size hypothesis: plants producing large fruit crops are likely to attract a greater number and variety of frugivores (and thus have increased seed dispersal success) compared with plants with small fruit crops	(1) Does the crop size hypothesis account for among-plant variation in the amount of crop removed away from the plant crown? (2) What are the spatial scales at which bird and ant seed dispersal operate? (3) What is the role of each dispersal vector for plant regeneration?
Guo et al. (2012)	Populations and taxa living at high elevations should exhibit a lower log reproductive versus log vegetative mass slope than those in more benign environments	(1) What are the relationships between reproductive (R) and vegetative (V) biomass within populations and among species of *Pedicularis* growing at different elevations? (2) How do the slope and intercept of the log R–log V relationships change with increasing elevation among populations and among species?
Bode and Kessler (2012)	Long-term exclusion of herbivores would result in differential natural selection on plant resistance traits that are expressed by marked differences in mean resistance between plants from populations with and without herbivores. (There were also four additional and more specific hypotheses)	Is there diffuse selection by multiple species on a generalized resistance to herbivory? Or, is resistance specific to individual herbivore species or small groups of species?
Hernández-Barrios et al. (2012)	Females have higher costs of reproduction, expressed as lower tolerance to defoliation, than males	(1) What are the effects of different levels of repeated defoliation on survivorship, growth, and reproduction of females and males? (2) Does the reproductive function of females imply higher costs in terms of growth, survival, and reproduction than that of males? (3) Are females less tolerant to repeated defoliation than males? (4) To what extent are there differences in these responses to defoliation among closely related species?

will not be found (Popper 1968, 1983; Loehle 1987). By contrast, deductive reasoning (deduction being 'inference from general to particular'; Sykes 1976) develops theory from a set of assumptions about the universe. Using this approach, we might deduce the similar theory that 'herbivory reduces plant fitness' from the general idea that hungry herbivores, such as beetles, might eat the reproductive structures and hence reduce seed set. We can arrive at this theory through deductive reasoning without any prior observations.

We need not worry about trying to prove or disprove theories at this stage. In practice, scientists use inductive and deductive reasoning to develop hypotheses as well as theories. The former is what we work with, on an everyday basis, in the lab or field. The latter are developed later in summarizing areas of the discipline as a whole (e.g. Tilman's models of plant community structure based upon plant life-history strategies; Tilman 1988). Whether derived through deduction or induction, hypotheses are the basis for developing specific research questions. Loehle (1987) presents a discussion of how hypotheses are developed and used in ecology. He notes, 'Dreams, crystal balls, or scribbled notebooks are all allowed. In fact, induction may be used to create empirical relations . . . even though induction cannot be used to prove anything'.

Let us now return to the use of hypotheses in classical hypothesis testing. With this approach we confront a model with our data. Hence, we begin with the following hypotheses (Hilborn and Mangel 1997):

H_o: model M_1 is true

H_a: some other model (M_2) is true

The alternative hypothesis (H_a) is what we believe to be true if our original hypothesis (H_o) is not supported (falsified). We derive the precise questions for our study from H_o, and if that does not work out then we move on to test H_a.

The phrasing and interpretation of H_o as 'model M_1 is true' can be problematic since it may blur phenomenon and mechanism. Suppose the model being tested is the –1.5 thinning rule. This rule predicts density-dependent mortality of the smallest individuals in the population as the cause of a commonly observed –1.5 slope relationship between mean plant density and mean plant yield (White and Harper 1970). Let us further suppose that the mortality patterns of seedlings planted at various densities in an experiment match the predictions of this model. We might reasonably conclude that we have proved, or at least supported, the model. However, it is possible that, unbeknown to the investigator, the actual cause of the observed pattern is a herbivorous insect that attacks only when a population reaches a certain biomass. However, if H_o is framed in a more general sense (e.g. mortality is higher in high-density populations) then we are forced to face the fact that we have seen the expected pattern (supported H_o) but have not still demonstrated the mechanism.

This conundrum suggests that the nature of a hypothesis needs to be considered carefully. As a supposition, or starting point for further investigation (Table 2.1), an ecological hypothesis is an educated guess about a process based on prior experience or previously published research, or both (that's why it is an 'educated' guess). However, notice the difference in the form of the hypothesis in Table 2.1 and those in Table 2.2. The hypotheses in Table 2.2 are written in a positive (or alternative; see below) form. They are in plain English and predict exactly what the authors think is going on. By contrast, the hypothesis in Table 2.1 suggests that nothing is going on; that vesicular–arbuscular mycorrhizal fungi have no effect on the competitive balance between the plant species. This hypothesis is written in the 'null' form. *Null hypotheses* are 'reference points against which alternatives should be contrasted' (Strong 1980). In statistical testing of experimental results, the null hypothesis is the hypothesis that is being falsified (Chapter 3).

This idea of attempting to falsify a hypothesis is derived from Popper's ideas on the philosophy of science (Popper 1968, 1983). Popper argued that a hypothesis cannot be proved, only disproved. The reason why it is impossible to prove a hypothesis to be true is that it would require every possible observation to be true (Underwood 1990, 1997). Take the simple hypothesis 'roses are pollinated by bees'. We cannot *prove* this to be true without observing the pollination of every rose; an impossible task. In the *null* form the hypothesis can become 'roses are not pollinated by bees'. The first observation of a bee pollinating a rose falsifies this hypothesis, giving support to the alternative hypothesis (our original hypothesis in this example). If, after repeated observations or experiments, the alternative hypothesis is still supported (i.e. we repeatedly see bees pollinating roses) then it is accepted as being essentially true. This is the case with evolution through the process of natural selection. Darwin's (1859) theory has not been proven correct, but, most importantly, it has not been disproved. All the known facts are consistent with the theory and no one has yet come up with a more parsimonious alternative.

Null hypotheses require some care in writing. Disproof of the null hypothesis should leave the alternative hypothesis as the only possible one and obviate the need to attempt to prove the hypothesis.

This basic view of the scientific process was described by Platt (1964) as 'strong inference' and was likened to climbing a tree, with the branch points representing an experimental outcome. Platt (1964) said that the scientific method consists of the following steps:

1. Devising an alternative hypotheses.
2. Devising a critical experiment (or several of them), with alternative possible outcomes, each of which will, as nearly as possible, exclude one or more of the hypotheses.
3. Carrying out the experiment so as to get a clean result.
4. Recycling the procedure, making sub-hypotheses or sequential hypotheses to refine the possibilities that remain, and so on.

As noted before, the hypotheses are born from inferences drawn from observations, knowledge of the system under study, and concepts and theories already established in the discipline. The idea of this general approach has been reproduced in countless textbooks and lab manuals as *the* process through which science is undertaken (e.g. Fig. 2.1). In reality, the approach taken by scientists can be somewhat more flexible than this (Hodson 1982).

Classical hypothesis testing can lead to problems, especially if models M_1 and M_2 are both rejected or both accepted (Hilborn and Mangel 1997). If both models are rejected then a third model has to be developed. This process of sequential hypothesis testing could take a long time. Equally problematic is the situation when both models are true. For many questions it may not be conceptually or logistically possible to devise clean experiments that are conclusive. The testing of alternative models is often essential and can be accomplished through likelihood and Bayesian analysis (Chapter 7).

The hierarchy of hypotheses (HoH) approach

Sometimes we are caught between making a hypothesis that is too general to be readily tested (e.g. 'fungal pathogens cause seedling mortality') and writing hypotheses that are essentially trivial or too specific ('the density of seedlings of species X will increase with distance from parent trees as the abundance of pathogen Y decreases in lowland areas of my favourite nature reserve'). The HoH approach (Jeschke et al. 2012; Heger et al. 2013) hierarchically structures major hypotheses into smaller sub-hypotheses. This approach has been developed for invasion biology (Fig. 2.2), but could also be easily applied to plant population ecology. Basically, this approach is saying that a general hypothesis (e.g. the lack of eco-evolutionary experience hypothesis in Fig. 2.2) is of interest and is tested through a number of sub-hypotheses, which themselves can be subdivided up into sub-sub-hypotheses. The value of this approach is that it allows a research programme to be effectively planned out by proposing specific variables to be measured.

2.3 What should be measured? Ecological variables

So far in this chapter the philosophical approach taken in science, and in plant population ecology specifically, has been discussed. We now consider how one should choose the variables to be measured; for this then determines the experimental design. Experimental design, the contrast between the experimental approach and observational, comparative, and modelling studies, and what each can and cannot deliver are considered in Chapter 3.

Ecology is an empirical science, and even modelling (Chapter 8) ultimately requires data from the real world. It is necessary to decide what need to be measured in order to address the hypotheses and questions posed. The *objects* are the samples,

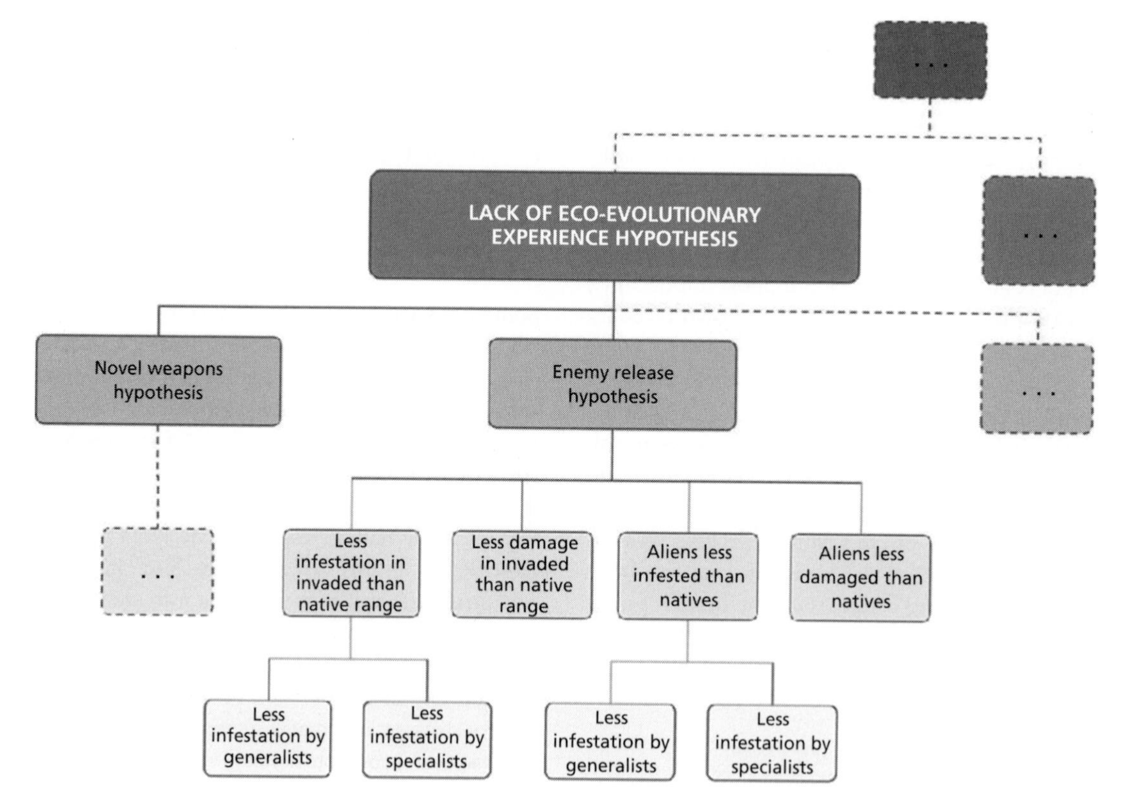

Figure 2.2 Example of a hierarchy of hypotheses for invasion ecology. Overarching ideas branch into more precise, better testable hypotheses at lower levels. Empty boxes indicate that the hierarchy may be extended. From Heger et al. (2013). With kind permission from Springer Science and Business Media.

locations, observations, sampling units, or subjects that are defined ahead of time by the sampling design and before making the observations (e.g. plant, leaf, seed, population, ramet). *Variables* are the actual properties measured by individual observations in a study (Sokal and Rohlf 1995) (e.g. length, colour, thickness, reproductive stage, density, height). More generally, the term 'variable' is also used to denote any factor that can affect an experiment. The *states* of a variable are the qualities observed, and they belong to various types as shown in Table 2.3 (e.g. x cm long/wide/thick/high, fertile, male, or yellow).

Variables fall into one of two categories (Table 2.3). *Dependent* (or *response*) *variables* are variables that we are seeking to predict and are uncontrolled in experiments. They may be mentioned in the hypotheses and include the plant variables measured by the investigator. *Independent* (*explanatory*) *variables* are variables that are free to vary if continuous, are causative (although not necessarily in relationship to the dependent variable), and may be set by the experimental treatments. The independent variables can be categorized into five classes (Table 2.4). There is nothing predictive about this list, rather it helps us isolate what should be measured (Peters 1980).

Identification of the dependent and independent variables and their states informs the experimental design and is important in data analysis (Chapter 7). Table 2.5 lists the objects, variables, and states for Case Study 4 in Chapter 1. The relationship between post-fire seedling density and the pre-fire tree (cone) density of the conifer *Widdringtonia nodiflora* in South African fynbos is shown in Fig. 2.3. In this example both variables are discontinuous, with seedling density being the dependent variable and cone density the independent variable

Table 2.3 Ecological variables (after Legendre and Legendre 2012). Note that, depending on the study question, most of the dependent variables listed here might on occasion be independent variables (e.g. fitness may depend on the number of flowers produced). Reproduced with permission from Elsevier.

Type of variable	Examples of dependent variables	Examples of independent variables
Binary	Presence or absence of flowering structures, rhizomes, leaf trichomes. Genetic data (RFLP, RAPD, and microsatellites)	Presence or absence of herbivory. Ambient versus elevated CO_2 or ozone
Multistate		
Non-ordered (qualitative)	Leaf and flower colour. Isozyme data	Species of herbivore. Soil type. Vegetation type
Ordered		
Semiquantitative (rank-ordered, ordinal)	Leaf position from bottom to top of flowering stem. Branching order. Ranking of herbivore damage (e.g. none, leaf damage, leaf removal)	Ranking of soil drought or fertility status (e.g. dry, moist, flooded)
Quantitative (measurement)		
Discontinuous (meristic, discrete)	Number of plant parts (e.g. flowering stems, flowers, pollen grains). Number of plants surviving or dying over a time period. Density measures	Number of herbivores per target plant. Number of species of pathogens
Continuous	Length, biomass, area. Fitness. Finite rate of growth. Physiological measures. Nutrient status	Temperature, soil moisture, light, humidity. Soil nutrients and pH

RFLP, restriction fragment length polymorphism; RAPD, random-amplified polymorphic DNA.

Table 2.4 Categorization of measurable independent variables. After Peters (1980).

Category	Examples
1. Characteristics of the plant itself	Biomass, age, foliar nitrogen content, stomatal conductance, leaf area
2. Characteristics of organisms of the same kind (i.e. intraspecific neighbours)	Density, mortality
3. Characteristics of organisms of other kinds (i.e. interspecific plant and animal neighbours)	Density, chemical composition, level of herbivory
4. Physical and chemical factors	Light, nutrient levels, frost frequency, mean annual temperature, etc.
5. Space and time	Seasonality, geographic distribution, phenology

Table 2.5 Ecological variables (from Table 2.3) in Case Study 4, a study of the effects of leaf damage on plant fitness (see Chapter 1).

Object	Variable	Type	State
Plant	Flower number	Dependent: discontinuous	Count
	Reproductive biomass	Dependent: continuous	Grams dry weight
	Total biomass	Dependent: continuous	Grams dry weight
Leaf	Leaf damage	Independent: non-ordered (qualitative)	Six different categories of damage[1]

[1]The six categories of leaf removal were: (1) control, (2) one mature leaf, (3) one new leaf, (4) 50% of two mature leaves, (5) 50% of two new leaves, (6) 25% of all leaves.

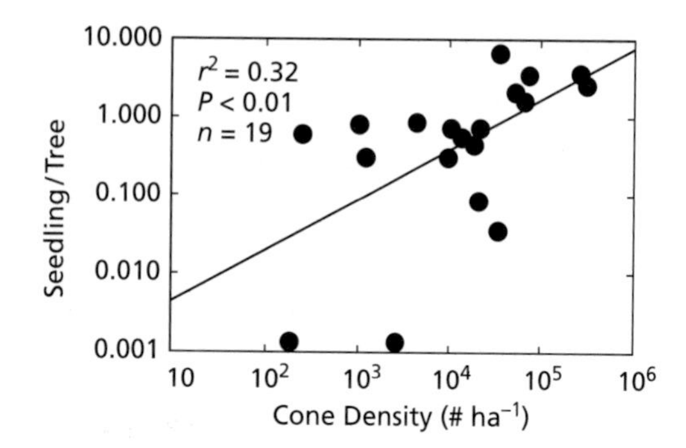

Figure 2.3 Density of post-fire seedlings of *Widdringtonia nodiflora* related to pre-fire cone density. From Keeley et al. (1998). Reproduced with permission from John Wiley & Sons.

(Keeley et al. 1998). The conifer has serotinous cones which open following fire, allowing seedling recruitment. The more cones there are before a fire, the more seedlings are observed afterwards. The seedling count must be the dependent variable since it is temporally constrained to come after tree density. By contrast, Fig. 2.4 shows the relationship between a continuous dependent (and derived) variable (root:shoot ratio) and a multistate, ordered, semiquantitative independent variable (soil patch type) from a greenhouse experiment on the clonal growth habit of the herbaceous plant *Glechoma hederacea* (ground ivy) (Wijesinghe and Hutchings 1997). In this experiment, the root:shoot ratio within clones of *G. hederacea* was dependent on soil quality (high versus low nutrients) and heterogeneity (patch size). These data are consistent with a theory of optimal foraging for this clonal plant.

Binary variables are simple presence/absence variables, whether we are seeking to predict them in a study (e.g. whether a plant flowers or not) or using them to code simple treatments (e.g. whether plants were fertilized or not). Statistically, these variables are coded as + or –, or as 1 or 0 to denote presence or absence. As data, binary variables can be quite easy to collect and may be as valuable as quantitative data for measures of plant abundance.

Other variables are referred to as multistate because they take more than two forms. Care has to be taken not to camouflage non-ordered or qualitative variables (e.g. different soil groups that have no order or sequence unless they can represent a fertility or moisture gradient) as ordered variables. Non-ordered variables have to be coded in such a way (e.g. A, B, C) as to avoid the temptation to order them

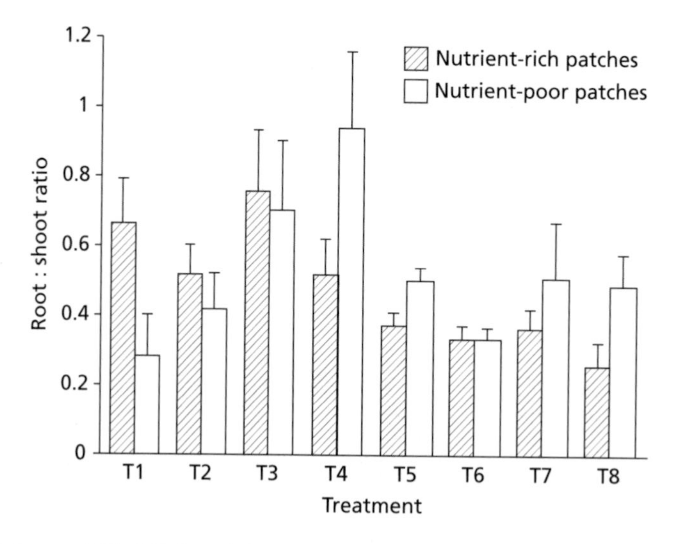

Figure 2.4 Root:shoot ratio (mean + 1 SE) of *Glechoma hederaea* in nutrient-rich and nutrient-poor soil patches declining in scale from 1250 cm^2 (T1), 625 cm^2 (T2), 312.5 cm^2 (T3), 156.25 cm^2 (T4 and T7), 56 cm^2 (T5) to 39 cm^2 (T6 and T8). The patch sizes in T7 and T8 were the same as those in treatments T4 and T6, respectively, but the nutrient-rich and nutrient-poor patches in T7 and T8 were physically separated by thin partitions so that roots were prevented from growing between adjacent patches (there were barriers between patches in T1–T6). See also Fig. 4.3. From Wijesinghe and Hutchings (1997). Reproduced with permission from John Wiley & Sons.

(e.g. do not code them as 1, 2, 3). These variables can also be reduced to a series of binary dummy variables if necessary for regression analysis (e.g. in soil group 1 or not in soil group 1; in soil group 2 or not in soil group 2, etc.).

Semiquantitative ordered variables can be used for the rapid enumeration of field data which are ranked. For example, the terrain in which a population occurs might be described as being flat, undulating, rough, hilly, or mountainous. These data are clearly ranked from flat to mountainous and can be coded as 1, 2, 3, 4, and 5 for analysis. Similarly, we may be able to rank a series of soil fertility treatments from a low to a high nutrient status (remember not to do this when such a ranking is not possible). Ranked responses (a few seeds, some seeds, or many seeds) can be used directly in non-parametric statistical analyses (Chapter 7). Quantitative data can also be readily transformed to ranked data for use in these procedures when necessary, such as when problems of non-linearity and normality arise (again, see Chapter 7).

Quantitative variables include all descriptions of abundance or other measurements that can be plotted on the continuous axes of a graph, and are of two types—continuous and discontinuous. Continuous variables can assume an infinite number of values between two end points (minimum and maximum). For example, there are an infinite number of possible lengths between 1.5 and 1.6 m. In practice, any reading taken of a continuous variable is an approximation to the exact reading, which is unknown. It is only as accurate as the resolution of the instrument; for example, most top-loading balances report values to only hundredths of a gram.

By contrast, discontinuous quantitative variables have fixed numerical values; intermediate values are not possible. For example, the number of seeds produced per fruit could take values of 0, 1, 2, 3, etc., but 1.5 seeds is obviously impossible. When these variables are analysed they frequently assume a continuous nature (e.g. the mean of 0, 1, 2, and 3 is 1.5). Caution needs to be used in reporting these averaged values. It is perfectly correct to report that a fruit produces an average of 1.5 seeds (plus or minus a specified range), but do not expect to ever find 1.5 seeds in a single fruit.

Derived variables

The variables described above are recorded as direct measurements or counts. However, a number of important variables such as the root-to-shoot ratio, the leaf area ratio (Chapter 5, Table 5.9), and several plant interaction indices (Table 5.1) are based on the relationship between two or more independently measured variables. These derived variables include ratios, percentages, indices, and rates. In some studies a ratio of ratios is even calculated; for example, the ratio of relative growth rates under two experimental conditions. While these measures can provide a useful indication of plant response, they have to be used with extreme caution and with reference to plots of the original variables. The following problems with derived variables have been identified (Sokal and Rohlf 1969; Green 1979; Jasienski and Bazzaz 1999):

1. They are inaccurate and biased estimates of the true mean value.
2. They have non-normal distributions (and hence are problematic for many forms of statistical analysis). For example, percentages have limits at 0% and 100% often leading to rectangular (flat) distributions.
3. They exhibit misleading dynamics, both temporally and ontogenetically.
4. They are liable to introduce spurious self-correlations between variables. This is a particular problem when the denominator variable is highly variable.
5. Their use leads to reduced statistical power and large Type II errors.

Derived variables can be used and may be appropriate for scaling data when treatment effects are large (so just about any analysis will work) or when the variability of the numerator variable is substantially greater than the variability of the denominator variable (Packard and Boardman 1988).

Allometric analyses (i.e. based upon differential growth of plant parts within a species) or analysis of covariance (Chapter 7) are generally preferred when dealing with the measurement of plant growth (Packard and Boardman 1988; Weiner and Thomas 1992). *Allometric analysis* is based upon the principle that there is not a simple, unchanging proportion between plant parts within populations either at one point in time or through time as individuals grow (this would be isometric growth). Furthermore, even when there is a linear relationship between two size variables (e.g. mass and height) it does not intercept the origin (i.e. at zero mass and zero height). Thus, the general form of an allometric relationship (the 'allometric equation') is

$$Y = aX^b$$

where X and Y are the measured plant variables being compared and a and b are constants (Weiner and Thomas 1992).

For example, extending the basic allometric equation to three variables, Shipley and Dion (1992) found that seed number per ramet (N_s) was related to average seed weight (W_a) and vegetative ramet weight (W_y) in 57 herbaceous angiosperms according to the allometric relationship $N_s = 1.4W_y^{0.93}W_a^{-0.78}$. The allometry of reproduction in plant populations is reviewed by Weiner et al. (2009).

2.4 The importance of pilot studies

Unless one is already very familiar with the plants, their habitat, and the organisms in the habitat then it is advisable to carry out a pilot study or at least collect some preliminary data. Doing so will almost certainly save a lot of time in the end. Pilot studies allow for a preliminary testing of the efficiency of field techniques and an assessment of experimental design. Modifications to techniques, sampling devices, etc., can be made at this stage. If this is not done then a whole field season or more could be wasted. Green (1979) provides a salutary example of how he wasted a field season because he tried to use an ineffective sampling device without first testing it.

Preliminary work also helps one gain familiarity with the study organisms and their habitat. Basic natural history data (e.g. habitat characteristics, associated plant taxa, phenology) and reading the literature on the system and species will help one to frame sensible and realistic hypotheses. Although not a botanical example, Dayton (1973) presents a good example of how insights from natural history allowed the correct mechanistic interpretation to be made from model predictions of an intertidal community. The models had been 'making the right prediction for the wrong reason'.

It is a useful exercise to ask how long it will take to collect the data that you are planning on collecting. If your design calls for each plant in your study to be screened for mycorrhizal colonization, then how long will it take you to collect and process each sample? How variable are your plants in their degree of mycorrhizal colonization? Would a simple presence/absence screening be sufficient, or do you need the percentage colonization of each root system? If the plants are to be collected from the field then how easy will it be to separate the roots of your target species from those of other plants? How easy will it be to remove the soil from the roots (clay is difficult, sand is easy)? Similar questions should be asked about all parts of your proposed study. A pilot study with preliminary data will greatly help in making this assessment. The logistics of planning fieldwork are discussed in detail in Chapter 3.

Preliminary data collected during a pilot study or the variability reported by others for the same system or similar species can allow an assessment of the variability to be expected in your own data. These data can be used to provide estimates of the optimal sample size and statistical power necessary to obtain the desired precision and accuracy (Chapter 3). Pilot studies can also be used to screen large numbers of populations for a trait of interest for more involved experiments. These screening studies have an obvious application in applied fields. N'Guessan and Quisenberry (1994) screened 50 lines of rice in order to find some that were resistant to the rice water weevil (*Lissorhoptrus oryzophilus*). Further field testing was then carried out on the resistant lines.

The data from pilot studies can be used in the construction of dummy data sets. These consist of data in the form that you expect to collect them. For example, if you plan on comparing the number of flower stems on 10 plants in two habitats each, then make up a data matrix of two columns (the two habitats) by 10 rows (the 10 plants in each habitat). Plug in random numbers and practice running the data through the proposed statistical analysis. This is an easy example, but try making the data matrix for the analysis reported for the

McEvoy et al. (1993) field experiment on *Senecio jacobaea* (Case Study 3 in Chapter 1). This dummy data set would be a lot more difficult to construct. Recall that they had 24 treatment combinations applied to 96 plots (four-way randomized blocks analysis of variance). Some variables are measured on every plant in a plot (e.g. reproduction), others were measured on a plot basis (e.g. ragwort cover). The exercise of setting up dummy data sets will help ensure that you know how to handle your data as you collect it, and to decide which statistical procedures to use. More importantly, it will ensure that you collect only those data that are useful for your study.

Pilot studies are critical, but the data cannot usually be combined with the *real* data from later on. Many funding agencies now expect to see the results of a pilot study in grant applications. Indeed, a common perception among many researchers is that you need to have practically completed the study before a funding agency will believe that you are capable of doing it! However, the data from a pilot study can certainly form an important part of any final report to an agency (indeed many first-year reports to funding agencies consist of little more than preliminary data) and may even be publishable in their own right.

Schmid (1994) provides a plant population example of a pilot study. He grew 24 clones and 24 half-sib families derived from 19 original parent clones of *Solidago altissima* (tall goldenrod) in field experiments to assess resistance to *Erysiphe cichoracearum* (mildew). This preliminary experiment revealed significant genetic variation to pathogen resistance in the source population. A subsequent field experiment on plant pathogen resistance assessed the effects of genetic diversity within plots on pathogen level and plant growth. The four least mildew-resistant and the four most mildew-resistant parents from the preliminary experiment were then used in this field experiment. Unpublished preliminary data were also alluded to in this paper. Schmid noted that morphological and quantitative genetic characters showed that individual genets identified in the field represented a single clone with a unique genetic identity.

Goldberg and Novoplansky (1997) used the results of a literature survey as a 'preliminary test' of their two-phase resource dynamics hypothesis. This hypothesis states that plants may experience two phases of resource availability: pulse periods when resource levels are high and growth occurs, and inter-pulse periods when resources are low and resource-induced mortality may occur. The authors hoped that publication of their idea and the support for their hypothesis from the literature would stimulate future critical experiments.

2.5 Follow-up exercises

1. Construct a table similar to Tables 2.1 and 2.5 for Case Studies 1–3 from Chapter 1.
2. Read the introduction to some papers in journals such as *Journal of Ecology*, *Ecology*, *Ecology Letters*, *Journal of Applied Ecology*, *Oecologia*, and *Oikos*. Identify the authors' hypotheses and research questions.
3. What sort of pilot study and preliminary data do you think might have been required for Case Studies 1–4? How about for your own research? How will, or how would, a pilot study have helped?

CHAPTER 3

Basic considerations in experimental design

I need hardly commence by telling you that my opinions are nothing, that the whole question wants working out by observation as you are doing, & by experiments which no one has attempted or even suggested that I know of. **(J. D. Hooker 1862, letter to H. W. Bates)**

The importance of field experiments cannot be overestimated. . . . The experimental method simply means the observation of processes under controlled conditions.

(Weaver and Clements 1938)

- **Design considerations: experiments and observations**
- **Where and how should you conduct your study? Choosing sites, plots, and plants**
- **Statistical issues in conducting experiments**
- **Darwin's 'vitality of seeds' experiment: the first plant population ecology experiment? (Box 3.1)**

Preamble

In this book it will be reiterated many times that the questions and hypotheses to be tested will direct the experimental design. In this chapter the important considerations that are used to test and address the study questions and hypotheses are described. Like the chicken and the egg (which came first?), the order in which these decisions are made is important, and will vary from one study to another. In some cases, particular hypotheses dictate a study organism, which then dictates the location of the field sites (wherever the plants can be found, perhaps). In other situations the identity of the taxa may be fairly irrelevant to the phenomena or mechanism of interest. In these cases the plants may be chosen later after aspects of the experimental design have been worked out (e.g. 'Three annual plant species are needed for this competition study').

An *experiment* has been defined as 'a deliberate action undertaken in order to interfere with phenomena in such a way as to compel a decision concerning a hypothesis' (Feibleman 1972). Generally speaking we can refer to any activity which confronts a hypothesis with relevant facts following a test or trial as an experiment. However, there is also an important place for observational studies, in which there is little direct interference with phenomena. The two approaches are complementary (Werner 1998). Indeed, we can classify some observational studies in which there is a clear perturbation of phenomena and a before–after structure as a type of natural experiment. Medawar (1969) said that as scientists we should be conducting 'demonstrative or Aristotelian' experiments. These are intended to 'illustrate a preconceived truth and convince people of its validity'. In other words your experiments should explicitly address your hypotheses.

In Section 3.1 of this chapter the advantages and limitations of different types of experiments are discussed. In Section 3.2 the topic of how to choose sites, plots, and plants is discussed, and in Section

Methods in Comparative Plant Population Ecology. Second Edition. David J. Gibson.

3.3 a number of statistical issues that are important for conducting experiments are introduced. These issues serve as an introduction to Chapters 7 and 8 that cover the use of statistical tests in plant population ecology. Finally, as an aside, one of the earliest plant population ecology experiments is discussed.

Whatever experimental design is used, remember that your hypotheses and questions should determine your design. Always think carefully about what you need to do to confront the hypothesis. Do not use a particular design just because it is commonly used (trendy). Remember also that you are as much concerned with discovering evidence for your hypotheses as against them. In fact, a negative result, if it is the result of a good study design, may be as important as a positive one (see Type I and Type II errors discussed later), and can make an important contribution (Loehle 1987). Negative results are difficult to publish, but can be useful for challenging established ideas. For example, ecotypes tolerant of heavy metals are typically found in populations growing on soils contaminated with high levels of lead, zinc, and cadmium. However, the failure to find these ecotypes developing in certain situations challenges the generality of this phenomenon (e.g. Egerton-Warburton 1995; Gibson and Risser 1982). Similarly, the production of potentially allelopathic compounds in members of the Brassicaceae does not have to mean that allelopathy is involved in the success of members of this family invading woodlands as exotic weeds (e.g. McCarthy and Hanson 1998). In accepting the conclusions from these studies we have to accept the author's contention that the experiments were an effective test of his or her hypotheses.

3.1 Design considerations: experiments and observations

Experiments can be categorized in two different ways. First, *mensurative* versus *manipulative* experiments—the contrast between whether the system is simply measured (the former) or manipulated and then measured (the latter) (Hurlbert 1984; Underwood 1997). Case Studies 1 and 2 in Chapter 1 are mensurative experiments because the investigators in each case simply measured the plants that were already growing at their field sites. They did not manipulate their system. By contrast, Case Studies 3 and 4 are manipulative experiments because in each the investigators manipulated the experimental system. In Case Study 3, cages were set up to keep out herbivores, the soil was disturbed, and background vegetation was removed around the target plants. In Case Study 4, different levels of leaf damage were applied to the target plants to investigate the effects of herbivory. Both mensurative and manipulative experiments are equally valid so long as the data collected provide a realistic test of some hypothesis.

A second way of categorizing ecological experiments recognizes four basic types (Diamond 1983). These are: (1) *lab*, (2) *field*, (3) *natural trajectory*, and (4) *natural snapshot* experiments (Table 3.1; and see examples in Table 3.2). For the plant population ecologist, lab experiments usually mean greenhouse and growth chamber studies. Lab experiments have the advantage that the investigator is able to regulate or hold constant across experimental units most biotic (e.g. plant age, herbivory, disease) and abiotic (e.g. temperature, light, moisture, humidity, soil) factors including, of course, the treatments (see Section 3.3).

Table 3.1 Different types of experiments. After Diamond (1983, 1986).

	Lab	Field	Natural trajectory	Natural snapshot
Control of variables	High	Medium	None	None
Spatial scale (ha)	<0.01	<1	Unlimited	Unlimited
Ability to follow changes	Yes	Yes	Yes	No
Realism	Very low	Medium	High	High
Generality	None	Low	High	High
Repeatability	High	Medium	None	None

Table 3.2 Selected literature for examples of plant population ecology studies using different types of experiments. All the lab experiments in these examples were conducted in the greenhouse.

Type of experiment	Topic	Reference
Lab	Genetic influences on competitive effect and response in *Arabidopsis thaliana*	Cahill et al. (2005)
Lab	Differences between clones in response to competition	Gibson et al. (2014)
Lab	Spatial scale and environmental heterogeneity effects on a clonal herb	Wijesinghe and Hutchings (1997)
Lab and field observations (snapshot)	Maternal determinants of seed dispersal in an annual plant	Donohue (1998)
Lab and field	Herbivore effects on volatile emissions from *Trifolium repens*	Kigathi et al. (2009)
Lab and field	Leaf litter effects on seedling establishment	Molofsky and Augspurger (1992)
Field	Deer effects on an understorey forb	Augustine et al. (1998)
Field	Differences in resource allocation driven by generalist versus specialist herbivory.	Huang et al. (2010)
Field	Growth and phenology of eight common weed species	Hegazy et al. (2005)
Field	Shrub protection against browsing of oak seedlings	Jensen et al. (2012)
Field	Population response to influence of light herbivory in an understorey shrub	Mooney and Niesenbaum (2012)
Natural trajectory	72-year changes in the demography of desert plants	Goldberg and Turner (1986)
Natural trajectory	Relationship of long-term population growth rates to climatic variability	Dalgleish et al. (2010)
Natural trajectory	85-year demographic study of saguaro cacti	Pierson and Turner (1998)
Natural trajectory	9-year study of the population dynamics of an annual grass	Watkinson (1990)
Natural snapshot/snapshot	Molecular analysis for understanding the long-term effects of herbivory on a wild population	Herrera and Bazaga (2011)
Natural trajectory/snapshot	Effects of environmental variability on the demography of three populations of a perennial herb over 7 years.	Toräng et al. (2010)
Natural snapshot	Comparative demography of semelparous and iteroparous *Lobelia* spp. on Mount Kenya	Young (1984)
Natural snapshot	Regeneration dynamics of subtropical thickets	Sigwela et al.(2009)
Natural snapshot	Life history and population structure of a cycad	Watkinson and Powell (1997)
Natural snapshot	Hurricane effects on regeneration of mangrove forest	Baldwin et al. (1995)

The number of test plants is carefully controlled and plants of a specified age or ontogenetic stage can be carefully placed in particular patterns or arrangements under controlled environmental conditions. Lab experiments should be repeatable if they are carried out carefully. Mechanisms of interaction such as competition can be readily studied in lab experiments (Gibson et al. 1999). However, the results of lab experiments have to be interpreted with care. Unless phenomena can be demonstrated under these highly controlled conditions then they are unlikely to be of importance in natural populations. But a greenhouse experiment has to be carefully carried out to ensure that phenomena can be demonstrated. For example, if leaf herbivory is important in the field, but only when plants are drought stressed, then a greenhouse experiment would have to include drought stress to demonstrate the importance of herbivory. Conversely, when phenomena are demonstrated under these artificial conditions, there is no certainty that they occur (at all, or in the same way) in nature. The disadvantages of lab experiments include the small spatial scale and a lack of reality compared with field situations.

Generalizations to nature can be made only with extreme caution. For example, based on greenhouse experiments using well-watered pots, plant competition below ground is often assumed to be symmetric (i.e. effects are in proportion to the relative size of the competitors). However, in the field, where there are likely to be strong moisture gradients, competition will tend to be more asymmetric (disproportionate to size, with the largest individuals often getting the lion's share of the resource). In this case the artificial nature of greenhouse conditions can lead to a misleading understanding of the process of belowground competition.

Field experiments are conducted outdoors and include artificial, sown populations as well as manipulations of natural populations. Common manipulations include the removal or addition of competitors, herbivores, soil nutrients, or moisture. Case Study 3 is a field experiment because treatments were applied to naturally sown plots of the study species. These experiments retain some of the control of lab experiments but are subject to the vagaries (reality) of the climate. It is because of year-to-year variation (e.g. climate, herbivore densities, disturbances, etc.) that field experiments may not provide the same results if they are repeated. A further disadvantage is that field experiments are a representation of the real world (nature) in which the investigator is trying to control factors that he or she is not interested in. Interpretation has to bear this caveat in mind. Field experiments thus have a high degree of realism but low generality. Generality is increased when repeated field experiments yield the same results, or at least the same interpretation. Using designs drawn from agricultural field trails (see Section 3.3), these experiments are very popular with plant population ecologists and allow both mechanistic and phenomenological issues to be investigated. Field experiments are extremely valuable and help bridge the gap between the study of processes in the lab and the study of patterns in natural experiments (Mertz and McCauley 1980). Reviews of field experiments in assessing the role of plant competition are provided by Aarssen and Epp (1990), Goldberg and Barton (1992), and Goldberg and Novoplansky (1997) (see Chapter 4).

There are two types of natural experiment. Neither of these involves any sort of manipulation by the investigator (hence the difference between natural experiments and field experiments described above). Therefore natural experiments are completely realistic (they are reality) and have the highest level of generality. In this sense natural experiments are equivalent to mensurative experiments. The investigator has no control over the regulation of the independent variables. The first type, natural trajectory experiments, are comparisons of plant populations before and after a perturbation, such as a storm, hurricane, volcanic eruption, or pollution event. These experiments can allow for the effects of perturbations to be followed for decades. Often the trajectory starts as or shortly after the disturbance happens and an undisturbed area is used for comparison. By contrast, natural snapshot experiments compare at one point in time plant populations that differ by factors such as the presence or absence of herbivores, pollution, fire or storm damage, or successional age. Obviously, natural experiments are wholly realistic since they are the real thing, but because of their design (or lack of one) they can be quite difficult to analyse statistically. There may be problems with adequate replication or pseudoreplication (see Section 3.3). Any correlations among factors that emerge from natural experiments provide only circumstantial, although perhaps compelling, evidence of a causative relationship. However, they may serve to suggest variables for subsequent manipulative experiments.

A natural experiment with a single treatment area (e.g. a single storm or pollution-impacted site) can be problematic to analyse, especially if compared with a single, untreated, control area. Replicated sampling in time before and after the impact is helpful but care has to be taken to avoid using either temporal samples or samples within the treatment area as replicates (see pseudoreplication in Section 3.3). Before/after, control/impact (BACI) designs are available to deal with these situations. It is recommended that with only a single treatment area, multiple untreated controls are established along with temporally replicated sampling before (if possible) and after the treatment or impact. Details of appropriate designs and statistical analyses for

BACI experiments are provided by Underwood (1992) and Smith (2002).

The four types of experiments described here are obviously part of a continuum. If the results of a perturbation on a plant are followed before and after the event in several populations then the experiment has attributes of both a natural trajectory and snapshot experiment. Case Studies 1 and 2 fall into this category. Similarly, a field experiment could be staged before and after a planned perturbation; or an investigator with a field experiment designed for something else may have the 'good fortune' to be in the path of a storm (e.g. You and Petty 1991). Since natural experiments address largely phenomenological issues, field or lab experiments are useful follow-ups for addressing the mechanistic aspects. Combinations of experiments, often of different types, are usually necessary in order to produce a full understanding of a particular issue.

Natural experiments which lack user-regulation of treatments are particularly appropriate for understanding long-term patterns in populations, comparative performance in different habitats, and the response to large-scale natural or anthropogenic events. Long-term studies, such as those by Goldberg and Turner (1986), Watkinson (1990), and Pierson and Turner (1998), are valuable for revealing population dynamics. It should be noted that these studies are mechanistic and not phenomenological (see Chapter 2). But there is, for example, no other way of revealing the patterns of population change noted by Pierson and Turner (1998) for the long-lived cactus (*Carnegiea gigantea*) that they studied. Watkinson (1990) studied *Vulpia fasciculata*, an annual grass, and his multi-generational 9-year study showed that population flux was related to sand movement in the dune habitat. Furthermore, the results of his study suggested that *V. fasciculata* was little influenced by interspecific competition, casting doubt on the relevance of earlier lab experiments. By contrast, studies such as those of Baldwin et al. (1995) and Brandrud and Roelofs (1995) illustrate the use of natural snapshot experiments to compare the population ecology of plants under different conditions but at a single time.

The above descriptions of the different experimental approaches should give an idea of the appropriateness and limitations of each and also the kinds of questions that can be approached with each type of experiment. There is an obvious contrast between the manipulative (lab and field experiments) and mensurative (natural trajectory and snapshot experiments) approaches. The observational or comparative methods of the latter do not make them any less 'scientific'. The key is to understand what each can and cannot deliver. As shown by the examples in Table 3.2, lab and field experiments allow quite precise and focused questions to be addressed. It would be very difficult, for example, to study the issues of soil heterogeneity and clonal growth investigated by Wijesinghe and Hutchings (1997) outside of a greenhouse (lab) setting, let alone without some sort of manipulation by the experimenter. Nevertheless, observational studies, either natural snapshot or natural trajectory experiments, may well provide reasonable grounds for conducting a lab experiment.

A review of long-term studies on plant populations is provided by White (1985a). An excellent discourse that considers the general value of experiments in ecology is provided by the contributors to the book by Resetarits and Bernardo (1998).

3.2 Where and how should you conduct your study? Choosing sites, plots, and plants

The following criteria should be used as the basis for deciding where to conduct your study:

1. The most appropriate place to address your hypotheses.
2. Logistical constraints (e.g. site accessibility, travel time, funding).
3. Personal preferences (e.g. love of the mountains, desert, coast, forest, prairie, etc.; loathing or fear of snake-infested swamps, tick-ridden old-fields, altitude, back-country bears, etc.).
4. Anywhere (basic processes occur in all habitats).

The first of the criteria sounds like a glib answer to the question of where you should conduct your study. However, if you are interested in a particular issue, such as the effects of heavy metals upon plant population ecology, then you obviously have to seek out soils that are contaminated with heavy metals. (The present author was interested in this topic for research for a master's degree and found that the nearest sites were 350 km from his university!) But this really is how study site selection should be made—on the basis of the goals of the study. If your hypotheses call for a particular species or group of species to be studied, then obviously you need to find sites with these plants. Logistical constraints are important too; if you cannot afford to travel to the tropics then do not plan your research there. If you need to make frequent measurements of your plants (e.g. measurements of growth through a growing season) then you need to be able to get to the site on a regular basis. You need to consider accessibility. Do you really want to walk several kilometres into a wilderness area carrying a 'portable' gas exchange system (they are heavy)? If your design calls for watering of plants you need to be sure that water is available on site or that you are prepared to carry containers of water (again, water is heavy). If your sample units are widely spaced you must allow for the time needed to move between plots. For example, it takes a long time to hike around several hundred hectares of wilderness area, especially if the terrain is heavily forested and rugged. Appropriateness and logistics aside, it is also true that most of us have distinct preferences about where we want to work (e.g. on the coast, in the mountains). If you don't like your study site at the start of your field work (for whatever reason; it's ugly, infested with ticks or snakes, too remote, too close to urban areas, etc.) then you are going to be really miserable by the end.

The contrasting view expressed by Criterion 4 'anywhere' is that if you are studying basic processes of population ecology then you should be able to address your hypothesis anywhere that plants grow. We would expect to find density dependence, herbivory, pathogen limitation, and resource competition in all communities. This was the view expressed by John Harper in choosing to work with his students at Henfaes field in North Wales for much of his career (see the publication list in White 1985b). He chose this 1-ha coastal grassland grazed by sheep and cattle precisely because it did not have any special, or particularly redeeming, features (J. L. Harper, personal communication).

How you conduct your experiment is important and subject to a number of considerations. Feibleman (1972) noted the following criteria for good experiments. A good experiment must be:

1. Isolated: uncontrolled variables should be eliminated or randomized.
2. Analytic: it should reveal properties beyond ordinary observational powers.
3. Repeatable: this may not be possible for some observational studies.
4. Crucial: it should render a verdict on the hypothesis.
5. Heuristic: it should open up new avenues of research, and pose more questions than it answers.

Life history

Choice of the appropriate study organism will, of course, depend upon the hypothesis to be tested. In some situations a particular plant or plant life form is called for to address specific questions, in which case some special considerations may be required (e.g. for vascular epiphytes see Mondragón 2011). In any event, a consideration of plant life histories is important. The life history of a plant describes (1) how long it lives and (2) how long it takes to reach reproductive size (Silvertown and Charlesworth 2001). There are number of classifications of plant life histories. Plants are typically classified as being either *annual* (having a life cycle that is complete within a single growing season), *biennial* (develops vegetatively during the first season, then flowers and dies in the second season), or *perennial* (lives for three or more growing seasons and may flower and fruit in the second and many succeeding seasons). Trees, shrubs, and lianas (woody vines) are considered as woody perennials. Plants are also classified solely on the basis of flowering frequency: *semelparous* (Latin: 'once' 'birth') or *monocarpic* (Greek: 'one' 'fruit') plants die after flowering once, while *iteroparous* (Latin: 'again' 'birth') or *polycarpic* (Greek: 'many' 'fruit') plants flower more than once. It is important to realize here that semelparous plants are not necessarily annual plants. Bamboos, for example, may live for more than 100 years before flowering once and then dying. The *life* (or *growth*) *form* of a plant depends upon its life history,

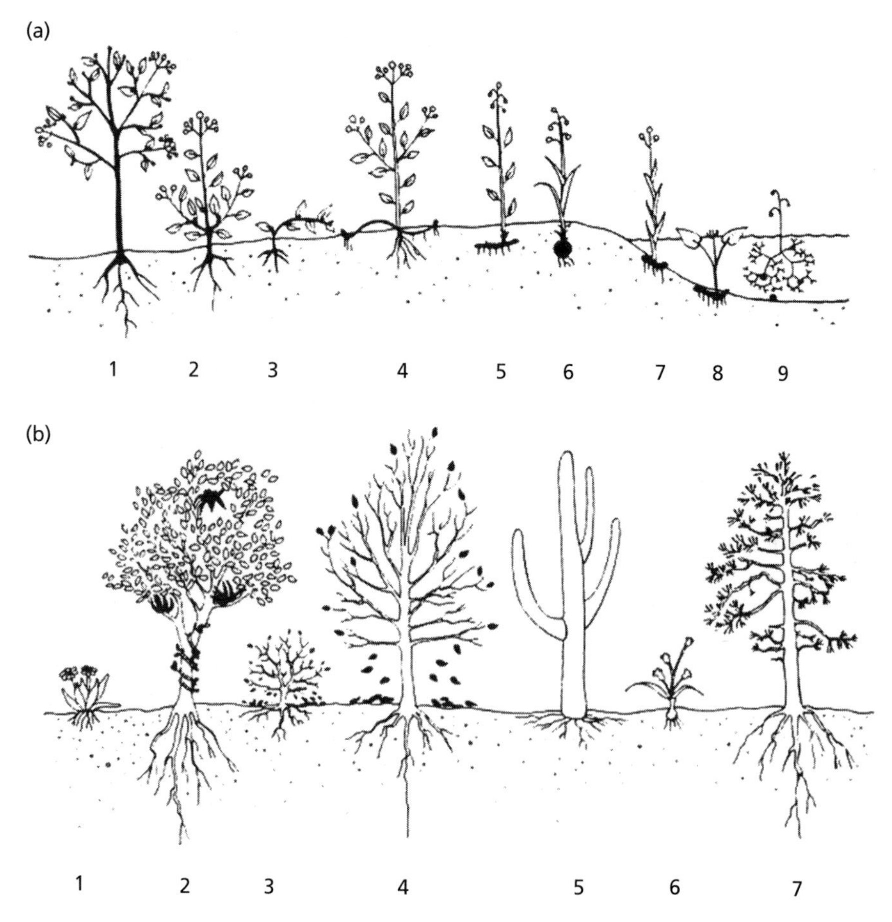

Figure 3.1 Plant life forms. (a) Raunkiaer's (1934) scheme according to the location of overwintering perennating buds, bulbs, and seeds (dark, solid areas): (1) phanerophyte, (2, 3) chamaephytes, (4) hemicryptophyte, (5) cryptophyte (4 and 5 also called geophytes as their buds are in dry ground), (7) helophyte, (8, 9) hydrophytes (4–9 can be referred to collectively as cryptophytes because their buds are below ground or under water). (b) Life forms based upon length of life, succulence, and leaf traits: (1) annual herb, (2) broadleaved evergreen tree with a liana and an epiphyte, (3) drought-deciduous shrub, (4) broadleaf deciduous tree, (5) stem succulent, (6) bulbous herbaceous perennial, (7) needleleaf evergreen. From Barbour et al. (1987).

location of perennating buds, degree of independence (rooted in the ground, parasitic, saprophytic, or epiphytic), and morphology, and is illustrated in Fig. 3.1. This scheme is based upon an older scheme of Raunkiaer (1934) and provides a ready comparison of the life history of a plant with the environment in which it occurs. Raunkiaer's scheme has been found to be useful in relating attributes of plant life history to the environment (Avinoam and Orshan 1990; McIntyre et al. 1995; Kos et al. 2012). For example, Kos et al. (2012) found significantly different representations of seed dormancy classes within Raunkier's life-form classes in a survey of 297 species representing 49% of the flora of the southern Kalahari.

The life history of a particular plant, whatever categorization scheme is used, is a function of the modular construction of the individual taxon. As noted in Chapter 1, a plant which arises as an individual from a seed is a *genet*. Plants are constructed from repeating subunits called modules. A *module* is an axis whose meristem creates all the differentiated structures of a shoot from inception to flowering (White 1980). *Meristems* are localized regions of dividing cells, and in plants they occur in discrete regions such as the root and shoot apices, certain areas of the leaves, and the cork and vascular cambium of woody plants. As defined, a module may be as small as a single apical meristem or as big as the branch of a tree. The size of a plant is a function of the number and size of its modules (e.g. the number of branches, leaves, meristems). This is in contrast to most animals which have the same *number* of modules (e.g. all rabbits have four legs, two eyes, and one tail) which differ in *size* (Jones 1985). The overall structure of a plant is determined by the number and distribution of meristems, and which of them develop into flower buds. Meristems which do not develop into flower buds allow the plant to continue to grow. *Clonal (vegetative) growth* occurs with the production of modules through horizontal growth. Semi-independent modules produced through clonal growth are known as *ramets*. Ramets are modular parts of plants arising through clonal growth; they have their own roots and leaves, and may flower. In grasses, these modules are referred to as tillers. The growth of ramets has important implications for the way in which plants acquire nutrients from the soil and the way in which they respond to competition for light from other plants (de Kroon and Hutchings 1995). The semi-independence or sometimes complete physiological independence of plant modules means that plant populations have to be considered at two hierarchical levels: that of the genet and that of the module (Jones 1985). For this reason, trees have been referred to as a population of buds.

A final life-history categorization that is relevant for choosing plants to study is the distinction between *'phalanx'* and *'guerrilla'* growth forms in clonal plants (Lovett Doust 1981). In phalanx plants the ramets are packed close to each other and the plant stays rooted in the same place on a year-to-year basis. Caespitose bunchgrasses (e.g. *Schedonorus arundinaceus*) offer a good example of the phalanx growth habit. Plants with the guerrilla growth form have ramets that are a long distance apart, often with long rhizomes or stolons. These plants can spread out across the habitat and infiltrate amongst the stems of other genets. White clover (*Trifolium repens*) is a well-studied example of a guerrilla plant. The effect of differences in growth form on the population ecology of two co-occurring grassland herbs, *Bellis perennis* (phalanx type) and *Prunella vulgaris* (guerrilla), was investigated by Schmid (1985a,b) and Schmid and Harper (1985) and high plasticity in *B. perennis* was ascribed to the relatively fixed position adopted by genets because of their phalanx growth form. By contrast, local specialization in the guerrilla plant *P. vulgaris* was argued to be a reflection of the ability of ramets to find favourable patches in the environment.

Popular or 'model' plants

A number of species have become quite popular in plant population ecology and you might wish to consider using these for your study (Table 3.3, Fig. 3.2 and Plates 2–10). The advantage of using 'popular' or 'model' species is the large background literature available as well as the wealth of knowledge acquired by researchers on how to use them in research. The diminutive annual *Arabidopsis thaliana* (thale cress; Brassicaceae) has become established as the botanist's 'fruit fly'. This plant is the subject of intense molecular interest (Koncz et al. 1992; Meyerowitz and Somerville 1994) with its genome having been completely characterized. A large number of different strains and ecotypes are available both free and commercially. The chronological 60-day seed-to-seed progression of the 30 growth stages of the commonly grown wild-type Columbia (Col-0) is well characterized (Boyes et al. 2001). *Brachypodium distachyon* is a self-fertile, inbreeding annual grass with a non-shattering seed head and a short life cycle of less than 4 months. Like *A. thaliana*, its genome has been sequenced (The International Brachypodium Initiative 2010), and with a small genome ($2n = 10$, with five chromosomes) it is being studied as a model system for functional genomics, including the development of new energy and food crops (Draper et al. 2001). Cytotypes with $2n = 20$ and $2n = 30$ have been proposed to be novel species (*Brachypodium stacei* and *Brachypodium hybridum*, respectively) (Catalán et al. 2012). Less frequently studied but similarly useful is the annual *Brassica rapa* (rape or field mustard). Rapid-cycling populations of *B. rapa* have been used extensively in plant breeding research and the teaching of biology and genetics (Tompkins and Williams 1990; Williams and

Table 3.3 Some herbaceous plants commonly used in population ecology research.

Name	Life form	Comments	Examples of applications
Arabidopsis thaliana (Brassicaceae)	Annual	Botanist's fruit-fly. Very small and easy to raise in large numbers. Well understood genetically. Many ecotypes and genetic mutants available	Baskin and Baskin (1972, 1983), Krannitz et al. (1991), Willis et al. (2010)
Brachypodium distachyon (Poaceae)	Annual	New model system for grass genomics. Small and easy to grow	Aronson et al. (1993), Liancourt and Tielbörger (2009)
Brassica rapa (Brassicaceae)	Annual	Often used in teaching. Short life cycle	Miller and Schemske (1990), Stowe and Marquis (2011)
Impatiens capensis, Impatiens pallida (Balsaminaceae)	Annual	Large, succulent. Horticulturally important relatives	Argyres and Schmitt (1992), Barto et al. (2010), Steets and Ashman (2010)
Kochia scoparia (Chenopodiaceae)	Annual	An 'annual tree'	Franco (1986), Franco and Harper (1988)
Trifolium repens (Fabaceae)	Stoloniferous perennial	Agriculturally important clover	Hutchings et al. (1997), Lötscher and Hay (1997), Gautier et al. (1998)
Lolium perenne (Poaceae)	Perennial grass	Agriculturally important ryegrass. Contains fungal endophyte	Wilson and Bell (1990), Marriott and Zuazua (1996), McNeilly and Roose (1996), Sugiyama (1998)
Schedonorus arundinaceus (syn. *Festuca arundinacea*) (Poaceae)	Perennial grass	Tall fescue. Agriculturally important forage species. Contains fungal endophyte and toxic alkaloids .	Clay (1998), Scheneiter and Améndola (2012), Rúa et al., (2013)
Glechoma hederacea (Lamiaceae)	Stoloniferous perennial	A stoloniferous, clonal perennial.	Widén and Widén (1990), Wijesinghe and Hutchings (1997), Roiloa and Hutchings (2013)
Solidago spp. (Asteraceae)	Clonal perennials	Well-defined ramets, attached to persistent, short and unbranched rhizomes	Schmid et al. (1988), Abrahamson and Weis (1997), Stoll et al. (1998)

Hill 1986). *Arabidopsis thaliana, B. distachyon,* and *B. rapa* are easy to grow, and by virtue of their small size and short generation time (30–35 days) can be raised in large numbers. Seed of *B. rapa* can be obtained commercially from Carolina Biological Supply, Inc., North Carolina, United States. Sometimes described as an annual tree, *Kochia scoparia* (summer cypress) is a broadleaf native to Europe and is a problematic weed in western North America. It is particularly suitable for addressing questions about the modular nature of plant growth. Perhaps because of their showy flowers and interest in the horticultural species, *Impatiens capensis* (orange touch-me-not) and *Impatiens pallida* (yellow touch-me-not) have been used in a large number of ecological and evolutionary studies. Native to North America, these two annual herbs are common in shady habitats of the eastern deciduous forest. *Impatiens pallida* grows rapidly in dense self-thinning stands (Thomas and Weiner 1989).

There are a number of perennial plants that are frequently the focus of population ecology studies. These include *Solidago* species (goldenrods), *Glechoma hederacea* (ground ivy), *Trifolium repens* (white clover), *Lolium perenne* (perennial ryegrass), and *Schedonorus arundinaceus* (syn. *Festuca arundinacea, Schedonorus phoenix,* and *Lolium arundinaceum*) (tall fescue). *Solidago* is a genus of largely North American species (100 species) that is common in grasslands and old-fields. A number of species were introduced to Europe as garden ornamentals and have since escaped cultivation, for example *Solidago altissima* and *Solidago canadensis*. A rhizomatous growth habitat makes *Solidago* species very useful for studies of clonal growth. *Glechoma hederacea* is smaller than the *Solidago* species, and is

Figure 3.2 Popular plants used in population ecology: (a) *Arabidopsis thaliana* (photo Rodney Mauricio), (b) *Kochia scoparia* (photo Miguel Franco), (c) *Brassica rapa* growing in dense herbaceous vegetation (photo David Gibson), (d) *Trifolium repens* clone growing in a lawn (photo David Gibson), (e) *Schedonorus arundinaceus*, endophyte-free plant on the left, endophyte-infected plant on the right (photo David Gibson), (f) *Glechoma hederacea* (photo Mike Hutchings), (g) *Solidago canadensis* clone growing on Konza Prairie, Kansas (photo David Gibson), (h) *Impatiens pallida* (photo Dan Nickrent), and (i) *Brachypodium distachyon* (photo Neil Harris). (See also Plates 2–10.)

a stoloniferous perennial that is commonly used for studies of clonal growth. Aspects of its ecology are reviewed in Hutchings and Price (1999). Two of the most widely studied plants are the agriculturally important *T. repens* and *L. perenne*. Beddows (1967) and Burdon (1983) provide reviews of their ecology. *Trifolium repens* is a nitrogen-fixing rhizomatous herb that has been widely used in studies of clonal growth, neighbour relationships, and genetic differentiation. *Lolium perenne* is the most important forage and turf grass in temperate regions and has been the focus of extensive studies. It is a caespitose perennial without rhizomes or stolons. It co-occurs with *T. repens* in managed and semi-natural pastures and the two species have been studied together in a mixture (Turkington and Jolliffe 1996). *Lolium perenne* is mycorrhizal and contains a fungal endophyte (*Neotyphodium lolli*). Both the mycorrhizae and the endophyte can affect its fitness and response to the environment (Cliquet et al. 1997; Newsham et al. 1998). The grass *S. arundinaceus* is similar to *L. perenne* in its agricultural importance, especially in the eastern United States where it is widely planted for forage and as a turf-grass. It is a larger plant than *L. perenne* and the presence of a fungal endophyte (*Neotyphodium coenophialum*) in many cultivars of *S. arundinaceus* has led to its use as a model species in the study of plant–endophyte interactions (Bacon and Hill 1997; Clay 1998). The ecology of *S. arundinaceus* is reviewed by Gibson and Newman (2001).

Woody plants are less amenable than the species mentioned above to direct, manipulative experiments and there does not appear to be a particular species that is commonly studied by plant population ecologists. However, there are numerous population studies of *Quercus* (oaks, 600 species; Fagaceae), *Pinus* (pines, 93 species; Pinaceae), and *Shorea* (357 species; Dipterocarpaceae) species. These are widespread and valuable lumber trees.

Kearns and Inouye (1993, pp. 407–409) provide a list of plants that are particularly well suited for pollination studies, including *Mimulus* (monkey flowers) and *Aquilegia* (columbines) species.

Ecological information about particular species can be found through literature or internet searches or by checking information contained in floras. Good sources are the *Biological Flora of the British Isles* series with accounts of over 300 species published since 1941 in the *Journal of Ecology* (see <http://www.britishecologicalsociety.org/publications/journals/journal-of-ecology/biological-flora-of-the-british-isles/>), the *Ecological Flora of the British Isles* (Fitter and Peat 1994) (<http://www.ecoflora.co.uk/>), accounts of Canadian weeds published in the *Canadian Journal of Plant Sciences* (129 accounts to February 2013, collected in a five-volume series, the *Biology of Canadian Weeds* published by the *Agriculture Institute of Canada*: <http://www.aic.ca/journals/weeds.cfm>), and the United States Department of Agriculture's PLANTS Database (<http://plants.usda.gov/java/>). The measurement of plant functional traits, which are not necessarily connected to plant species identity, is discussed in Chapter 5.

Phylogenetic considerations

If your research calls for a comparison of the population ecology of several species then you will need to consider their phylogeny, that is, their evolutionary relationships. Harvey et al. (1995) made the comment that ecologists need to be 'phylogenetically challenged'. By this they meant that it is important to consider evolutionary relationships in a comparative ecological study. The argument is that if we seek ecological similarities amongst species then we need to be very sure that these similarities are not due to a shared evolutionary history. Silvertown and Dodd (1997) gave the following example. A typical ecological question might be: do species found in one type of habitat have larger seeds than the species found in another type of habitat? We might find that species living in dry habitats have larger seeds than those in wetlands. We might also find that light-demanding pioneer species have smaller seeds than species in a mature tropical forest. There are three possible reasons for these findings:

1. Ecological sorting: only large-seeded species can colonize shaded habitats.
2. Natural selection: shade selects for species with large seed size.
3. Phylogeny: species able to grow in the shade are related and all have seeds of a similar size.

In options (1) and (2) the species are counted as independent data points. But, as suggested by option (3), species with a common descent (shared phylogeny) may confound such an analysis. The tendency for closely related species to share similar trait values is known as the *phylogenetic signal*. The strength of the phylogenetic signal can be quantified by calculating metrics such as Blomberg et al.'s (2003) K and Pagel's (1999) λ (see phytools in the Appendix). There are two ways to deal with the phylogenetic signal problem. One way is to ensure that you study only congeneric or confamilial groups of species (e.g. Boutin and Harper 1991; Schmid 1984a). A second solution is to use a statistical approach that takes the phylogenetic signal among species being compared into account. To do this, Felenstein (1985) proposed *phylogenetically independent contrasts* (PICS).

The basis for PICS is a phylogenetic tree of the species concerned which has been derived independently of the characters of interest. Molecular DNA data are used in the construction of the phylogenetic tree, and there are a number of software tools available for this such as the R package phyloGenerator (Pearse and Purvis 2013). Consider the hypothetical phylogenetic tree for four species shown in Fig. 3.3. Differences between species A and B are independent of differences between species C and D because each pair (A and B, and C and D) arose after the common ancestor of all four species (node '*g*'). These two sets of differences are referred to as contrasts and are phylogenetically independent of one another. The trait values at A (11) and B (9) can be averaged to obtain a hypothetical trait value for

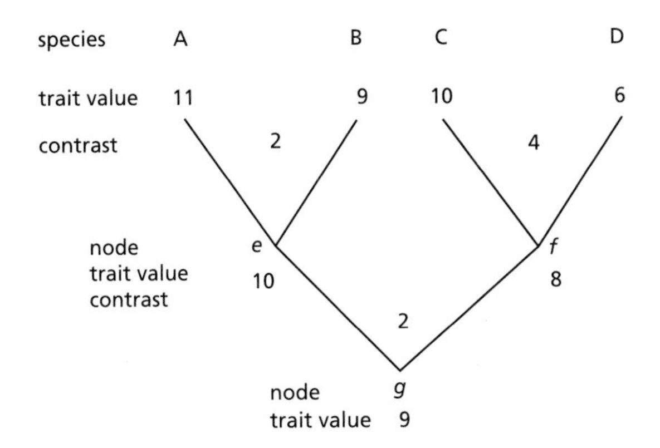

Figure 3.3 Phylogenetic tree for four hypothetical species (A–D). Contrasts are calculated as the difference between the trait values of the two species that branch from a common node. Trait values are assigned to internal nodes of the tree (nodes '*e*' and '*f*') by assuming an evolutionary model of character evolution. A node's trait value is the mean of the values for the two species that branch from it. From Silvertown and Dodd (1997). Reprinted with permission of Cambridge University Press.

their common ancestor '*e*' of 10. Similarly, the trait value for the common ancestor '*f*' of C and D is 8. The difference between nodes '*f*' and '*g*' is a third contrast. Through this process, an entire phylogenetic tree yields $n - 1$ contrasts for n species. The trait values for the three PICS in this hypothetical example can be correlated with the values of the three PICS for any other trait (e.g. seed weight with plant biomass). Association analysis can be used for comparing qualitative traits (e.g. seed weight with habitat type). The correct transformation of the original data has to be used to ensure a valid analysis (Freckleton 2000). PICS and other phylogenetic statistical methods can be undertaken using the R packages ape and phytools (see the Appendix) and are described in Paradis (2006). An accessible introduction to the literature and some of the issues related to phylogenetics and comparative methods is provided by Swenson (2014).

The assumptions of PICs are that continuous characters evolve randomly, equivalent to Brownian motion (or parsimonious change for discrete characters), rates of evolution are constant, and speciation and extinction rates are unrelated to trait values.

The hypothetical example described here should give an idea of how PICS are used. Silvertown and Dodd (1997) went on to show how this approach could be used to address questions about the reproductive effort of plants colonizing old-fields. Eriksson and Jakobsson (1998) conducted 43 PICS in a comparison of the life histories of 81 grassland species. Using this approach they found that the most abundant species had the smallest seed size and the smallest deviation in seed number. A comparison of 297 species of Kalahari desert plants showed no association between the kind of seed dormancy (dormancy present versus absent) and sand content of soil when potential phylogenetic relatedness among taxa was accounted for using PICs (Kos et al. 2012). A discussion on the need to consider phylogeny in comparative studies of plant life histories is provided in the book by Silvertown et al. (1997) and a series of eight forum papers in the 1995 issues of the *Journal of Ecology*.

Plant phytometers

In much the same way that canaries were used in mines, plants can be used as indicators of habitat conditions and in an assessment of environmental suitability for population growth. The use of plants in this way, as phytometers, was first suggested and used extensively by Clements and Goldsmith (1924). They advocated the use of a few standard plants that would grow vigorously and could be handled easily. The growth of a phytometer integrates environmental microsite variability into a suite of readily measured growth parameters that are important for the population biology of the organism. Variability among individuals should not mask the effect of the habitat, and where possible plants of low or uniform genetic variability should be used. Thorough replication of phytometers placed in the habitat of interest should ensure that the effects of the environment on phytometer growth are clear. For the most part Clements and Goldsmith used agricultural plants (especially sunflower, beans, oats, and wheat) in a series of extensive investigations across the plains and montane regions of the United States. Most investigators now use plants native to the habitat under study (e.g. Miller et al. 1995; although see Laliberté et al. 2012). Use of native rather than agricultural species affords a better and more ecologically relevant bioassay of the effects of the environment on plants

Table 3.4 Examples of phytometers used in plant population ecology research.

Species/family	Life form	Role of phytometer	Reference
Agrostis capillaris (Poaceae)	Perennial grass	Bioassay to assess soil fertility	Lee and Fenner (1989)
Bouteloua rigidiseta (Poaceae)	Perennial grass	Microspatial performance within native sites	Miller et al. (1995)
Chlorella spp. (Chlorellaceae)	Single-celled green algae	Suspended in flasks to assess forest microclimate	Hopkins (1965)
Dimorphotheca sinuata (Asteraceae)	Annual herb	Assessment of competitive hierarchies versus 15 pioneer species	Rösch et al. (1997)
Eucalyptus delegatensis (Myrtaceae)	Tree	Effect of fire on seedbed	O'Dowd and Gill (1984)
Lolium perenne (Poaceae)	Perennial	Assessment of competitive hierarchies versus 20 dicot grassland species	Wardle et al. (1996)
Penthorum sedoides (Saxifragaceae)	Perennial	Assessment of competitive hierarchies versus 26 wetland species	Keddy et al. (2000)
Phlox drummondii (Polemoniaceae)	Annual	Intraspecific competition	Clay and Levin (1986)
Poa alpina (Poaceae)	Perennial grass	Bioassay of plant-available soil nitrogen	Keuper et al. (2012)
Ranunculus repens (Ranunculaceae)	Perennial herb	Reciprocal transplants of genets in woodland and grassland sites	Lovett Doust (1981)
Raphanus sativus (Brassicaceae)	Annual	Indication of forest herb layer productivity	Axmanová et al. (2011)
Sinapis arvensis (Brassicaceae)	Annual	Landscape and local effects on parasitism and herbivory rates	Gladbach et al. (2011)
Trichostema brachiatum (Lamiaceae)	Annual	Root and shoot competition	Belcher et al. (1995), Keddy et al. (2002)
Triticum aestivum (Poaceae)	Annual	Proxy for primary productivity along a dune chronosequence	Laliberté et al. (2012)

that are presumably adapted to the environment under investigation. Non-native species may have different nutritional requirements from native plants in the soils of the habitat being tested. By contrast, agricultural plants may grow faster than native plants, affording a more rapid and pragmatic choice of phytometer (see the contrasting views on this in Laliberté et al. (2013) and Uren and Parsons (2013)). The choice of life form is important. Annual plants may offer the opportunity for rapid growth to reproduction, whereas clonal perennials are often easily propagated *en masse* to obtain genetically identical plants. Genotypic versus phenotypic and environmental versus genetic influences can be maintained and readily studied through an appropriate choice of phytometer species. Selected examples that illustrate the range of applications and species of phytometers used are provided in Table 3.4.

The 'plants as phytometers' approach can also be used to help assess the plasticity of a genotype's response to the environment through reciprocal transplants of genets across sites with different environmental conditions (see Section 4.3 and Doust 1987).

3.3 Statistical issues in conducting experiments

The statistical tests that are appropriate for testing different experimental designs are discussed in Chapters 7 and 8. However, in planning the design there are a number of important statistical issues that require some careful consideration (see Hairston (1989) for additional discussion of some of these points).

Controls, replication, accuracy, precision, and bias

A *control* is a baseline experimental condition against which one or more treatments are being compared. The purpose of a control is to remove the effect of all other factors apart from the one under investigation (Dytham 2011). A *factor* is the experimental intervention that we impart on the system under study (e.g. a shading factor consisting of three *treatments* or *levels*: no shading, 50% shading, and 75% shading). The nature of experimental treatments is discussed in detail in Chapter 5. In most experiments the control will be the unmanipulated condition; for example, the plant that is not grazed in a herbivory experiment. In most experiments, a comparison with a valid control is necessary to demonstrate an effect (Green 1979). A valid control is a treatment that differs from the treatment of interest only by the nature of the intervention (i.e. the treatment itself). Other confounding effects (i.e. other influences) should not differ between the treatment and the control. When confounding effects are present then there is no way to determine causality (i.e. the process operating according to the model and hypothesis; Underwood 1997). For example, if you were interested in testing the effects of addition of soil nutrients on the fitness of an annual plant in the field, you may decide to add aliquots of a liquid fertilizer to sample plots in which the plant is present. A valid control would be a series of the same number of plots to which is added the same volume of water balanced to the same pH as the fertilizer. Note that the control plots are not necessarily unmanipulated. An unmanipulated control group in this case would not allow differentiation between the effects of added nutrients and added water. Water is added to the control plots because the fertilizer is in liquid form and you do not want a confounding moisture effect. The plants that are watered but not fertilized represent a *procedural control* group. Similarly, the added water should be at the same pH as the liquid fertilizer because you do not want a confounding pH problem. Needless to say the water and fertilizer are added at the same time and in the same quantity. Look back at Case Studies 3 and 4 in Chapter 1 and identify the control treatments in those experiments.

A control is not relevant or required in all experiments, especially many mensurative experiments. For example, there were no controls used or necessary in Zuidema et al.'s (2010) comparison of tree recruitment in tropical forest habitats (Case Study 1 in Chapter 1). In manipulative experiments a control may not be necessary when the goal is a comparison of treatments of which none could be designated as a baseline or unmanipulated. For example, in Case Study 3 in Chapter 1 the authors created two experimental disturbance factors with two treatments (autumn 1986 and spring 1987). Neither the autumn nor the spring disturbance can be regarded as a control, and a control was not necessary in this situation. Similarly, when the effects of soil moisture on population growth are being compared, for example, there may not be a defined control soil moisture level per se.

One of the first questions that students seem to ask is, 'how many samples do I need?'. This is an important question, but not one that can be answered satisfactorily until after the study is done, or at least a pilot study completed (see Section 2.5). All too often it seems we hear at professional meetings how 'the data were too variable for a significant effect of the treatments to occur', or 'the difference between the two treatments would have been significant if we had used a larger sample size'. Hindsight is always valuable, but there are some steps that can be taken at the planning stage that might help alleviate some of these problems.

Replication is the taking of multiple, independent observations or measurements of a variable on different sampling units. In experimental terms, replication refers to the repetition of experimental units or observations under similar conditions, for example designation and subsequent measurement of X plants to receive treatment A and X plants to receive treatment B (where X is the level of replication and is greater than or equal to three). Replication reduces the effects of noise or random variation and increases the precision (see below) of an estimate. Variability in measurements arises from the intrinsic and extrinsic properties of the plant population under observation. For example, the height of a plant and its growth rate are subject to intrinsic genetic variability from one individual to another (unless clonal fragments from the same genet are being measured). Thus, there is no possibility that sizes and growth rates will be identical unless individuals are themselves genetically identical. Even then, genetically identical individuals will differ because each lives in a different environment, even if they are neighbours or seeds in the same pod. Somatic mutation can also cause plant modules from a single genet to differ genetically and hence differ in morphology. Extrinsic causes of variation are properties outside the unit being measured, and include errors in making measurements. The methods that we use to make measurements are not entirely accurate (see below) and introduce variability. If you were to weigh an acorn on a top-loading balance you would probably come up with a different figure every time because the third decimal place on these machines is sensitive to the slightest movement of air in the lab.

It is the inherent variation in natural systems that necessitates the taking of replicate samples. The statistical tests we use are an attempt to allow us to make statements about potential differences in dependent variables in response to independent variables (factors) given this variability. Decisions on what constitutes a valid replicate sample or measurement require some care, as discussed below.

Accuracy is the closeness of the sample mean to the true population mean. The sample mean represents the measurements that we make of a variable. How close are they to the 'real' value (the population mean)? Imagine shooting arrows at an archery target aiming for the centre bull's eye. An arrow landing in the bull, which is analogous to the 'real' or true value, is considered to be very accurate. Arrows landing at the edge of the target would be considered inaccurate (Fig. 3.4).

Precision is how close the replicated sample values are to each other, irrespective of the true population mean. Precise samples are similar to each other but may or may not be accurate. Again, imagine if you shot three arrows and they all landed within the centre bull ring; they would be considered to be both precise and accurate (Fig. 3.4). Arrows landing in a tight cluster but in the edge would be considered to be precise, but inaccurate if aiming for the bull (centre). Arrows landing in a wide scatter around the target would have low precision and low accuracy.

Bias is a consistent under- or over-estimation of a true population parameter and is usually the result of a consistent inaccuracy in a sampling procedure. If you were trying to get an estimate of seedling numbers in a plot but consistently missed the smallest seedlings because of dense vegetation then your measure would be inaccurate and biased (too low).

Sample adequacy, Type I and Type II errors, and power

Consideration of the need for replication along with the issues of precision, accuracy, and bias

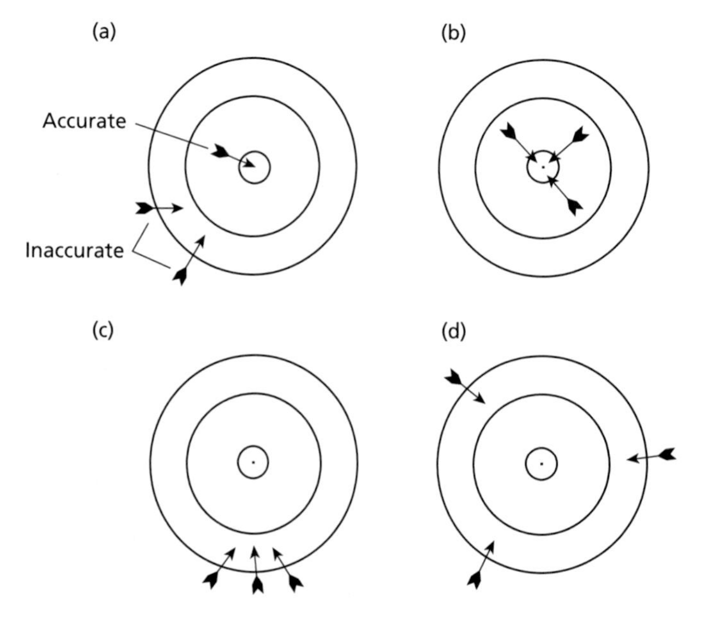

Figure 3.4 A shot of three arrows aimed at the bull (the centre of the target) as an illustration of accuracy and precision: (a) one accurate and two inaccurate arrows, (b) all highly accurate and precise, (c) inaccurate but precise, and (d) low accuracy, low precision.

lead to the issue of adequacy of the sample size. In other words, 'how many samples do you need to take?' and 'how many plants should you measure?'. The trite answer is that the best sample number is the largest sample number. The precision of an estimate of the mean increases with sample size. However, there is obviously a point of diminishing returns above which your estimate becomes only marginally more precise. In simple statistical terms, the standard error of the mean decreases in proportion to the square root of the sample number. Note that accuracy may not be improved with increased sample size because of methodological problems (e.g. always missing those small seedlings in the plots).

The sample number (n) required to achieve a desired precision can be estimated by solving for n:

$$x \pm t_{1-(1/2)_\alpha} \frac{s}{\sqrt{n}}$$

where x is the sample mean and s is the sample standard deviation (Green 1979). For example, you may wish to estimate the mean density of tillers of a grass population within ±10% (your level of precision), at a 90% confidence level of doing so correctly. Assuming random sampling and a normal distribution of the sample data, then 10 preliminary values may yield a sample mean of x = 244.2 and s = 79.16. Applying the formula above:

$$244.2 \pm t_{1-(\frac{1}{2})\alpha} \frac{79.16}{\sqrt{n}}$$

The only unknowns are n and the t-statistic. The latter is a function of its degrees of freedom, that is, $n - 1$. However, we can estimate t as being close to 2. With 0.95 confidence limits $\alpha = 0.05$, ±10% of the mean, gives

$$24.42 \approx (2)\frac{79.16}{\sqrt{n}} \text{ and } n \approx 42.$$

For $n - 1 = 41$ degrees of freedom, $t_{0.975} = 2.02$. Solving for n again using this t-value yields 42.9, which is very similar. Thus, a sample size of 43 or more will allow an estimate of tiller density to be made that is within ±10% of the mean, within a 1 in 20 chance of being wrong.

Examples in which investigators actually use this approach are difficult to find in the literature because when used it is part of a preliminary study that may not be published. Table 3.5 lists the sample sizes necessary to take measurements for some morphological attributes for some graminoids, herbs, woody species, and a succulent. Note that the sample size ranges from 28 to 1025 depending on the species and the attribute. Sample sizes in the complete data set of 16 species and eight attributes were unrelated to growth form, physiognomic type, or physical size (Croy and Dix 1984).

Table 3.5 Sample sizes required to bring the 95% confidence interval to within 10% of the mean for three morphological attributes of select plants. Abridged from Croy and Dix (1984).

Species (family)	Life form	Plant height	Horizontal distance1	Leaf blade length
Bouteloua gracilis (Poaceae)	G	101	440	102
Agropyron smithii (Poaceae)	G	47	162	44
Stipa comata (Poaceae)	G	41	236	119
Antennaria parviflora (Asteraceae)	H	126	89	28
Achillea lanulosa (Asteraceae)	H	51	63	67
Lactuca serriola (Asteraceae)	H	138	690	43
Sphaeralcea coccinea (Malvaceae)	W	55	411	40
Pinus ponderosa (Pinaceae)	W	135	1025	38
Opuntia polyacantha (Cactaceae)	S	83	484	149

G, graminoid; H, herbaceous plant; W, woody; S, succulent.
[1] Horizontal distance measured from the plant stem at the soil surface to the plant apex projected vertically onto the soil surface.

Notice the two parameters that we set in the estimation of n in the example above. We decided that we wanted our estimate to be within ±10% of the mean, and that we wanted to have only a 1 in 20 (95% confidence) chance of being incorrect. Different values for these parameters would have yielded different estimates of n. For example, if we had wished to estimate n ± 5% of the mean with 0.999 confidence limits $\alpha = 0.001$ then the equation above yields $n \approx 455$; that is,

$$12.21 \approx (3.29)\frac{79.16}{\sqrt{n}} \text{ and } n \approx 455$$

where $t_{0.999} \approx 3.29$. The values that we assign to these parameters relate to the level of Type I and Type II errors that we are willing to accept (see below).

Several times already we have alluded to the confidence that we have in our tests of hypotheses. When we believe that we have falsified a hypothesis we do so with a certain level of confidence. This is especially true in ecology where we cannot directly observe many of the processes that occur. We cannot observe density-dependent thinning in a population, but we can observe effects that are consistent with such a process (e.g. a negative relationship between density and stand yield). In making these statements we are accepting a certain degree of confidence in the experimental tests of a hypothesis of, in this case density-dependent thinning. Thus, when we test a hypothesis we have two sources of error that we have to assign a confidence level to.

A *Type I error* (α or P-level) is the probability that we will reject a null hypothesis when it is in fact true. For example, if your null hypothesis stated that male and female holly bushes are the same size but the results of your statistical test led you to believe they are of different sizes (a rejection of the null hypothesis), then you would be committing a Type I error if the null hypothesis were actually true. By convention, but for no particular reason, the Type I error level, is set at $\alpha < 0.05$. This is the level of probability reported in the results sections of papers. The null hypothesis is not rejected at what is known as the '5% level' if the probability that the data could have arisen from the null hypothesis is greater than 0.05 (see Chapter 7). For the holly bush null hypothesis above, a significant result at $\alpha < 0.05$ would mean that the data suggest there is a less than 1 in 20 chance that male and female bushes are indeed the same size. The probability of making a Type I error is increased if multiple non-independent statistical tests are made. This can happen, for example, if the complete set of 105 correlations among 15 variables was calculated. Even if there were no true correlations among the variables, with a Type I error of 5% at least five correlations would produce a significant result by chance alone. In this case the Bonferroni correction can be used to designate a new critical level (see Chapter 7).

A *Type II error* declares a null hypothesis to be true when it is not; that is, the probability of failing to reject a false null hypothesis or not finding a 'real' significant difference. The Type II error, β, is related inversely to the Type I error level. For any statistical analysis or test, lowering the Type I error level increases the Type II error. Type II error levels are often not considered in population ecology studies, but they can be quite important. For example, a test of two treatment regimes for managing the population of a rare plant might find no significant difference between the two treatments; that is, at an α set at $P < 0.05$ the null hypothesis of a difference between the two treatments is not rejected. Given this result, the recommendation would probably be to use the least expensive and easier regime (perhaps doing nothing versus burning the sites). However, burning may be just what this rare species needs if the Type II error for this test is unacceptably high. The lack of a significant difference between the two treatments may be due to a poor experimental design, insufficient sampling, or unreliable measurements, that is, an inadequate test of the hypothesis. In this case, accepting the hypothesis when it is in fact not true could lead to the loss of the rare species.

The *power* of a test is the probability of getting a significant result (rejecting the null hypothesis) at a given level of α. More precisely, power is equivalent to $1 - \beta$. Power provides the probability of correctly rejecting the null hypothesis in favour of the alternative (Underwood 1997). Power can be calculated in a prospective power analysis either using preliminary data as part of a pilot study (Chapter 2) or in a retrospective analysis. Prospective power analysis, such as the determination of sample size (see above) allows decisions to be made about the α level to use (it does not have to be 0.05) for desired effect sizes (how large a difference there is between the null hypothesis and the alternative one). Retrospective power analysis can be used to determine the power of hypothesis tests already conducted (Thomas and Krebs 1997). Reviewers and journal editors can sometimes ask for the results of a retrospective power analysis to assess the adequacy of sample size when non-significant results are being presented. Calculation of power depends upon three parameters: sample size (n), intrinsic variance in the population being sampled, and effect size. While prospective power analysis can be useful in the planning of an investigation, care must be taken in retrospective power analysis depending on the goal of the analysis. If the goal is to determine whether a study design can detect biologically meaningful patterns Thomas (1997) recommends calculating power using pre-specified effect sizes. A good discussion of power analysis is provided in Underwood (1997) and a description of statistical tests in Cohen (1988), Steidl and Thomas (2001), and Ellis (2010). A review of software suitable for power analysis for ecologists appears in Thomas and Krebs (1997). The R package pwr can be useful for these analyses (see the Appendix).

Power analyses are considered to be important in wildlife biology and animal conservation ecology (Gerrodette 1987; Zielinski and Stauffer 1996; Thomas 1997; Nickerson and Brunell 1998), and can be valuable in plant population ecology, although applications are less frequent. Savolainen and Hedrick (1995) provide an example which shows that fitness in Scots pine (*Pinus sylvestris*) is not related to intrinsic heterozygosity. Retrospective power analysis showed that the power of their tests to detect a difference between heterozygous and homozygous individuals (determined using isozyme analysis of 12 polymorphic loci) depended upon the trait measured (Table 3.6). For tree height and diameter and pollen shedding the power of their test exceeded 0.5 (maximum possible value 1.0), whereas for pollen weight and cone production the power was quite low at <0.1. Karban (1993) and Karban and Niiho (1995) used retrospective power analysis to test the efficacy of herbivory experiments. Analysis showed that their experimental design had sufficient power to detect an immunological memory in the response of cotton (*Gossypium* sp.) to spider mites.

Table 3.6 Power analysis of the hypothesis that there is no difference between heterozygous and homozygous individuals of Pinus sylvestris grown in Viitaselkä, Finland. Power is the proportion out of 1000 samples where the calculated t-statistics exceed the tabular value for a given α value. α is equivalent to the Type I error rate. From Savolainen and Hedrick (1995). Reprinted with permission of the Genetics Society of America.

Trait	$\alpha = 0.05$	$\alpha = 0.1$
Diameter	0.569	0.976
Height	0.727	0.999
Pollen weight	0.079	0.155
Pollen shedding	0.928	1.000
Cone production	0.090	0.115
Seed weight	0.171	0.395
Mean	0.427	0.607

Experimental designs and avoiding pseudoreplication

Pseudoreplication is 'the use of inferential statistics to test for treatment effects with data from experiments where either treatments are not replicated (though samples may be) or replicates are not statistically independent' and was introduced by Hurlbert (1984). Although controversial (Hurlbert 2009), pseudoreplication is an important aspect that has to be avoided in planning the replication of samples and measurements in an experimental design. In other words, you have to consider carefully what a bona fide replicate is, otherwise it does not count as a replicate. Pseudoreplication is an issue for both mensurative and manipulative experiments and a simple example illustrates this issue. Suppose that you were interested in the following null hypothesis: 'elevated atmospheric CO_2 had no effect on plant growth' and you placed 10 plants in environmental chamber A maintained at ambient CO_2 and 10 plants in environmental chamber B maintained at double ambient CO_2. A statistical test (e.g. a t-test; see Chapter 7) to see if there were significant differences in growth between the plants grown in the two chambers would be inappropriate because the so-called replicate plants (10 per chamber) are not statistically independent (anything that affects one plant in a chamber is very likely to affect them all, regardless of CO_2 status). Any difference in growth would provide evidence only of a difference between the two environmental chambers, which happen to be at different CO_2 levels but also differ in location and a whole host of other unmeasured factors. To avoid pseudoreplication, several environmental chambers would be required, with equal numbers at ambient and at double ambient CO_2, each containing one or more plants. The performance of plants in each chamber would be averaged, with the average values representing the true replication and experimental units (n being the number of chambers, not the number of plants). Indeed, growth chambers and FACE (free air carbon dioxide enrichment; see Section 4.4) experiments are a very common source of pseudoreplication.

Three forms of pseudoreplication that can (and often do) occur in manipulative experiments are simple, sacrificial, and temporal pseudoreplication (Fig. 3.5). Simple pseudoreplication occurs when there is a single 'replicate' per treatment despite multiple samples being taken from each (Fig. 3.5a). The data from the samples are not independent. An example would be measuring the amount of damage done by a herbivore to 50 leaves on one oak tree and comparing it with the damage done to 50 leaves on a beech tree. In this situation there is a sample size of one tree per treatment (species of tree), not 50. The problems in using growth chambers described above are an example of simple pseudoreplication.

Sacrificial pseudoreplication (Fig. 3.5b) occurs when there is true replication of treatments but either the data for the replicates are pooled prior to analysis or two or more measurements are taken from each experimental unit and treated as independent replicates. In Fig. 3.5(b) X_1 and X_2, for example, are not independent measurements of the first replicate. The means of X_1 and X_2, X_3 and X_4, Y_1 and Y_2, and Y_3 and Y_4, respectively, should be taken and these submitted for analysis (for a total sample size of four, not eight). Similarly, $X_1, \ldots, X_4$ and $Y_1, \ldots, Y_4$ should not be pooled into two sets prior to analysis as this results in a loss (sacrifice) of information. Continuing the example from before, we can imagine here four trees represented by two oaks

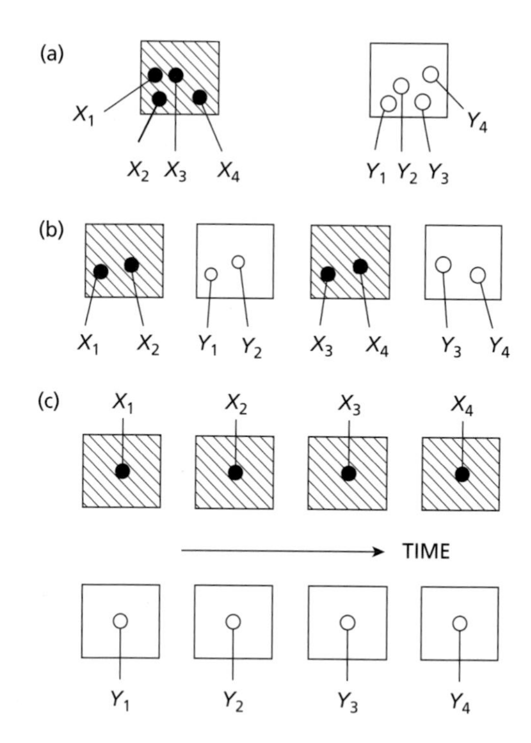

Figure 3.5 Pseudoreplication in manipulative experiments. $X_1, \ldots, X_n$ and $Y_1, \ldots, Y_n$ are replicate samples of two treatments (shaded and open). After Hurlbert (1984). (a) simple pseudoreplication, (b) sacrificial pseudoreplication, and (c) temporal pseudoreplication. Reprinted with permission of the Ecological Society of America.

and two beeches with two leaves being sampled from each. The correct analysis is one experimental factor (tree species) with two treatments (oak and beech) and a sample size of two in each (which incidentally is insufficient replication for an appropriate analysis). Similarly, measuring multiple plants within the same sample plot and treating them as separate data points is a form of sacrificial pseudoreplication common with beginning students doing field work.

Temporal pseudoreplication (Fig. 3.5c) can occur when multiple samples from each experimental unit are taken sequentially over several dates and the dates are taken to represent replicated treatments. The point here is that the successive samples are not independent because they were taken from the same sample unit. Using our herbivory example again, we might wish to sample each tree several times over the growing season. Such repeated sampling is perfectly good, but the successive dates should not be treated as replicates when they are not (the same tree is sampled each time). The correct analysis would be to use repeated measures of analysis of variance (Chapter 7) with four sample dates and one factor (of two treatments: an oak tree and a beech). Again, this example is simplistic and one would really need to have three or more oaks and beeches, each resampled over time. Note that if *different* sample units are sampled each time (e.g. a different oak is chosen on each date) then it is appropriate to treat the data from each date as an independent sample unit, although they cannot be treated as replicates.

A 'trick' that has been used to alleviate the problem of pseudoreplication in unreplicated growth chamber experiments has been to alternate the treatments and experimental units among the chambers. For example, the elevated CO_2 treatment may be allocated to chamber A one week and to chamber B the next week. The experimental plants assigned to the control and treatment conditions are transferred from one chamber to the next every time the CO_2 level is changed between chambers (e.g. Dale and Press 1998).

Randomized intervention analysis (RIA) is an elegant approach which allows unreplicated control and treatment samples to be statistically analysed (Carpenter et al. 1989) (see also BACI designs in Section 3.1). This procedure is gaining popularity in ecosystem ecology and requires paired, time-series data for both treatment and control conditions before and after manipulation. RIA avoids issues of simple and temporal pseudoreplication (see below) because it uses non-parametric randomization tests that are not subject to the assumptions of normality, independence, and equal variance implicit in parametric statistics (Chapter 7). Beerling (1999) used RIA to test whether differences in the leaf gas exchange rates of three woody species grown in control and treated sections of an experimental climate change greenhouse could have arisen by chance. The greenhouse was divided into two sections providing unreplicated ambient and elevated temperature and atmospheric CO_2 units.

Interspersion refers to the spatial and temporal placement of experimental treatments in a manipulative experiment and is a key feature to avoiding pseudoreplication. Treatments must be interspersed with each other to avoid pseudoreplication. In mensurative experiments the 'treatments' are isolated, which can cause problems in the correct application of inferential statistics. There are five forms of spatial interspersion of which two commonly lead to pseudoreplication (Fig. 3.6) (Hurlbert 1984).

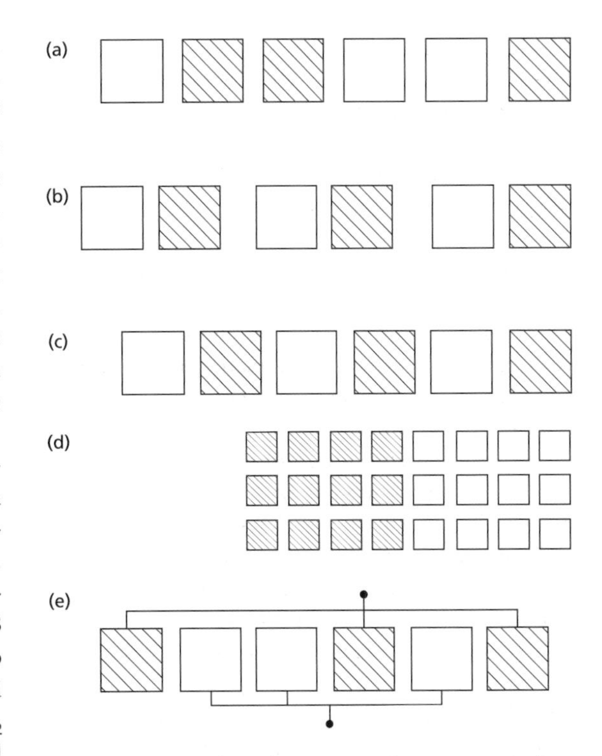

Figure 3.6 Modes of spatial interspersion of two treatment levels (shaded and open): (a) completely randomized, (b) randomized block, (c) systematic, (d) simple segregation, and (e) randomized with interdependent replicates. After Hurlbert (1984). Reprinted with permission of the Ecological Society of America.

A *completely randomized design* is the most straightforward and basic approach to interspersion. Experimental treatments are applied at random to the units of study (plants, plots, etc.) using something like a random numbers table, coin flips, etc. While there is nothing statistically wrong with this approach, it can be problematic. Space limitations in the field or greenhouse may require the placement of sample units (pots, plants) in different areas (e.g. different greenhouse benches). Logistical reasons may make it more convenient to place sample plots in groups. A block design (below) should be used when this occurs. In addition, a completely randomized design assumes environmental homogeneity. This is unlikely since even within a greenhouse some areas will be warmer than others, pots placed on one bench may inadvertently get more water than those on another, etc.

A *randomized block design* is commonly used and is a very good design. Note in Fig. 3.7(b) how the pairs of treatment units are separated by a space. Each pair of experimental units is a block and may represent different greenhouse tables, for example. This design is particularly appropriate when you have to deal with a pre-existing gradient in the location of your experiment, such as a cold end of a greenhouse (Potvin 2001). There are several different ways in which experimental units can be blocked, and some of these are illustrated in Fig. 3.7. A good ecologically orientated discussion of blocking is provided by Potvin (2001) and the statistical issues of blocking are covered in detail by Mead (1988). Case Study 3 in Chapter 1 used a four-factor randomized block design (see details under factorial designs below).

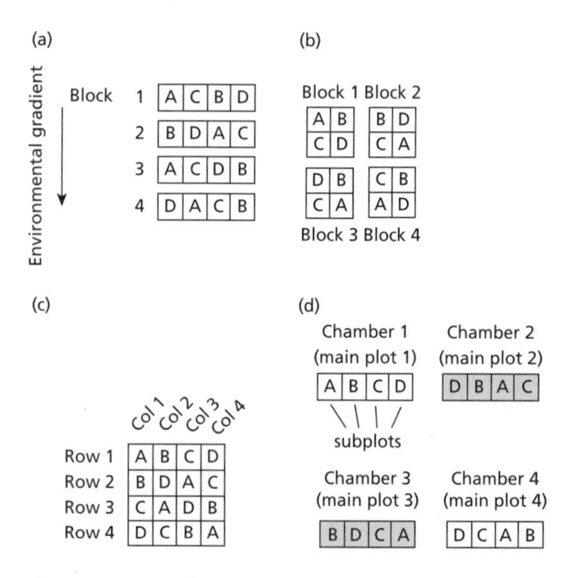

Figure 3.7 Types of blocking designs to compare four treatment levels (A–D) of a single factor: (a) randomized block in four rows, (b) randomized block with four blocks, (c) Latin square, and (d) split-plot. The Latin square design is a double-blocking system with each treatment occurring once in each row and each column. The split-plot design compares the four treatment levels (A–D) nested as subplots within two main treatments (shaded and non-shaded) of a second factor associated with each main plot [e.g. four nutrient levels (subplots A–D) nested within four CO_2 chambers (1–4)]. After Mead (1988) and Potvin (2001).

A *systematic design* (Fig. 3.6c) involves a very regular interspersion of treatments. This design should be used with caution because it is not random, randomness being one of the criteria for inferential statistics (Chapter 7). There is a risk of the spacing coinciding with the periodicity of a pre-existing pattern in the field or greenhouse. For example, an old ridge and furrow pattern may occur in a field—you would not want all of one set of treatment plots to be in the ridges and the others in the furrows.

Segregated designs (Fig. 3.6d) involve the placement of the different treatment plots in separate locations. This design involves simple pseudoreplication (Fig. 3.5a) and can lead to spurious treatment effects. Pre-existing differences between locations may exist and confound potential treatment effects. Chance effects might affect one location and not another. This is the experimental design equivalent of 'putting all your eggs in one basket'.

Interdependent replicates (Fig. 3.6e) can occur when so-called replicate experimental units are linked together. For example, this might occur in the field using open-topped growth chambers to study the effects of elevated CO_2 if all the chambers with elevated CO_2 were vented from one fan and the ambient chambers were vented from a separate fan. In this case the chambers with elevated CO_2 cannot be considered independent units, and neither can the ambient level CO_2 chambers. This practice leads to statistical confounding of variables. If the fan venting one set of chambers, for some unknown reason, introduces a contaminant into the gas lines then that set of chambers will have an unwanted, confounding treatment. A more appropriate design would have separate fans for each chamber, or perhaps pairs of elevated and ambient level chambers in blocks; for example, Ashenden et al.'s (1992) design used by Applebee et al. (1999) to study the effects of elevated CO_2 upon allelopathic interactions on alfalfa (*Medicago sativa*) seedlings (see Fig. 4.5).

Nesting uses replication of experimental units in at least two levels of a hierarchy (Underwood 1997) and may be necessary for logistical reasons. This design results in one experimental factor having different treatments at each of the levels of another factor. In Fig. 3.7(d) the four nutrient levels are nested as subplots within the CO_2 chambers.

Designs incorporating nesting are often required when there are two or more factors, and it is logistically impossible to have a separate experimental unit for each treatment interaction. For example, it is very expensive and laborious to establish chambers with elevated CO_2 and so these tend to be quite large. As suggested by the design in Fig. 3.7(d) it is then advantageous to test other treatments within the CO_2 factor. For example, Firbank et al. (1995) used a split-plot design with nutrient levels and plant density nested within CO_2 and temperature treatments to study the density dependence of the annual grass *Vulpia ciliata*. Nesting also happens in situations when seeds from individual maternal plants are used to rear 'families' that are all from the same population. The families are then nested within the population.

A *factorial design* involves an experiment in which two or more factors are considered and analysed simultaneously, for example the effects of temperature and the presence of pathogens on fitness. These designs are quite difficult to handle because of the large number of experimental units required and can be prohibitively expensive or logistically difficult. For example, two factors each divided into two treatments require $2 \times 2 = 4$ experimental units to allow for each combination of treatments before allowing any replication. Three factors each divided into four treatments would require $4 \times 4 \times 4 = 64$ experimental units before replication. A further problem is with the analysis and interpretation of interactions (see Chapter 4). A two-factor factorial design has a straightforward interaction ($A \times B$), but a three-factor factorial design has three first-order interactions, $A \times B$, $A \times C$, and $B \times C$, and a second-order interaction, $A \times B \times C$. As factors are increased the number of interactions increases rapidly: for a k-factor factorial, the number of factorial combinations of m-order interactions $= k! = [(m + 1)!\ (k - m - 1)!]$ (Sokal and Rohlf 1969). The experiment described for Case Study 3 in Chapter 1 involved a four-factor factorial design established in the field in randomized complete blocks (four blocks × two disturbance treatments × three treatments of background vegetation × two treatments of exposure to cinnabar moth giving 96 plots with 11 factorial combinations plus the effects of blocking). In this example, the second- and third-order interactions were not significant, meaning that interpretation could be made on the basis of the significant main effects of time and clipping (see Table 7.2). If a second-order (two-way) interaction had been significant, it would have meant that the dependent variable was affected by both treatments in the interaction. Interpretation of the results from factorial designs should be made with care (see Section 7.6).

Box 3.1 Darwin's 'vitality of seeds' experiment: the first plant population ecology experiment?

There is some debate about what the first plant population ecology experiment was and who carried it out. Harper (1977) suggests that the first *significant* paper was that of Nägeli (1874), and indeed that is an unusually detailed study of plant competition for the time. White (1985a) notes the early studies by Sinclair (1824) who censused the numbers of distinct plants and species in 1 ft^2 (~0.09 m^2) squares of turf from several natural and artificial pastures in England. Sinclair also carried out experiments to determine the 'produce and nutritive qualities' of pasture grasses, but his data are wholly agricultural. Darwin (1859) in his *Origin of Species* cited Sinclair's work. As pointed out by Harper (1967), the *Origin of Species* itself contained much of interest to plant population ecologists. It was also in 1856 that the famous Rothamsted Park grass experiment was set up in England (Lawes et al. 1882). This experiment was designed to test the effect of agricultural fertilizer on the botanical composition of a hay meadow, and over the long term has revealed some interesting plant community results (Dodd et al. 1995). In addition, studies on individual species have revealed population differentiation in response to the soils (Snaydon 1970; Snaydon and Davies 1972). However, as noted in Chapter 1, the mantle of 'first documented plant population ecology experiment' goes to Darwin's earlier study of charlock (*Sinapis arvensis*; Brassicaceae) (Darwin 1855a).

Darwin had long maintained an interest in the longevity of seeds, and in 1844 he was the ghost writer of an account for a correspondent, Mr William Kemp, on the recovery of germinable seed that had been buried at a depth of 25 ft (7.6 m) in a sand quarry (Burkhardt and Smith 1986). Later, in 1855, he began a series of experiments in which he immersed the seeds of various species in barrels of salt-water

continued

Box 3.1 *Continued*

to see how long they continued to float and remain alive. Since plate tectonics and continental drift were still unknown at that time, Darwin speculated that long-distance drifting of seed across the sea was the best explanation for the distribution of the world's flora. Unfortunately, he ultimately found that of 30 to 40 herbaceous plants and branches with ripe seed placed in salt-water, only the fruit of *Euonymus* remained floating for more than a month. This was despite the observation that the seeds of many of the species remained alive (Darwin 1855b).

With this background interest in seed longevity in mind, Darwin's experiment with charlock was prompted by his observation that seedlings of the plant had established in four patches of ground. Earlier, in the spring of 1855, some thorn bushes (*Crataegus* sp.?) had been pulled up in that part of the woodland near his home (Fig. 3.8 and Plate 11). On 1 July he recorded six, three, three, and one individuals of charlock in flower on the four patches of ground, respectively. This observation interested Darwin, because although charlock is a common weed in the cornfields of England, it had not been seen for 9 years in that particular woodland. The woodland had been an arable field 15 years previously, and then in pasture before being ploughed and planted with trees in the spring of 1846. Charlock plants were abundant during the summer after ploughing the field, but were not seen thereafter in the ensuing 9 years. Darwin set up his experiment as a result of these observations. He cleared the vegetation away and dug the earth over to 'one spit deep' (a spit is a spade-depth) in three 2-ft^2 (~0.19 m^2) plots on 21 July 1855. On 1 August he recorded eleven, six, and five charlock seedlings on the three plots, respectively. These data allowed Darwin to make the following conclusion, '. . . this seems good evidence that the Charlock seed had retained its vitality within a spit's depth of the surface during at least eight or nine years'. He further went on to generalize his results to seeds as a whole, '. . . the power in seeds of retaining their vitality when buried in damp soils may well be an element in preserving the species, and, therefore, that seeds may be specially endowed with this capacity'. Modern studies have corroborated Darwin's conclusions regarding the longevity of charlock seeds in the seed bank (McCloskey et al. 1996).

It is interesting to consider Darwin's use of the scientific method for this experiment. He made some observations (charlock seedlings on disturbed soil) and followed these up with a field experiment (digging over plots of ground). We can assume that he had some sort of working hypothesis (seeds can remain dormant in soil), although this is not stated in his article or correspondence. His data (numbers of established seedlings) support the hypothesis and allow him to generalize outside of the bounds of his experiment, but within the scope of the topic. Darwin is reported as saying that he could not resist forming a hypothesis on every subject (Medawar 1969), and clearly his experiment here is an example of classical hypothesis testing in the best Aristotelian sense.

Figure 3.8 View from nearby woodland of Darwin's home, Down House, Kent, UK, as it appears today. (Photo David Gibson.) (See also Plate 11.)

3.4 Follow-up exercises

1. What sort of pilot study or preliminary data would have been useful for Case Studies 1–4 in Chapter 1?
2. Look through the plant population ecology papers for one year of a journal. Identify any cases of pseudoreplication.
3. Read some of the papers listed in Table 3.2. How did the investigators of these studies intersperse their experimental treatments?
4. Go down to your department or college's greenhouse. Using one of the bays as a potential location decide upon the best arrangement for an experiment involving three treatments and 90 sample units (e.g. plant pots). Assuming the pots are too large to fit onto a single bench, how would you best arrange them to accommodate the concerns of pseudoreplication and treatment interspersion?
5. Read Hurlbert's (1984) paper. It should be essential reading for any ecologist. Does he take the issue too far?

PART 3

Doing the Study

CHAPTER 4

Experimental treatments

In 1779 a flower of Papaver orientale *when swollen with pollen was inserted in a flower of* Papaver somniferum *from which the stamens had been removed, and was held there. The seeds of* Papaver somniferum *matured and these seeds gave rise to plants very similar to* Papaver orientale *except that the stem was almost always many-flowered. So great was the similarity that I kept quiet about this experiment in case some error had crept in, until it could be repeated.*

(John Hope, FRS, 1781, in Morton 1986)

To trace the effect of variations in water supply on vegetation, two different methods may be adopted. One may increase the supply where it is low, or diminish it where it is high.

(Jeffreys 1917)

- **What are experimental treatments?**
- **Growing plants**
- **Biotic treatments**
- **Abiotic treatments**
- **Interactions between plants**

Preamble

This chapter considers some of the experimental treatments that are used to address questions in plant population ecology studies. For the most part these are applied in manipulative experiments (Chapter 3), but in several cases they will have relevance for mensurative (unmanipulated) experiments too. The emphasis here is to describe the different types of treatments that are used and provide advice about their advantages and limitations, but details of standard procedures are not discussed. The best sources for this information can be found in the accompanying references. Numerous examples from the literature are cited, so it can be clearly seen how others have set up treatments to address their specific hypotheses. Each population ecology investigation is unique and you will probably need to combine and adapt the treatments that are discussed. The procedures described here will have to be adapted to best suit the hypotheses and questions to be addressed (e.g. see Table 2.2). In many cases you will need to invent your own treatments, guided by the questions that you are asking. On the other hand, never do an experiment simply because the treatment is convenient. This is a seductive trap and usually leads to mushy results: the question must always lead to the method, not the reverse.

In Section 4.1 some caveats regarding the setting up of experimental treatments are discussed. You should refer back to Chapter 3 where a number of important statistical issues were covered which have relevance for experimental treatments. The statistical tests themselves are discussed in Chapters 7 and 8. In Section 4.2 the factors involved in choosing a particular plant species to study and the factors that affect the growth of plants are discussed. Experimental treatments are divided into two major categories—biotic and abiotic—and these are discussed in Sections 4.3 and 4.4, respectively. In Section 4.5 issues relevant to establishing experimental treatments for studying plant interactions are discussed. Throughout, and as noted in Chapter 2, it is important not to lose sight of the

Methods in Comparative Plant Population Ecology. Second Edition. David J. Gibson.

questions and hypotheses being addressed (review Table 2.2 again).

4.1 What are experimental treatments?

It was noted in Chapter 3 that a *factor* is an experimental intervention (e.g. herbivory) that is made on/to an experimental unit. A factor may have one or more *treatments* or *levels* and include a control; for example a herbivory factor with three levels may be (1) some leaves removed, (2) many leaves removed, and (3) a control with no leaves removed. The *experimental unit* (or sample unit) is the object (plant, genet, module, plot, etc.) of interest in the study.

The length of time for which a treatment is maintained is important. The response of a plant to a one-time treatment is likely to be quite different from its response to repeated application of the treatment. For example, the effects of a single defoliation versus repeated defoliation, or which leaves are removed, may have quite different effects on a plant's fitness (Doak 1992). An experiment involving a single application of a treatment is referred to as a 'pulse' experiment, whilst repeated or sustained application of a treatment is referred to as a 'press' experiment (Bender et al. 1984). The impact might also depend on how long after treatment measurements are taken. The effect could be small at first and grow, or it could vanish with time. Mortality might ensue if you wait long enough, and its timing might not be witnessed if you wait too long.

Care must be taken to ensure that the treatments used do not result in any additional, confounding '*hidden treatments*' (Huston 1997). Hidden treatments are indirect factors that intentionally or unintentionally interfere with the course or interpretation of the experiment. Huston (1997) discusses an example in which different levels of nitrogen fertilizer were the experimental treatments in a grassland experiment (Tilman 1996). However, the experimental plots were later put into eight groups according to differences in species richness. These groups were subsequently used as treatments to test the effect of species richness on ecosystem stability. Levels of species richness were not the true, original, treatments. Instead, differences in species richness became established following a biomass gradient produced in response to the nitrogen treatments. We could imagine an analogous example in population ecology occurring where investigators were interested in the effects of aphid herbivory and fertilized their plants at different levels to increase their palatability to aphids. Plants with a higher leaf nitrogen content might attract more aphids and this would probably have a greater effect on the plants, but this is an indirect and inappropriate way of manipulating herbivore numbers. In addition, it confounds the effects of plant nutrients on plant growth with the response to herbivory. A more appropriate experiment might be to establish treatments in which different numbers of aphids are placed or caged onto plants that do not differ in tissue quality or size.

4.2 Growing plants

The factors listed below should be considered if your experimental design requires you to grow plants, whether in the greenhouse, growth chamber, or in the field. All of these factors affect plant growth and potentially the responses of plants to experimental treatments. Plants respond simultaneously to many aspects of the environment, many of which are not independent in their effects. As noted in Chapter 3, Type II errors need to be minimized by ensuring that your experimental design is an appropriate test of your hypothesis. The effects of the following factors on plant performance need to be considered in the experimental design:

- plant age
- ontogeny
- sex and mating
- density
- genetic identity
- space available for above- and belowground plant parts
- appropriate and realistic above- and belowground environmental conditions (light, soil moisture, nutrients, and biota)
- protection from unplanned disturbances.

A plant's response to the environment depends upon both its age and ontogenic stage. Ontogeny refers to the sequence of developmental phases through which an individual passes as it grows. Although ontogenetic change is a continuous process (e.g. directional changes in the root/shoot ratio as plant mass increases), Gatsuk et al. (1980) defined several consistently recognizable ontogenic stages including:

- seed
- seedling
- juvenile
- immature
- virginile
- reproductive (young, mature, and old)
- subsenile
- senile.

Other classifications of ontogenic stages have been proposed for forage grasses and crop plants. For example, five main primary ontogenic growth stages of perennial forage grass tillers are recognized (Moore et al. 1991):

- germination
- vegetative—leaf development
- elongation—stem elongation
- reproductive—floral development
- seed development and ripening.

Similar ontogenic growth stages have been identified for the annual herb *Arabidopsis thaliana* including vegetative rosette stages based upon leaf number (Boyes et al. 2001).

The ontogenic stage of a plant is probably more important than its age in the structure and dynamics of populations. For example, tree saplings can remain in an immature growth stage for many years when suppressed under a thick forest canopy. These ontogenically immature plants may be better able to respond rapidly to the growth opportunities afforded by a canopy gap than more mature individuals. The results and interpretation of an experiment with seedlings may be quite different from the same experiment involving mature individuals. This means that the results of an experiment are generalizable only to individuals of an equivalent ontogenic stage. Ontogeny is of particular importance in experiments involving plant interactions (see Section 5.5) and in the classification of individual plants when using analytical methods such as population transition matrix models (Chapter 8).

The importance of density-dependent effects on plant behaviour is well known and needs to be considered when deciding upon the experimental design. Plants cannot perceive density directly—they respond to density effects indirectly. A plant grown in isolation may respond quite differently to a treatment from a plant grown with neighbours; in other words, plants are more sensitive to local density than mean density. The response will differ depending upon the identity and density of neighbouring plants. Depending on the purpose of the experiment, individuals should generally be planted equidistant from each other (e.g. in regular arrays) when there is more than one plant per pot. A regular arrangement equalizes density-dependent effects among experimental units. It is also important to recognize that replacing one individual of species A with one individual of species B is going to change the impact on a focal plant.

Only a small percentage of the world's flora bear male and female flowers on different individuals (~4% are dioecious and ~7% gynodioecious, with androdioecious plants being very rare; Silvertown and Charlesworth 2001), but this may be an important consideration in experimental design if you are studying one of these species. Dioecious species have male and female flowers on separate plants (from the Greek for 'two houses'), and include hollys (*Ilex* spp.), maples (*Acer* spp.), willows (*Salix* spp.), cedars (*Juniperus* spp.), and cannabis (*Cannabis sativa*). Gynodioecious species bear either hermaphrodite or female flowers, and include many species within the Lamiaceae, such as ground ivy (*Glechoma* spp.) and thyme (*Thymus* spp.). Androdioecious species have individuals that are either male or hermaphrodites and include the herbaceous *Mercurialis annua*. Sex biases can occur in the number of male and female plants in populations of dioecious species. There can be spatial segregation of the sexes, aggregation of seedlings around female plants, and differences in size and reproductive effort between the sexes. Sex expression can be labile, and in some species, for example jack-in-the-pulpit (*Arisaema triphyllum*), an individual changes sex as it ages (Policansky 1987) or, as in the normally gyndioecious *Glechoma hederacea*, if environmental conditions change (Hutchings and Price 1999). Different sexes have different resource requirements and different schedules of resource use. By mixing sexes in an experiment, you are introducing a new physiological variable, exactly as if you had used two species.

The breeding system of a plant is also important. Seed set is going to be severely reduced for self-incompatible species if pollinators are not present. In a greenhouse setting this may mean that artificial pollination will be necessary. This can often be accomplished using a brush to transfer pollen from the anthers of one plant to the stigma of another (for details see Kearns and Inouye 1993).

The genetic identity of a plant can affect its response to experimental treatments. What is most appropriate for your experiment: is it better to have the diversity of genotypes that is represented in a mixed seed lot or the genetic uniformity that comes either from seed from the same mother (half-sibs) or from cloned vegetative shoots (full-sibs)? Knowledge of the genetic constitution and origin of your experimental populations is important. For example, individual genotypes of the herbaceous annual *Polygonum pensylvanicum* respond differently to the range of conspecific densities encountered in the field (Thomas and Bazzaz 1993).

The rooting volume determined by pot size is an important issue for pot-based studies. Performance declines and the response of a plant to the environment changes when it becomes root bound. Asking the question 'Is physical space a soil resource?' McConnaughay and Bazzaz (1991) showed that plants grown with larger rooting volumes had greater vegetative growth, different root:shoot ratios, and often a higher reproductive output than ones with smaller rooting volumes, even when provided with equal amounts of nutrients and adequate water. If plants are going to be grown to a fairly large size then it is sensible to grow them as seedlings in flats or small pots before transplanting them into larger pots. The pots should be sufficiently large that the plants will not become root bound during the course of the experiment. To avoid rooting volume problems it has been advised that the pot size should be chosen so that plant biomass does not exceed 1 g L^{-1} (Poorter et al. 2012).

If environmental factors such as light (intensity, quality, and day length), soil moisture, nutrients, texture, and soil biota (e.g. mycorrhizae) are not part of the experimental treatments, then their levels need to be standardized and appropriate for the species concerned throughout the duration of the experiment. If environmental factors are limiting to plant growth and are not standardized then they can confound the experimental results. It may not be possible to control all environmental factors in a field experiment, especially with self-seeded plants in undisturbed soils. Either way, whether controlled or not, the levels of environmental factors may need to be measured and reported (see Sections 4.3 and 4.4 and Chapter 6).

Unplanned events and disturbances can disrupt or ruin the best-planned experiment. At worst, unobserved disturbances can completely confound results and render an entire experiment worthless. However, if they are carefully recorded such events sometimes offer serendipitous insights. For example, Alexander Fleming did not plan to introduce the *Penicillium* fungus into his bacterial cultures, but he was quick to recognize the value of the failure of his original experiment. It is not easy to predict disturbances, but some careful planning will not go amiss. For example, unless the effects of herbivory on seedlings are being studied, seedlings need protection from rodents, slugs, grasshoppers, etc. (see Section 4.3). The same seedlings may require supplemental watering for the first few weeks in order to help them establish. An ecological example of an unplanned intrusion is provided in Case Study 3 in Chapter 1. The experiment in this case study intended to study the effects of moths on ragwort, but ragwort seed fly larvae (*Botanophila seneciella*; Diptera) infested flower heads of the target species. The effect of this unplanned experimental disturbance was monitored carefully. The difference in capitula per study plot between plants protected from moths and unprotected plants (a planned treatment) was increased by 4% as a result of the fly larvae. Fortunately, this was considered to introduce only a negligible bias in estimates of the effects of the moth.

4.3 Biotic treatments

Biotic experimental treatments include factors that alter the effect of other organisms on the target plants. Some of these treatments mimic interactions between biotic factors and plants. In this section, the source locations of populations, the effect of neighbours, manipulation of reproduction, herbivory, pathogens, and allelochemicals are considered.

Treatments involving the source location of populations

Since the pioneering field experiments of Turesson (1922, 1930) and Clausen et al. (1948) (see Fig. 4.1) it has been clear that the characteristics of transplanted populations often differ by virtue of their source location. These characteristics are due to the occurrence of plant ecotypes. An *ecotype* is a genetically distinct population of a species that is partially reproductively isolated from other populations and has evolved and adapted to the local environment (definition based upon Lowry 2012). Ecotypes represent the genotypical response to a particular habitat and are designated according to the habitat type they occupy (e.g. coastal ecotypes, heavy-metal tolerant ecotypes). An ecotype may or may not be morphologically distinct from other populations of the species and may or may not correspond to a subspecies or race (Briggs and Walters 1984). The term ecotype has to be used with some caution, as there is a tendency among many investigators to use it in a narrow sense and synonymously with 'locally adapted population'. It is argued that this narrow view, in which each population is considered an ecotype, is of limited value (Quinn 1978, 1987).

Ecotypes can be problematic in experimental design. Results of an experiment based on a single source population cannot be generalized with confidence to the species as a whole, especially if a known ecotype (or a particular cultivar, variety, or subspecies) is used. Conversely, an investigation of ecotypic differentiation between source populations can provide valuable information on the responses of populations to the environment.

Identification and characterization of ecotypes in response to habitat characteristics may be the focus of investigation. Experimental treatments to test the effect of ecotypic differentiation range from hydroponic greenhouse experiments to reciprocal field transplants. In these experiments the source population is treated as a categorical (multistate, non-ordered) independent variable. Source material can be seed, genets, or cloned modules (Hume and Cavers 1981). If whole plants are used care must be taken to remove soil and other possible contaminants from the roots. The growth of genets transplanted from one environment to another may reflect the

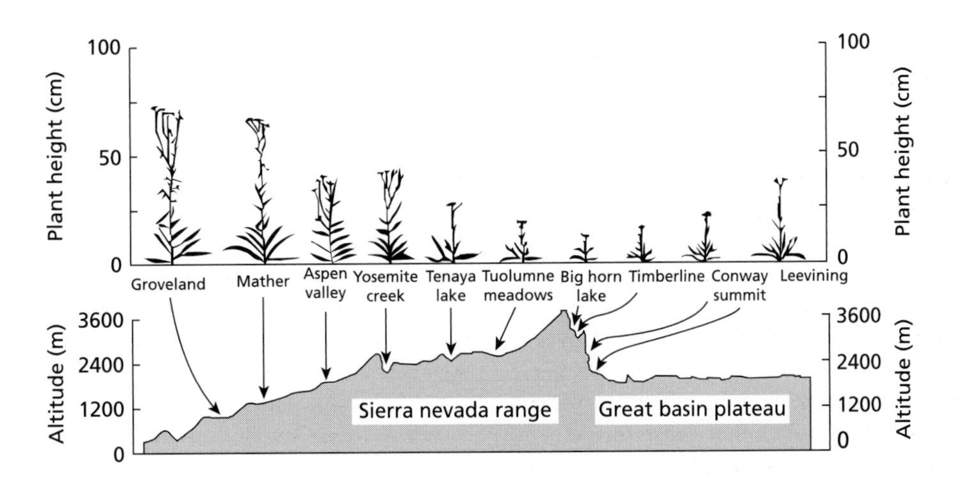

Figure 4.1 Results of Clausen et al.'s (1948) common garden experiment with *Achillea lanulosa* demonstrating distinct ecotypes across a wide topographic gradient. Reprinted with permission of the Carnegie Institution for Science.

original environment in which they (acclimation, a within-generation effect) or their parents (carry over, a between-generation effect) developed (McGraw and Antonovics 1983). Seedlings and adult plants should be allowed to acclimate to their new growth environment before measurements are taken. Better still, experimental plants can be allowed to acclimate to a standard environment before they are assigned to treatments. Assignment to treatments can be random (if there are a very large number of replicates) or paired by source population. Many recent population studies of ecotypes include measurements of the genetic makeup of the population (see Chapter 5) to allow differences between and within populations to be expressed on a genomic basis.

Common garden experiments, either in the greenhouse or the field, allow for potential morphological differences between putative ecotypes to be evaluated. Plants are grown together under uniform conditions from seed or as transplants from source populations. Measurements are taken either through time or at final harvest. The uniformity of soil and climate afforded by a common garden experiment can reveal the presence of ecotypes when populations differ under these conditions. Experiments of this form are widely used in forestry to assess differences between provenances (e.g. Kuser and Ledig 1987). The extensive nature of ecotypic differentiation in many grasses was shown by McMillan (1959), who grew clonal transplants of 12 prairie grass species collected from 39 sites across the North American Great Plains region in a common garden in Lincoln, Nebraska. Phenological patterns of growth and flowering of the source habitats were retained for 9 of the 12 grasses in the common garden. A limitation of this approach is that the growth of a population on a single soil (that the plants may not have had prior experience of), at a single location perhaps far removed from its source, and in the absence of natural competitors, herbivores, and pathogens, may not be reflective of its growth in its native habitat (Antonovics and Primack 1982). There is therefore no certainty that plants from different source populations that grow in the same way under one set of uniform conditions are not ecotypically distinct, because they may all respond in the same way to the new environment (i.e. there is a risk of a Type II error for failing to reject the null hypothesis that there is no difference between two populations).

Common garden experiments can also be used to investigate both genetic differences between individuals from within a single population and *reaction norms*, i.e. the phenotypic expression of given genotypes across a set of environmental states. To plot a reaction norm, the performance of each individual (or genotype or clone) is plotted against an experimental treatment, for example density as a continuous variable (Fig 4.2) or, say, soil moisture as an ordered categorical variable (Gibson et al. 2009). The slope of a reaction norm reflects an individual plant's capacity to adapt to different environments, i.e. its phenotypic plasticity, and can be analysed using a random regression model approach in R with the odprism and pamm packages (see the Appendix). As a rule of thumb, measurements should be made on at least 40 individuals with a data set comprising at least 1000 observations to analyse reaction norms with sufficient accuracy, precision, and power (van de Pol 2012). For example, Thomas and Bazzaz (1993) used this approach to investigate variation between 25 genotypes from within a single population of the herbaceous annual *Polygonum pensylvanicum*. The genotypes were cloned to obtain enough plants for growing at three densities per population in the greenhouse. Genotypes that grew rapidly early on had the greatest advantage in the highest-density treatments (Fig. 4.2).

Reciprocal transplant–replant experiments using phytometers (Chapter 3) provide a realistic setting for the assessment of the effects of the local environment upon genetic variation in plant populations. In these experiments plants from two or more source populations are removed (or seed is germinated) and reciprocally replanted back into both their sites of origin and each other's site(s) of origin. The mortality of transplants and the variation in growth of surviving transplants may both be high, necessitating large numbers of transplanted replicates. However, the realistically obtained magnitude of genetic and environmental variation makes these experiments valuable (e.g. Antonovics and Primack 1982; van Tienderen and van der Toorn 1991; Waser and Price 1985). Similarly, within-site reciprocal transplant–replant experiments allow for characterization of

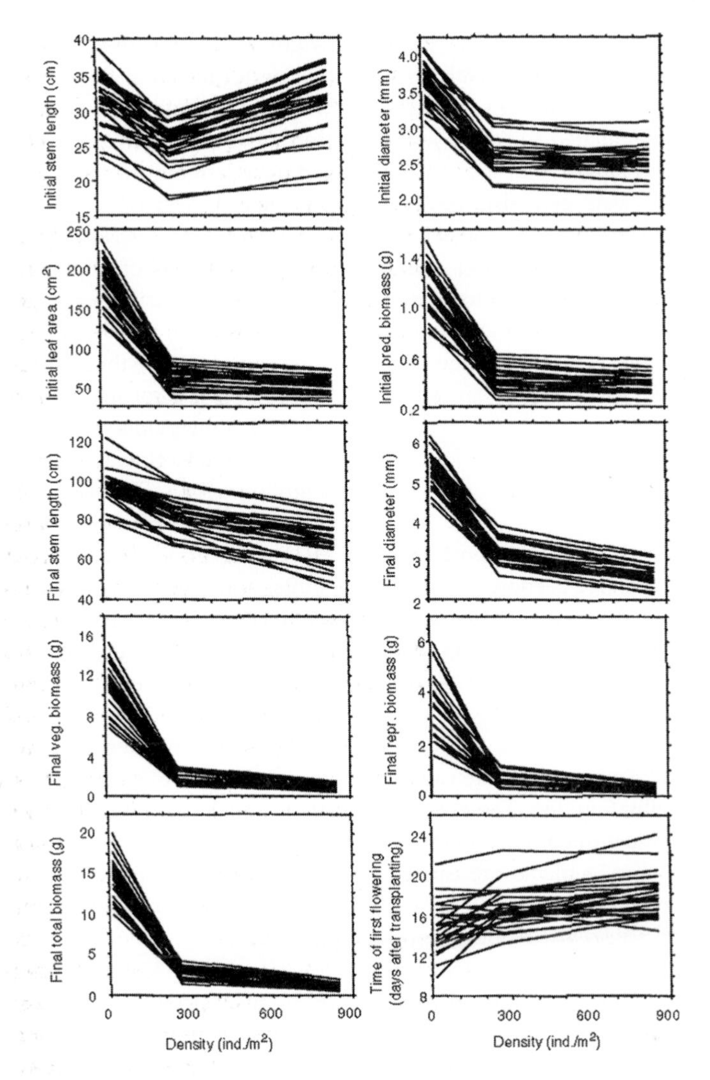

Figure 4.2 Reaction norms for 25 genotypes of *Polygonum pensylvanicum* in relation to density. Republished with permission of the Ecological Society of America from Thomas and Bazzaz (1993).

local adaptation and for genotype–environment interactions to be determined (Antonovics et al. 1988).

Heavy metal ecotypes. Ecotypes that are tolerant of heavy metals develop naturally on metalliferous soils in a wide range of plant populations, especially grasses (Antonovics et al. 1971). The standard test for the development of a heavy metal-tolerant ecotype involves root tillers of suspected ecotypes in hydroponic culture. Aliquots of heavy metals are added to the hydroponic solution and the root length of the tillers is measured through time. Root growth relative to plants grown without metals provides an index of heavy metal tolerance where (Jowett 1964)

$$\text{Index} = \frac{\text{mean length of longest root in solution with heavy metal}}{\text{mean length of longest root in solution without heavy metal}}.$$

Values of the index will range from 0 to 1, where a value of 0 would indicate that the plant did not grow at all, or died, in the heavy metal solution, and values close to 1 indicating a high level of tolerance to the concentration of heavy metal in the treatment solution. Lead-tolerant populations of the grass *Agrostis tenuis* from abandoned mines in Wales had index values ranging from 0.52 to 0.73, whereas populations from uncontaminated soils had index values of 0.27–0.35 (Jowett 1964).

Standard toxicological tests can also be performed to assess the toxicity of heavy metals, including determination of LD_{50} levels. LD_{50} stands for 'lethal dose 50' which is the dose of a toxin that kills (or reduces the growth) of 50% of the test population after a specified time. For example, the LD_{50} for nickel was determined as 100 mg L^{-1} in canola (*Brassica napus*) (Ali et al. 2009).

Treatments affecting neighbours

The potential effects of neighbours are important in studies of plant interactions (see Section 4.5 where experimental designs are discussed). Neighbouring plants affect target plants by intercepting light and moisture above ground and nutrients, space, and moisture below ground and through the production of allelochemicals ('Treatments involving allelopathy', below, and Section 6.1). There can be positive interactions (facilitation) too. The effect of these factors needs to be quantified when manipulations involving neighbouring plants are carried out as part of an experimental treatment.

Clipping or tying back neighbours. Aboveground treatments include clipping neighbours back partially or to ground level. This approach can completely eliminate aboveground effects of neighbours but requires frequent maintenance. A disadvantage is that the root system of clipped neighbours may alter belowground interactions, or the roots of clipped individuals may die and decompose, resulting in a flush of nutrients. Laboratory bioassays of the performance of the target species with decaying roots prior to removal experiments may help assess the importance of these side effects (McLellan et al. 1995). As an alternative to physical removal, neighbouring plant stems can be held away from target plants or plots using string or plastic mesh. This non-destructive approach has the advantage of allowing neighbours to remain alive so root interactions are maintained, although the growth of the tied back plants may change. Skeel and Gibson (1998) used this approach to investigate the effect of neighbours on the photosynthetic rates and vegetative performance of the grass *Sorghastrum nutans*. The aboveground parts of all neighbouring species were held back with a piece of vinyl screen which had a circle removed through which a target plant was allowed to grow.

Weeding of non-target plants. This approach allows complete removal of neighbour effects, both above and below ground. However, weeding involves disturbing the soil, which, among other potential effects, can stimulate the germination of seeds from the seed bank. Repeated removal of resprouting or re-establishing plants is usually necessary.

Herbicides. Herbicides can be used to remove neighbours (e.g. Bertness and Shumway 1993; Silander and Antonovics 1982; Suwa and Louda 2011). Care is needed to avoid killing the target plants, and so systemic herbicides that lose their activity in the soil are often used. Glyphosate [*N*-(phosphonomethyl) glycine; Roundup®, Monsanto Inc.] is a broad-spectrum, non-selective, non-residual, post-emergence herbicide that is applied to the leaves (Smith and Dehme 1992; Székáacs and Darvas 2012). Upon absorption into the leaves glyphosate is translocated via the phloem to the roots and rhizomes. It is commonly used because it rapidly binds with soil particles, rendering it unavailable for uptake and no longer phytotoxic. It only affects plants that receive a direct application of the herbicide. Glyphosate is most effective in rapidly growing plants, as translocation within plant tissues is impaired in dormant plants. The primary mode of action in plants is through the blocking of the synthesis of aromatic amino acids and metabolism of phenolics, leading eventually to a cessation of growth, cellular disruption, and plant death. Herbicidal effects generally take 2–4 days in annuals and up to 7 days in perennials and are preceded by the gradual wilting and yellowing of treated plants.

Belowground root trenching and exclusion tubes. Reduction or removal of the belowground effects of neighbours is problematic, especially in a field setting (see Section 4.5). Root trenching can be used to sever roots from neighbouring plants, decreasing their effects on plants inside the trenched area. This method involves the cutting of vertical incisions in the soil. Robertson (1947), in a study of the effects of competition with sagebrush (*Artemisia tridentata*), showed that trenching to a depth of 25 cm increased the yield and size of individual range grass plants. Important considerations are the depth to which trenching is done (it needs to be below the depth that the roots are growing in the soil), trench width (to decrease the probability of ingrowth), the frequency of trenching (to prevent root regrowth), and the physical constraint placed on the rooting volume of the target plants within the trenched area. The need for repeated trenching can be avoided by burying plastic or metal sheets or tubes in the soil (Wilson 1993); however, the effects of trenching with or without the installation of barriers may not be the same (Powell and Bork 2006). Trenching can also promote mineralization of nutrients if roots are severed and subsequently die. While trenching may reduce the belowground effects of neighbouring plants, it may also sever roots and restrict the soil volume available to the roots of the target plant. A review of root competition studies in forests, including the efficacy and effects of root trenching, is provided by Coomes and Grubb (2000), who noted that soil fertility and shade tolerance affected species' responses.

The use of root exclusion tubes involves planting seedlings or saplings of target plants inside soil-filled bags or tubes with drainage holes in the habitat of interest (e.g. Gunaratne et al. 2011). This approach excludes neighbours from around the root zone of the target plants and does not have the problem of potentially increasing the supply of nutrients from decaying roots that is associated with trenching. However, a limitation of this approach is that the soil conditions inside the tube do not adequately mimic the conditions under which a seedling invades soil containing active fine roots (Coomes and Grubb 2000).

In situ root chambers. A novel approach for investigating competition between fine roots was proposed by Hertel and Leuschner (2005) using small (90 mm × 70 mm × 30 mm) plastic chambers containing field soil. Intra- and interspecific competition can be addressed by measuring the length of in situ fine roots from one or two plant species before and after they have been placed into a chamber in which they are allowed to grow for a short period of time (see 'Roots and belowground structures' in Section 5.3).

Treatments manipulating reproduction

Reproduction, through seed production, is a measure of plant fitness. Treatments involving the manipulation of structures related to reproduction can provide information on fitness costs to the plant. The field of reproductive biology is extensive and well beyond the scope of this book (Lovett Doust and Lovett Doust 1988). The discussion here is limited to treatments that may affect the population ecology of plants directly through effects on reproduction. Examples are selected to illustrate the range of treatments that can be used to test hypotheses regarding plant reproduction. Specifically, the following topics are covered in this subsection:

- defoliation
- pollination and pollinators
- seed addition experiments
- maternal effects
- clonal growth and the integrated physiological unit.

Defoliation. The removal of plant parts can allow an assessment of their reproductive costs or benefits to a plant. Leaves are a source of carbohydrates that may be critical for developing floral buds or maturing seed. For example, Lehtilä and Syrjänen (1995) observed decreased seed mass per plant and mass per seed in the herbaceous *Primula veris* when leaves were removed. Similarly, experimental defoliation of fruiting branchlets of the understorey tree *Cornus florida* increased fruit abortion and decreased fruit size (Boyd and Pitzer 1991). In some plants the reproductive structures themselves may also produce carbohydrates to compensate for the loss of leaves. By contrast, reproduction may incur a cost on future reproduction. This was demonstrated in the herbaceous perennial *Ranunculus acris*, where fecundity was reduced in unmanipulated plants compared with plants in which the flower buds had been removed over four previous years (Hemborg 1998). Although not considered in these studies, the timing of removal, the number of leaves removed, and which leaves are removed are also likely to be important and should be considered in such experiments. The authors of these studies emphasized that their treatments were not testing the effects of herbivory. As noted later in this subsection, artificial defoliation treatments are frequently used in herbivory studies, although the effects of removing leaves with scissors, for example, may be quite different from the effects of leaf removal by a herbivore because of positive or negative effects of saliva and other factors.

Pollination and pollinators. Pollinators are essential except in self-compatible or apomictic species (i.e. those that can produce seed without the requirement for sexual reproduction). Treatments to test the importance of pollen limitation on a species or group of species (a pollinator network) include the following (with a representative example study in parentheses):

1. Supplemental hand pollination (Sletvold et al. 2012).
2. Preventing the visitation of pollinators (e.g. placing bags over flower heads; Hornemann et al. 2012).
3. Comparing the pollination efficiency of different pollinators (Arena et al. 2013).
4. Testing the effects on pollination of the quality and quantity of nectar (Real and Rathcke 1991).
5. Manipulating the foraging behaviour of pollinators (Kunin 1993).
6. Disrupting the integrity of pollinator networks (e.g. by removing floral resources of target plant species; Lopezaraiza–Mikel et al. 2007).

Detailed methods for conducting these and other treatments can be found in Dafni (1993) and Kearns and Inouye (1993).

Some highly original experimental treatments have been used to mimic pollinators or flowers to gain insight into the issues related to the energetics of plant–pollinator interactions. Two examples will serve as an illustration. The influence on pollinator behaviour of variability in nectar reward and flower colour was tested using artificial blue and yellow flower morphs constructed of cardboard squares centred under plastic wells filled with honey water for nectar (Real 1981). This study showed that plants are most likely to benefit from pollinator visits when they can offer a constant nectar reward compared with a variable reward. Molau and Prentice (1992) also conducted a manipulative field experiment in which they assigned plants from three species of arctic saxifrages (*Saxifraga cespitosa*, *Saxifraga tenuis*, and *Saxifraga cernua*) to one of three treatments: hand-selfed individuals, where the pollen from an anther was transferred to the stigma of the same individual; autodeposition, in which individuals were caged to prevent insect pollination; and control individuals which were uncaged, unmanipulated, and open-pollinated. These manipulations of the three saxifrage species were conducted as part of a larger study to investigate the extent to which self-pollination or vegetative spread along with phenotypic diversity in leaf shape (a measure of population differentiation) contributed to or maintained population differentiation.

Experiments on the resilience and integrity of pollinator networks are logistically difficult because of the large number of (generally unknown) interacting partners (pollinators and plants). Nevertheless, experimental treatments involving the removal of a partner can be carried out. For example, Lopezaraiza–Mikel et al. (2007) investigated the impact of an alien plant (*Impatiens glandulifera*) on a native plant–pollinator community by experimental removal of the floral resources of the alien plant (leaving the rest of the plant) from treatment plots and comparing the structure of the plant–pollinator community with unmanipulated control plots.

Seed addition experiments. Are populations limited by seed input and seedling establishment? Seed limitation can affect population size, especially in annual species. Addressing this issue can be important for understanding the metapopulation dynamics of a species. Measuring the proximate cause of seed limitation (e.g. frugivory, shading) is discussed later when we consider the design for herbivory, litter, and space treatments. However, testing the immediate effects of seed limitation on population size and population growth rate can be assessed by seed addition experiments. Care is needed in conducting these experiments as the recruitment function (i.e. the density-dependent rate of seedling establishment) and ambient seed density need to be known. Depending on how a seed addition experiment is carried out, it can address issues related to scale (local versus regional limitation) and population growth rate or carrying capacity (Table 4.1). Recommendations for seed addition experiments include the following (Clark et al. 2007):

Table 4.1 Inferences possible from seed addition experiments. With kind permission from Springer Science + Business Media: from Münzbergová and Herben (2005).

Scale		Limitation by	
		Population growth rate	Carrying capacity
Local	Term and definition	Seed limitation: population size is limited by seed availability at the local scale	Microsite limitation: population size is limited by availability of opportunities for establishment and growth
	Evidence	Number of seedlings increases after adding extra seeds or excluding herbivores feeding on seeds or seedlings	No increase in population size after seed addition
Regional	Term and definition	Dispersal limitation: distribution of species is limited by seed dispersal	Habitat limitation: distribution of species is limited by the availability of suitable habitats
	Evidence	Adding seeds to unoccupied habitats results in successful germination and establishment	Species does occur in all suitable habitats that are determined by a seed sowing experiment or an alternative method

1. Measure the ambient seed rain and seed bank.
2. Add seeds at a range of densities.
3. Monitor supplemental seed production from individuals that become established from the added seed.

The addition of 1000 seeds of *Aristida longiseta* or 1000 spikelet heads of *Bouteloua rigidiseta* to 45 cm × 90 cm plots in a Texas grassland led to six to nine times the number of one- and two-tillered plants after 5 years compared with plots that did not receive additional seed (Fowler 1995). Some caution is needed in interpreting the results of an experiment like this because of the cumulative nature of the increase in plant numbers in the seed addition plots. Plants that established from the additional seed themselves produced extra seed, which increased the magnitude of the effect of the treatment. We do not know from this experiment why seeds were limiting the size of the populations of these species, although to be fair the experiment was designed to address somewhat different issues of density dependence. Seed production by these species is apparently abundant in the grassland (Fowler 1995). Were levels of granivory limiting? Were the seeds succumbing to pathogens? Additional experiments would be necessary to answer these questions. For example, populations of annuals were limited by seed number in granite outcrop communities dominated by *Sedum smallii* (Houle and Phillips 1989). The population size of the annual community did not differ between treatments where (1) seeds were added in the presence of the dominant species and (2) seeds were added and the dominant *S. smallii* was removed. In both cases population size was larger than in controls where no seeds were added and no plants removed. D'Antonio (1993) used a seed addition treatment to assess the factors limiting the invasibility of an exotic species, *Carpobrotus edulis*, in California. Populations of *C. edulis* were found to be seed limited, but the response to seed additions was context specific, varying from one habitat to another and with levels of herbivory.

The source of seed in the seed-limitation experiment (discussed above) can be a treatment itself. Seed collected from across the range of a species can be used to investigate the influence of ecotypes on seedling establishment, as discussed earlier in this chapter. Such variation can directly influence the reproductive output of populations (Aarssen and Clauss 1992; Pigliucci and Schlichting 1995a). This can be investigated by germinating and growing seedlings of each genotype under controlled conditions. Pigliucci and Schlichting (1995b) tested the effect of geographic origin on the plasticity of 26 seed stocks of *A. thaliana* by growing plants in a growth chamber through maturity and senescence. In this study the source population was considered as a treatment in the statistical analysis.

Seed retrieval studies are also an important corollary to seed germination studies as they allow an understanding of seed longevity in the field. In these studies, seed is placed in mesh bags in the field for a predetermined time. At specified intervals, perhaps every month, a number of bags are retrieved from the field and the seed tested for viability and germinability (Chapter 6). Time is the independent variable in these studies.

Maternal effects. As with testing plants grown from seed with different geographic origins, plants grown from seed derived from maternal parents raised under different environmental conditions may also differ. The contribution of the maternal parent to the phenotype of its offspring beyond the equal chromosomal contribution expected from each parent is referred to as a *maternal effect* (Roach and Wulff 1987). Generally, environmental maternal effects are most pronounced early in a plant's life and decrease in later developmental stages. Differences in seed germination, seedling growth, and leaf area and plant height may be revealed when the maternal origin of plants is considered a treatment. At least two generations of plants are necessary to conduct these experiments, and so annuals or early flowering perennials are often used. To avoid maternal effects in experiments it is necessary to sample seeds randomly from individuals in the source population. Bulk seed collections, in which seeds from many plants are mixed, are intended to overcome potential maternal effects. To account for maternal effects some investigators include seed mass or seedling size as a covariate, although this 'correction' may underestimate genetic variation. To avoid maternal effects confounding an experiment it is generally best to grow at least one generation of plants in a standard environment and use their progeny (Bischoff and Müller-Schärer 2010). Conversely, to investigate maternal effects the parental origin of each seed needs to be recorded and treated as an experimental unit or treatment (parent A, B, C, etc.).

Clonal growth and the integrated physiological unit. Clonal growth and vegetative 'reproduction' is, strictly speaking, not reproduction since recombination does not take place but is a form of plant propagation. Nevertheless, the vegetative growth of plants can be manipulated to assess its importance to population dynamics. Anatomically connected subunits of a clonal plant are referred to as an integrated physiological unit (IPU). For example, excavation revealed up to 18 IPUs in tussocks of the grass *Schizachyrium scoparium* (Derner et al. 2012). For clonal plants, many research questions revolve around establishing the extent of physiological and developmental integration between ramets. It is important to know when young shoots become physiologically independent and the extent to which they remain integrated with the rest of the clone after rooting. Manipulative experimental treatments include the removal of clonal fragments, branches, or modules from individual genets. These experiments need to be more sophisticated than simply seeing if plants grow differently when parts are removed. For example, Hartnett (1993) used a sharp blade driven to a depth of 30 cm into the soil to sever rhizome connections between halves of mature clones of switch grass (*Panicum virgatum*). There was no effect of rhizome severing, suggesting that the size of the IPU was smaller than that of the clones themselves (1.0–1.6 m diameter). In a more complex experiment with the clonal shrub *Gaylussacia baccata*, Matlack (1997) established 15 experimental treatments involving excavating rhizomes, clipping stems and rhizomes in several combinations, and reburying. The clipped treatments included isolated rhizomes and shoots, isolating pairs of shoots connected by a rhizome, and isolating rhizomes cut to various lengths. Procedural controls were established to measure the effect of excavating, severing, and reburying rhizomes, and included the excavation and reburying of rhizomes with further manipulation, clipping of stems and reburying of rhizomes, and clipping stems without excavating the rhizomes. The experiment showed that movement of resources between a shoot and its supporting rhizome was more important than movement of resources between clonal shoots via rhizomes. This pattern allows rapid regrowth following widespread destruction of ramets by fire. Other experimental manipulations of the vegetative structures of plants to assess their role in the population ecology of the species include complete separation of all ramets, and removal of leaves, tubers, and inflorescences.

Treatments involving herbivory

Herbivores can have demonstrable effects on populations at all scales of analysis. At the level of the individual plant, the effects of herbivores range from partial defoliation to complete consumption of the plant, and trampling of plants and surrounding soil. Some herbivores are specialists, eating only certain parts of the plants of one or a limited range of species (e.g. giant pandas eat only bamboo shoots), while other herbivores have a varied diet, for example goats, which seem to consume just about anything. Herbivores act on plants above or below ground, or both, and the herbivore and plant can interact (Johnson et al. 2012b). As a result of herbivory, plants may change the allocation of resources to shoots and roots, flowering may be suppressed or restricted to parts of the plant which escape herbivory, and plants will almost inevitably alter their growth form. Seed production and survival, survival of seedlings and mature plants, and vegetative growth can all be affected by herbivory (Crawley 1988). Indeed, the easily observed effects of herbivory can be misleading. Demonstration of damage to a plant by herbivores does not necessarily mean that there is an effect on plant population dynamics. For example, excluding insects for 6 years through spraying of oaks yielded large effects on seed production despite no effect on shoot production (Crawley 1985).

The design of effective treatments to assess the effects of herbivory should start with the question of whether you wish to keep existing herbivores away from target plants (i.e. exclusion) or whether you wish to keep specific herbivores on or around the plants (i.e. inclusion). The latter can be more problematic and may require the introduction and maintenance of populations of the herbivore for the duration of the experiment. In either case, population levels of the herbivore need to be monitored to measure the effectiveness of your treatments. Indirect measurements of herbivore abundance are sometimes necessary and may be the most practical method, for example counting rabbit droppings per unit area as a surrogate measure of rabbit abundance. Clear demonstrations of the importance of herbivory on plant populations require well-replicated manipulative field experiments. It is also important that the appropriate dependent variables are measured.

Belowground herbivory is affected by a variety of insects, nematodes, rodents, and fossorial mammals (pathogens are considered separately later in this subsection). Treatments centre on controlling their abundance through either physical or chemical exclusion. Chemical treatments can range from non-specific killing of most soil organisms (e.g. by chloroform fumigation; Sarathchandra et al. 1995) to the use of specific treatments. Examples of biocides targeted toward specific organisms include organophosphates such as isofenphos (Oftanol®, Mobay Chemical Co.) and ethoprop (MoCap®, Rhone-Poulenc Ag. Co.) to reduce numbers of soil invertebrates (Gibson et al. 1990). These are contact and gut toxins with effective activity against soil insects for 1–3 months following application. Isofenphos is recommended specifically for control of coleopteran and lepidopteran larvae (Evans 1991). Chlorpyrifos (also called Dursban and Lorsban) (Kenaga et al. 1965) is a broad-spectrum organophosphate insecticide that has been used to suppress both above- and belowground invertebrates with no evidence of phytoxicity; it adds negligible amounts of nitrogen and phosphorus to the soil (e.g. Coupe et al. 2009). By contrast, a non-chemical treatment, freezing soil to −20 °C, can be used to reduce the activity of soil nematodes (Sarathchandra et al. 1995). Whatever approach is taken, the experimental setting needs to be considered when interpreting the results of belowground herbivory studies on plants. Meta-analysis of multiple studies has shown that significant effects are more likely to be found in controlled greenhouse or laboratory settings than in field settings, and simulated herbivory treatments can overestimate the effects of natural herbivory (Zvereva and Kozlov 2012). Indeed, it is difficult to make estimates of herbivory comparable with these simulated approaches.

Aboveground herbivores include a variety of insects, molluscs, birds, and mammals. As with the belowground case, aboveground invertebrates can be controlled experimentally through the application (usually spraying) of pesticide. Carbaryl (Sevin 1, Union Carbide Co) photo-oxidizes rapidly without residual effects and has been used in a number of studies (Gibson et al. 1990; Boyd and Barbour 1993).

Carbaryl is insoluble in water with little infiltration into the soil. Other insecticides that have been used in ecological studies include Dursban, dimethoate, and malathion (Brown and Gange 1989; Brown et al. 1987).

Unforeseen indirect effects are a problem with many of the chemical treatments used to reduce above- or belowground herbivory. For example, when Seastedt et al. (1987) used organophosphates to reduce the numbers of belowground arthropods, application of the pesticide resulted in an increase in the abundance of earthworms (*Diplocardia* spp.). This response was thought to be due to a release from predation by the target arthropods that were negatively affected by the biocide. In the same study, a decrease in abundance of western ragweed (*Ambrosia psilostachya*) in plots treated with organophosphate was thought to be due to its sensitivity to herbivorous nematodes that had increased in response to this treatment (Gibson et al. 1990). An observed increase in leafhoppers and planthoppers (Auchenorrhyncha) in this study may have occurred through release from competition with the belowground arthropod fauna (Evans 1991). Chemical treatments like these may also have undesirable and unintended effects on the chemical and physical properties of soils, including cohesiveness, soil aggregates, and wettability.

Lists of pesticides that may be useful in ecological research are available in a number of guidebooks, for example Martin (1972) and Page and Thomson (1997). These should be consulted carefully to ensure that the chosen chemical will effectively target the organisms of interest. It is also critical that the biocide does not stimulate plant growth through the addition of water or nutrients, or inhibit plant growth through phytotoxic effects (either directly or indirectly through long-term accumulation of toxic by-products). A test of this possibility should be an integral part of any such study. If the pesticide is applied in liquid form, then it is important to apply an equal amount of the carrier (usually water) to the control plants. A useful general reference is *Haye's handbook of pesticide technology* (Krieger 2010).

Physical methods of altering aboveground herbivory range from laboriously removing each insect from plants (Whittaker 1982) to the use of exclosures and enclosures. An investigator will often be faced with damage to a plant caused by an unknown species of herbivore, but identification of the specific herbivore can be made through careful combinations of selective exclusion treatments. Bhadresa (1997) provides some guidelines for the design and use of exclosures. It is important that the fencing used is appropriate for the type of animal (and it may not be fencing for exclusion of small animals such as aphids, Lepidoptera and hemipterans). Fence supports need to be strong enough to withstand damage from large animals such as cattle or bison when they are present. Cattle and sheep can be controlled quite easily with lightweight single-strand electric fencing. Conversely, bison require more substantial fencing. Deer are problematic because they can jump over fences of at least 2 m in height. For smaller mammals a small mesh (size 3–4 cm for rabbits, 0.5 cm for voles and mice) is necessary, and netting should be buried in the soil to prevent access through burrowing. Insects can be retained on or around target plants, or excluded, using mesh, paper bags, or cages. A potential problem is that the bag or cage itself may affect the response of the plant. For herbivore inclusion studies, a cage or bag effect can be measured by having control plants that are not caged/bagged as well as control plants that have cages/bags but no herbivores. For herbivore exclusion studies the problem is more difficult to address, because while bagging, for example, excludes the herbivore it may affect the plant. A partial solution is to include a set of plants with bags that have holes to allow herbivore access. For example, in Case Study 3 in Chapter 1 'sham cages' were used as part of the insect herbivory treatment. These cages had insect netting covering the top of the cages but the netting was rolled half-way up the sides to allow access for herbivores. Hulme (1994) provides a good example of a study design with treatments to assess the impact of vertebrate and invertebrate herbivory. A combination of exclosures with 3.0 or 0.64 cm wire mesh, insecticides, and a molluscicide were used to control small mammal, arthropod, and mollusc herbivory, respectively, on seedlings of 17 grassland species.

Surrogate treatments for herbivory (herbivory simulations) are often used and may be quite convenient, especially in greenhouse studies where the use of live herbivores can be prohibitive or in the field where there may be logistical problems preventing the use of live herbivores. These treatments include clipping or cutting leaves or plant parts, often with scissors or hole punches (see Case Study 4 in Chapter 1). Nevertheless, there are limitations to this sort of approach as herbivores may not cut leaves in the precise manner of a pair of scissors. Incremental manual removal of leaves or leaflets from a set of test plants can allow assessment of the effects of herbivory on plant performance. For example, Koptur et al. (1996) carried out a greenhouse study where leaflets of the annual herb *Vicia sativa* were removed in 25% increments by clipping. They observed a significant reduction in the number of pods and seeds, total seed mass, and individual seed mass with leaf area lost. Similarly, experimental defoliation reduced reproduction of herbaceous woodland plants in a comparable manner to the effects of herbivory by deer and ground hogs (Rockwood and Lobstein 1994). Herbicides can also be used to defoliate target plants. While these studies can be used to answer quite precise phenomenological questions, their limitations have to be kept in mind. The saliva that herbivores deposit on plant tissues may also affect their regrowth and scissor treatments cannot substitute for these effects (Detling et al. 1980).

Treatments involving pathogens

Plant pathogens include above- and belowground viruses, bacteria, and fungi and have been the focus of much ecological and agricultural research. Treatments result in plants or soil in which the pathogen is rendered inactive or has reduced activity. In addition, plants and soils can be inoculated with the pathogen. These treatments of pathogen reduction or addition are qualitative (the pathogen is either present or absent, or at ambient, reduced, or increased levels). Quantitative differences in pathogen activity are more difficult to obtain, except perhaps in the case of the inoculation times for pathogen vectors. Host–pathogen interactions are frequently complex and experimental designs need to consider a range of causal factors, including the mosaic nature of natural communities. Distance to neighbours, the identity of neighbours, and plant ontogeny affect the effectiveness of pathogen infections and need to be controlled or accounted for in experiments.

There is a danger of treating the belowground soil community as a 'black box'. Some of the techniques by which the soil is sterilized make this assumption. There can be nutrient flushes following sterilization or different nutrient uptake patterns during the conditioning phase as a target plant is grown in a soil to prepare the soil for later experimental use—both need to be accounted for (Hendriks et al. 2013). These experiments have to be interpreted with caution because it may not be clear which soil organisms and which interactions are ultimately responsible for effects on plant performance (Newsham et al. 1994). Nevertheless, Bever et al. (1997) describe a two-step protocol designed to evaluate the feedback between the microbial soil community and interacting plant species. With this approach the role of the soil community is evaluated in a general sense without the need to identify the microbial agents involved in the feedback. First, soils are 'cultured' by growing the plant species of interest in the soils in pots. The plant species are then grown in pots inoculated with each type of cultured soil. The difference between the growth of plant species in soil in which the same species had previously been cultured provides a measure of the magnitude of the feedback between the soil community and the plant species. The relationship can be represented by an interaction coefficient

$$I_s = \alpha_A - \beta_A - \alpha_B + \beta_B$$

for two plant species, A and B, with α and β representing the change in the soil community resulting from the presence of A and B, respectively (Bever et al. 1997). Statistically this is tested using 'home versus away' or 'leading diagonal' contrasts (Turkington and Harper 1979; Bever 1994). The implications of plant–soil feedbacks based upon interaction strengths are summarized by van der Putten et al. (2013), with a graphical analytical methodology in Revilla et al. (2013). Culturing soil in this way can also be used to develop soil treatments for allelopathy and root exudates (see 'Treatments involving allelopathy').

One of the principal methods for controlling fungal pathogens is through the use of chemical fungicides. Paul et al. (1989) provide a review of fungicide use in ecological studies and note that selection of a fungicide should consider four criteria: (1) spectrum of activity, (2) efficiency and persistence, (3) toxicity to non-target organisms, and (4) direct effects on the plant species present in treated vegetation. Of particular importance is their observation that because of the wide taxonomic diversity of fungi there is no single systemic fungicide that is effective against both the subdivision Mastigomycotina (which includes the important downy mildews, Peronosporaceae, and *Phytophthora* and *Pythium*) and the higher fungi (Zygomycotina, Ascomycotina, Basidiomycotina). Most fungicides do not show basipetal transport and so separate foliar and soil sprays may be necessary. Repeated treatments may be necessary as the persistence of treatment effects may be short-lived. Fungicides can be used to reliably assess the effect of saprophytic fungi on buried seed (Mitschunas et al. 2009).

For the control of soil-based fungi, liquid fungicides can be sprayed directly onto the surface of the soil in treatment plots with equal amounts of deionized water sprayed on control plots. The carbomate benomyl [active ingredient 50% Benlate®, methyl 1-(butylcarbamoyl)-2-benzimidazole carbamate, Du Pont Ltd] was for many years the most widely used fungicide in ecological studies (Paul et al. 1989; West et al. 1993) and was frequently used in studies of mycorrhizal fungi. Plant population ecology studies using benomyl indicate that interactions with mycorrhizal fungi can affect the competitive relationships of plants and their demography and fitness (Hartnett et al. 1994; Newsham et al. 1994). Benomyl is most effective against Zygomycotina and least effective against Basidiomycotina and Ascomycotina. Benomyl is considered to be generally non-toxic to plants but it has been shown to increase the fecundity of plants in some situations. In addition, benomyl has been shown to be effective in reducing the levels of some other rhizosphere and root-infecting fungi (West et al. 1993; Newsham et al. 1995). Commercial production of benomyl ceased in 2001, although ecological studies using the fungicide continue, presumably using old stock (e.g. Yu et al. 2012). Topsin-M [thiophanate-methyl: dimethyl (1,2-phenylene)-bis(iminocarbonothioyl) bis(carbamate), Cerexagri Inc., Philadelphia, Pennsylvania] has been proposed as a suitable substitute for benomyl in ecological studies (Wilson and Williamson 2008) and appears to be suitable for establishing mycorrhizal-suppressed treatments in individual grass genets (e.g. Collins et al. 2010).

Other soil fungal treatments include fumigation with methyl bromide and chloropicrin. Both of these are insidious poisons and have to be handled with great care. Methyl bromide (MeBr) is a colourless gas that was first introduced as an insecticide in 1938 (Martin 1972). Chloropicrin is also an insecticide, but is added to fumigants as a warning agent because of the intense irritation it causes to eyes and mucous membranes. An advantage of soil fumigation is that fairly large areas of ground can be cleared of the soil biota for use in experiments. Shumway and Koide (1994) fumigated the soil of a 0.4-ha field 2 weeks prior to establishing an experiment to test the effects of the presence of mycorrhizae and soil phosphorus level on reproduction of the annual *Abutilon theophrasti*. Although fumigants have been used in plant ecology studies in the past (Wallace 1987; Wallace and Svejcar 1987) they are less often used today because of the availability of easier and more efficient methods. Moreover, use of MeBr is heavily regulated and it is being phased out because of its role in the depletion of stratospheric ozone (Johnson et al. 2012a). An alternative to chemical fumigation that is being promoted in the crop production industry is biofumigation where plants of the mustard family (Brassicaceae) are grown in soil before a conventional crop. The roots of these mustards release glucosinolates that break down to produce isothiocyanates that serve as biochemical fumigants (Reddy 2013).

Non-chemical options for soil sterilization include soil solarization and the baking or autoclaving of soil for use in greenhouse experiments. Soil solarization involves heating the upper soil layers through radiant heat that accumulates under clear plastic film (solar tents) covering the soil surface (Stapleton 2012). However, there is the danger that all the soil biota are killed, not just the fungal pathogens. Microwaving the soil for a short period (e.g. 2.5 min kg^{-1} at 1500 W full power; Mihail et al. 1998) reduces the soil mycoflora with little effect on the other soil biota and without disrupting the soil structure (Ferris 1984). For example, microwaved soil was used in an assessment of coexistence between prairie plants to obtain a soil treatment with reduced soil fungal activity (Holah and Alexander 1999). Alternatively, mycorrhizae-free soil can be obtained by passing a blend of non-sterile (field collected) soil and sterile distilled water through a 38-μm sieve. This sieve size retains the spores of indigenous mycorrhizal fungi (Wilson et al. 1988). The sievings contain active bacteria and non-mycorrhizal fungi that can be added to experimental soils as a non-sterile soil amendment.

Mycorrhizae can be added to experimental treatments by inoculation with fungal spores. The mixture of mycorrhizal species retained as spores by a 38-μm sieve can be added directly to soils. Alternatively, spores of specific mycorrhizal fungi can be first isolated and collected from plant pot cultures by wet sieving, decanting, and sucrose gradient density centrifugation (Pacioni 1994). Mycorrhizal inoculum mixes can also be purchased commercially. This inoculum can then be added to the soil used in an experiment (e.g. Shumway and Koide 1994; Veiga et al. 2012). Mycorrhizal inocula can also be added to soils in the form of coarsely chopped roots from plants grown in pot culture containing mixed fungal cultures (Kormanik et al. 1982). Mycorrhizal spores are easily dispersed in water and by animal vectors (e.g. small mammals and earthworms) and therefore a great deal of care is needed to avoid contamination of non-mycorrhizal treatments. Non-mycorrhizal plants can be obtained by growing seedlings in sterile growth media (e.g. sterilized sand; Buwalda 1980). Mycorrhizal plants can then be obtained as necessary by inoculation of the soil with fungal spores as described above.

Inoculation to produce fungal-infected plants can be achieved by using fungal spores or by placing infected plants in close proximity to uninfected treatment plants (Dhingra and Sinclair 1995). There are four methods of inoculation using spores:

1. An atomizer can be used to uniformly deposit urediospores of the rust fungi (*Puccinia* spp.; Uredinales, Basidiomycetes) on seedling and adult cereal plants (Andres and Wilcoxson 1984). This approach can be used to study the relationship between inoculum concentration and plant performance (e.g. Welty and Barker 1993).
2. Healthy plant leaves can be dipped into spore suspensions obtained by crushing sporulating lesions from infected plants into deionized water. For example, Morrison (1996) dipped clipped leaves of healthy *Juncus dichotomus* into a teliospore suspension of the parasitic smut fungus *Cintractia junci* (Ustilaginales, Basidiomycetes) prior to planting. The rest of the spore suspension was then poured into the soil plug around the target plant roots.
3. Fungal spores can be dusted onto plants. For example, Alexander and Antonovics (1988) used a toothpick dusted with fungal spores of the anther-smut fungus *Ustilago violacea* to artificially inoculate naturally grown plants of *Silene alba*. The toothpick was prepared by being gently inserted into the floral tube of healthy flowers. The authors showed that the probability of a plant producing diseased flowers (and hence having reduced fecundity) increased with the numbers of flowers that were inoculated.
4. Diseased plants can be placed in close proximity to healthy target plants. For example, Alexander (1990) placed diseased flowers in jars of water at disease-free study sites. By monitoring the number of diseased flowers on naturally grown individuals surrounding the diseased plants, she was able to show that the likelihood of spore deposition decreased with increasing distance from the inoculum source.

Fungal endophytes. Systemic fungal endophytes in the Clavicipitaceae (Ascomycota) infect a wide range of grasses and can dramatically affect their population dynamics (Siegel et al. 1987; Clay 1990, 1998). Depending upon the nature of the relationship the fungus may sporulate in the host and spread contagiously from plant to plant or remain entirely within the plant host throughout its life cycle. The latter relationships are mutualistic, and an infected plant cannot infect another plant. The fungus is passed on maternally through the seed from one plant generation to the next. Examples of this mutualistic relationship include endophyte-infected tall fescue (*Schedonorus phoenix*) and perennial ryegrass (*Lolium perenne*). There has been a wealth of ecological and agronomic studies of these two grasses and their endophytes (Clay 1998), but there is concern that the preponderance of studies on the agronomic cultivars of these two taxa of forage grasses may not be representative of wild-grown grasses (Gundel et al. 2012). Endophyte-infected and endophyte-free treatments can be established by either using an uninfected seed source or through vegetative cloning of uninfected tillers. There are a large number of cultivars of both *S. phoenix* and *L. perenne* and it is necessary to use infected and uninfected plants of the same cultivar in experiments. The fungicides benomyl and propiconazole can be used to remove, or at least reduce, the fungus in grasses (Bacon and White 1994; Cheplick 1997). The sterol inhibitor triadimenol can eradicate the endophyte from treated seed (Williams et al. 1984). The viability of the endophyte can be reduced in seed by storing for 7–11 months at 21 °C or by short-term heat treatments (57 °C for 40 minutes or 49 °C for 7 days), but with some loss (up 16%) of seed viability (Siegel et al. 1984). Hot-water treatments are also effective (Williams et al. 1984).

Soil bacteria. Treatments to increase the general activity of soil bacteria include application to the soil of carbon sources (sugar, sawdust, straw, or grain hulls) (Seastedt et al. 1988; McLendon and Redente 1992) or crabshell chitin (poly-β-1,4-*N*-acetylglucosamine) (Sarathchandra et al. 1996), both of which will stimulate the microflora. The soil microbes accumulate soil nitrogen, lowering its availability to plants. Reever-Morghan and Seastedt (1999) used this approach to study proposed restoration efforts on the non-native *Centaurea diffusa* and observed a decrease in biomass per plot following carbon supplementation but no effect on seed production. The effects of specific bacteria on plant performance can be tested by directly inoculating soil with bacterial cultures. For example, seedling emergence of *Agropyron cristatum* was found to be enhanced after soils were treated with a solution containing the free-living soil diazotroph *Bacillus polymyxa* (Holl et al. 1988). The bacterium was isolated from a pasture soil and cultured in nutrient broth.

Symbiosis between most legumes (Fabaceae), 158 species from 14 non-legume genera (Bond 1976), and nitrogen-fixing bacteria (*Rhizobium* and *Bradyrhizobium* in legumes, *Frankia* in non-legumes) is thought to bring mutual benefit to each participant and can markedly affect plant population dynamics (Chanway et al. 1991). Treatments to assess the effects of these bacteria on plant growth can be established by inoculating the soil with a bacterial suspension. The inoculum can be poured directly onto the soil surface and watered in. Bacterial samples can be obtained from isolated nodules using standard methods (Vincent 1970; Chanway et al. 1989; Food and Agriculture Organization of the United Nations 1993; Somasegaran and Hoben 1994) and stored in culture until required. Measures of soil nitrogen are important in these experiments because of the nitrogen-fixing ability of the rhizobia.

Viruses. Plant viruses are predominantly transmitted from plant to plant via insects, although around 12% are spread by other vectors including mites, nematodes, and fungi (Power 1996). However, while it is well known that viral infections can reduce agricultural yield, the frequency and abundance of viruses and their effects on plant fitness are poorly known (Prendeville et al. 2012). Viruses can be detected in plant tissues using an enzyme-linked immunosorbent assay (ELISA) or reverse transcriptase polymerase chain reaction (RT-PCR); the latter is generally the more sensitive method (Hu 1995). Experimental treatments involving viruses have to consider the three-way interaction between virus, host plant, and vector. For example, the barley yellow dwarf virus (BYDV) is one of the most important diseases of grasses worldwide and has been the subject of considerable study (D'Arcy and Burnett 1995). To test for the effect of the virus upon barley (*Hordeum vulgare*) and oats (*Avena sativa*), Harper et al. (1976) infested test plants with viruliferous or non-viruliferous aphids (*Rhopalosiphum padi*). *Rhopalosiphum padi* is one of the primary vectors of BYDV, and in Harper et al.'s experiment it was used to produce virus-infected plants. Virus vectored by the viruliferous aphids reduced the height of barley and oats, the leaf width of oats, and the number of barley tillers. Differences in virus content in test plants can be obtained by varying the inoculation times (i.e. the time spent feeding, and hence transmitting virus, by the viruliferous aphids) (Pereira et al. 1989). For *R. padi*, inoculation times differed for different strain–vector combinations, ranging from 15 minutes to 48 hours (Power and

Gray 1995). Another way to assess the effects of a virus on plants is to infect test plants directly with virus-infected cell sap prepared by grinding virus-infected plant tissue in a suitable buffer solution, and then applying it to leaves as a viral suspension (Molken et al. 2012).

Treatments involving allelopathy

Allelopathy is 'any direct or indirect harmful or beneficial effect of one plant (including microorganisms) on another through production of chemical compounds that escape into the environment' (Rice 1984). The chemical compounds involved in this interaction are termed allelochemicals (see Section 6.1 for the methodology for chemical identification). The relevance of allelopathy to plant interactions is unclear, not least because of methodological and logistical difficulties. Allelopathy has been implicated as a mechanism in the 'novel weapons hypothesis' for understanding alien plant invasion (Inderjit et al. 2008). Experimental treatments to test the role of allelochemicals upon plant populations can range from comparatively simple bioassays in the laboratory to more complex field manipulations. However, as noted in more detail below, it is difficult to equate the results of experimental allelopathy treatments with the effects that might apply in the field because concentrations of active allelochemicals are rarely known and difficult to mimic.

Laboratory bioassays. Bioassays measure the germination or growth of a test plant, often seedlings, against extracts, exudates, leachates, or decomposition products of suspected allelopathic plants (Romeo and Weidenhamer 1998), or growth in 'cultured' soils in which allelopathic root exudates have accumulated (see 'Treatments involving pathogens'). In particular, laboratory bioassays can help to identify and explore the toxicity of putative allelochemicals and the fate, role, movement, and interactions of allelochemicals in soil and with soil microbes (Inderjit and Nilsen 2003). The following commercially available plants have been proposed as standard target species for allelopathic investigations (Macías et al. 2000):

- lettuce (two varieties *Lactuca sativa* cvs. Nigra and Roman)
- Onion (*Allium cepa*)
- Cress (*Lepidium sativum*)
- tomato (*Lycopersicon esculentum*)
- barley (*Hordeum vulgare*)
- carrot (*Daucus carota*)
- wheat (*Triticum aestivum*)
- corn (*Zea mays*).

However, others argue that native plants from the field setting of the presumed allelopathic target plant allow greater realism (Inderjit and Nilsen 2003).

Leachates can include solutions of rhizochemicals extracted from soils or washed from the roots in which suspected allelopathic plants are grown (e.g. Hagan et al. 2013). Effective treatment concentrations of the allelochemicals are obtained by varying the concentration of extract or leachate that is applied to the test plant, or the amount or condition of shoots, roots, or leaves that are incorporated into the growing media. Concentrated extracts can be obtained by freeze-drying the plant material and grinding it in a Wiley® grinding mill, although such extracts are likely to be more concentrated than in field situations. Nevertheless, the ground material is agitated in distilled water, filtered, and the filtrate diluted to the required strength. Equal amounts of distilled water should be added to control plants to avoid confounding moisture effects. The use of organic solvents should be avoided. Example applications include Wardle et al. (1996), Applebee et al. (1999), Mudrák and Frouz (2012), and Viard-Crétat et al. (2012); Tang (1986) reviews the methodology for obtaining allelopathic root exudates and Inderjit and Nilson (2003) discuss protocols for laboratory bioassays.

Despite the value of laboratory bioassays, there are several problems with generalizing their results to the field (Romeo and Weidenhamer 1998):

1. The results of bioassays are determined to a large extent by the bioassay technique.
2. Determining the appropriate concentration of an allelopathic chemical for a bioassay is difficult because in nature it depends upon release rate, the water content of soil and plants, runoff, leaching, and spatial relationships. Allelochemicals may interact with each other and with other

plant chemicals, and uptake and response may be density dependent.
3. Physiochemical and microbial activities alter the toxicity of allelochemicals after they are released into the environment.
4. Positive bioassay results may not translate to ecologically relevant effects.

Leather and Einhellig (1986) suggest that several different bioassay methods should be used for each case of a suspected allelopathic interaction. A combination of terbutryn and triasulfuron has been proposed as an internal standard of known phytoxicity for comparison against new allelochemicals (Macías et al. 2000).

Field studies. Laboratory studies of allelopathy cannot demonstrate an unequivocal role for the compound in natural systems. Rather, evidence to support a role of allelopathy in a field setting needs to include the following information on the putative allelochemicals (Callaway 2003):

- concentrations
- release rates
- longevity in the soil
- chemical transformations involving soil microbial communities and organic and inorganic carbon components of the soil.

Field experiments to establish allelopathic treatments generally involve the application of leachates or plant tissues to the soil around target plants. It is exceptionally difficult to design a field experiment that unequivocally tests the effect of a presumed allelopathic compound without involving confounding effects from competition for water, light, or soil nutrients or the involvement of herbivory and soil microbes (Williamson 1990). Nevertheless, Nilsson and Zackrisson (1992) demonstrated an allelopathic role of the dwarf shrub *Empetrum hermaphroditum* in the establishment of *Pinus sylvestris* seedlings. Laboratory bioassays using foliar leachates had shown the presence of an allelopathic inhibitor (3-methoxy-5,3'-dihydroxy, dihydrostilbene, or batatasin III). Bioassays with freshly collected soil samples demonstrated that the inhibitor accumulated in the soil as a leachate from the foliage of *E. hermaphroditum*. A field experiment involved the addition of activated carbon to the soil around pine seedlings which were watered with foliage leachate. The activated carbon either removed or deactivated the inhibitor, resulting in a lack of inhibition compared with control plots without the activated carbon, in which there was significantly higher mortality and slower growth of the pine seedlings. However, while activated carbon may be valuable for neutralizing allelochemicals in both greenhouse and field settings, it has the drawback that it can increase plant growth because of increased nutrient availability, especially nitrogen (Lau et al. 2008), and can reduce mycorrhizal colonization of plants, again affecting performance independently of allelochemicals (Wurst et al. 2010). An experimental treatment to test the effect of activated carbon with and without fertilizer on target plants can help address this limitation.

4.4 Abiotic treatments

Abiotic treatments involve the manipulation of non-living components of the environment in which plant populations are growing. Treatments to manipulate soil nutrients and moisture, temperature, light, atmospheric changes, and disturbance regimes are considered.

Treatments to manipulate the abiotic components of soil

Soil is an ecological paradox. On the one hand this complex structure is so commonly limiting that experimental manipulations demonstrating its effect upon plant populations abound in the literature. Conversely, soil is so complex that we often do not know the extent of the interactions between the biotic and abiotic components that are altered following a manipulation. Treatments involving the biotic components of soil have already been discussed.

Treatments to manipulate the abiotic soil components include increasing, decreasing, or otherwise altering levels of the following:

- soil nutrients (e.g. Gough et al. 2012; Santiago et al. 2012)

- water relations, ranging from flooding to the imposition of drought stress (e.g. Middleton 1990; Levine et al. 2011)
- pollutants ranging from heavy metals to acid precipitation effects (e.g. Gartside and McNeilly 1974; Dean et al. 1998; Wierzbicka and Panufnik 1998; John et al. 2009)
- plant litter (e.g. Molofsky and Augspurger 1992; Schramm and Ehrenfeld 2010; Kalamees et al. 2012)
- temperature (discussed below in 'Treatments to manipulate temperature') (e.g. Landhäusser and Lieffers 1994; Lee 2011)
- allelochemicals (see 'Treatments involving allelopathy' above)
- soil heterogeneity (e.g. Fitter 1982; Bell and Lechowicz 1991; Brandt et al. 2013)
- trampling effects (e.g. Barrett and Silander 1992; Kobayashi et al. 1999; Sunohara et al. 2002).

Whatever type of treatment is used on the soil it is necessary to include measurements of the effect on the soil itself to ensure efficacy of the treatments. For example, measurements of gravimetric soil moisture status in drought stress experiments will allow the effectiveness of the treatment to be assessed and reported. If the effects of the treatment are not monitored then there is no certainty that the intended manipulation has been made.

Soil nutrients. Soil nutrient levels are commonly altered in field or greenhouse experiments through the addition of fertilizer. Fertilizers are commercially available in a wide variety of forms ranging from slow-release granular formulations to fast-release liquids. The effects on plant populations will depend not just on the amounts of nutrients added but also on the timing of the release of the nutrients. Fast-release nutrients from a liquid fertilizer are easily leached and lost from the soil but have the advantage that the plant response may be fast as the nutrients are readily available to plant roots. Equal amounts of water (preferably distilled and pH-balanced to that of the fertilizer) should be added to control plots when using liquid fertilizer for nutrient additions. Conversely, slow-release nutrients supplied in a granular form can provide a steadier supply of nutrients, perhaps more similar to the natural release of nutrients through microbial degradation of organic matter. Granular fertilizers are most suitable for long-term experiments on plant populations (e.g. John and Turkington 1997).

Precise concentrations of nutrients required for plant growth experiments can be achieved by growing plants either hydroponically or in artificial soils. The latter are frequently made by mixing various amounts of vermiculite (a clay mineral), perlite (a naturally occurring light-weight, insoluble, volcanic glass), and *Sphagnum* peat moss, which holds water but also acidifies the soil. The exact proportions of these materials have to be considered carefully as they determine the texture of the soil, which itself determines the water-holding capacity. The use of pure sand as a growing medium is useful if measurements of plant roots are required but it has a poor water- and nutrient-holding capacity. Other materials will make it difficult to separate soil from the roots, and to separate two or more root systems. Nutrient concentrations in artificial soils or water culture can be achieved using liquid fertilizers or nutrient stock solutions. Hoagland's solution is a nutrient stock solution that provides a full complement of plant macro- and micronutrients. Variations allow specific solutions to be made up lacking one of the following micronutrients: Ca, S, Mg, K, N, P, or Fe (Machlis and Torrey 1956, pp. 41–45; Kaufman et al. 1975, p. 130). Smith et al. (1983) provide a comparison of nutrient solutions that can be used for the growth of plants in sand culture. Hewitt (1966) should be consulted for descriptions of sand and water culture methods and the over 100 nutrient solutions used in the study of plant nutrition. Booth et al. (1993) provide procedures for growing plants in Rorison nutrient cultures (similar to Hoagland's solution) according to the integrated screening program (ISP), a standardized comparative program for comparing the ecological traits of British vascular plants (Hendry and Grime 1993). Rorison and Robinson (1986) provide a good discussion of all aspects of the mineral nutrition of plants, including methods for various experimental techniques.

Soil water. Treatments to manipulate soil water can be as simple as varying the watering frequency of plants or plots. This approach is commonly used in both the greenhouse (e.g. Marks and Strain 1989) and the field (e.g. Sans et al. 1998), and provides a qualitative treatment (e.g. some plants received some water, others received a lot).

Watering pots to a specified weight can allow a constant and uniform level of water stress to be maintained through an experiment. Water is added to pots when weighing them reveals that a certain amount of water has been lost. A large number of pots can be maintained at specified soil moisture levels by combining an automatic watering system with modified balances (Schwaegerle 1983). This type of device was used as part of an experimental investigation into the response of nine populations of the annual *Phlox drummondii* to environmental gradients that included six levels of soil moisture (Schwaegerle and Bazzaz 1987).

By altering the watering regime, soil moisture levels can be established at saturation, field capacity, or the permanent wilting point. *Saturation* is the point at which all soil pores are filled with water. *Field capacity* is the amount of water in a soil that has been allowed to drain. At this point, the small pore spaces are still filled with water but the large pores are filled with air. The *permanent wilting point* is the point at which available water in the soil has been taken up by plant roots or drained away. Below this point plants will loose turgor as their roots are unable to take up the small amounts of remaining water, and they are likely to die.

Water stress treatments can be imposed by flooding pots with solutions of a high-molecular-weight osmoticum such as polyethylene glycol (PEG). The physiological effects of PEG are relatively minor, at least in the short term (Lawlor 1970), although it can limit ion transfer and the diffusion of oxygen to the roots. Musselman et al. (1975) used PEG to establish two levels of water stress in an experiment to investigate the adaptation of seedlings of northern white cedar (*Thuja occidentalis*) ecotypes to their local environment. NaCl or other mixed salts that cause osmotic stress can be used as a short-term surrogate for water stress treatments, although concentrations should not be too high in order to avoid toxic effects (Munns et al. 2010).

In the field, natural rainfall can be excluded using rain shields, although these may introduce additional unplanned effects such as increased humidity and changes in the light regime (quality and intensity of photosynthetically active radiation, PAR) (Pake and Venable 1995). Studies involving wetland plants will frequently include a flooding treatment. For example, the surface of a substrate within trays was maintained at either 0, −3, or −13 cm below the water surface in a study on the effects of water depth on the growth and survival of three emergent wetland species, *Paspalidum punctatum*, *Nymphoides cristatum*, and *Ipomoea aquatica* (Middleton 1990). In the field, knowledge about natural flooding levels can be used to locate sample plots and allow the effect of a range of flooding frequencies to be investigated. Losos (1995) used this approach to study seedling establishment and survivorship of two palm species in an Amazonian forest. More ambitious approaches to carrying out moisture treatments in the field include large precipitation arrays coupled with rainout shelters to reduce natural precipitation. A solar powered system that allowed up to 80% reduction and 80% addition to ambient precipitation is described by Gherardi and Sala (2013) (Fig. 4.3 and Plate 12). For example, Fay et al. (2000) established rainfall manipulation shelters in tallgrass prairie that allowed rainfall redistribution treatments of 30% rainfall quantity and 50% inter-rainfall dry periods over 36-m^2 sample plots to investigate the potential effects of future changes in seasonal rainfall patterns. Effects of these treatments included changes in seedling emergence and survival. Such rainfall manipulation experiments can investigate both 'trend'- and 'event'-based changes and 'extreme' versus 'average' scenarios (Jentsch et al. 2007) allowing a high degree of realism with regard to future climate change scenarios.

Chemical pollutants. Treatments to test the effect of chemical pollutants in the soil on plant populations involve experiments similar to those that manipulate soil nutrients. Airborne nitrogen pollutants (NO_y and NH_x) act principally on plants through accumulation of soil nitrogen or soil acidification, and can be studied through manipulation of the soil (Bobblink et al. 1998). Chemical pollutants can be added to the soil experimentally in liquid form, or applied as some sort of powder or granular solid. Treatments involving contaminated soils can be established by growing test species in greenhouse bioassays (e.g. Redente and Richards 1997) or by monitoring plants in the field in contaminated versus uncontaminated soils (e.g. Zvereva and Kozlov 2001). Many of the studies of heavy metal tolerance (see Section 4.2) involve growing plants on

Figure 4.3 Automated rainfall manipulation system in the Chihuahua Desert, New Mexico, United States: (A) interception and irrigated plots, (B) a view of the interception component and its connection to its water storage tank, (C) four modules of rainout shelters being put together, including a removable panel with hinges that allows walking access to the centre of the plot. From Gherardi and Sala (2013). (See also Plate 12.)

contaminated soils that have been brought into the greenhouse. For example, Gartside and McNeilly (1974) examined the germination and growth of nine species in response to various levels of copper contamination in potting compost/copper mine waste mixtures. The addition of chemical pollutants to hydroponic solutions can provide a way of establishing treatments without the complexity of confounding soil interactions. For example, Shupert et al. (2013) added automotive brake pad dust to hydroponically grown *Salvinia molesta* to investigate the potential toxic effects of copper and iron dissolution products on this aquatic plant.

Litter. The soil litter layer comprises the recently deposited leaves and other dead and decaying organic materials on the soil surface. Germinating seeds or newly arising vegetative shoots have to grow through the litter layer. Consequently, the litter layer is a physical barrier that also markedly alters the microclimate at the soil surface, decreasing the red/far-red light ratio and thereby significantly affecting seedling survival and plant growth (Vazquez-Yanes et al. 1990). Treatments to manipulate the litter layer include additions, removal, or reduction to specified depths. Litter from different species decomposes at different rates, reflecting intrinsic chemical variability, and hence differentially affects nutrient input to the underlying soil and species performance (Coq et al. 2012). Unless the effects of litter from individual species are being tested or the natural community of interest is a near monoculture, mixed litter treatments are more representative of natural systems than treatments involving manipulation of litter from a single species. Molofsky and Augspurger (1992), for example, established four litter depths (0, 1, 6, and 12 cm) in experiments both in a greenhouse and in the field in semideciduous tropical forest on Barro Colorado Island, Panama. They found that the litter layer in this forest system increased seedling diversity by allowing seedlings of some species to survive in light gaps. By contrast, small-seeded shade-intolerant species survived poorly in the presence of litter.

Soil heterogeneity. Soils are inherently variable in the field at many scales. Spatial variability in soils occurs in terms of nutrient levels and soil microbial communities, both of which can affect plant performance and population growth (Ettema and Wardle 2002).

Many studies of this variability are correlative, seeking to establish the relationship between existing levels of soil variability in the field and plant performance. Experimental manipulations of the soil to produce treatments that vary in the level of heterogeneity are uncommon in the field (although see Baer et al. 2005). In the greenhouse, however, different soil types can be manipulated quite readily. At least two approaches have been used to assess the role of soil heterogeneity on plant populations (see Rorison and Robinson (1986) for other examples). The first approach makes direct assessments of the effects of heterogeneity by growing plants in samples of soil collected from the field while the second approach assesses the response of a plant to soil heterogeneity using artificial soil constructed with particular degrees of heterogeneity:

1. Bell and Lechowicz (1991) sought to assess the degree of environmental heterogeneity in an area and its relationship to plant fitness. They used explant trials where plants were grown in 'samples of the environment' (soil cores) under controlled conditions. In this study, two genotypes of *Arabidopsis thaliana* and *Hordeum vulgare* were grown as bioassays in cores of soil removed from a 50 m × 50 m plot in the field. By taking a hierarchical series of soil cores at different spatial scales this method allowed an investigation of the overall environmental variance within the sample area. Different plant growth forms may experience very different scales of heterogeneity in the same area.
2. Wijesinghe and Hutchings (1997) established eight different treatments, each with a different number and size of patches in a greenhouse study which showed that resource foraging by *Glechoma hederacea* depended upon the scale of spatial heterogeneity (see Figs. 2.4, 3.2, and 4.4). Patch types provided a contrast in nutrient availability, with all treatments providing the same quantity of nutrients. A similar approach is described by Grime et al. (1993).

Trampling. Animals, including humans, frequently trample vegetation, compacting the soil and creating microsite heterogeneity in the topography (e.g. hoof prints) that can markedly affect seedling recruitment and plant performance. Subjecting sample plots to specific frequencies and intensities of trampling can experimentally assess the effects of trampling on the demographics of target plants. Ikeda and Okutomi (1995) subjected sample plots sown with either monocultures or mixed cultures of seven species to four levels of trampling plus an untrampled control. The levels of trampling varied from one trampling per day every 2 weeks to four tramplings per day for 4 days a week. A single trampling involved one 65-kg person walking over the entire sample plot with rubber-soled shoes. This trampling imposed a downward force of approximately 0.34 kg cm^{-2}. The trampling treatment significantly increased soil hardness, and plant leaf and shoot lengths decreased while tiller and leaf number increased. Barrett and Silander (1992) established treatments to assess the effects of animal hooves on recruitment of *Trifolium repens* seedlings in a mown lawn and a grazed pasture by removing above- and belowground vegetation in 10 cm (approximate cow hoof print size) or 25 cm (approximate size of mole disturbed areas) diameter plots. At both sites, there was strong limitation on seedling recruitment that was reduced in the 10 cm diameter vegetation removal plots.

Treatments to manipulate temperature

In the laboratory or greenhouse, soil or air temperature can be experimentally manipulated using climate-controlled growth chambers, temperature-gradient tunnels (Grime et al. 1989), heated water baths (Hobbie and Chapin III 1998), or, for testing seed germination, thermogradient bars (see Chapter 5). Treatments to manipulate temperature in the field can be established using passive devices to raise the temperature through a greenhouse effect (e.g. polyethylene covered cages; Hobbie and Chapin III 1998) or open-topped chambers (Marion et al. 1997). Raising the temperature in plots in the Arctic tundra using cages increased the survivorship and growth of the treeline species *Picea glauca* (Hobbie and Chapin III 1998). Although there are concerns that these passive devices may do more than raise temperature (e.g. they may alter the availability of light, access of pollinators and herbivores, the daily temperature range, soil–air relationships, wind, and soil microbial activity), they may provide

Figure 4.4 Creating a heterogeneous environment at different scales. From Wijesinghe and Hutchings (1997); photograph used with permission from John Wiley & Sons. (See also Plate 13.)

reasonable analogues of natural temperature variability (see discussion and evaluation in Hollister and Webber 2000; Frenne et al. 2010; De Boeck et al. 2012). There is also concern that the short-term responses of plants to the use of passive devices may be different from their long-term responses (Hollister et al. 2005). More direct temperature treatments can involve placing infrared heating lamps over plots (Wan et al. 2002; Kimball 2005). In a 4-year study in an alpine meadow, heat lamps increased the heat flux by approximately 3% over total average ambient levels (Harte and Shaw 1995). Consequently, snowmelt was earlier and soil moisture lower in the heated plots. Survivorship of sagebrush (*Artemisia tridentata*) seedlings was significantly enhanced in the warmed plots compared with the controls. An alternative approach that avoids the 'chamber effects' noted above is to place heating/cooling elements directly on or just below the soil surface to raise/lower soil temperatures (Hillier et al. 1994).

Snow melt or snow removal and snow addition experiments allow the effects of future changes in the snow pack and winter climate to be addressed. The relatively simple use of strategically placed snowfences or open-top chambers can allow changes in winter snow conditions to be tested (Walker et al. 1999; Wipf and Rixen 2010). Snow removal accomplished physically (e.g. with a shovel) or through warming (e.g. open-topped chambers or the use of reflective materials placed on the snow surface) can also increase soil freezing ('colder soil in a warmer world') and leaching losses of carbon, nitrogen, and phosphorus (Henry 2007). Generally, snow removal advances phenology, especially of dwarf shrubs (Wipf and Rixen 2010).

Treatments to manipulate light

As with manipulations of temperature, light treatments are readily established in the laboratory or greenhouse using lamps of different light intensities to manipulate PAR (i.e. PAR in the range 400–700 nm). The main precautions when using lamps are to ensure that they illuminate evenly and throughout the PAR range and that they do not produce infrared radiation which causes confounding temperature effects. Additional manipulations include setting the red/far-red (R/FR) ratio (the ratio of irradiance at 660 and 730 nm) to either a shade (~0.32) or a sun (~1.2) condition (Campbell et al. 1993), or something in between, and setting the daylength. Grime and Booth (1993) describe a procedure using growth chambers to establish the response of flowering and vegetative growth in plants to day length.

Shade cloth or mesh can be used both in the greenhouse (e.g. Petit and Thompson 1997; Gibson et al. 2009) and in the field (e.g. Pigliucci and Schlichting 1995a; Munguía-Rosas et al. 2012) to reduce light to desired levels. It is important to measure the levels of PAR achieved under different degrees of shading to provide a standardized measure of the effectiveness of the treatment. The material used should be tested to see if it depletes all wavelengths relevant to plants to a similar extent. Other parameters are also likely to be altered by the shade treatment, especially temperature, and these should be monitored as well.

Plastic films or filters can be used to alter light quality, the R/FR ratio, or both (Novoplansky 1991). For example, innovative experiments can mimic light conditions characteristic of sunny periods when plants are actually growing in crowded environments. Under these conditions the plants can be 'fooled' into growing as if they were in isolated conditions, i.e. the effects of light competition can be minimized and investigated (Novoplansky et al. 1994).

Experimental treatments to manipulate levels of ultraviolet light are discussed in the next subsection.

Treatments to investigate the effects of atmospheric change

The Earth's atmosphere has been changing due to anthropogenic activities since the beginning of the Industrial Revolution 250 years ago (Intergovernmental Panel on Climate Change 2013). In particular, levels of atmospheric carbon dioxide (CO_2), reduced (NH_x) and oxidized (NO_y) nitrogen, and sulphur dioxide (SO_2) are known to be increasing, as well as levels of ultraviolet radiation (UV-B) reaching the Earth's surface. Ozone (O_3) levels are decreasing in the upper atmosphere (allowing increases in UV-B) but increasing in the lower atmosphere. There is a pressing need to assess the effects of these atmospheric changes on plant populations.

A number of experimental approaches have been developed to investigate the effects of changing concentrations of atmospheric gases (for review see Allen et al. 1992). When only two experimental concentrations are possible the de facto standard has been to set the treatments at ambient and twice ambient levels (ambient atmospheric levels are ~350 mg L^{-1} for CO_2 and ~18 μg L^{-1} for O_3, with wide diurnal and seasonal variation in the latter).

Treatments to manipulate atmospheric constituents in the laboratory revolve around the use of environmental chambers, either specially built (Hill 1967) or adapted for the purpose (e.g. Reekie et al. 1994). Environmental chambers must be airtight and may involve internal recirculation when toxic gases are being used. It is important to avoid pseudoreplication (see Section 3.3) when using environmental chambers. Placing 'replicate' plants into each of two chambers when one serves as the treatment and the other as the ambient control is statistically inappropriate because the treatment variable is inseparable from the chamber variable. Alternating ambient and treatment conditions between chambers with relocation of the plants can be used to redress this problem when only a single chamber is used per treatment level. Replication of chambers per treatment level or RIA (Section 3.3) is a better approach where possible. Large phytotrons have been constructed especially for environmental research (e.g. the Duke Phytotron; Kramer et al. 1970), but pseudoreplication remains a problem. The former UK Ecotron (discontinued in mid 2013) was a unique facility with 16 controlled environment chambers that was particularly suited for experiments involving manipulations of atmospheric constituents (Lawton et al. 1993; Jones et al. 1998; Milcu et al. 2012), although its use for addressing questions in plant population ecology was limited. The European Ecotron is a similar facility (<http://www.ecotron.cnrs.fr/>). Johnson et al. (1993, 2000) describe a 39-m growth tunnel for studying the responses of plants to superambient to subambient levels of CO_2. A blower forces air or CO_2 through the tunnel, and a gradient of decreasing CO_2 is established as photosynthesis by the plants closer to the input end progressively removes CO_2 from the air circulating above plants at the far end of the tunnel. This tunnel can be set up over naturally established stands of vegetation. A combined CO_2–temperature gradient chamber is described by Lee et al. (2001).

In the field there are two approaches to establishing atmospheric treatments other than the placement of environmental chambers over the plants: these are open-topped chambers (OTCs) and free air carbon dioxide enrichment (FACE) experiments (Dahlman 1993). Many researchers use OTCs, and generally each OTC consists of a circular or square metal or plastic frame covered by plastic sheeting. As the name suggests, OTCs are open at the top, although a frustum (cone-shaped flange) may be added to decrease the internal recirculation of gases. Gas is pumped into the OTC at the base, or sometimes half-way up the sides, and allowed to vent through the open top. Ashenden et al. (1992) provide a design for an inexpensive OTC system that is useful for testing the effects of elevated CO_2 on plants (see Fig. 4.5). Wulff et al. (1992) provide a design for comparable experiments to manipulate atmospheric O_3 concentrations. The spatial variability in CO_2 concentration within chambers is lowest in round chambers with a frustum (Hileman et al. 1992). However, the 'chamber effect' is a problem; that is, the alteration to the physical environment caused by the chamber itself regardless of treatment. There may be higher temperature and humidity inside an OTC compared with outside, and lower light levels inside (see discussion under 'Treatments to manipulate temperature' above). This problem can be addressed by including non-chamber controls in addition to ambient gas-concentration controls (Fischer et al. 1997). A design in which the chamber walls are left 7 cm above the ground may also ameliorate this problem, as well as providing access for smallish(!) animals to move into and out of the chambers (Leadley and Körner 1996; Leadley et al. 1997). Another solution is to replace the fans which blow 'air' into each OTC with evaporative coolers and in-line heaters so that the air temperatures within the chambers can be controlled (Norby et al. 1997). OTCs are also limited in size and are unable to hold very large numbers of plants. OTCs have been used to study the effects of CO_2 (Fischer et al. 1997; Jongen and Jones 1998; Newman et al. 2003; Vannette and Hunter 2011), O_3 (Ho and Trappe 1984; Pfleeger et al. 2010), and SO_2 (Garcia et al. 1998) on plant populations, including competition, allelopathy, plant–insect interactions, and reproductive allocation.

FACE experiments can be used to study the effects of elevated CO_2 on plant populations (Hebeisen et al. 1997; Bucher et al. 1998; Lüscher et al. 1998; Way et al. 2010), and involve the release of CO_2 from

Figure 4.5 Open-topped chamber system for investigating the effect of elevated CO_2 on plants. Design according to Ashenden et al. (1992). (Photo David Gibson.) (See also Plate 14.)

a central array of standpipes (Hendry 1992; Norby and Zak 2011). Local wind and turbulence create a desired 'sweet zone' in the centre of the circle. With no walls, the chamber effect of OTCs is avoided, and the treated areas are much larger (up to 30 m in diameter). Large numbers of mature plants, including trees, can be included in experiments. The disadvantages of FACE experiments are the initially high construction costs and the ongoing expense of using large amounts of CO_2. Smaller, less expensive mini- and mid-FACE systems designed to expose smaller areas to experimental treatments have also been described (Miglietta et al. 1997; and see photo in Lee 1998). FACE experiments are inappropriate for experiments with noxious gases.

Experimental treatments to investigate the effects of acid precipitation generally involve the spraying or misting of plants. The test plants may either be growing directly in the soil or in pots in the greenhouse or field. Sheppard et al. (1998) placed pots of *Picea abies* seedlings in OTCs before spraying. Acidified solutions are made from deionized water mixed with sulphuric or nitric acid to the desired pH. As with natural acid precipitation, there are the confounding indirect effects in which the availability of essential minerals and toxic metals in the soil is altered and minerals are leached from foliage. Many effects of acid precipitation are the result of a combination of agents, including pollutants and climatic and soil factors. The testing of all possible interactions between these agents is likely to be outside the scope of most investigations. Furthermore, one of the major effects of acid precipitation may be to increase the susceptibility of affected plants to damage from other factors, such as frost (Fowler et al. 1989).

Ultraviolet B radiation (UV-B; 280–315 nm) is increasing due to decreases in stratospheric ozone. Elevated UV-B can inhibit plant growth, development, and physiological processes, eliciting the production of UV-B absorbing compounds (Newsham and Robinson 2009). UV-B was shown to affect seed and berry production and chemistry in sub-arctic heathland plants (*Vaccinium myrtillus* and *Empetrum hermaphroditum*) with implications for human and animal consumers (Gwynn-Jones et al. 2012). Experimental treatments involving UV-B can be established using a series of fluorescent lamps and filters (see Flint et al. 2003 for procedures; McLeod 1997). Hendry (1993) describes the procedure advocated as part of the ISP and recommends using white light sources filtered to remove UV-C at wavelengths < 280 nm and UV-A wavelengths > 320 nm (the UV-A spectrum is 315–400 nm). The biologically effective UV-B dosage needs calculating and reporting along with the equivalence of stratospheric ozone reduction (see procedures in Caldwell 1971; Caldwell and Flint 1997). The type of filter used is relevant, as energized Mylar-filtered fluorescent UV lamps have been shown to affect plant growth compared with unenergized lamps, and so should only be used when a comparison between cellulose acetate-filtered and glass- or Mylar-filtered lamps has been made (Newsham and Robinson 2009). Other precautions include ensuring that shadows are not cast by support structures for the lamps. The UV intensity should not be varied simply by placing racks of lamps at different heights. Rather, PAR from lamps at a single height can be controlled through the use of filters or electrically modulated lamp control systems (Flint et al. 2009). Although many of the early studies of the effects of UV-B on plants have been conducted in the lab or greenhouse, emphasis has shifted to field responses (see papers and photographs in the 1997 special issue of *Plant Ecology*, volume 128). Pseudoreplication has been a problem in several field studies of UV-B and needs to be guarded against in setting up arrays of lamps. In addition, lamp output (PAR, spectral composition, and dose) needs to be monitored carefully and adjusted as necessary to ensure uniform irradiation within arrays and between control and treatment lamps (McLeod 1997) and to account for natural radiation fluctuations (Santas et al. 1997).

Treatments to manipulate abiotic disturbance regimes

Disturbance is defined as 'any relatively discrete event in time that disrupts ecosystem, community, or population structure and changes resources, substrate availability, or the physical environment' (White and Pickett 1985). This generalized definition can apply to many of the biotic effects that we have already discussed, such as herbivory and trampling. Abiotic disturbances are exogenous from plant populations, i.e. they arise outside the community. White (1979b) lists 13 different types of natural, exogenous disturbance. These can be placed into three main categories, reflecting the effects of (a) fire, (b) weather (wind, ice, temperature, precipitation), and (c) soil disturbance (erosion, deposition, flooding, and movement). The effects of anthropogenic disturbance and management can be added to this list as these may, in some cases, be similar to the natural disturbances (e.g. prescribed burning that is used to control woody plants that are not subject to natural fires). In a sense, treatments to address the effects of these disturbances have been discussed earlier. For example, treatments to manipulate flooding effects were described in the discussion on treatments to manipulate the soil.

Treatments to investigate the effects of specific forms of disturbance upon plant populations will be unique for each situation. It is important to consider the temporal and spatial scales at which the treatment is applied, whatever disturbance regime is being experimentally tested. To allow generalization of the results from an experiment in which disturbances are manipulated or mimicked, it is important that the imposed disturbance occurs at the same temporal frequency (has the same return time), with the same intensity, and at the same spatial scale (patch size) as that at which the natural disturbance typically occurs. It is also important to consider whether a particular disturbance might be acting directly or indirectly upon plant populations. Several examples are now described to illustrate questions that can be addressed using a range of approaches.

Mowing: how important is clonal integration for ramet performance in a herbaceous perennial? Mowing is an anthropogenic disturbance that can be used to retard succession or simulate grazing. Meyer and Schmid (1999a,b,c) conducted a 5-year experiment to assess the effects of mowing as a control on the demography of the perennial *Solidago altissima* invading old-fields in Switzerland. The mowing treatment consisted of annual late-summer mowing to a height of 5 cm and there were unmown controls. Experimental plots were 16 m × 8 m, arranged in three blocks of two plots each according to a split-plot design, with the whole plot treatment factor assigned according to the randomized complete block design (see Chapter 3). Mowing was found to reduce the production of seeds and shoots, thus preventing spread of this species. No effects of mowing were observed on seed germination or seedling survivorship.

Fire: what effect does fire have on the growth and reproduction of a perennial herb? Fire is a natural component of the disturbance regime of many ecosystems, especially grasslands, and is often used as a management tool. The effect of prescribed fire, set to mimic the natural fire regime of tall grass prairie, was used to investigate the perennial *Ratibida columnifera* (Hartnett 1991). As part of a larger, multiple-scale study, fire treatments were applied, in April, to 20–30 ha watershed units. Two watersheds were burned in the growing season of the study, two were 4 years post-fire, and there was one watershed representing each of 9, 10, and 15 years since the last fire. Time since fire was found to reduce plant size, stem production, and seed production. An important conclusion was that the effects of fire on plants appeared to be indirect. The plants were thought to be responding to changes in post-fire abiotic conditions (fewer competitors) rather than the direct effects of fire on buds and plant growth. When conducting an experiment in large landscape units such as this it is important to ensure that each population studied is on the same soil type and topographical position (elevation, slope, and aspect). One could also approach the problem through a stratified random design with a large number of replicates, or a randomized block design. The latter gives valuable insight into potential interactions between fire and underlying site factors. Whatever design is used it is important to understand the underlying influence of site factors, for example soil types.

When using fire treatments it is important to ensure that the study plots containing target plants are actually burned and mimic what would happen in a natural fire. Fires can burn unevenly across the landscape as a result of variations in fuel load and microclimate. Great care has to be taken in the use of prescribed fire for ecological experiments. In many locations, prescribed burns require the expertise of licensed personnel and permission from local authorities. At the very least, precautions have to be taken regarding smoke drift and fire escape. Guidelines for the use of prescribed fire are widely available (e.g. Chandler et al. 1983; Biswell 1989) and produced by many agencies, including the United States Forest Service (Wade 1989) and The Nature Conservancy (2012).

Burial of aboveground parts: what is the physiological basis for burial-stimulated increase in performance of a dune grass? Burial is a frequent disturbance in which plants are buried due to deposition of sediment in an unstable substrate (reviewed in Kent et al. 2001) including coastal and fluvial habitats, around volcanoes (Tsuyuzaki 2010), and following dust storms, landslides, and storms (Affandi et al. 2010). For example, adult plants of two grasses, *Ammophila breviligulata* and *Calamovilfa longifolia*, were buried beneath 0, 20, and 40 cm of sand in a field study on a sand dune on the shore of Lake Huron, Canada (Yuan et al. 1993). The treatments were randomly assigned to 12 50 cm × 50 cm plots within areas of relatively even shoot density for each species (a total of 24 plots). Wooden frames were placed around the plots for the two burial treatments and a control to contain the sand. A similar experiment was established in the greenhouse where germinating seedlings of each species were planted individually in 15-cm diameter pots. When the seedlings were 2 months old they were buried to depths of 0, 3, 6, 9, and 11 cm of sand, with six replications per treatment of each species (a total of 60 pots). Plastic piping was placed around the pots to retain the sand. Both the field and greenhouse experiments showed that plants buried at depth may emerge later than shallower-buried plants. Thus in these sand dune species, burial up to 40 cm stimulates growth due to higher assimilation rates in emerging shoots. The density of shoots of both species in the field was, however, reduced when buried by 40 cm of sand compared with 0 and 20 cm.

Storms: what are the effects of storms on the environmental conditions of microsites and subsequent recruitment of forest trees? Storms are a major source of natural disturbance, especially in forest and coastal systems, where physical damage (blowdown) can occur due to high winds or ice accumulation or burial from sediment accretion (see 'Burial', above). Manipulative experiments testing the effects of damage by ice storms are infrequent because of logistical difficulties. However, a novel experiment in which water was sprayed over a forest canopy during the winter led to accretion of an ice glaze allowing damage effects to be assessed (Rustad and Campbell 2012). Experimental treatments associated with disturbance from hurricane blowdown were studied in a hardwoods–white pine–hemlock stand in Massachusetts, United States (Carlton and Bazzaz 1998). Pulling down randomly selected canopy trees created experimental blowdowns. Uprooted trees created three types of microsite; tip-up mounds, pits, and north-facing vertically oriented forest floor on the old tree base. In addition, microsite changes associated with level open areas and areas covered with understorey vegetation were also identified. Regeneration of three sympatric birch (*Betula* spp.) species on these microsites was studied by monitoring natural seed dispersal, recruitment, survival, and seedling growth, as well as the growth of transplanted seedlings. The results of this experiment showed that forest floor heterogeneity associated with the five types of microsite affected all aspects of establishment among the three birch species, including initial seed dispersal patterns. A suite of environmental characteristics associated with each type of microsite affected seedling survival and growth (Fig. 4.6).

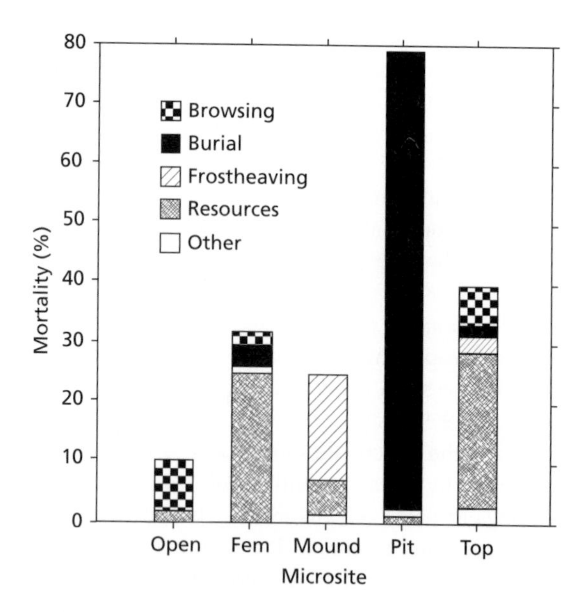

Figure 4.6 Mortality rate and cause of death for birch seedlings during three seasons of growth on five microsites of experimentally simulated hurricane blowdown. Reproduced with permission of the Ecological Society of America from Carlton and Bazzaz (1998).

4.5 Interactions between plants

There are many forms of interaction between plants. An interaction is any effect one plant has on another, whether it be competitive (at least one plant loses out), facilitative (one plant does better in the presence of another), or mutualistic (both gain). Investigation of interactions between plants can take place either in the greenhouse or the field, and preferably in both locations. Experiments and designs for studying plant interactions have been the subject of several comprehensive reviews (Connell 1983; Schoener 1983; Aarssen and Epp 1990; Goldberg and Barton 1992; Cousens 1996; Gibson et al. 1999; Jolliffe 2000; Connolly et al. 2001; Goldberg and Scheiner 2001). The basic types of design that can be used are described here along with some general cautions in regard to the conduct of these experiments. These experiments are referred to in terms of interactions between species, although population ecologists are traditionally also interested in intraspecific interactions between genotypes, cultivars, subspecies, or varieties within species.

Designs

Three main types of design are commonly used for comparing interactions between pairs of species: simple pairwise (SP), additive (AD), and replacement series (RS; also called substitutive designs) (Fig. 4.7). SP designs usually maintain a 1:1 ratio of species. The simplest AD design (the partial additive) holds the density of one species (the focal or target species) constant while the density of the associated (or neighbour) species is varied. In RS, species are grown in different proportions and performance is compared with their growth in monoculture, with a constant total density across all mixtures. More complex designs involve simultaneously varying the proportions of focal and associate species (i.e. addition series; Fig. 4.7d). Interactions between several species can be investigated by using multiple comparisons of these designs. A diallel design uses all possible combinations of n species, each represented by one equiproportioned mixture and two monocultures, again, at constant total density (Fig. 4.7e). When a design includes a range of densities and relative frequencies of species then

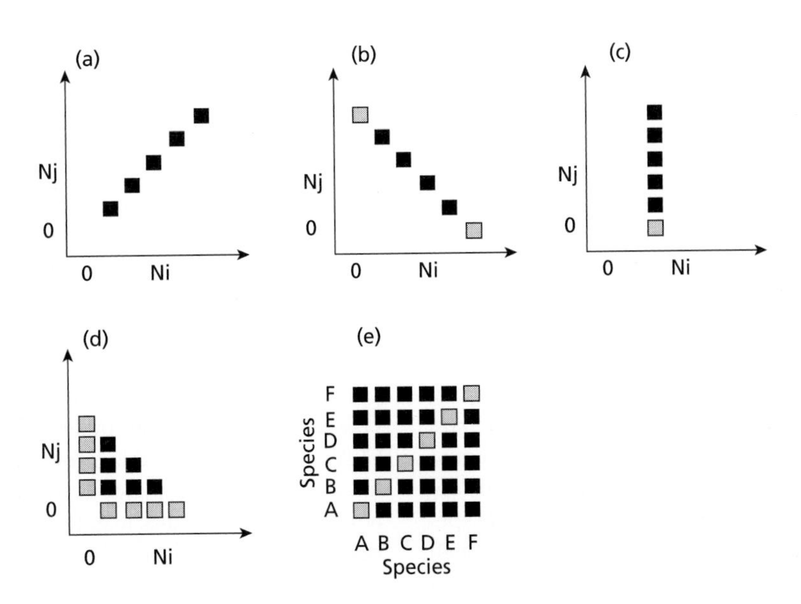

Figure 4.7 Five designs for competition experiments plotted on joint-abundance diagrams for two species (designs a–d) and five component species in a diallel design (e). In (a)–(d) lightly shaded symbols represent monocultures. (a) Simple pairwise design at multiple densities and without the monocultures included by some investigators. (b) Replacement series at a single total density. (c) Target-neighbour or partial additive form of an additive design with a constant density of component i. (d) Additive series and (e) diallel design including redundant intraspecific mixtures (lightly shaded symbols). From Gibson et al. (1999). Reproduced with permission from John Wiley & Sons.

regression-style response surface models can be fitted to performance measures (Freckleton and Watkinson 2000). Response surface models allow measures of per capita species performance to be related to the density of each species, and represent the most powerful method of analysis of interactions between pairs of species (e.g. Connolly and Wayne 1996). All of these designs suffer from limitations that need to be carefully considered in light of one's original research hypotheses before they are used (Gibson et al. 1999). Use of the RS design has been questioned because of problems related to the inclusion of monocultures, size bias, and the use of fixed total density (Gibson et al. 1999; Jolliffe 2000).

Neighbourhood interactions can be included in the design of plant interaction experiments through the use of a hexagonal fan design (Boffey and Veevers 1977). These designs utilize a plant spacing pattern such that each individual is surrounded by zero to six intraspecific neighbours and zero to six interspecific neighbours. This array of hexagons can be arranged to vary density or frequency along a plant spacing gradient. Schmid and Harper (1985) used a fan design in an investigation of competitive interactions between two grassland species with phalanx and guerrilla growth forms (*Bellis perennis* and *Prunella vulgaris*, respectively; see 'Life history' in Section 3.2). A complete reversal of competitive superiority was found to occur with changing density as the phalanx life form (*B. perennis*) was the aggressive competitor at high densities but a weaker competitor at low densities, and vice versa with respect to the guerrilla species (*P. vulgaris*).

Designs to investigate multispecies interactions include multispecies mixtures (e.g. Grime et al. 1987), removing all or selected species from around focal plants (e.g. McLellan et al. 1997a), or enlarging one of the designs discussed above to include extra species (e.g. Austin 1982; Fowler 1982; Turkington 1994). The limitations of the different types of design still apply when they are used for multiple species interactions. The simplex design (Cornell 1990) allows multispecies mixtures to be established in which a relative frequency of species is established to cover the range of possible species proportions for the number of species concerned. For example, centroids with equal proportions of each species, corner points dominated by one species with all other species having equal initial numbers (e.g. density, biomass, or canopy cover), and midpoints with one subordinate species and equal numbers of the other species. The simplex design has been used to investigate the effect of elevated CO_2 on multispecies mixtures of five annual old-field weed species (Ramseier et al. 2005), to assess the effects of agricultural weeds on soybean yield (Gibson et al. 2008), and to demonstrate that the number and strength of pairwise interspecific interactions among species drives the yield of experimental grassland mixtures (Kirwan et al. 2007).

There are several issues that need to be considered in planning an investigation of plant interactions concerning the choice of the most appropriate design. These are discussed below. Additional discussion can be found in Gibson et al. (1999), Connolly et al. (2001), Keddy (2001), and Trinder et al. (2013).

Greenhouse, field plots, or natural stands?

The advantages and disadvantages of laboratory versus field settings discussed in Chapter 3 are important here. Laboratory studies may allow for the demonstration of plant interactions, but it is only in the field that their true relevance can be fully realized (if at all). Field studies also allow for interactions to be studied over several seasons, although most investigators do not take advantage of this opportunity (Goldberg and Barton 1992). Field studies within natural populations have to account for inherent variability in ages, ontogeny, and the spatial distribution of individuals. Dealing with this variability requires large appropriately located numbers of sample units (see Section 3.3), careful tracking of biotic and abiotic variability (Chapter 6), and the application of appropriate statistical methods that can take into account unmanipulated factors (e.g. ANCOVA, see Chapter 7). It is recommended that plant interaction studies combine both laboratory (greenhouse) and field studies (e.g. Berendse 1983) to take best advantage of the benefits of each (Table 3.1).

Measuring appropriate dependent variables at the right time

To be of evolutionary relevance, studies of plant interactions should measure fitness (see Section

5.8) or something directly related to it. However, in many cases this is not possible as plants may not flower or otherwise reproduce within the realistic time frame of the experiment. In addition, other dependent variables such as biomass, numbers of leaves, etc. may be relevant for particular questions (see Chapter 5). Only taking measurements at the conclusion of the experiment may appear expedient, but must be avoided. Final measurements can miss dynamic changes in species interactions during the course of the experiment, and only allow a limited inference about which species dominates a particular mixture (Gibson et al. 1999; Connolly et al. 2001; Trinder et al. 2013). The initial size of individuals in the experiments also needs to be determined and accounted for. Initial size differences can severely bias the eventual outcome of any competitive interaction. Therefore it is recommended that dependent variables be measured at the start, at regular intervals throughout the study, and at the end. Time (date of measurement) should then be included in the statistical analysis as an independent repeated measure. Consideration of the method of statistical analysis must be made early on, and preferably before a final design is decided upon (Freckleton and Watkinson 2000).

A distinction has to be made between addressing questions on the *outcome* of competition (the end point for the community) versus the *effects* of species on each other (their impact or response) (Goldberg 1990). Intermediaries such as resources, pollinators, dispersers, herbivores, or microbial symbionts mediate the interactions between species. These indirect interactions consist of two processes: the *effect* of the plants upon the intermediary (and hence the *effect* on the competing plant), and the subsequent *response* of the plants to changes in abundance or impact of the intermediary. The choice of experimental design should reflect these distinctions, and reflect the questions being addressed; i.e. are you interested in competitive response, competitive effect, the outcome of competition, or some combination of all of these? Measurements of suspected intermediaries should be included. For example, if you suspect that the effect of one species upon another will be mediated through competition for light, then light should be measured over the course of the experiment. Trinder et al. (2013) recommend that light, water, soil and plant tissue nitrogen, phosphorus, and other nutrients in general are important resources to measure in plant competition studies. Obviously, a balance has to be struck to ensure that the right intermediaries are measured—not just a smorgasbord of measurements because certain instruments happen to be available. Not everything has to be measured.

Separating above- and belowground interactions

In a very general sense, interactions above ground primarily involve light and those below ground involve moisture and soil nutrients. There is a large potential for confounding these effects. For example, a superior competitor for soil nutrients may produce more biomass above ground, and in turn may shade its neighbours, thereby increasing its competitive edge (e.g. Aerts et al. 1990). Separating aboveground and belowground interactions is problematic and requires the shoot and root systems of plants grown in the mixture to be kept apart from each other. This separation can be achieved by a combination of appropriate design (see above) and appropriate means of physical separation of the shoots or roots, or both, of interacting neighbours. Physical separation can be achieved using partitions placed between neighbouring plants above ground and/or below ground. Partitions should be constructed of material that does not allow the roots of neighbours to intermingle or soil water to move from the volume occupied by the root system of one competitor to that of another (Wilson 1988). Above ground, the partitions should prevent light competition between neighbours without reducing ambient light intensity or interfering with the direction of incident light. White or reflective materials are the best for this purpose. Transparent materials or mesh can be used if the neighbouring plants are held back from each other sufficiently to stop mutual shading. In pot studies, the root systems of plants in a mixture can also be restricted to grow in separate, adjacent pots. With designs of this sort, it is important to maintain equal soil volumes for all treatment levels. The separation of roots and shoots using the constraints described above is somewhat artificial. Isolated plants with and without partitions can be included as controls to separate biotic interactions from the constraint effect. A review and critique of methodology for separating above- and belowground competition is presented by McPhee and Aarssen (2001) who recommend an additive design (see above) with a complete set of controls to assess the effects of partitions.

A salient example of how design issues can haunt one's best intentions to address an important concept is in the debate on the importance of self/non-self discrimination between the roots of plants competing intraspecifically (i.e. kin recognition, one manifestation of the so-called tragedy of the commons phenomenon in plants). Following the proposal that roots of plants grow to exploit more than their share of resources in the presence of roots of a different individual of the same species (Gersani et al. 2001), much of the ensuing debate centred on methodological issues revolving around whether or not an appropriate experimental design had been used (Hess and De Kroon 2007; Semchenko et al. 2007; O'Brien and Brown 2008). One outcome of the debate is the need to carefully design studies and effectively measure appropriate traits before and during the competitive process (Dudley et al. 2013). Moreover, it has been argued that these types of competitive interactions may not be manifest directly in fitness consequences; it may be necessary to determine how neighbours affect natural selection through the measurement of multiple correlated fitness-related traits (File et al. 2012).

Four examples are now described to illustrate the variety of approaches that have been used to assess questions about the importance of above- and belowground interactions; additional examples are discussed in Wilson (1988) and Silvertown and Charlesworth (2001).

1. *Which plant properties determine the success of a perennial plant species under nutrient-poor and nutrient-rich conditions, respectively?* Aerts et al. (1990) used four experimental treatments to assess the effects of nutrient conditions on competition between three heathland species, *Erica tetralix* (Ericaceae), *Calluna vulgaris* (Ericaceae), and *Molinia caerulea* (Poaceae). In a fully isolated treatment, plants were grown in 13 cm × 13 cm × 13 cm pots. Belowground interactions were allowed by growing five plants in containers of the same depth but with five or six times the root volume (65 cm × 13 cm and 78 cm × 13 cm pots, respectively). This design assured an equal mean rooting volume per plant for all treatments. In treatments where aboveground interactions were prevented, white nylon stockings were placed over a vertical framework of sticks separating each plant. Each plant was thus guaranteed an equal volume of aerial space. Fertilizer treatments (no nutrients or addition of nitrogen, phosphorus, or potassium) were implemented in a simple additive design where each species was grown either in monoculture, or in a 1:1 density mixture with each of the other two species in turn. Belowground interactions predominated in this experiment, but only in fertilized pots.
2. *Does the intensity of competition differ above and below ground and does it change along an environmental gradient?* In a field experiment, Belcher et al. (1995) separated the effects of root and shoot competition along a soil depth gradient. The annual *Trichostema brachiatum* (Lamiaceae) was used as a phytometer (Chapter 3) in treatments without neighbours, with competition from neighbour roots only, and with competition from neighbour roots and shoots (Fig. 4.8). Single seedlings of the phytometer were transplanted into the centre of 25 cm × 25 cm plots. The no-neighbour (NN) treatment was established by applying glyphosate to the plots 10 days prior to transplanting the phytometers. The competition from neighbour roots only treatment (NR) was established by placing a steel mesh cone (2.5-cm mesh), narrow end down, in the centre of each plot. Shoots of aboveground vegetation were trained to the outside of the cone through weekly checking. This arrangement allowed the phytometer to grow in the presence of neighbour roots and in the absence of neighbour shoots. The treatment allowing competition from both neighbour roots and shoots (NRS) was established in intact vegetation. The mesh cones were placed into both NN and NRS treatments to allow for potential cone effects. Shoots of the neighbouring vegetation were allowed to grow through the mesh. A treatment to test the effects of only aboveground interactions was not established. The results of this experiment suggested that competition was predominantly below ground, but competition

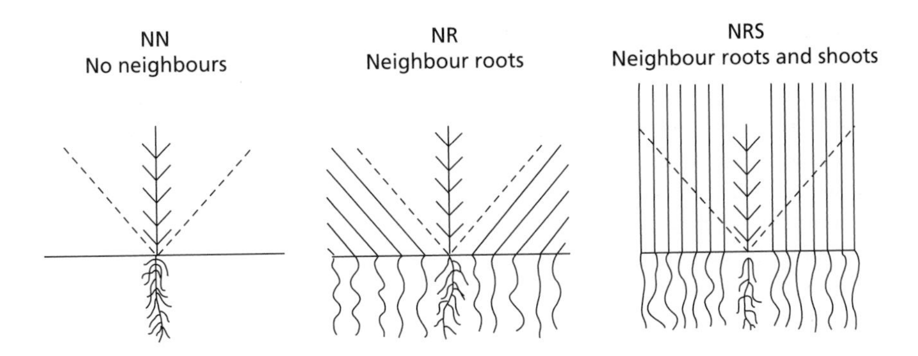

Figure 4.8 Experimental design used to measure above- and belowground interactions. Reproduced from Belcher et al. (1995) with permission from John Wiley & Sons.

intensity did not change along the soil depth gradient. Since this design involved monitoring the response of a single target plant against the natural density of neighbours, it is a multispecies design but does not correspond to one of the designs illustrated in Fig. 4.7. Perhaps one lesson that can be drawn is that new experimental methods may need to be 'invented' to answer specific questions.

3. *What is the relationship between competition, productivity, and the interaction among early successional species of contrasting growth forms?* Cahill (2002) used an additive design similar to that of Belcher et al. (1995) with PVC root exclusion tubes and thin tree netting to tie back neighbours to investigate the effects of root and shoot interactions on four species in an early successional community in response to a fertilizer treatment. Fertilizer altered the relative strengths of root–shoot interactions, but not the total strength (measured as competitive response, see Chapter 6) of root plus shoot competition. The labile nature of the root–shoot interactions exhibited in this study reinforces the need to measure the strength of both above- and belowground competition to fully understand how plant interactions vary in response to productivity gradients.
4. *What are the morphological responses to competition of a clonal perennial when limited by either soil-based resources or light and soil-based resources?* Price and Hutchings (1996) investigated the morphological response of *Glechoma hederacea* clones to belowground or above- and belowground competition. Their experimental design involved allocating replicate clones (each of two primary ramet-producing stolons at the start of the experiment) of *G. hederacea* at random to each of five treatments (Fig. 4.9) in 7-cm diameter pots in the greenhouse. The five treatments included tall grass (TG) and tall grass split (TGS) treatments, where one or both ramets were

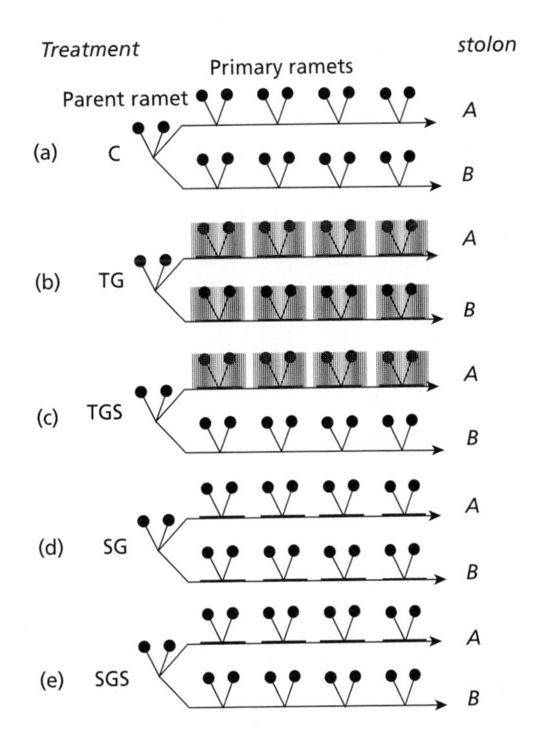

Figure 4.9 Schematic diagram of five experimental treatments used to compare the growth of clones of *Glechoma hederacea* in competition with *Lolium perenne* for light (treatments TG and TGS—long swards) or soil resources (treatments SG and SGS—clipped swards). From Price and Hutchings (1996). Reproduced with permission from John Wiley & Sons.

competing for light and soil resources with uncut grass (*Lolium perenne*). In the short grass (SG) and short grass split (SGS) treatments, competition for light still occurred, but was reduced by clipping at 1 cm above ground level compared with treatments TG and TGS, and one or both ramets were competing primarily for soil-based resources. A control treatment (C) consisted of both ramets growing without competition from the grass. This experiment was carried out at a single density and corresponds to a simple pairwise design (Fig. 4.7a). Competition for light had a larger effect on performance than competition for soil resources. Under the highest levels of light competition (treatments TG and TGS), *G. hederacea* stolons had the longest internodes and highest petiole length and mass, suggesting that this clonal plant grew aggressively to avoid light competition. Increased biomass and petiole length enhances light acquisition under these conditions. A limitation of this experiment was that while light levels were measured under the different treatments, measurements of soil resources (nutrients and water) were not made. Competition for soil resources between *G. hederacea* and *L. perenne* is suspected on the basis that significant short grass (SG and SGS) effects occurred.

4.6 Follow-up exercises

1. Design a greenhouse experiment to investigate the importance of a biotic and an abiotic treatment upon an annual and a perennial plant from Table 3.4. How should the experiments differ because of the differences in the plants' life histories?
2. What field-based experiments would you conduct in parallel with the experiment that you designed in Question 1?
3. Extend part of the experiment from Question 1 to include competition with another plant with the same life history (i.e. another annual or another perennial plant).

CHAPTER 5

Measuring individual and population parameters

The winged seeds, where they lie cold and low, Each like a corpse within its grave, until Thine azure sister of the spring shall blow . . .

(P. B. Shelley, *Ode to the West Wind*)

Plants stand still and wait to be counted . . .

(Harper 1977)

. . . it is obvious, however, that during the daylight period the plant is adding new material continuously.

(Blackman 1919, The compound interest law and plant growth)

- **Plant traits and measurements**
- **The value of making repeated measurements**
- **Locating and marking population units**
- **Morphological measurements**
- **Measurements of seeds and other dispersal units**
- **Physiological measurements**
- **Molecular measurements**
- **Indices for expressing plant growth**
- **Fitness, fecundity, and reproductive effort**

Preamble: plant traits and measurements

The quotes above exemplify in different ways the goal of this chapter, which is to describe the measurements that can be made on individual plants or populations. While individual plants do not move, their units of dispersal and populations as a whole certainly do; and the plants themselves are continually growing and changing shape and size. This dynamic characteristic of individuals and populations provides the challenge of measurement. There are two aspects to plant traits and their measurement that require consideration. First, there is the taking of the measurement, and second there is the interpretation of the measurement to express characteristics of the plant or population.

The measurements described here are critical as they provide the data that will be used to directly test the hypotheses posed at the start of the study (Chapter 2). Because of inherent problems with the precision and accuracy (Chapter 3) of the instruments and tools used in taking measurements (including observer bias), data represent a summary or estimate of reality and encompass a loss of absolute information. It may not seem so when starting out, but it is also easy to accumulate many measurements and data, some of which are more useful than others. Such an accumulation of data can make the eventual analytical process drawn out and perhaps frustrating, as 'wheel-barrow' loads of data have to be summarized and analysed. The temptation to collect data simply because of the availability of

Methods in Comparative Plant Population Ecology. Second Edition. David J. Gibson.

an instrument or the apparent ease of making the measurement should be avoided. As emphasized in Chapter 7, advance planning of the analysis will ensure that adequate data are collected and that unnecessary measurements are avoided.

A list of the data required to test your hypotheses should be drawn up as the experiment is being planned. Hypothesis-driven data collection will streamline field work in the same sense that hypothesis-driven experimental protocols streamline the complete study.

A *trait* is defined by Violle et al. (2007) as any morphological, physiological, or phenological feature measurable at the individual level, from the cell to the whole organism, without reference to the environment or any other level. The value or modality of a trait at a particular time or place is an '*attribute*'. Traits are measured for the purpose of testing a hypothesis. As an example, consider the list of traits measured by the investigators of the four case studies described in Chapter 1 (Table 5.1) (note that they may have measured additional traits that they chose not to report). There is some redundancy in these lists as many traits were measured for each study. This redundancy arises in part from imprecise original hypotheses, and also because of the difficulty of identifying the best traits to measure. Some plant traits are hard to measure or require surrogate or indirect measurements. For example, the number of flowers borne by a plant was used by the investigators of Case Study 4 as a surrogate estimation of fitness. Among the many assumptions associated with this measurement is that a plant producing a large number of flowers will eventually produce a correspondingly large number of viable seeds that will survive, germinate, establish, and reproduce. The investigators hedged their bets somewhat by also using the dry mass of flowers and buds, and total plant biomass as additional estimators of fitness.

Table 5.1 Traits measured in the case studies described in Chapter 1. A ✓ indicates that the trait was measured. Traits used as surrogates for fitness or fecundity (authors' definitions) are indicated.

Trait	Case study			
	1	2	3	4
Mapped plants	✓		✓	
Plant origin	Habitat of likely mother			
Date of birth/death	✓			
Number of:				
Flowers				Fitness
Seed (including number viable)			Fecundity	
Stems		✓		
Flowering whorls		Fecundity		
Rosette leaves				✓
Stem leaves				✓
Stem height			Estimation[1]	
Stem basal diameter		✓	Estimation[1]	
Flowering state		✓	Estimation[1]	
Leaf area				✓
Length of longest leaf			Estimation[1]	
Dry mass of:				
Flowers and buds				Fitness
Stem				✓
Rosette leaves				✓
Stem leaves				✓
Roots				✓
Total				Fitness

[1] Stem height, stem basal diameter, flowering state, and leaf length were used to estimate the number of capitula, dry weight of capitula, foliage, and total plant biomass, respectively, by regression.

In general, what traits are likely to be important ones for population ecologists to measure? Weiher et al. (1999) approached this question by considering the challenges that a plant faces in its environment. A plant needs to disperse to new sites, become established at these sites, and then tolerate the biotic and abiotic conditions of the site in order to reproduce and produce new units of dispersal. Each of these challenges has one or more components that can be identified, for example dispersal in space and time. Hard and easy to measure traits are then identified for the components of each challenge (Table 5.2).

A global archive of trait values for over 69,000 plant species is available through the TRY database (<http://www.try-db.org/TryWeb/Home.php>; Kattge et al. 2011). These trait values are useful for making species comparisons (e.g. Gallagher and Leishman 2012; Loranger et al. 2013).

Table 5.2 Hard and easy traits (from Weiher et al. 1999). Hard traits are the most obvious and direct traits but may be difficult to measure, involving measurements over a long period of time or requiring experimental manipulation. Easy traits are less direct surrogates of hard traits that are easier to measure. Some easy traits such as seed mass, specific leaf area (SLA), and leaf water content (LWC) may be surrogates of more than one hard trait. Reproduced with permission from John Wiley & Sons.

Challenge	Hard trait	Easy trait
1. Dispersal		
(a) Dispersal in space	Dispersal distance	Seed mass, dispersal mode, seed rain
(b) Dispersal in time	Propagule longevity	Seed mass, seed shape, seed rain
2. Establishment		
(a) Seedling growth	Seed mass	Seed mass
	Relative growth rate	SLA
		LWC
3. Persistence		
(a) Seed production	Fecundity	Seed mass, aboveground biomass
(b) Competitive ability	Competitive effect and response	Height, aboveground biomass
(c) Plasticity	Reaction norm	SLA, LWC
(d) Holding space/longevity	Life span	Life history, stem density
(e) Response to disturbance	Resprouting ability	Resprouting ability
(f) Response to stress and disturbance avoidance	Phenology, palatability	Onset of flowering, SLA, LWC

It has generally been assumed, especially in community ecology applications, that traits vary more between than within species (i.e. interspecific variability > intraspecific variability). However, the degree of intraspecific variability varies among traits (e.g. leaf dry matter content is less variable than specific leaf area, SLA) and this variability can be relevant—especially at local scales, across short environmental gradients, and when measuring relatively plastic traits. Under these situations, intraspecific variability needs measuring and accounting for (Violle et al. 2012; Auger and Shipley 2013). A decision tree and methodology for dealing with intraspecific variability were proposed by Albert et al. (2011) which involve testing the sensitivity of intraspecific variability of traits using simulation studies.

The obvious and most direct traits may be difficult to measure and are referred to as 'hard' traits. Conversely, 'easy' (also called 'soft') traits are analogues or surrogates of the hard traits that we may be forced to use because of logistical constraints. Alternatively, if a statistical relationship between a hard and an easy trait can be demonstrated through the use of regression then the hard trait can be estimated directly. For example, the length of the longest leaf of *Senecio jacobaea* plants was found to give an acceptable estimate of plant biomass in Case Study 3 of Chapter 1 (Table 5.1).

Two additional categories of traits are functional and performance traits (Violle et al. 2007). Functional traits are any trait which has an indirect impact on fitness through its effects on growth, reproduction, and survival. Performance traits are direct measures of fitness and are of three types—vegetative biomass, reproductive output (e.g. seed biomass, seed number), and survival. A framework illustrating the relationship between functional traits, performance traits, plant performance, and individual fitness is illustrated in Fig. 5.1. In a set of detailed measurement protocols, Pérez-Harguindeguy et al. (2013) related the value of plant functional traits to environmental change, plant competitive strength, defence, and fitness, and the effects of plant on biogeochemical cycles and disturbance regimes. The 43 plant functional traits that they described were

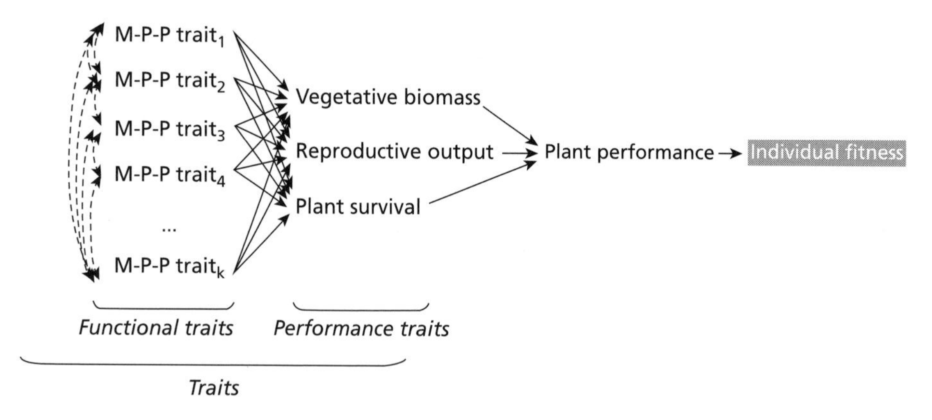

Figure 5.1 Framework illustrating the relationship between functional traits, performance traits, plant performance, and individual fitness. Morpho-physio-phenological (M-P-P) traits (from 1 to *k*) modulate one or all three performance traits which determine plant performance and individual fitness. M-P-P traits may be interrelated (dashed arrows). Performance traits may also feed back to M-P-P traits (not shown for clarity). From Violle et al. (2007). Reproduced with permission from John Wiley & Sons.

placed into four categories: whole-plant traits (13 traits; e.g. plant height), leaf traits (16 traits; e.g. SLA), stem traits (5 traits; e.g. stem-specific density), belowground traits (3 traits, e.g. specific root length), and regenerative traits (6 traits; e.g. seed mass). Recommended preferred sample size varied among traits from $n = 3$ for photosynthetic pathway traits to $n = 25$ for plant height. Some traits such as growth form, plant height, and seed mass can be measured on any plants in the population that meet the trait criteria. Other traits such as SLA and leaf nitrogen concentration may be measured on the same individuals. A generally accepted minimal set of relevant plant traits includes plant size (usually measured as plant height), seed size (usually seed mass), and leaf structural tissue (usually SLA or dry matter content). The smallest possible number of uncorrelated traits from multiple plant organs should be measured (and it may be only one) (Laughlin 2013).

This chapter is divided into sections dealing with the value of making repeated measurements (Section 5.1), locating and marking population units (Section 5.2), morphological measurements (Section 5.3), measurements of seeds and other dispersal units (Section 5.4), physiological measurements (Section 5.5), molecular measurements (Section 5.6), the calculation of common indices for expressing plant growth (Section 5.7), and the calculation of fitness, fecundity, and reproductive effort (Section 5.8). The indices and calculations described in Sections 5.7 and 5.8 comprise part of the evaluation of the *vital rates* of a population, that is, the statistics concerned with or essential for population growth, fecundity, or mortality.

5.1 The value of making repeated measurements

The plant population literature is replete with studies in which measurements were taken only 'at the end of the experiment', 'upon final harvest', or 'at the end of the growing season'. In a sense, if we are interested in just the final outcome of a process or series of interactions then a single measurement in time might be appropriate. This would include certain questions dealing, perhaps, with fecundity. In addition, some traits can only be measured once, since they occur once or represent the outcome of a process. The number of seeds produced by an annual plant in its single season is an example. However, there are many traits that can be measured several times over the time course of an experiment. These include, for example, plant height, leaf number, and flower number (in plants with indeterminate reproductive structures). The disadvantage of taking measurements only at the end of an experiment, or at the end of the field season, is that only relationships apparent at that one time will be expressed. This may or may not correspond to the potential

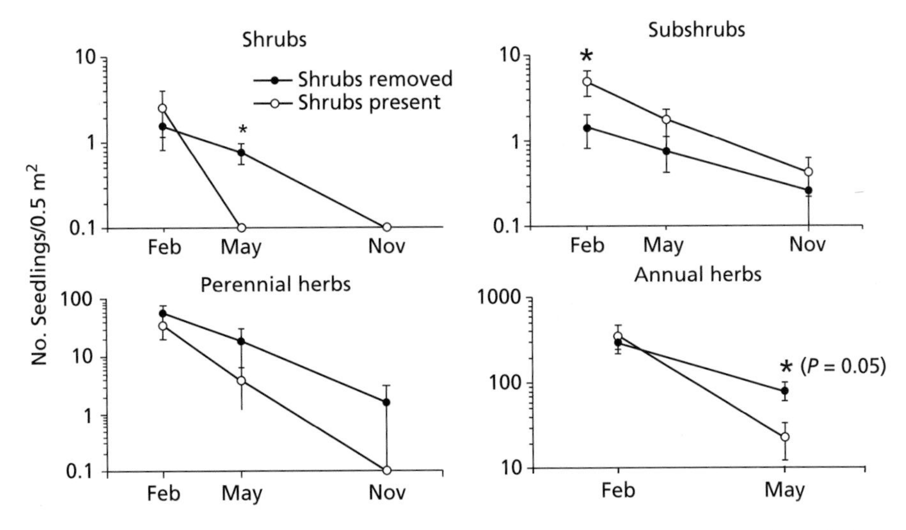

Figure 5.2 The value of repeated measurements. Seedling establishment of shrubs and subshrubs in Californian chaparral when surrounding adult shrubs are removed or present. The asterisk (*) indicates a significant difference between treatments ($P < 0.05$). From Tyler (1996). Reprinted with permission of the Ecological Society of America.

final outcome that would occur. The 'end of the experiment' may be an ecologically irrelevant point in time corresponding to experimental logistics or limitations (e.g. no more money to purchase supplies to maintain an experiment, or someone else needs the greenhouse space). The advantage of taking repeated (serial) measurements of a trait throughout time is that the temporal dynamics of experimental patterns can be observed and quantified. Repeated measurements are particularly important for following the dynamics of plant competition studies (Gibson et al. 1999; Trinder et al. 2013). Inclusion of time in the analysis of an experiment can be accommodated through a repeated measures ANOVA or by treating time as a block (Chapter 7). A clear example of the value of making repeated measurements is shown in Fig. 5.2. If seedling densities had only been measured at the end of the season in November then the earlier effects of shrub removal on shrub and subshrub seedling densities would have been missed.

If the goal of a study is to determine population flux through time then repeated measurements will be necessary. The frequency of sampling depends upon the life history of the species and the rate at which changes are occurring (or are expected to occur). Annual or biennial populations need to be censused every year. Less frequent sampling may be adequate for perennial plants. The sampling interval will determine the minimum time-step available for calculating population transition matrices (Chapter 8). Estimates of λ to quantify population growth on a yearly basis requires an annual census.

A disadvantage of making repeated measurements on the same plant is the risk of damage to the plant. Even minimal handling of leaves to take measurements, for example, can lead to a marked loss of mass through increased respiration rates (Evans 1972).

5.2 Locating and marking population units

In plant population ecology the units to be studied include three basic levels of organization: genets, ramets, or modules. The *genet* population comprises genetically distinct individuals each derived from a single seed. The *ramet* population includes tillers, shoots, or rosettes in clonally growing species. *Modules* are units of construction such as leaves or flowers which may be repeated within genets and ramets (Hutchings 1986). *Metamers* are basic growth units consisting of a piece of stem with its internode complex, leaf or leaves, and associated lateral (axillary) meristems (Watson and Casper 1984). The organization of growth units is nested or hierarchical; for example, the module is an aggregate of metamers formed by a single apical meristem.

Clonal plants form a special case. Harper (1977) euphemistically defined a clone as 'a single plant that falls to pieces as it grows!'. He was referring to those plants that comprise ramets; the unit or module of clonal growth that may follow an independent existence if severed from the parent plant. The growth of these modules is referred to as *clonal growth*. About 70% of the plant species occurring in the temperate, deforested areas of the Earth are clonal. Because of this, and the dominance of clonal species in pastures, there has been a great deal of research by population ecologists to try and understand the nature and consequences of clonal growth (van Groenendael and de Kroon 1990).

The measurements required to better understand the growth of clonal plants revolve around the need to determine the number of '*integrated physiological units*' (IPUs) into which a plant can be resolved. IPUs consist of 'identifiable arrays of morphological subunits' (Watson and Casper 1984). For any particular plant, this depends upon a number of factors including age, size, and structure of the plant. Understanding clonal plant growth requires the measurement of basic morphological variables (e.g. relative growth rate, see Section 5.7) to assess important factors such as resource foraging, branch autonomy, and vegetative reproduction (Hutchings and de Kroon 1994). Determining the extent and role of IPUs requires a careful assessment of the integration of both morphological and physiological parameters.

For the sake of simplicity, the units of study are referred to here as individuals or plants in the following discussion regardless of the actual level of organization that may be the focus of study.

The number of plants that need to be sampled cannot be prescribed because of wide differences between and within populations. Indeed, this variation, especially in relation to the environment, may be the focus of a study. As noted earlier in Section 2.5, preliminary estimates of variation through use of a pilot study will assist greatly in devising an appropriate level of sampling intensity. An important complication is that the demographic behaviour of a population may change markedly as the environmental conditions change through time. For example, the variation shown by a particular plant trait may be comparatively small at the beginning of an experiment, but become very large following the imposition of experimental treatments.

Regardless of the units to be studied, data need to be collected from either individual plants or groups of plants. The former will be necessary when hypotheses require the retention of the identity of individual plants. This requires methods of precise relocation of individuals. Measuring groups of individuals under the same experimental treatments is less exact and in many respects will be easier, with fewer logistical requirements. The disadvantage of measuring groups of individuals is that there is no link between the plants recorded at one census and those at another.

Plot-based approaches

Groups of plants can be marked for repeated experimental treatments and measurement within sample plots. Plots (also referred to as quadrats) are more or less equal-sided sample areas and range in shape from circles or squares to rectangles. Circular quadrats have the smallest perimeter to area ratio and hence will incorporate the smallest amount of internal heterogeneity and the fewest edge (or boundary) effects. Square plots are often easier to lay out and mark in the field. Plot size can vary over several orders of magnitude from 1 cm^2 to 1 ha depending upon the size and density of the individuals to be studied. The morphology of the species concerned is the principal consideration in choosing an appropriate quadrat size. The following empirical sizes have been suggested (Cain and Castro 1959):

- moss layer: 0.01–0.1 m^2
- herb layer: 1–2 m^2
- low shrubs and tall herbs: 4 m^2
- tall shrubs: 16 m^2
- trees: 100 m^2.

The 100-m^2 quadrat is equivalent to the 0.1-ha plot size commonly used in forest ecology. Quadrats of size 1 m^2 (with sides of either 1 m × 1 m or 0.5 m × 2 m) are perhaps the most commonly used for

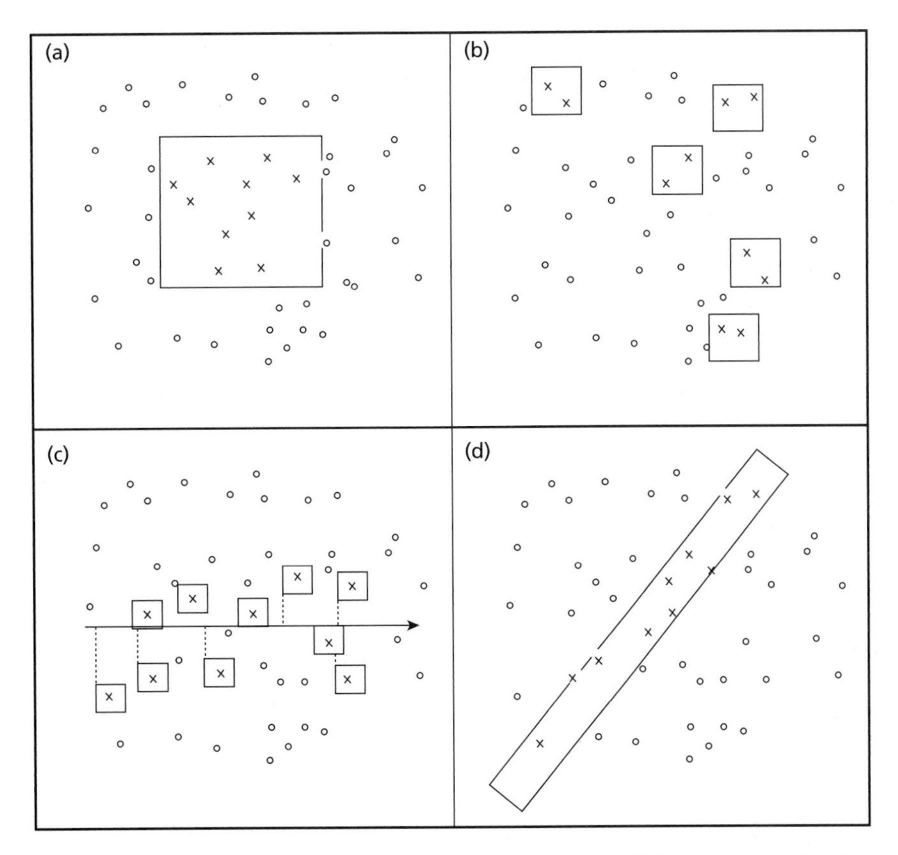

Figure 5.3 Use of sample plots to locate 10 individual plants for study: (a) single sample plot, (b) five randomly placed small sample plots with two plants sampled in each, (c) sample plots located with reference to a line transect with one plant per plot, and (d) plants sampled within a belt transect. X, sampled plants; O, unsampled plants of the same species.

herbaceous layer plants allowing comparisons to be readily made on a unit area basis. Plot size can influence estimates of frequency and are affected by the level of spatial aggregation exhibited by species. A plot size that will result in a mean density of less than 10 is recommended when aggregation is suspected to be high (Hayek and Buzas 1997).

Since in population ecology we are concerned with measurements of or on individuals, the size of a plot is important only in as much as it allows the identification of (a sufficient number of) individuals. At the largest scale, we may wish to delimit a single sample plot within the bounds of a population. We would then sample a number of individuals at random from within this single sample plot (Fig. 5.3a). Alternatively, small sample plots can be randomly located within a population (Fig. 5.3b). In this case, the plots are small enough to allow only a small number of individuals to be sampled in each plot. The plots (and plants within them) could be assigned to specific treatments, in which case a more systematic arrangement of the plots may be necessary (Fig. 5.3c or Figs 3.5 and 3.6).

Transects are elongate sample areas, either rectangular or lines of zero width that run through a study site. In Fig. 5.3(c) plots are located with reference to a line transect. The line transect is, as the term suggests, a line passing through the population. Sample plots can then be located using a stratified random design as follows:

1. Starting a point X, a set distance *a* (e.g. 5 m or five paces) is travelled along the transect. The value of *a* depends on the length of the transect and the number of sample plots that one wishes to sample.

2. A random number is used to designate an additional distance, less than *a*, to be travelled along the transect.
3. A second random number is used to make a left or right decision.
4. A third random number is used to provide a distance to move to the left or right (according to Step 3) perpendicular to the line of the transect. A plot is established at that point providing it contains the desired number of individuals of the species to be studied. If too few individuals are present in the plot then it is discarded.
5. Return to the location identified prior to Step 2 and the procedure starts again from Step 1.

Care has to be taken when using this procedure to ensure that the distance *a* and the distances obtained using the random numbers allow for the desired number of plots to be established. If values are too large you may stray outside the bounds of the population or the transect may cross the population before a sufficient number of plots have been established. If values are too small sample plots may overlap or an unrepresentative portion of the population may be sampled. Individual plants rather than sample plots can also be located this way by tagging for study the closest individual to the point reached at Step 4.

A belt transect involves placing a long, narrow sample plot through the population of interest and sampling the plots (or a portion of the plants) contained within (Fig. 5.3d). Like the size of a square or rectangular plot, the width of a belt transect will depend upon the size and density of the plants to be studied.

Relocating individual plants

Being able to accurately relocate plants within a sample plot is necessary (a) when repeated measurements of individual plants are required or (b) when repeated censuses of the population are necessary. Techniques suitable for recording the location of plants in small plots, for example <1 m^2, include the use of pantographs, mapping tables, bar plotters, field digitizers, and photographs (Hutchings 1986). For example, Sarukhán and Harper (1973) used a pantograph in their demographic study of buttercups and McEvoy et al. (1993) mapped seedlings of *Senecio jacobaea* by circling the position of each seedling on a piece of acetate placed over sample plots (see Case Study 3 in Chapter 1).

In larger plots individuals can be mapped with respect to their distance from two corners of the plot. Rectangular coordinates within the plot can be obtained using the cosine rule (Fig. 5.3). Gibson and Menges (1994) used this approach to map bushes of *Ceratiola ericoides* (see the worked example in Chapter 8). Hutchings (1987) similarly recorded the position of orchids in a 20 m × 20 m plot over 11 years.

Difficulties in using this approach include:

1. The corners of the plots (*A* and *B* in Fig. 5.4) have to be permanently marked. Any change in the location of these reference points translates directly into inaccurate coordinates and potential confusion with the identification of plants.
2. Species which perennate below ground, for example bulbous species, may appear to move through time as the exact location in which they appear above ground varies each year. Some flexibility in making records of a plant from one census to another is necessary otherwise it might appear that an unusually large number of plants are both dying and being born.
3. Germinated seedlings may move because of rain splash, frost heave, or movement by the soil fauna (Liddle et al. 1982).
4. Some perennial, non-woody species do not re-emerge every year. In noting this problem with orchids for demographic studies, Hutchings (1987) considered as dead an orchid that failed to reappear for three consecutive years. Similarly, the appearance of individuals after the start of a study may not indicate recruitment, but merely the reappearance of plants that have lain dormant for a period.

Marking plants

Relocating plants by their location in sample plots is greatly facilitated by some sort of tag on, or adja-

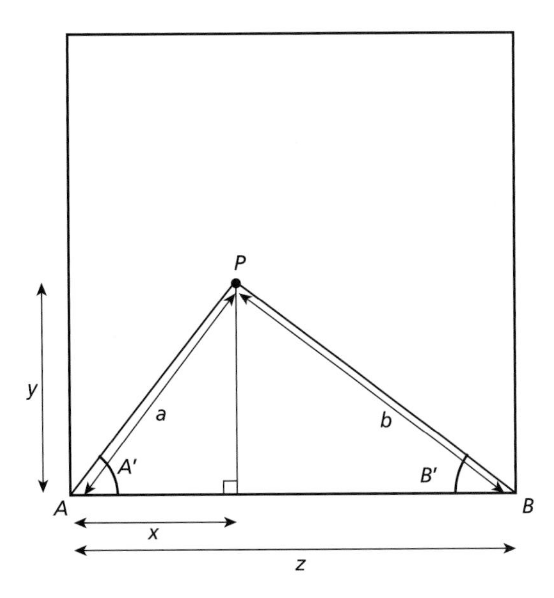

Figure 5.4 Obtaining rectangular coordinates (x and y) of a plant (P) from measurements a and b, taken from corners A and B of a study plot using the cosine rule: $\cos A' = (z^2 + a^2 - b^2/2za)$; $x = a \cos A'$; $y = \sqrt{(a^2 - x^2)}$. From Hutchings (1986). Reproduced with permission from John Wiley & Sons.

cent to, each plant. In this respect, plant ecologists usually tend to be quite inventive, often making use of inexpensive materials. The important criteria are twofold: first, the label must unambiguously identify the unit of study (genet, ramet, or module); second, the label must remain on the plant and be identifiable from one census to the next (Fig. 5.5 and Plate 15). Nearly all markers have the propensity to attract something, be it people (and worst of all vandals) or animals (which may gnaw, eat, or collect markers), or become lost or destroyed following storms, flooding, fire, etc. Markers should be only as visible as is necessary to find them again, without attracting unwanted attention. It is a good idea to check markers between census periods to guard against loss. The advantages and disadvantages of some different types of markers are indicated in Table 5.3.

In some areas it will not be practical or feasible to mark individual plants. In these situations, individuals can be identified with respect to their distance and direction from reference points such as large trees, gate posts, boulders, etc. Permission should always be obtained from site owners or managers before placing any markers in the ground or on plants.

Figure 5.5 Individual shrub of *Ceratiola ericoides* marked with an aluminium tag. (Photo David Gibson.) (See also Plate 15.)

Table 5.3 Markers and tags used in plant population ecology.

Marker/tag	Advantages	Disadvantages
Placed next to plants		
Wooden stakes	Easily constructed, and inexpensive	Will rot, markings will fade
Plastic stakes	Durable	Lightweight and may be easily dislodged
Metal stakes	Durable. Numbers can be welded onto stakes. Can relocate with a metal detector	Expensive
Stake wire flags	Inexpensive. Easy to drive into soil. Different colour flags assist in identification	Plastic flags fade and rip quickly
Toothpicks	Inexpensive	Do not allow numbering system, easily lost
Placed on the plant		
Wire loops	Inexpensive. Durable. Can be cut to fit around branches, petioles, etc.	Colour coding necessary to identify individuals
Plastic rings	Inexpensive	Colour coding necessary to identify individuals
Plastic straw collars	Inexpensive	Colour coding necessary to identify individuals
Steel or aluminium tags	Durable, can be stamped with unique identifier	Can be expensive
Paint or ink marks on leaves or stem	Easy to apply, can allow unique colour coding.	Can fade or wash off. Potentially toxic

Herbarium specimens

In addition to using living plants, many concepts in population ecology can be addressed using measurements and observations obtained from plants preserved as herbarium specimens. Applications include determining changes in plant range and size in response to climate change (Leger 2013) and human harvesting (Law and Salick 2005). Introduced species were shown to exhibit morphological change over 150 years based upon herbarium specimens of 23 species in New South Wales, Australia (Buswell et al. 2010). Categories of threat for Red Listed taxa can be derived from herbarium specimen data (Willis et al. 2003). Limitations of using herbarium data include an assumption of no collector bias in terms of location or specimen selection. Some herbarium specimens have limited or inadequate labels lacking accurate locations, dates, or habitat information. Validation for using herbarium specimens for these sorts of investigation was provided by Robbirt et al. (2011) who matched field and herbarium data as part of a study investigating phenology of the terrestrial orchid *Ophrys sphegodes* and climate change.

5.3 Morphological measurements

This section describes how the traits such as those listed earlier in Table 5.2 can be measured or estimated. The distinction between measurement and estimation is important. Traits such as plant height can be measured directly, whereas traits such as seed number may have to be estimated through subsampling. The 'hard' traits in Table 5.2 most likely have to be estimated.

Units of measurement should conform to the Système Internationale d'Unités (the International System of Units, abbreviated SI). The SI system of units is constructed from seven base units each precisely defined (Table 5.4). All other units for physiochemical quantities are derived from these. For example, 1 ha is 1000 m^2 and is derived from the metre base unit. The derived unit of force is the newton (N) and is 1 kg m s^{-2} (i.e. it is a compound unit, being the product of three SI base units). Care should be taken to use units that are appropriate to the scale of work rather than the SI base unit multiplied by some factor of 10; for example, cm rather than 10^{-2} m. However, Imperial units such as feet, inches, pounds, and ounces are still used routinely in the

Table 5.4 SI base units and their symbols. From Style Manual Committee Council of Science Editors (2006), reprinted with permission of Cambridge University Press.

Quantity	Unit	Symbol[1]
Length	metre	m
Mass	kilogram	kg
Time	second	s
Electric current	ampere	A
Thermodynamic temperature	kelvin	K
Amount of substance	mole	mol
Luminous intensity	candela	cd

[1]Capitalization in this column reflects the correct usage of upper- and lowercase letters.

United States in agriculture and forestry. Full details on the use of SI units can be found in author's style manuals such as the *Scientific style and format: the CSE manual for authors, editors, and publishers* (Style Manual Committee Council of Science Editors 2006) and in Salisbury (1996).

Counts

As shown in Tables 5.1 and 5.2 there are many plant traits for which it may be useful to obtain a numerical count. Numbers of plant parts such as stolons, stems, leaves, inflorescences, and seeds are commonly counted. It is important to have a clear working definition of the unit that is being counted. For example, when counting leaves it is important to decide ahead of time whether to include (or exclude) from the count immature leaves (e.g. for grasses, leaf blades still rolled in the leaf sheath), damaged, diseased, or dead leaves. It is important to decide how much damage, disease, or necrosis defines a leaf as being in the 'dead' category for example. It might also be important to make separate counts of basal rosette and cauline (stem) leaves.

Numbers of fruits or seeds may be particularly difficult to count, especially in plants such as composites with compound inflorescences. In these situations it may be necessary to take a representative subsample and estimate the total by multiplication.

Linear measurements

The following traits are readily measured using a ruler, tape measure, or callipers: plant (stem) height, internode length, leaf length and width (leaf area), and inflorescence length (in graminoids). Leaf area is an important measure that is readily determined using a leaf area meter. Care has to be taken to keep the leaves fresh as the slightest drying out will lead to leaf curl, roll, and shrinkage with consequent reductions in apparent leaf area. If a destructive harvest is not possible, for example with plants that are to be measured repeatedly, then leaf area can be estimated using regression from leaf length and width measurements. Some leaf area meters can be used on leaves that are still attached to the plant. In graminoids and plants with linear leaves, length alone may be a sufficient estimator of leaf area. The size of shrubs can be estimated by calculating volume from measurements of height, length, and width. In this case, length is determined as the longest linear extent of the aboveground shrub canopy and width is the maximum extent of the canopy perpendicular to the orientation of the length measurement. Estimating volume as a rectangular box from these measurements is likely to overestimate canopy volume and an inverted cone (or fustrum) may be more appropriate.

The height of trees above 2 m (head height) can be obtained using a clinometer. The top of the tree is sighted at a known distance (L) from the bole of the tree. The angle between the ground and the top of the tree (α) is recorded. The height (X) is calculated as (McLean and Ivimey Cook 1968):

$$X = L \tan \alpha$$

Accommodation needs to be made for uneven or sloping topography in making the calculations by calculating the average angle from two measurements made at equal distances on either side of the tree (Fig. 5.6).

Roots and belowground structures

Measurements of belowground structures can be informative for plant population ecology, especially in studies of resource acquisition and plant interactions. For example, the monitoring cohorts

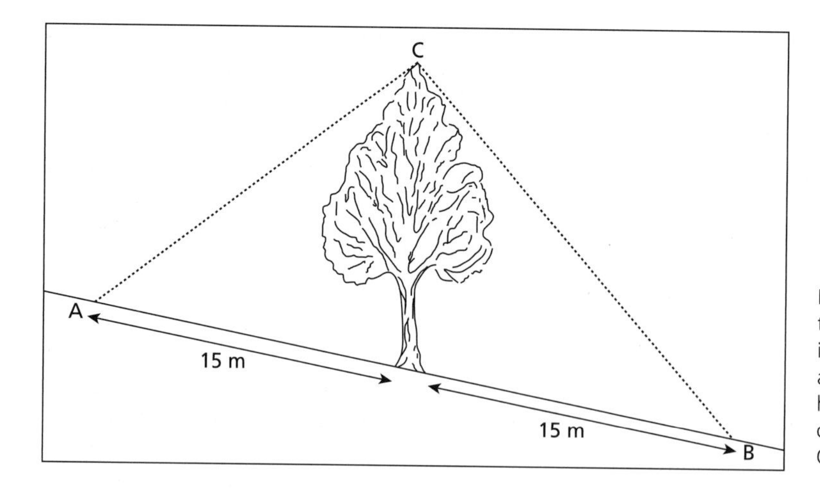

Figure 5.6 Measuring the height of a tree on sloping ground. If the angle CAB is 48° and the angle CBA is 38° then the average angle (α) is 43°. Therefore, the height $X = L \tan \alpha$, where $L = 15$ m in this case. Adapted from McLean and Ivimey Cook (1968).

of roots can provide demographic assessments of root longevity related to the biotic and abiotic components of the soil environment (Hendrick and Pregitzer 1992). The 'why and what' to measure for root studies is detailed in Atkinson (2000). For these studies it may be necessary to determine one or more of a number of belowground parameters including:

- root length or surface area
- root topology or branching
- root turnover and longevity
- biomass (see next subsection), including belowground net primary production (BNPP)
- mycorrhizal or rhizobial infection levels (see Section 6.1)
- species identity in mixed systems (see 'Molecular methods', section 5.6).

In many cases studying roots involves either destroying them or disrupting the delicate soil environment in which they grow, or more likely both. Methodological studies of plant roots were pioneered by Weaver (1919, 1920), but are still the subject of debate as different approaches provide sometimes widely divergent estimates of root production and demography. A meta-analysis indicated that indirect estimates of fine root production and longevity (including estimates obtained from sequential cores; Yuan and Chen 2013) were 87 and 124% higher than direct methods (ingrowth and minirhizotrons), respectively (Yuan and Chen 2012). Detailed descriptions of methodology for studying plants roots can be found in Böhm (1979), with more recent treatments available in chapters in Smit et al. (2000) and Mancuso (2012). A wealth of information is available from these sources.

In brief, some approaches for studying plant roots that are useful for plant population ecologists include the following:

1. Excavation: pits, profiles, direct excavation, monoliths, and the use of soil augers to obtain (sequential) cores. Excavating roots or viewing them *in situ* in pits or profiles is destructive, time consuming, and subject to loss of small and fine roots. Estimates of BNPP using these methods can be skewed and are subject to numerous biases. However, decision matrices using data from sequential cores can provide acceptable estimates of fine root production and mortality (Yuan and Chen 2013, who also provide R code for making calculations).
2. Ingrowth cores and doughnuts—roots are allowed to grow into sifted, root-free soil encased in a mesh bag of some sort for a period of time before the bag and soil is removed (or cores removed from within the bag) and the roots separated from the soil and processed. This is a popular, relatively easy, and inexpensive method. Difficulties can arise in removing the bag and in reinstalling subsequent bags in the same location. Underestimates can arise due to gaps between the soil and the bag, mesh sizes that limit

root ingrowth, and roots that grow and senesce between sample intervals. Overestimates can arise from root proliferation in the absence of existing roots in the soil in the bag. A full critique is provided by Milchunas (2012).

3. Transparent soil interfaces: root windows and minirhizotrons. Images of roots are captured as roots grow against transparent surfaces placed into the soil. The older use of root windows (literally, planes of glass or plastic placed against a soil profile with roots drawn onto acetate sheets or photographed) has been superseded by minirhizotrons—images of roots in the soil are digitally recorded from recording devices placed inside narrow clear plastic tubes inserted into the soil. Biases and problems with minirhizotrons arise from the methodology of field installation, image collection, data processing, and conversion of raw data to meaningful estimates of root production and demography (Milchunas 2012). Image analysis of data obtained from minirhizotrons is described in Richner et al (2000). While minirhizotrons are perhaps the best method for assessing fine root volumes, a serious problem with minirhizotron sampling can be that the small sample volumes encountered can lead to non-representative samples and substantial underestimates of dynamic root volumes (Taylor et al. 2013). A combination of methods (monolith sampling, soil cores, allometric calculations, along with high-volume minirhizotrons) may be necessary.
4. Electrical capacitance. This rapid and non-destructive field method involves measuring the electrical capacitance when electrodes are placed in the soil and on the stem of a target plant. According to the model of Dalton (1995), roots can be considered a cylindrical capacitor with the epidermis and xylem as external and internal electrodes and internal roots tissues as a dielectric insulator. Capacitance varies in proportion to root surface area, mass, length, and the permittivity and density of root tissue. Use and applicability of this technique is, however, debated and is an area of active research. Ellis et al. (2013) (see commentary in Phillips 2013) found this method to be useful and were able to distinguish intermingled root systems of herbaceous plants and trees over several orders of magnitudes across several ecosystems. However, Dietrich et al. (2013) tested a new physical interpretation of plant capacitance and reported that capacitance was not always directly proportional to root mass but was more related to plant tissue between the surface of the substrate and the electrode attached to the plant.
5. Three-dimensional imaging and analysis of whole root systems in transparent growth media. A number of optically transparent growth media [e.g. Phytagel (Sigma), gellan gum (Sigma), granulated agar (DIFCO), and Gelzan CM (Sigma)] have been developed that allow real-time visual imaging of whole plant root systems. These approaches maintain the topology of the root systems allowing root system architecture to be captured (Iyer-Pascuzzi et al. 2010). For example, growth of *Medicago truncatula* in a structurally heterogeneous hydrogel allowed three-dimensional time-lapse imaging to follow primary root growth (Silverberg et al. 2012). A number of software platforms have been developed to facilitate handling and analysis of these data (Clark et al. 2013).
6. Isotopes and tracers introduced into the soil or as labels in plants allow detection of root activity (Bingham et al. 2000). Stable isotopes, as isotopically distinct and chemically indistinguishable tracers (e.g. ^{15}N), are particularly suitable for field situations as they are non-hazardous and can be used without special precautions. Radioisotopes (e.g. ^{3}H, ^{14}C, ^{35}S, ^{32}P) can be detected in smaller quantities than stable isotopes and readily visualized using autoradiography or electronic imaging systems, but are potentially toxic. The amounts of radioisotopes used are typically below the exposure limits for adverse effects and their half-life would lead to a low retention time in the soil; for example the half-lives for ^{35}S and ^{32}P are short and decay would remove either isotope in 2.5 or 0.4 years, respectively. In addition, radioisotopes require proper disposal, and often large quantities of contaminated soil may need to be remediated. Chemical cation tracers of elements that are generally rare in the natural environment can be used as analogues of common cations, for example Li^{+} as an analogue for

K^+ (Gibson 1988) and Sr^{2+} as and analogue for Ca^{2+} (Veresgolou and Fitter 1984).
7. Computer-assisted tomography (CAT) and magnetic resonance imaging (MRI) of roots. A variety of energy sources (X-rays, gamma rays, electrons, protons, alpha particles, lasers, radar, seismic, and nuclear magnetic resonance) can be used in conjunction with CAT scanners to produce multiple slices of two-dimensional images of horizontal slices of a plant–soil system which can be combined to produce three-dimensional images (Asseng et al. 2000). These methods are largely limited to greenhouse and laboratory investigations because of the size of the equipment, although portable units are becoming available. The advantage of these approaches is the ability to obtain non-destructive, highly detailed, high-resolution three-dimensional images of intact root systems (Mairhofer et al. 2012; Mooney et al. 2012; Helliwell et al. 2013).
8. Ground penetrating radar (GPR). This method provides imagery allowing rapid field estimates to be made of root biomass of coarse roots (>5 mm in diameter) using portable, readily available instrumentation. When combined with other methodology such as minirhizotrons, reliable estimates of BNPP among different species and genotypes of trees across ecosystems under a range of soil conditions can be obtained (Butnor et al. 2012).

Biomass

Measurements of plant production are commonly made for assessments of yield and fitness. As with counts and linear measurement just about any plant genet, ramet, or module can be separated and weighed. *Fresh weight* (FW) refers to biomass measured prior to any evaporative or senescent water loss from the plant tissue. Plants in this condition contain all the water present in the functioning living plant at the time of sampling. Plants should be at full turgor (i.e. fully watered to field capacity) if it is desired to later obtain the fresh weight/dry weight (FW/DW) ratio (see below). To obtain FW, plants should either be weighed immediately after harvesting, or else stored at below 8 °C in watertight and sealed plastic bags. Storage time should be kept to a minimum to avoid decomposition and the growth of fungi on the plant tissues. If samples are collected from the field for FW analysis then storage in plastic bags is necessary. Fresh material is also required for certain chemical analyses, for example some carbon, nitrogen, and phosphorus fractions (Chapin et al. 1986). *Dry weight* (DW) refers to biomass measured after drying to a constant weight (for at least 24 h) at 80–105 °C in a forced air oven. Samples should be weighed straight out of the oven to avoid uptake of moisture from the atmosphere (dried plant tissues are very hygroscopic). Ideally, the plant materials should be placed back in the oven for a further 24 h and then weighed again. If the two weights agree then it can be taken as final; if not, drying for a further 24 h may be necessary. Air drying of samples does not remove all the moisture and sample biomass will be dependent upon local humidity. The *FW/DW ratio* is a characteristic of species when young and represents the extent to which a species invests photosynthate into each unit of leaf tissue (Hodgson and Booth 1986). The FW/DW ratio also allows for an estimation of total dry weight when subsamples have to be removed for other analyses prior to drying.

Non-destructive estimates of biomass are frequently used to avoid harvesting plants. Measurements of cover repetition or cover abundance involve recording the number of times that plants touch a thin pin placed or held vertically in the vegetation. Frames for holding the pins can be used where a systematic sample is required. The pins should be as fine as possible to avoid exaggerating the estimates of fine-leaved species (Goldsmith et al. 1986). Samples of 100 measurements per sample area are recommended. While useful as a repeatable method for estimating biomass, this method is extremely tedious, time consuming, and can be impractical in tall vegetation and under windy conditions.

Phytometric equations allow biomass to be estimated from repeatable, non-destructive measurements. These involve calculating the statistical relationship between biomass and some combination of other plant variables such as tiller number, leaf size, and stem height from a sample of plants that are not otherwise included in the experiment. For example, the stem dry mass (in grams) of the grass *Spartina alterniflora* before flowering was estimated as $0.147H \times D - 0.309$ ($r^2 = 0.94$, $P < 0.001$, $n = 49$), where H is the height (in cm) of the stem from the base to the top below the unexpanded leaves and D is stem basal diameter (in cm) (Dai and Wiegert 1996). Flowering stem dry mass was estimated using the same relationship after first decreasing plant size ($H \times D$) by 84%. These estimates allowed the authors to relate seasonal biomass production to environmental factors. Similarly, McEvoy et al. (1993) (Case Study 3 in Chapter 1) estimated the dry mass of food resources (foliage and capitula) available to cinnabar moths using regression equations incorporating stem basal diameter and leaf length of *Senecio jacobaea*.

5.4 Measurements of seeds and other dispersal units

The post-dispersal movement and fate of seed is an important component of a plant's life cycle. In this section, the methodology for estimating the magnitude and consequences of dispersal is described. All units of dispersal (strictly called diaspores), including fruit and vegetatively produced clonal fragments, are included. For the sake of simplicity, these reproductive propagules are generally referred to throughout as seeds.

Successful dispersal occurs through the movement of seeds to an environment where they may germinate (if seeds), establish roots, grow, and reproduce. Dispersal occurs in two phases: *primary dispersal* is the movement of a seed from the inflorescence on the parent plant to its first settling point (e.g. an acorn dropping off an oak tree and settling on the soil surface); *secondary dispersal* is any subsequent movement prior to germination (e.g. seeds being blown across the ground). Different vectors may act as either primary or secondary dispersers or as both. Potential vectors include animals (including humans, birds, various grazers, and insects) (i.e. *zoochory* with more specific terms for certain vectors such as *aviochory* for birds), wind (*anemochory*), and water (*hydrochory*). These terms are based on the suffix '-chory' derived from the Greek verb *chorein* (χωρεῖν) meaning to go. An understanding of the importance of seed dispersal for a species is likely to involve estimating the role of more than one vector. For example, Matlack (1989) showed that wind was the primary dispersal vector of birch seeds, but there was significant secondary dispersal across snow. An extensive review of various dispersal mechanisms is provided by Van der Pijl (1982).

Methods for monitoring seed movement can focus on either tracking dispersal or determining how many seeds arrive at a particular site (Table 5.5). Ultimately, the quantitative descriptor of seed dispersal is an estimate of the *seed dispersal kernel*—the probability density function of the dispersal distances relative to a source point (e.g. the mother plant) (reviewed by Nathan et al. 2012).

Seed dispersal

Measurements of seed dispersal are obtained through one or other of direct observation, seed markers, and seed traps. These approaches are described below. The data from these assessments can be used to plot seed dispersal curves and describe the seed shadow (see below).

Rates of fall and terminal velocity. The potential movement of anemochorous seed is assessed by determining the rate of fall or terminal velocity of seeds in a column of still air. A long, transparent plastic pipe held vertically and treated to remove static can be used. The time taken for the seed to drop a set distance is observed and measured with a stopwatch. The data from seeds that touch the sides of the tube are rejected. Several replications are needed, and the values obtained will depend not only on the type of seed (shape, size, mass, volume, presence of wings, plumes, and other dispersal aids) but also upon relative humidity. More accurate estimates, and dispersal rates for smaller seeds, can be measured by timing the rate of fall between two laser lights (Askew et al. 1997). True measures of terminal velocity depend upon determining when a dropped seed reaches a constant velocity and are best achieved using stroboscopic photography (Green 1980). For this reason, the data obtained from measurements made as described here by an observer with a stopwatch should use the term 'rate of fall' or 'rate of descent'. These data provide a useful general idea of seed fall, but under field conditions terminal velocity will be greatly

Table 5.5 Methods of tracking dispersal units or determining how many arrive at a site. See text for details. With kind permission from Springer Science+Business Media, after Bullock et al. (2006).

Method	Type of propagule/problems
Tracking dispersal units	
Following wind-dispersed propagules	
Abscission in the field	Large and easily visible, slow
Hand release in the open or in the field	Large and easily visible, slow, difficult to follow over large distances
Following animals carrying seeds, e.g. ants, vertebrates	Slow, observer may affect vector behaviour
Relocating marked populations	
Fluorescent powder, paint, radioisotopes, thread	May need to search large areas. Loss of marker can be a problem. Marker can interfere with dispersal
Locating germinants/seedlings	Any. May need to search large area
Tracking prologues on arrival	
Sticky traps	All, but not very large seed
Containers on or in the substrate	All. Containers attract or trap non-target organisms
Netting on a frame	Large. Mesh size issues
Netting or traps in water	All. Loss of netting or traps during flood events
Soil in seed trays	All (germinants are counted)
Artificial turf to trap seed containing sediment	Not large
Trapping sediment	All. Traps need anchoring to substrate

altered by the wind conditions. Very few seeds will be dispersed under the still air conditions obtained in a vertical tube.

Lateral movement. The lateral movement of seed can be determined by measuring the lateral distance moved following the dropping of seed in still air, but without the vertical tube. Lateral movement of wind-dispersed diaspores is most affected by biomass and morphological features of the wing or plume (e.g. wing area) (Matlack 1987). However, lateral movement of seeds in still air does not provide an indication of dispersal capacity since wind dispersal adaptations slow the rate of fall but may impart little lateral movement.

Experiments in wind tunnels can provide supplementary estimates of the probability of entrainment as a function of wind speed (Greene and Johnson 1997). Entrainment is important because the potential lateral movement afforded by plumes and wings may be of little relevance if the wind speed is not great enough to remove the seed from the parent plant in the first place. This concern is of less consequence for species that release seed directly into the atmosphere following dehiscence or decay of floral tissues.

Secondary seed dispersal. Assessing secondary seed dispersal in the field is an important corollary to the laboratory experiments described above. The usual approach is to place or release seeds on or close to the maternal parent in caches and track their movement. Single, naturally isolated individuals may be used, flowers may be removed from other plants, or a plant may be transplanted or moved in a pot to a site with no other conspecifics. Seeds can be visually followed from source to landing site, although this will usually be difficult and time consuming, if not impossible. Individual seeds can be more efficiently tracked by marking them in some way that allows them to be later identified and distinguished from other seeds not forming part of the experiment. The small size of most seeds can make marking individuals somewhat tricky, and care is needed to ensure that the marker itself does not impair or otherwise alter seed movement. Investigators have used a variety of approaches including spray painting (Carey and Watkinson 1993; Greene and Johnson 1997), spray painting with fluorescent paint and later searching with a UV lamp (Bossard 1990), attachment of marked tags, thread, or wire (Thomas et al. 2005; Xiao et al. 2006), radioactive tagging (Winn 1989; Kalisz et al. 1999; Pons and Pausas 2007), still photography and automated remote video recording (Jansen and den Ouden 2005), and the use of molecular markers (Ouborg et al. 1999). With all methods, recovery of marked seed and attraction or repulsion of seed predators or dispersers is a concern. Attachment of thread or tags can damage the seed, reducing seed survival and germination. It is generally assumed that the choice of methodology for tracking seeds does not influence the results. However, this may not be the case; for example, tagged seeds of European beech (*Fagus sylvatica*) were easier for investigators to relocate compared with seeds marked with UV-fluorescent powder, but the tagged seeds were also consumed by rodents at a much higher rate (Wróbel and Zwolak 2013). Recovery rates of marked seeds can be high. For example, Carey and Watkinson (1993) recovered 68% of painted diaspores falling from infructescences of the grass *Vulpia ciliata*, and (coincidentally) Bossard (1990) recovered 68% of the seed of ant-dispersed *Cytisus scoparius* within 56 h of placing marked seed out in depots. Radioactive labelling is a useful approach if a gamma-emitting radionuclide in solution is injected into parent plants before or while seeds are developing. The radionuclide is incorporated into the developing seeds and after the seeds disperse and germinate they can be located using a Geiger counter which detects the label (Winn 1989). Kalisz et al. (1999) recovered up to 76% of radionuclide-labelled *Trillium grandiflorum* seed. An advantage of radionuclide labelling is that a range of potential gamma-emitting labels are available which can be used singly or in combination to simultaneously label uniquely the seeds from several parent plants. Screening naturally established seedlings for the presence of molecular markers can provide valuable information on parentage and seed dispersal rates. However, this indirect technique is based upon a number of assumptions and requires considerable resources to identify and utilize a suitable marker. The use of molecular markers is discussed in more detail later in this chapter.

Seed rain

In addition to monitoring seed movement, measurements of the density of seeds comprising the *seed rain* may be important. The seed rain refers to the input of seed to the soil surface and can reflect primary dispersal from local plants and secondary dispersal from plants further away. Recovery of seed to allow expression of the seed rain on an area basis is achieved using a seed trap. Whatever sort of seed trap is used, the following attributes are important: (a) it must be possible and practical to separate seeds from litter, soil particles, and insects, (b) there must be protection against seed predation, and (c) dry preservation of seed material is necessary—the seeds must not rot or decay prior to examination (Kollmann and Goetze 1998). As with other methods of sampling, the spatial heterogeneity of the measured variable, seed rain in this case, is likely to be considerable under field conditions and must be considered in the establishment of an appropriate sampling regime. The following seven types of seed traps have been used (Kollmann and Goetze 1998):

1. Flat traps. Tissue or foil is spread out on the ground to catch seed. This method is very inaccurate since losses of seed due to rain, wind, or predation will be high. Nevertheless, Greene and Johnson (1997) used an appropriate modification of this approach. They dug circular holes of radius 1.0 m in snow, placed the excavated snow from the plot on a sheet, and counted the seeds of *Betula alleghaniensis*.
2. Soil traps. Pots or trays of sterilized soil are placed out in the field. The number of germinating seedlings is taken as a measure of the density of seed rain. Differences in germination rates, mortality of seeds and seedlings, seed predation, and secondary dispersal make this method unreliable.
3. Sticky traps. Boards, pans, or filter paper painted or sprayed with non-drying glue, for example Tanglefoot®, TangleTrap®, or grease, can be exposed horizontally or vertically. Seeds landing on the glue become tightly held and the glue-covered portion of the trap can be removed to the lab for seed identification and counting. This design is very efficient, especially for anemochorous

species, but may also collect a high density of insects, dust, and debris making the traps difficult and time-consuming to check. The glue-covered portion of the traps may need to be replaced frequently. This design is inefficient when the traps are frozen or covered with snow or overhanging vegetation.

4. Wet traps. Seeds are caught in basins or pans filled with water or kerosene. If water is used then a layer of paraffin oil may help reduce evaporative losses. Depending upon the design used, problems with this type of trap include losses due to water runoff after rainfall, predation by birds attracted to the water, build up of insects in the water, and rotting of the seeds in warm weather. The use of kerosene instead of water in these traps helps avoid freezing in cold conditions (Matlack 1989).
5. Bucket traps. These can be fixed to the ground with a nail, or recessed into the soil. A drainage hole is required at the bottom of the bucket. Placing clamps upside down around the rim of the buckets can prevent birds from perching. These traps are particularly suited for monitoring the seed rain of fleshy fruited zoochorous species.
6. Funnel-shaped or cylindrical dry traps. These have been widely used (e.g. Hughes et al. 1987; Schott 1994) and are regarded by Kollmann and Goetze (1998) as the most efficient type of seed trap. Care has to be taken in the design to ensure that seeds stay in the trap after being caught, especially in windy conditions. Wire gauze stretched over the top of the trap can be used to keep predators out. A collection vial or bag at the base of the funnel/cylinder allows samples to be collected at frequent intervals. Very small seeds may be difficult to detect and can be caught in coarse tissue or wood-wool inside the funnel. A sophisticated version of this, the 'Melbourne trap', was found to be more effective than bucket traps and sticky traps (Morris et al. 2011). The hydrochorus seed in aquatic systems can be captured using floating funnel traps which can stay on the surface of the water with rising and falling water levels (Middleton 1995). Densities obtained from aquatically dispersed seed in such a device must account for the rate of water movement which effectively increases the surface area sampled over the sampling period.
7. Natural traps. Samples of zoochorus seed can be obtained by collection and separation from animal faeces, hair, feathers, or fur. For example, endozoochory (dispersal of seed in animal guts) by frugivorous birds was estimated by counting the density of seeds in 'droppings traps' (nylon sheets) placed at random under trees (Izhaki et al. 1991). Similarly, endozoochory of exotic species by horses was assessed through counting seedlings emerging from samples of horse dung placed both in the field and maintained in the greenhouse (Campbell and Gibson 2001).

The seed shadow

Data from measurements of seed dispersal and the seed rain allow quantification of the movement of seeds from parent plants. The *seed shadow* is the post-dispersal spatial distribution of seeds around their maternal plant source (Willson 1992). Except for a number of epiphytic plants, seed shadows exist in two horizontal dimensions and are usually asymmetrical due to the wind (Fig. 5.7). A *seed dispersal curve*, describes the frequency distribution of dispersal distances, and is used to relate the number of seeds deposited at different distances from the parent plant. The *dispersal kernel* is the probability density function of the dispersal distance or the dispersal end point (Nathan et al. 2012). The shape of the dispersal curves is generally leptokurtic, right-skewed, and fat-tailed for anemophilous species with a high peak close to the parent plant and a long monotonically decreasing tail. The monotonic decrease often fits a negative exponential curve or a negative power function. These phenomenological models are converted to a straight line when plotted on a log–log or semilogarithmic scale, respectively, with the negative slope providing a measure of the rate of decline in seed density with distance. A more mechanistic approach is provided by fitting these data to a tilted Gaussian plume model in which dispersal is shown to be proportional to wind speed and to the height of release (Okubo and Levin 1989).

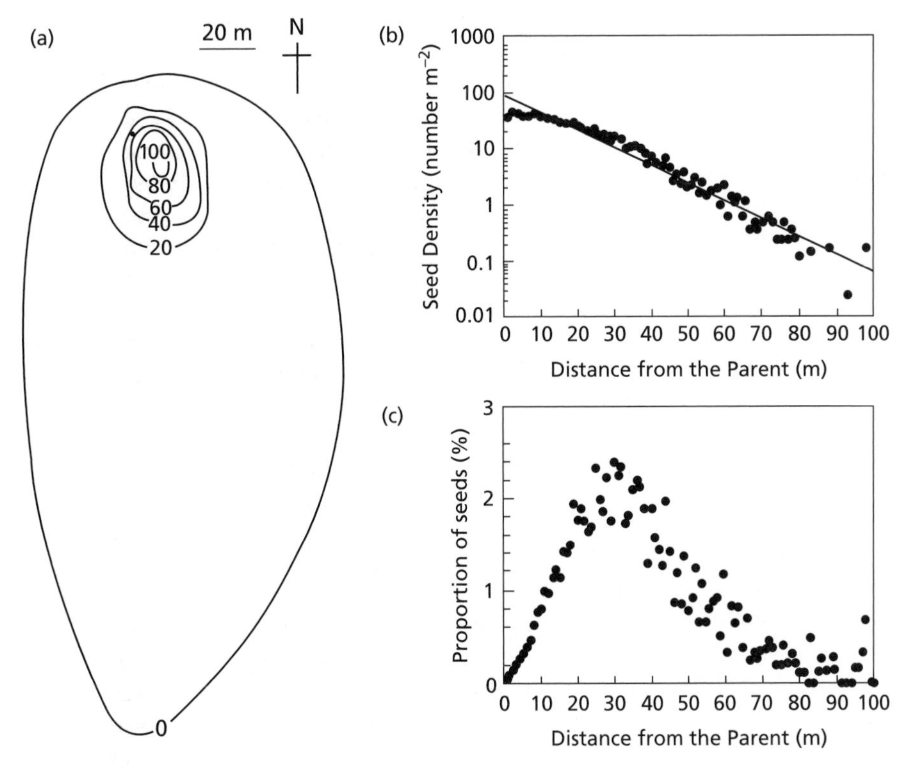

Figure 5.7 Seed shadow around a parent of the anemophilous canopy tree *Tachigalia versicolor*. (a) Isoclinic lines of density (number per m^2) of dispersed seeds, with the black dot showing the location of the parent tree. Winds are predominantly from the north. (b) Density of seeds at distance from the parent averaged over all compass directions. The line indicates the exponential fit: $y = 95.0\ e^{-0.073x}$. (c) Number of seeds at 1-m intervals from the parent plant, expressed as the proportion of the entire seed shadow. From Fenner and Kitajima (1999), originally adapted from Kitajima and Augspurger (1989) and Augspurger (unpublished). Reprinted with permission of Taylor and Francis Group LLC Books.

Many dispersal studies fail to measure adequately the tail of the seed distribution. To do so properly, the trap area needs to be increased so that the same proportional area is sampled at increasing distances from the parent plants (Bullock and Clarke 2000). Negative exponential or power models poorly describe the tail even when adequate measurements are made. An empirical mixed model based on the two functions was found to be a significant improvement:

$$S_N = T_A\left[a_3 \exp(-b_3 D) + c_3 D^{-p_3}\right]$$

where S_N and T_A are, respectively, the total number of seeds trapped and the total area of traps at distance D from the centre of the plant, a, b, and c are constants, and p is the inverse power parameter.

Dispersal distances should be reported as the range and mean of the measured dispersal distances (e.g. see data for several vectors in Hughes et al. 1994). Mean dispersal distances depend upon the seed biomass and morphology, dispersal vector, parent plant height, climatic conditions, and structure of the surrounding vegetation. The length of the tail of the fitted dispersal curve can be reported as the distance to which 84% (one standard deviation to the right of the mean trajectory) of the dispersed seeds fall (Greene and Johnson 1989).

The shape of the seed shadow and of seed dispersal curves for zoochorous species may be substantially different from that described above for anemochorus seed. The seed distribution of zoochorus species is often highly clumped. Frugivorous birds, for example, may deposit seed around perch sites distant from feeding sites; likewise, seed- or fruit-eating mammals may defaecate seed in latrines.

The seed bank

The reserve of persistent seeds in the soil is referred to as the *soil seed bank* and is assessed as the number of seeds in a given volume of soil (or under a given area of ground). In many ways these seeds represent the future of the plant community. There is, however, no connection between persistence in the soil and seed dormancy (Thompson 2000). The seeds in the seed bank may persist in the soil for years or decades, but often not in a dormant state. To fully understand the population ecology of a species it is important to determine the proportion of viable seeds stored in the seed bank. A review of the ecology of soil seed banks is provided in Leck et al. (1989).

When it is necessary to measure the *total seed bank* (i.e. the total number of seeds in the soil), an estimate can be obtained by counting the seeds extracted through sieving and floatation of seeds from the soil. For example, the dynamics of the seed population in the soil formed an important component of a study of the demography of three herbaceous *Ranunculus* species (Sarukhán 1974). In this study, seeds were recovered from soil samples through sieving and flotation. Sieving was done using a mesh size small enough to retain seeds. The seeds were then recovered from the material retained by the sieve that subsequently floated in water. The recovered seeds were characterized as: (a) dormant-enforced (*ED*; germinated under a fine mist in the lab), (b) dormant-induced (*ID*; tested as viable using the tetrazolium test described below, but would not germinate), or (c) dead (*D*; empty, non-living or failing the germination and viability tests). Coupled with the germinated fraction (*G*), these data allowed the total seed production (*S*) to be estimated as

$$S = \mathrm{G} + \mathrm{ED} + \mathrm{ID} + \mathrm{D}.$$

This is similar to a model used by Schafer and Chilcote (1969) to parameterize the persistence and depletion of seeds in the soil as

$$S = P_{ex} + P_{end} + D_g + D_n.$$

where S is the total number of seeds buried in the soil, P_{ex} are persistent seeds prevented from germinating by exogenous or environmental factors (e.g. darkness), P_{end} are persistent seeds prevented from germinating by endogenous factors (i.e. they are dormant and will not germinate even under ideal conditions), D_g are seeds germinating *in situ* and no longer part of the seed bank, and D_n are dead seeds.

Price and Reichman (1987) used a sieve with a 3.35-mm mesh and then floated the sievings in a saturated solution of K_2CO_3 (specific gravity 1.56, allowing more seeds to float than water). The floating organic fraction was then decanted through coarse filter paper or fine cloth. After drying for over 3 hours the seeds were counted under a dissecting microscope at 10× magnification.

In other studies, interest lies in measuring the *germinable* (or *viable*) *seed bank*. This constitutes the seed that will germinate given the appropriate environmental conditions. This seed population may be quite a bit less than the total seed population described above. The germinable seed bank is obtained by counting the number of seedlings which germinate from a soil sample after it has been spread out thinly on a seed tray and kept under conditions favourable for germination. A layer of vermiculite in the tray below the soil sample helps provide sufficient depth for seedling root development and allows the soil sample to be spread out thinly (1 cm or more). Few deeply buried seeds will germinate. It is also useful to reduce the volume of soil by washing it over a fine sieve and spreading the concentrated sievings as a 3–5 mm thick layer over sterilized potting compost (Ter Heerdt et al. 1996). This procedure promotes rapid germination and reduces the area of greenhouse required. In practice, 'conditions favourable for germination' usually means greenhouse conditions or a sheltered location outdoors, free from other dispersing seeds. The temperature and light conditions should include a cold stratification period, otherwise only seeds that are not dormant at the time of sampling are likely to germinate and the persistent seed bank (see below) may be underestimated. The environmental conditions should be recorded and reported. The soil in the trays should be kept moist and fertilized as necessary to improve seedling growth. Experimental treatments such as the inclusion of drought or flooding conditions may be useful (Middleton et al. 1991). Seedlings are removed and voucher specimens made when they can be identified growing up from the soil samples. Seedlings that grow too large for the trays before identification is possible should be transplanted into separate pots. Trays of sterilized soil should be interspersed among the sample trays to allow assessment of greenhouse contaminants. Species which appear in these control trays have to be judged with caution if they also emerge from the soil samples. After the initial flush of seedlings germinating from the soil has finished, then the soil should be stirred to stimulate further germination, especially of any deeply buried seed. Thompson (1993) recommended following seedling emergence for no more than 6 months and stirring up the soil only once; however, Baskin and Baskin (1998) suggested that such a short period of time would be likely to underestimate the persistent seed bank. Sieving and hand-sorting can be used to determine the presence of remaining seeds when no further germination is recorded. A thorough knowledge of the floristics of the field site will greatly aid in seedling identification, and it is common for many annual species, otherwise rare at the field site, to emerge from the seed bank samples.

The number of samples needed to adequately characterize the seed bank of a species should accommodate the large vertical and horizontal spatial heterogeneity of seeds in the soil. Most short-lived seeds are confined to the upper soil layers and it may be useful to separate the upper 5 cm of soil and the 5–10 cm layer. The ratio of seeds in the upper and lower layers provides information about seed longevity (Thompson et al. 1997). Thompson (1993) recommended a minimum sample size of 50 for reasonable accuracy in estimating the density of the common constituents of the seed bank. Fewer samples may be adequate to merely detect the presence of most species in a seed bank. A preliminary survey to construct a curve relating the number of species recovered to the soil volume sampled will help determine the necessary amount of soil required. This is analogous to a species–area curve and yields estimates in the range of 400–6000 cm^3 depending upon the vegetation type (Hutchings 1986). Many small samples composited together better allow spatial variability to be accounted for than do a few large samples.

The integrated screening programme (ISP; Hendry and Grime 1993) recognizes three types of seed bank (Thompson 1993; Thompson and Grime 1979; examples from Bakker et al. 1996; Fenner 1985):

1. Transient. Seed bank of species with seeds that persist in the soil for less than 1 year. This category can be subdivided into autumn (Type I) and spring (Type II) germinators. Type I species are often large-seeded grasses which germinate over a large range of conditions, for example *Festuca arundinacea* and *Lolium perenne*. Type II species often have seed with a requirement for chilling that imposes winter dormancy; for example *Heracleum sphondyllium* and *Mercurialis perennis*.
2. Short-term persistent. Species with seeds that persist in the seed bank for more than 1 but less than 5 years. This category can be subdivided into seed banks where a small (Type III) or large (Type IV) fraction of the annual seed production enters a persistent seed bank. Species with a short-term persistent seed bank often have seeds which are light demanding and germinate only within a narrow range of temperatures; for example, Type III species include the winter annuals *Poa annua* and *Arabidopsis thaliana*, and Type IV species include *Calluna vulgaris* and *Juncus effusus*.
3. Long-term persistent. Species with seeds that persist in the soil for more than 5 years. These species may be the most important for regeneration of the vegetation and include *Achillea millefolium* and *Trifolium repens*.

The timing of collection of soil samples affects the proportion of transient and persistent species in the seed bank. To obtain an estimate of only the persistent seed bank, samples should be collected before seed dispersal of most species or after the germination season is completed. Again, knowledge of the flora of a site is important. To obtain an estimate of the persistent seed bank of a single species, sample collection should be timed to occur after the germination season is over and before new seeds for the season are dispersed (Baskin and Baskin 1998). If the species of interest germinates throughout the growing season then sites of future soil collection can be marked and covered with a fine mesh to prevent new seed dispersal before collecting the soil at the end of the growing season.

Germination tests

Measurements of seed dormancy and germination requirements provide important data that may characterize a species, different populations within a species, or even seed from different types of flowers within a single individual. *Seed polymorphism* is the production of distinctly different types of seed on the same plant. Mean seed weight is usually invariant within and among populations. However, individual seed weights and seed polymorphisms can exhibit high levels of plasticity. An example is the seed from chasmogamous (open-flowered and cross-pollinated) and cleistogamous (closed-flowered and self-pollinated) flowers on the same individuals of the perennial *Lespedeza cuneata* which differ in size and viability (Donnelly and Patterson 1969). At the larger scale, numerous studies have shown differences in germination requirements of seed collected from geographically widespread populations (see the review in Baskin and Baskin 1998).

Germination requirements are measured using a germination test (Bradbeer 1988). The test should be carried out on samples of 100 seeds allowed to germinate for 28 days on moist filter paper placed in Petri dishes. An experimental protocol is provided in Table 5.6. Seed that does not germinate under the conditions listed under Points 5a–f may

Table 5.6 Experimental protocol for seed germination tests. From Bradbeer (1988) with kind permission from Kluwer Academic Publishers

1. Collect ripe seed
2. Identify, sort and clean seed
3. Air-dry the seed in the open laboratory for 2–3 days
4. Split the seed sample into three equal samples and pre-treat as follows:
   (a) Either test germination immediately as in (5) below or place the seed in a sealed container in the −20°C deep freeze for later testing
   (b) Store air dry seed in an unsealed container at laboratory temperature for 3 months
   (c) Chill at 5°C in moist sand for 3 months
5. Test the germination of each of samples 4a, 4b, and 4c under the following conditions:
   (a) 5°C unilluminated
   (b) 10°C unilluminated
   (c) 15°C unilluminated
   (d) 20°C unilluminated
   (e) 20°C under continuous illumination at 33 $\mu mol\ m^{-2}\ s^{-1}$
   (f) 20°C under total darkness
6. Wherever appropriate, carry out germination tests under the following treatments:
   (a) After removal from the fruit
   (b) With the naked embryo
   (c) After scarification[1]
   (d) During exposure to fluctuating conditions of light and temperature
   (e) During treatment with chemicals[2]
   (f) During treatment with gases[3]
   (g) After leaching
   (h) In response to illumination treatments

[1]Rupturing of the seed coat with a scalpel, needle or concentrated H_2SO_4.
[2]For example 10^{-4} M GA_3 (gibberellic acid), 10^{-4} M kinetin, 0.1 M thiourea, and 0.2% KNO_3.
[3]For example CO_2 and ethylene.

be under deep or enforced dormancy and the tests listed under Point 6 may be necessary. The results are expressed as the cumulative percentage of seed germinating (appearance of the radicle) at 24-h intervals. Statistical differences between treatments can be tested using chi-square or ANOVA on arcsin transformed data (Chapter 7; Bradbeer 1988). Seeds that germinate are considered as a portion of the total viable seeds. Viable seeds are those that are capable of producing normal seedlings, without reference to whether they actually will germinate or not. Viability can also be tested by placing or painting the bisected embryo from fully imbibed seeds in a 0.1% aqueous solution of 2,3,5-triphenyltetrazolium chloride, pH 7 (Grabe 1970). A 1.0% solution is recommended if uncut seeds are tested. After 24 hours in total darkness, living cells stained with tetrazolium develop a red/purple precipitate indicating that the seed was alive and germinable. Care has to be taken in the interpretation of this test because the spores of pathogenic microorganisms will also take up the stain and may give a false positive result. Also, as noted above, the viability test may overestimate potential germination. The value of this test, however, is that a large number of seeds can be readily screened for viability without having to worry about whether the seeds are dormant or not.

Detailed investigations to measure germination response to temperature may be carried out using thermogradient bars. These allow seed to be placed at specific points along a gradient of temperature. Seed germination is observed at each temperature. Plans for constructing a thermogradient bar are provided by Thompson and Fox (1971) and Larson (1971) with applications in Thompson (1975, 1977) and Danielson and Toole (1976). Detailed guidelines for experimental treatments involving temperature and the germination of seeds are provided in Baskin and Baskin (1998).

Seedling distributions: the population recruitment curve

The seed shadow and the distribution of recruited seedlings around parent plants (the population recruitment curve, PRC) are often discordant. The PRC is a plot of seedling density (y-axis) against distance (x-axis). Distance can be represented as a series of classes (e.g. Fig. 5.7) or with continuous units allowing the determination of the best-fitting regression models. Theoretically, seedling density should decline with distance in a simple exponential manner, but this does not always occur and has been the subject of much investigation (e.g. Augspurger 1983; Barot et al. 1999). The so-called Janzen–Connell escape hypothesis (Connell 1971; Janzen 1970) stated that PRCs in tropical forests would be shifted outwards relative to the seed shadow because of high density-dependent or distance-dependent predation or pathogen probabilities close to target trees (Fig. 5.8).

5.5 Physiological measurements

Plant physiological ecology (or plant ecophysiology) considers the relationship between the environment and plant processes related to survival, growth, and reproduction. Specifically, physiological ecologists seek to understand the importance and role of processes such as carbon acquisition, energy and water loss and gain, nutrient uptake, and the response of these processes to environmental factors. As noted by Harper (1986), we know a lot about the physiology of individual plants, or plant parts, outside their communities. Rather less is known about the physiology of the individual in the population. Physiological measurements are important because they may be the first, or most immediate, response of a plant to changing environmental conditions (beyond gene expression changes; see Section 5.6). By contrast, changes in whole plant morphological parameters may take days, weeks, or even years to be expressed.

There are several important considerations relevant to the measurement of ecophysiological parameters that plant population ecologists should bear in mind.

1. The goals of population ecology are to determine individual/population success and, in particular, fitness.
2. Physiological measurements generally show high levels of temporal (especially diurnal) variation and may provide only an instantaneous measure of individual plant performance.

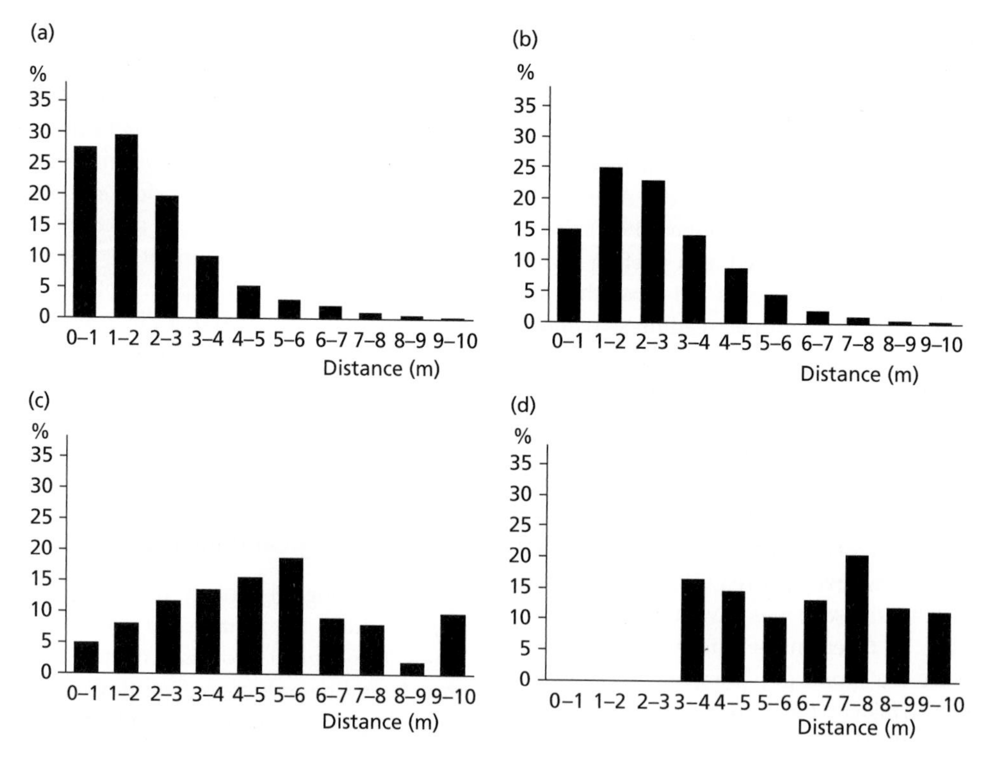

Figure 5.8 Population recruitment curves for the palm *Borassus aethiopum* shown as the distribution of (a) fruits and (b)–(d) seedlings in three successively older stages: (b) entire-leaved seedlings, (c) split-leaved, and (d) juveniles around female trees (Barot et al. 1999). Reproduced with permission from John Wiley & Sons.

3. Individual physiological measurements on their own may be poor predictors of plant performance, but can be important when summed or in combination with other factors.
4. Measurements made on a leaf area basis must be scaled to the entire leaf area of an individual, and take into account variations among leaves. Furthermore, measurements of belowground processes are also important determinants of carbon budgets. This includes not only root and rhizome respiration but also losses of carbon to symbionts such as mycorrhizal fungi and nitrogen-fixing prokaryotes.
5. There are many physiological measurements that *can* be made. Do not be tempted to take measurements simply because of the availability of a piece of sophisticated instrumentation.

It is sometimes implied that physiological measurements should be taken under ideal, laboratory conditions. This is misleading as it fails to consider the impact of the natural environment on plant function and performance. It is often important to make physiological measurements to understand, or predict, performance and fitness. Plant ecophysiological traits can be used to understand species interactions at the whole plant level and at community and ecosystem scales (Bazzaz 1996; Wardle et al. 1998). One of the most exciting current areas of ecophysiological research is the balance between genetic and environmental control of processes and the implications for plant fitness (Bazzaz and Stinson 1999).

The data obtained from physiological measurements may vary substantially depending upon the environment of the plant. Plants grown indoors under constant, uniform conditions, with few or no neighbours may yield very different data from plants growing in the field under the vagaries of natural conditions. Physiological measurements can also vary significantly depending upon the time of the day and the season. Gas exchange rates, for example, exhibit circadian rhythms and diurnal changes reflecting the daily pattern of sunlight. Hence, measurements taken on the same plant at 10 a.m. may be very different from those taken at 2 p.m. Physiological measurements also vary significantly from one part of the plant to another, and depend upon both the age of the plant and the age of the organ measured. Again, considering gas exchange, the highest rates are normally obtained from the youngest, most recently expanded leaves near the top of the canopy. By convention it is these leaves that are usually measured for gas exchange, but an understanding of the role of other leaves may have relevance for some situations.

Make only the measurements that you need. Physiological measurements often involve the use of expensive equipment that may take time to calibrate and learn to use. These instruments may measure several parameters in a short period of time. While there are numerous physiological measurements that *can* be made, it is important to identify early in your study, preferably in the planning stages, the critical measurements necessary. Ecophysiological measurements are rarely a surrogate for direct measurements of individual plant fitness (Section 5.8), and may indicate very little, if anything, about population growth. Physiological measures do provide an indication of the underlying biochemical and biophysical processes that determine and limit growth. Leaf-level measurements of photosynthesis, for example, have rarely been shown to be directly correlated with plant success, except perhaps among similar life forms within habitats (e.g. McAllister et al. 1998). The importance of ecophysiological measurements of carbon and water exchange is considered further below.

Four physiological processes often measured by population ecologists are energy, net carbon uptake, water relations, and mineral and nutrient uptake. In the rest of this section we focus on the basic methodology for making these measurements and their relationship to plant performance and fitness—the detailed methodology is described elsewhere (Pearcy et al. 1989; Reigosa 2001). There are many other physiological processes that are less frequently measured. General treatments of physiological ecology are provided in Bazzaz (1996), Prasad (1997), Larcher (2003), and Lambers et al. (2008) and for desert and tropical plants in Smith et al. (1997) and Lüttge (1997), respectively. The matching of molecular markers with ecophysiological measurements using quantitative trait loci is described in Section 5.6.

Energy

Energy provides the ability to do work. An understanding of energy transfer and flow among and within components of the plant and its environment is of fundamental importance (Ksenzhek and Volkov 1998). This reductionist view is a scaled-down contrast to the macroecology of ecosystem energetics in which interactions among trophic levels are studied.

The relevance of measuring plant energy costs and budgets in plant population ecology is in determining and assessing energy tradeoff for performance and fitness. For example, Harper et al. (1997) addressed the hypothesis that the energy and resources needed for plants to exhibit heavy metal tolerance are diverted away from other essential traits, resulting in reduced fitness. Freeman et al. (1993) measured the biomass, nitrogen, and energy content of male and female plants of *Atriplex canescens* and concluded that the genders were approximately equal on an energy and resource basis.

The energy budget of a plant may be viewed in a hierarchical manner related to the flow and use of energy within the plant. The main source of energy for plants is solar radiation, hence *quantum energy* refers to the energy exchanged between a plant and its environment by radiation, convection, conduction, and transpiration. In turn, radiation flux, air temperature, wind speed, and water vapour density describe the energetically relevant environment in the proximity of a plant (Gates 1980). By contrast, the *chemical energy* of a plant refers to physiologically based energy currencies, principally adenosine triphosphate (ATP) and NADPH. Both are produced in chloroplasts and mitochondria associated with the electron transfer pathways of photosynthesis and respiration, respectively, and are the two main classes of energy currency (Nobel 1999). Thus, chemical energy is derived from the harnessing of quantum energy. The *total energy* contained within the molecular bonds of a plant is referred to as the calorimetric energy content and is determined as the units of heat liberated when a unit weight is burnt in oxygen.

Determining the quantum energy budget of a leaf requires the measurement of several parameters related to radiation absorption and loss; that is, solar radiation absorbed, long-wave (infrared) radiation absorbed, long-wave radiation loss (chlorophyll fluorescence), latent heat loss, and sensible heat loss. Methodologies for making these measurements are described in Ehleringer (1989). Leaf temperature is the principal factor affected by changes in the leaf energy budget and a determinant of virtually all plant processes including enzyme catalysis, membrane transport, and transpiration. Plant fitness can be directly affected by the radiation and temperature environment of an individual (e.g. Noy-Meir and Briske 1996; Conner and Zangori 1997).

Measurements of chemical energy provide an indication of the short-term energetic status of the plant. For example, ATP was undetectable in air-dried *Allium cepa* cv. *Wadenswil* seed, but increased rapidly following imbibition, reflecting an increased metabolic rate (Siegenthaler and Douet-Orhant 1994). Levels of ATP extracted from plant tissues can be determined using the *luciferin–luciferase ATP assay*. The principle of this assay is that luciferase from the American firefly catalyses the reaction between ATP and D-luciferin in the presence of oxygen. The intensity of light emitted is proportional to the concentration of ATP and permits direct quantification of ATP (McElroy and DeLuca 1983). Despite the simplicity of this assay, measurements of chemical energy do not appear to have been widely used in plant population ecology studies. Instantaneous measurements of chemical energy may bear little or no relationship to plant performance and fitness because of the high turnover (synthesis and use) of ATP in tissues (Poorter and Bergkotte 1992).

Total energy budgets of plants, and other organisms, are given in units of watts per square metre (1 kg m^2 s^{-3} = 1 W), where a watt (W) is a measure of power describing the rate of working of one joule per second. Work energy is measured in joules (1 J = 1 kg m^2 s^{-2}) where one joule is the work done by the force of one newton moving an object one metre in the direction of the force or the work done or heat generated by a current of one ampere flowing for one second against a resistance of one ohm. The calorimetric energy content of biological material can be measured using a bomb calorimeter which determines the units of heat liberated when a unit weight is burnt in oxygen. The sample is ignited inside a thick, pressurized container (the bomb) containing oxygen. The increase in temperature is measured and compared with a standard such as benzoic acid (26 447 J g^{-1}) (Chapman 1986; Paine 1971). The term calorimetry refers to the old unit of the calorie (1 joule = 1 thermochemical calorie). The total energy of plant tissues grown under a wide range of environmental conditions has been determined using calorimetry and widely reported.

Caution has to be used in reading the literature on plant energy budgets because many researchers use biomass as a measure of total energy (e.g. Korpelainen 1992; Tang 1990). While this may be acceptable in a general sense, the limitations have to be acknowledged (see discussion on 'Reproductive effort' in Section 5.8). Total energy refers to the energy accumulated over the entire life cycle of the plant, or at least to the time of sampling. Measurements in dry weight terms instead of joules may provide a reasonable estimate when the main reserves are carbohydrates, for example in cereal grains. However, when oils constitute a significant fraction of the plant tissue then biomass may underestimate calorimetric measurements (Harper and Ogden 1970). Furthermore, it is perhaps worth bearing in mind that a plant uses only about 1% of incident (incoming) energy in building new biomass (roots, stems, branches, and leaves). The greatest share of the energy obtained is used to perform intrinsic metabolic functions such as water and nutrient uptake and transport (Ksenzhek and Volkov 1998).

Carbon uptake: photosynthesis

Comprising approximately 45% of a plant's dry mass, carbon is one of the most important elements in a plant. Green plants utilize light energy from the sun to generate ATP and NADPH in a series of light-dependent reactions to provide the energy necessary to fix atmospheric CO_2 into simple, energy-rich sugars in a coupled series of light-independent reactions. Understanding and measuring the rate of carbon uptake, that is, net photosynthesis, has become central to understanding plant growth (see Farrar (1999) for discussion of the importance of additional post-photosynthetic processes). Notwithstanding the importance of measuring single-leaf rates of photosynthesis, these rates are not well correlated with plant growth rate. Single-leaf and short-term measurements of photosynthesis are important, but very limited, measurements of a plants ecophysiological status. It is not appropriate to scale from single-leaf to canopy-level photosynthetic rates, or from short-term to long-term rates.

There is no direct technique for measuring photosynthesis. The most common integrative measure of photosynthesis is to determine the rate of CO_2 uptake at the leaf level. Portable gas exchange systems allow the uptake of CO_2 to be determined in relation to leaf and air temperature, humidity, and light intensity on intact plants growing in the field. Generally, a living leaf, or a whole shoot if it is small enough, is enclosed in a sealed chamber and the change in atmospheric CO_2 is measured over the course of a few seconds or minutes using an infrared gas analyser. If the system is 'closed', the CO_2 concentration in the recirculating air will decrease and the rate of decline is proportional to the photosynthetic rate of the leaf. If the system is 'open', then the CO_2 concentration of fresh air entering the chamber will be higher than that leaving the chamber. This difference in CO_2 concentration can be used to determine net photosynthetic rate as CO_2 uptake per unit leaf area (A_n, μmol CO_2 m^{-2} s^{-1}) (Field et al. 1989). Net photosynthesis is calculated as:

$$A_n = \frac{u_e c_e - u_o c_o}{L}$$

where u_e and u_o are the air flow entering and leaving the chamber, respectively, c_e and c_o are the CO_2 concentration entering and leaving the chamber, respectively, and L is leaf area. A clear description of the open and closed systems is provided by Tamayo et al. (2001).

Loss of water from plants or leaves (transpiration rate) and leaf stomatal conductance to water vapour diffusion are measures of plant water relations (see below) and can be obtained simultaneously. A_n is considered to be a direct measure of photosynthetic rate and increases with increasing irradiance until an asymptote is reached due to light saturation of the photosynthetic apparatus. Generally, net photosynthesis in C_4 plants saturates at higher light levels than in C_3 plants, but this can depend strongly on the developmental environment of the plant (e.g. Chen et al. 1995). Measurements of A_n can be made under ambient light; however, maximum A_n (A_{max}) for a plant is measured under saturating light conditions (bright sunlight or with supplemental lighting). Plants in the field often experience considerable variations in A_n as a result of fluctuating light levels, either through variable cloud cover or through the passage of sunflecks under the canopy of other plants (Knapp and Smith 1990; Zipperlen and Press 1997; Horton and Neufeld 1998). Indeed, A_n is directly sensitive to many environmental parameters including temperature, soil moisture, and atmospheric CO_2 (Lambers et al. 2008), and indirectly to factors such as crowding (Gibson and Skeel 1996). A_n is sensitive to atmospheric CO_2 concentration, increasing in C_3 plants in particular. Many reports, however, indicate that A_n acclimates or down-regulates under constant high levels of atmospheric CO_2 (Sage 1994).

Measurements of A_n require concurrent measurements of air and leaf temperature in addition to CO_2 concentration. As noted above, the physiological status of the plant greatly affects A_n as does the health, age, maturity, and position in the canopy of leaves being measured. Generally, the youngest fully expanded leaves on a plant have the highest A_n and these are the ones that should be measured. The instrumentation available to determine gas exchange rates readily makes these measurements.

Photosynthetic rates can also be estimated based upon oxygen evolution, uptake, and liberation using a 'Clark-type' silver–platinum oxygen electrode (fully described in González et al. 2001). For these measurements, the sample has to be in a closed chamber. For example, Gylle et al. (2013) used this method to compare photosynthetic rates among laboratory-grown samples of marine and brackish water ecotypes of the algae *Fucus vesiculosus* and *Fucus radicans* (Phaeophyceae). The use of the oxygen electrode is particularly useful for measuring photosynthetic rates in small-scale laboratory settings, even week-old seedling cultures of *Arabidopsis thaliana* seedlings (Benamar et al. 2013).

Chlorophyll fluorescence measurements allow quantification of the health (size and functional integrity) of the photosynthetic apparatus in intact leaves under field conditions. Measurement of chlorophyll fluorescence may be included as an option on portable photosynthesis systems or via separate instruments, and can be used to quantify plant stress in response to environmental factors. Specifically, the emission of radiation at wavelengths of approximately 680–760 nm following irradiation of chlorophyllous tissue with photosynthetically active radiation (PAR; 400–700 nm) or <400 nm originates mainly from chlorophyll on photosystem II (PSII; Krause and Weis 1991). Three fluorescence parameters are frequently measured using a fluorometer (descriptions in Reigosa and Weiss 2001; Weiss and Reigosa 2001). Minimal fluorescence (F_o) is the brief level of fluorescence emitted when a dark-adapted leaf is subjected to a strong white light (>5000 µmol m^{-2} s^{-1}, $\lambda > 680$ nm) or a weaker intermittent (modulated) light. Under saturating irradiance provided by a flash of actinic light, fluorescence rises to a maximum value (F_m) in about 200 ms. The difference between maximal and minimal fluorescence is the variable fluorescence ($F_v = F_m - F_O$). F_O reflects the size of the PSII chlorophyll antenna and the functional integrity of the PSII reaction centres. The F_v/F_m ratio, referred to as the fluorescence ratio in some studies (Smith and Moss 1998), measures the maximum photochemical yield of PSII (i.e. the efficiency of primary photochemistry). Under the Q_A model (Q_A referring to the primary quinone electron acceptor in PSII), the fluorescence ratio can be used to quantify photoinhibition as the ratio correlates with the number of functional PSII reaction centres. However, current views suggest that this interpretation of the fluorescence ratio may need some re-evaluation (Schansker et al. 2014) Fluorescence ratios from ~0.67 up to 0.83 are considered representative of unstressed plants (Castillo et al. 2000). Low fluorescence ratio values indicate photoinhibition or the activation of mechanisms for protecting the light-harvesting apparatus. For example, fluorescence ratio values of 0.54–0.65 in transplants of *Spartina densiflora* at low elevations in a salt marsh reflected low photochemical efficiency due to the decreasing redox potential of the sediment (Castillo et al. 2000). Fluorescence ratios were 0.589 ± 0.034 for seedlings of *Pilicourea riparia* growing in a tropical forest canopy gap compared with values of 0.635 ± 0.030 under the gap edge, indicating photoinhibition or light-induced damage to the photosynthetic apparatus (Fetcher et al. 1996).

Water relations

Water is critical for all physiological processes in plants and is required in large amounts. While 80–95% of the biomass of non-woody tissues may be water, turnover is high as less than 1% of the water taken up by plants is retained. Measurements of plant water status and flux are critical for an understanding of plant status.

Water potential (ψ_w, measured in MPa) describes the status of water in a plant and is a comparison of the potential energy of water in the plant with that of pure water (reviewed in González and Reigosa 2001). At standard temperature (298 K) and atmospheric pressure, pure, free water has, by definition, a water potential of 0 MPa and has the greatest potential energy. Thus, since all metabolic water is impure (i.e. has various solutes dissolved within it), values of water potential in a plant are lower. Increasingly negative water potential values indicate greater water stress. Halophytes and plants in arid climates can be adapted to water stress and may have a very low water potential (i.e. less than –2.5 MPa). Passive movement of water occurs from areas of more positive to more negative water potential; that is, under non-drought conditions from the soil into plant roots, from cell to cell within the plant, and from the plant cells through the stomata into the atmosphere. The steepest gradient in water potential occurs from just inside the leaf to the atmosphere, reflecting the very negative water potentials of the atmosphere and the major regulatory role of the stomata in governing plant water relations.

The two principal methods for measuring ψ in plants growing in the field are thermocouple psychrometric and pressure chamber techniques (González 2001; Koide et al. 1989). Psychrometric methods work on the principle that ψ in a plant sample can be related to the equilibrium relative humidity (h) in a psychrometric chamber held at constant temperature

$$\psi = \frac{RT\,\text{In}(h)}{V_W}$$

where R is the universal gas constant, T is the temperature in kelvin, and V_w is the partial molar volume of water. *In situ* continuous monitoring of ψ is possible using thermocouple psychrometers subject to three problems discussed in detail by Koide et al. (1989): (a) vapour pressure disequilibria, (b) thermal gradients and instability, and (c) changes in ψ due to growth of the sample.

The pressure chamber technique is the most widely used method for measuring ψ. Briefly, a freshly cut petiole or plant stem is placed into a pressure chamber with the cut surface protruding slightly out from a rubber sealing gasket. Pressure inside the chamber is gradually increased by adding nitrogen gas or air until a droplet of fluid starts to exude from the cut surface. At this point, the positive balance pressure in the chamber equals the negative apoplasmic (cell wall) value of water in the cells of the leaf, which in turn equals the symplasmic (vacuolar and cytoplasmic) value of ψ The main assumption is the equality of apoplasmic and symplasmic ψ, an acceptable assumption under most conditions (Koide et al. 1989). Precautions in using this technique include: (a) rapid transfer of excised leaves from plant to chamber often with leaves enclosed in a plastic sheath for some species; (b) not recutting excised petioles to avoid erroneously high ψ measurements; (c) minimizing the length of petiole protruding from the pressure chamber; (d) ensuring a tight-fitting rubber gasket around the petiole to avoid gas leaks from the chamber; and (e) slow rates of pressurization in the chamber to prevent ψ disequilibria.

Because of diurnal fluxes in temperature and soil water potential, ψ is conventionally measured both immediately before dawn (pre-dawn, when water stress is minimal and the plant is assumed to be in equilibrium with soil water potential) and at midday (when water stress is maximal).

Leaf stomatal conductance to water vapour (g_{lw}, mol H_2O m^{-2} s^{-1}) is a measure of the regulatory control that a plant has over the flux of gases through the leaf stomata. Stomatal conductance is the inverse of resistance, but the former is presented because it is proportional to flux and better expresses the degree of regulatory control. Total stomatal conductance to water vapour (g_{tw}) is measured using the same closed-chamber equipment used to determine photosynthetic rates in which changes in the gradient in water vapour concentration from the intercellular species to the atmosphere are compared. Leaf conductance (g_{lw}) is subsequently calculated from g_{tw} and estimates of boundary layer conductance depending upon whether the leaf has stomata on one or both sides (equations in Pearcy 1989a). Stomatal conductance to CO_2 (g_c) can be similarly calculated.

Through regulation of stomatal opening, photosynthesis becomes co-limited by CO_2 availability and light-driven electron transport. As shown in Fig. 5.9, the highest rates of A_n correspond to high stomatal conductance. Under dry conditions, for example, g_{lw} is likely to be low as stomata are closed to reduce water loss.

Transpiration of water vapour (E, mol H_2O m^{-2} s^{-1}) is a direct measure of water loss incurred by the plant and can be measured with the same instrumentation used to determine A_n and g_{lw}. Leaf energy balance and plant water status are primarily determined by E, which in turn determines water use efficiency (below). Leaf transpiration can be calculated as

$$E = \frac{u_e(w_o - w_e)}{L(1 - w_o)}$$

where w_e and w_o are the mole fractions of water vapour (mol mol^{-1}) entering and leaving the sample chamber, u_e is the total molar flow rate (mol s^{-1}) entering the chamber, and L (m^2) is the leaf area (Pearcy 1989a).

Water use efficiency (WUE) describes the relationship between carbon or biomass gain and water loss during photosynthesis. It is calculated as either (a) *water use efficiency of production*, the ratio between (aboveground) gain in biomass and water loss, or (2) *photosynthetic water-use efficiency*, the ratio between carbon gain in photosynthesis and transpirational water loss (A_n/E). (Photosynthetic WUE can also be calculated as the ratio of leaf conductance for CO_2 and water vapour, respectively (i.e. g_c/g_{lw})

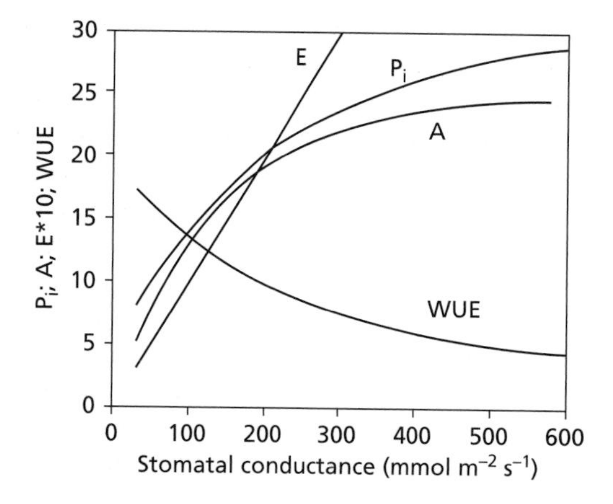

Figure 5.9 Relationship between stomatal conductance and transpiration rate (*E*), rate of CO_2 assimilation (*A*), intercellular CO_2 partial pressure (p_i), and water use efficiency (WUE). From Lambers et al. (2008) with kind permission of Springer Science+Business Media.

(Lambers et al. 2008) although the ratios of conductance are very seldom used and A_n/E is usually preferred.)

WUE is highly dependent upon stomatal conductance and decreases as g and A_n increase (Fig. 5.9). Among species, WUE differs with functional type, that is, WUE for C_4 plants is typically 4–12 mmol mol^{-1} whereas for herbaceous C_3 plants it is 2–5 mmol mol^{-1} (Lambers et al. 2008). Plants with crassulacean acid metabolism (CAM plants) have very high WUE (4–20 mmol mol^{-1}) because they minimize water loss during the day by keeping their stomata closed. Within populations, WUE is particularly sensitive to light levels, water stress, and soil nitrogen.

Relative water content (RWC; relative turgidity technique) is an indicator of the state of water balance of a plant, and is expressed as the percentage of plant tissue that is water (often a leaf, hence leaf RWC) relative to water content at full turgor (González and González-Vilar 2001):.

$$RWC = \frac{\text{fresh weight} - \text{dry weight}}{\text{saturated weight} - \text{dry weight}} \times 100\%.$$

Fresh weight should be obtained on tissue samples immediately after removal from a plant, or on samples that have been stored in a cooler in a sealed plastic bag. The saturated weight (= turgid fresh weight) is determined by placing the tissue into deionized water for a period of time at 4 °C until it reaches constant weight. The cool temperature while saturating the tissue is necessary to reduce physiological activity, including growth and respiration, that might otherwise introduce an error. The period of time necessary for water imbibition to obtain the saturated weight can vary among species, tissue age, and environmental conditions during development (generally 4 hours for herbaceous plants, 12–48 hours for conifers; Barrs and Weatherley 1962; Yamasaki and Dillenburg 1999). Dry weight is obtained following oven drying at 70 °C for at least 48 hours. RWC can be used to test the efficacy of water treatments on plant water status, and has been used to distinguish between species and among populations within species under different moisture conditions (Gratani et al. 2012). Fully turgid plant tissues can have a RWC approaching 98%, whereas when plants are severely desiccated RWC values may be only 30–40%. CRWC is the critical relative water content and is the RWC above which photosynthetic rates are insensitive to water stress. In a comparison of 10 species, CRWC was 81 to 90% of RWC. Below the CRWC, the net photosynthetic rate decreases linearly due to stomatal closure (Anderson and McNaughton 1973).

Leaf water content (LWC) is a similar measure of plant water status to RWC expressed instead as the percentage of water in fresh leaf tissue relative to tissue dry weight:

$$LWC = \frac{\text{leaf fresh weight} - \text{leaf dry weight}}{\text{leaf dry weight}} \times 100\%.$$

LWC can also be estimated instantaneously and non-destructively using remote sensing techniques based on canopy reflectance (Gente et al. 2013; Jin et al. 2013). Some investigators estimate LWC by expressing tissue water as a proportion of fresh leaf weight rather than dry weight (i.e. g H_2O/g fresh weight) (e.g. Mooney and Niesenbaum 2012). Regardless of whether LWC is expressed relative to fresh or dry weight, diurnal and/or seasonal changes in both dry and fresh weight can confound and affect the values of LWR that are obtained. RWC expressed relative to water content at full turgor overcomes these problems and is to be recommended where possible.

Nutrient uptake

There are 17 metabolically essential elements required by all plants: C, O, H, N, P, K, Ca, S, Mg, Mn, Mo, Fe, B, Cu, Zn, Ni, and Cl. With the exception of carbon and oxygen (obtained from the atmosphere) and hydrogen (as H_2O), these elements are absorbed from the soil solution. The importance of soil nutrients to plant growth and reproduction has been clearly established (Barker and Pilbeam 2007). The outcome of interactions among neighbours, for example, is frequently based upon competition for soil nutrients (Goldberg and Novoplansky 1997). Availability of these nutrients depends upon a suite of soil factors including soil pH, mineralogy, temperature, and moisture, all of which influence the solubility and speciation of the nutrient. Plants allocate resources to roots, stems, leaves, and other parts based in large part on the uptake of soil nutrients (Bazzaz and Grace 1997). It is often important, therefore, in plant population studies to measure rates of nutrient uptake and the concentration of nutrients in the soil and plant tissues (Chapter 6). Rorison and Robinson (1986) list six factors important in investigations of mineral nutrition:

- nutrient availability
- nutrient sources
- uniformity of nutrient supply
- effects of nutrients upon specific plant processes
- effects of specific environmental factors upon plants, in relation to their nutrition
- community–nutrient relations.

In this subsection discussion is focused upon the use of soil and plant nutrient data through the calculation of a number of nutrient and productivity indices.

Nutrient uptake can be determined through one of three procedures: direct harvest of plant tissues, radioisotopes, or non-radioactive tracers.

Nutrient levels obtained from sequential harvests of plant tissue provide an exact measure of nutrient acquisition. However, problems with this approach include: (a) removal of tissues (leaves usually) from individuals may affect subsequent plant growth, (b) nutrient levels must be corrected for losses to herbivory, reproduction, and senescence, (c) incomplete harvesting (e.g. just leaves) does not allow distinctions to be made between soil uptake and translocation, and (d) plants or tissues (i.e. experimental units) being compared must be within the same replicate block. Ideally, all plant parts, including roots, are harvested to avoid problem (a). However, this obviously may not always be possible, and would certainly not be desirable when marked individuals are being monitored. Detailed laboratory procedures for determining the nutrient concentration in plant tissues are available elsewhere (Allen et al. 1986; Chapin and Van Cleve 1989; Rorison et al. 1993).

Stable isotopes (e.g. ^{15}N), radioactive isotopes (e.g. ^{32}P), and non-radioactive tracers allow the movement of ions to be followed from the soil into the plant. Non-radioactive tracers involve the use of cation analogues of the essential nutrients such as Li, Sr, or Rb that are typically present in the soil and in plant tissues in small amounts (Veresgolou and Fitter 1984; Gibson 1988). In either case, the tracer is placed in the soil at a specific location and depth. The appearance of the tracer in the plant is later determined using conventional means (allowing a determination of rooting depth and uptake kinetics). Difficulties with this approach can include: (a) uniform labelling of the soil, (b) bioavailability of the labels to the plant, (c) equivalency of non-radioactive tracers to essential nutrients, (d) potential toxicity of non-radioactive tracers if used in high concentrations, (e) interaction of tracers with the soil, and (f) discrimination between nutrients and their analogues. Nevertheless, these approaches can allow a sensitive assessment of resource allocation patterns both from soils to plants and among plant parts. For example, Derner and Briske (1998) used ^{15}N pulses in the soil to assess the equitability of resource acquisition among ramets within clones of perennial grasses. Disproportionately greater amounts of the isotope in ramets closest to the label compared with more distant ramets established the potential for asymmetric intraclonal competition below ground. Similarly, Lötscher and Hay (1997) used ^{32}P to assess genotypic variation in physiological integration in *Trifolium repens*. Chapin and Van Cleve (1989) provide further discussion on the methodology of assessing nutrient uptake.

Nutrition productivity (NP or P; mg mol^{-1} day^{-1}) measures the efficiency of nutrient use to produce new biomass, and is the ratio of relative growth rate (RGR; Section 5.7) to whole *plant nutrient concentration* (PNC; mol g^{-1}, frequently measured in terms of nitrogen) in the plant tissue; that is, NP = RGR/PNC. NP is referred to as nitrogen productivity with the same acronym when tissue N is being considered. In this case, NP is a measure of the investment of nitrogen into plant growth and is defined as the increase in plant dry mass per unit time (dM_P/dT) and per unit plant N content (N_P) (Garnier et al. 1995):

$$NP = 1/N_P \times dM_P/dT.$$

Under conditions of optimal supply of N, high values of NP reflect rapid growth, efficient and high use of leaf N in photosynthesis, and a small use of N in respiration. Fast-growing species have a higher NP than slow-growing species (Poorter et al. 1990).

The PNC itself is a useful measure for specific nutrients and provides information on plant status. PNC varies widely among nutrients and concentrations in plant tissues depend upon the environment, allocation to woody and herbaceous tissues, developmental stage, and species (Chapin 1980; Lambers et al. 2008). Tissue concentrations of nutrients in plant ash range from < 0.1 μg g^{-1} for Ag, Co, and Li to $> 10\,000$ μg g^{-1} for macronutrients such as K, Ca, Na, Mg, and P (Lambers et al. 2008).

Nutrition-use efficiency (NUE; g g^{-1} N $year^{-1}$, usually characterized in terms of nitrogen) represents the time during which nutrients remain available in the plant to support growth. NUE = NP × MRT where MRT (in years) is the mean residence time of the nutrient in the plant reflecting the time that N is in the plant before being lost to senescence, herbivory, etc. Important differences in NUE, or its components, occur between plant functional groups. For example, similarity in NUE between evergreen and deciduous species was attributed to low NP and high MRT in the former and the opposite in the latter (Aerts 1990).

Photosynthetic nitrogen use efficiency [PNUE or Φ_N; μmol CO_2 (mol leaf N)$^{-1}$ s^{-1}] is the ratio between the rate of photosynthesis (A_n) and leaf N content (N_L) and provides an instantaneous measure of the efficiency of photosynthesis per unit leaf N. PNUE typically ranges between 60 and 130 mmol mol^{-1} s^{-1} (Lambers and Poorter 1992). Interspecific differences in PNUE can be attributed to SLA and organic N content per unit leaf area (Poorter and Evans 1998). Generally, values of PNUE are low for slow-growing species, although the reason for this is unclear and may be related to the allocation of N to the photosynthetic apparatus (Lambers and Poorter 1992; Poorter and Evans 1998). Low PNUE may also reflect a high WUE. There is, of course, great agricultural interest in improving NUE and PNUE among crop cultivars. Within natural plant populations, NUE and PNUE are useful measures of instantaneous nitrogen use. PNUE was significantly reduced in *Liquidambar styraciflua* saplings by root competition with vines, alone or in combination with aboveground competition, indicating a shift in allocation in response to competition (Dillenburg et al. 1995).

The value of ecophysiological measurements in population studies

As noted above there are many physiological measurements that can be made. Individually or together these can provide important information about plant status and competitive abilities. Numerous studies, for example, attest to the ecophysiological response of individuals to environmental factors. But, how useful are these measurements for addressing issues relating to plant population ecology? Several years ago Mooney (1976) noted that '... physiological responses of plants ... can be utilized in an integrative manner to predict potential success in habitats where resources are closely defined'. Physiological measurements provide the means to determine the immediate, and often real-time, response of an individual to its biotic or abiotic environment. For example, at a fairly coarse scale of investigation clearly distinct differences in physiological attributes have been demonstrated between early and late-successional plants (Table 5.7) and between fast-growing and slow-growing species (Lambers and Poorter 1992; Chapin et al. 1993).

Table 5.7 Physiological attributes of early and late-successional plants. From Bazzaz (1996), reprinted with permission of Cambridge University Press.

	Early successional	Late successional
Light saturation intensity	High	Low
Light compensation point	High	Low
Efficiency of carbon gain at low light	Low	High
Photosynthesis rates	High	Low
Respiration rates	High	Low
Transpiration rates	High	Low
Stomatal and mesophyll conductances	High	Low
Resistance to water transport	Low	High
Allocation flexibility	High	Low
Acclimation potential	High	Low
Resource acquisition rates	High	Low

Ecophysiological traits have also been found to correspond highly to the life-history characteristics of closely related species (Bazzaz 1996). For example, a continuum of variation related to A_{max}, g_s, and leaf mass per unit area was related to the life-history characteristics of nine pioneer *Macaranga* species in forests in Borneo (Davies 1998). Within species, the incorporation of ecophysiological measurements can improve the development of plant–plant interaction models. For example, Vargas-Mendoza and Fowler (1998) found that interspecific competition models for *Ratibida columnifera* were significantly improved by the inclusion of measures of plant water uptake and usage. On a comparative basis, ecophysiological traits can provide a useful conceptual link between individual leaf/whole plant measurements and ecosystem-scale processes, such as rates of plant litter decomposition (Wardle et al. 1998).

At the individual scale, ecophysiological measurements can provide insight into the mechanisms responsible for reproductive success and fitness. For example, Farris and Lechowicz (1990) used multiple regression analysis and a path diagram (Chapter 7) to investigate the contribution of individual morphological, physiological and phenological traits on fruit number and size in *Xanthium strumarium*. No single trait determined reproductive success, but of the physiological traits measured, pre-bud and seasonal WUE were shown to be biologically significant (Fig. 5.10). This analysis emphasized the integrative nature of the traits that we often measure and the danger in attempting to relate individual performance to a single factor. In a related study, variation in photosynthetic rate was found to affect fitness in *Amaranthus hybridus* by approximately 17% (Arntz et al. 1998, 2000). Photosynthetic rate indirectly contributed to fitness by affecting the linkage between plant size and reproduction.

Cautions in using ecophysiological traits for population level studies are: (a) instantaneous, one-time measurements are unlikely to provide a good relationship to plant performance, (b) scaling-up from individual-based measurements to higher

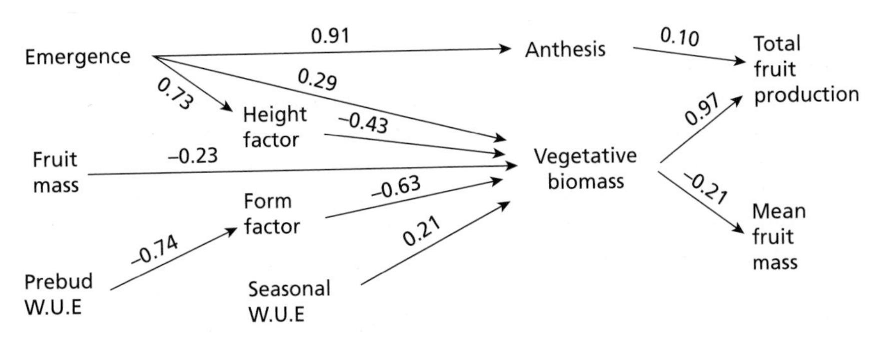

Figure 5.10 Path analysis of biologically significant character traits related to fitness in *Xanthium strumarium*. The numbers above each arrow are standardized partial regression coefficients (WUE, water use efficiency). From Farris and Lechowicz (1990). Reprinted with permission of the Ecological Society of America.

community or ecosystems can be problematic (Ehleringer and Field 1993), and (c) remember that correlation does not demonstrate causation—unmeasured environmental effects may be affecting both the physiological and the performance trait, and may do so via other unmeasured traits.

5.6 Molecular measurements

Ecologists are making increasing use of molecular markers to assist in understanding a number of important aspects of plant populations. Molecular markers allow aspects of the genetic composition of individuals within and among populations to be estimated. This estimation is important since the genetic composition of mature individuals may reflect selection on early stages by local microenvironments (Cruzan 1998). The most frequently addressed issues are related to within- and among-population variability (e.g. Knapp and Rice 1998; Gustafson et al. 1999). Comparative assessments of genetic diversity based upon molecular markers have allowed for general patterns in life history to be examined (Godt and Hamrick 1998). Additional topics include issues related to adaptive evolution (Anderson et al. 2013; Thurber et al. 2013), plant breeding systems (e.g. Loveless et al. 1998; Procaccini and Mazzalla 1998), clonality (e.g. McLellan et al. 1997b; Syde and Peakall 1998), population history (e.g. Dietz et al. 1999), gene flow, especially seed dispersal (Ouborg et al. 1999) and metapopulation dynamics (He et al. 2004), and genet identity in mixture studies (e.g. Hazard and Ghesquière 1995). Molecular methods are particularly useful for addressing issues of taxonomic definition, hybridization, and population structure in conservation biology (Haig 1998; Hamilton et al. 2013). In addressing these topics, molecular markers may also reveal subtle adaptive changes made, for example, in response to environmental adaptation or in alterations to metabolic pathways and processes. These changes may only be evident at the molecular scale, especially ecotype-specific changes to gene expression or coding sequences. Changes to gene expression (see 'Gene expression', below), may reveal adaptive strategies including interorganismal interactions that have a molecular basis. Gene expression precedes the appearance of a trait, and can be measured as an attempt by the plant to engage a strategy even if it is ultimately a failure (no gain in fitness).

Molecular markers are generally assumed to be neutral, i.e. not to change in response to selective pressure. Neutrality is thus an important assumption in the interpretation of results from allele population studies. Neutral markers make it possible to examine drift and gene flow without having to make estimates of selective pressures. On the other hand, it is often selection which drives the structure of populations, communities, and ecosystems. Thus, it is desirable to explore both neutral and trait-based markers at the molecular level at the same time in order to separate these processes. The neutrality of several molecular markers such as microsatellites has been called into question, and examples of non-neutral microsatellites have been shown which may give some genes/traits the ability to evolve rapidly. However, while non-neutral markers might represent rare exceptions rather than the rule, statistical methods have been developed to estimate selective pressure on microsatellites and should be used whenever possible (see Haasl and Payseur 2012).

A large number of molecular markers are available, and each has to be regarded as a tool with both limitations and strengths. These markers can be absolute and discrete (i.e. the presence or absence of a marker) rather than continuous when variables such as biomass, height, or other aspects of size are measured. The basic molecular approaches that are used are described here, the types of molecular markers available, and the use and limitations of each (summarized in Table 5.8). Detailed methodologies can be obtained elsewhere (Baker 2000) and there is an extensive review on the use of molecular markers in ecology in Freeland et al. (2011).

There are two measures of molecular variation that can be estimated (McLellan et al. 1997b):

1. Genotypic variation: the number and frequency of individual genets.
2. Genetic variation: the number and frequency of alleles, that is, the expected heterozygosity.

Table 5.8 Comparison of molecular methods most frequently used in plant population ecology (adapted from Newton 2007). Reproduced with permission from Oxford University Press

	Isozymes	RFLP	SSR (microsatellites)	RAPD	AFLP	SNP
Methods	Gel electrophoresis and histochemical staining of cellular enzymes and proteins	Total genomic DNA digested with restriction endonucleases then probed with specific DNA fragments by Southern blotting and hybridization	Specific PCR primers used to amplify previously characterized hypervariable repeat motifs in nuclear or organelle genomes	Short sequence primers (usually 10-mers) used to PCR amplify random loci throughout the entire genome	Total genomic DNA digested with two restriction enzymes, DNA adaptors fitted to cut sites, and products selectively amplified by using PCR primers	Dideoxy or HTP sequencing of DNA followed by sequence comparison
Type of marker (polymorphisms in...)	Molecular weight of proteins	Restriction sites in genomic DNA	Length of a repetitive region in genomic DNA[1].	Position of primer sites in genomic DNA	Restriction sites in genomic DNA (amplified)	Variation of a single nucleotide base in DNA
Advantages/disadvantages	Well-documented enzyme systems can provide unequivocal measures of allele frequencies. Fresh and often specific tissues required, e.g. buds, germinated seeds	Same probes/methods applicable to different taxa. Usually requires radiolabelled probes and large amounts of sample DNA	Unequivocal single-locus alleles can be scored. Microsatellite-containing regions differ between taxa, therefore expensive and laborious development required for each new species	Coding and non-coding DNA of potentially all three plant genomes, randomly analysed. Can give low reproducibility and artefactual markers. Genomic location unknown without controlled crosses	More reproducible than RAPD, but more expensive. Automation of marker scoring available. Radioactive labels may be required	Methods can be automated. Requires use of high throughput RNA-seq. Usually only two alleles per SNP (C or T)
Evolutionary window (polymorphism rate over time)	Intermediate	Intermediate	High	Intermediate	Intermediate	Low
Number of loci typically obtained	30–50	100s	10s	1000s	1000s	4
Degree of polymorphism	Low	Moderate	High	Moderate	Moderate	Extremely high
Dominance	Co-dominant (usually)	Co-dominant, occasionally dominant	Co-dominant	Co-dominant, occasionally dominant	Co-dominant, occasionally dominant	Co-dominant
Reliability (reproducibility)	High	Very high	High	Low to medium	Medium to high	High
Amount of sample required per assay	mg of tissue	2–10 μg DNA per lane	25–50 ng DNA	5–10 ng DNA	25 ng DNA	< 100 bp
Ease of assay	Easy	Difficult	Moderately difficult	Easy	Difficult	Moderately difficult
DNA sequence information needed	No	No	Yes	No	No	Yes
Cost of equipment	Low	Moderate/high	High to very high	Low/moderate	Moderate/high	High
Cost of development	Low	Moderate/high	Very high	Low/moderate	Moderate	High
Cost of assay	Low, although some reagents expensive	High	Moderate to high	Moderate	Moderate	Low

RFLP, restriction fragment length polymorphism; SSR, simple sequence repeat; RAPD, randomly amplified polymorphic DNA; AFLP, amplified fragment length polymorphism; SNP, single nucleotide polymorphism; PCR, polymerase chain reaction; HTP, high-throughput profiling; RNA-seq, RNA sequencing; bp, base pairs.

Molecular markers provide an indirect measure of the plant genome. Because of this there are two main assumptions (sources of sampling variation) inherent in these methods:

1. The genotypes assessed are a sample of the population.
2. The molecular markers examined are a sample of all potential markers in the genome.

There are two main types of molecular marker: protein markers (isozymes) and DNA markers. Isozymes have been extensively used for several decades, whilst the development of DNA markers is a rapidly moving area of research. Of the methods described below, microsatellites, single nucleotide polymorphisms (SNPs), and expressed sequence tag (EST)-simple sequence repeats (SSRs) are currently the most widely used. Rapid advances in sequencing technology are allowing the development of methods of gene expression, and functional metagenomics is a fast developing area (Orsini et al. 2013).

Protein-based markers

Isozymes and allozymes are protein-based markers representing different molecular forms of an enzyme that migrate different distances across a starch or poly-acrylamide gel in response to an electric field (i.e. the electrophoresis method) due to differences in weight or charge. Staining specificity makes it possible to distinguish particular enzymes in a tissue extract. If the polypeptide constituents of the enzyme are coded by more than one gene then it is referred to as an *isozyme*. Conversely, when coded for by a single gene, they are referred to as *allozymes*. Thus, different banding patterns between individuals are assumed to represent the presence of alternative, and neutral (or nearly neutral), alleles at a given locus. This formerly widely used method allows genetic variation within and among populations and species to be readily and cheaply compared. The major advantage of isozymes, and the reason why they are still used, is that the bands represent Mendelian loci. Their co-dominant expression means that estimates of genetic diversity and gene flow can be readily calculated (i.e. using *F*-statistics; Wright 1978). Assumptions of the method include: (a) differences are assumed to be genetically based and heritable, (b) homology, i.e. that bands migrating to the same position on a gel represent the same molecular form, and (c) selective neutrality of alleles (assumed despite some known exceptions). Isozymes can underestimate true genetic variation for three reasons: (a) only a very small part of the entire plant genome is being considered even though more than 100 loci can sometimes be screened, (b) parental enzymes may not be expressed, and hence not scored as present in the earliest stages of embryo development, and (c) in any given organism only a fraction of genes are expressed, and this is often dependent on the developmental stage and is strongly influenced by environmental factors such as temperature and other abiotic stressors. Moreover, within species, more than half of all loci may be monomorphic, and loci with more than three alleles are uncommon (Parker et al. 1998). Thus, isozymes cannot be used as molecular markers in some population ecology studies. Another disadvantage of isozyme/allozyme studies is the potential for false positives or false negatives, as genes coding for proteins may be differentially expressive and turned on or off because of environmental stress. In addition, within-species paralogues of a constituently expressed protein can occur, leading to two or more functionally different forms of the same protein.

Detailed methodology for conducting isozyme analysis can be found in Soltis and Soltis (1989). Isozyme and allozyme methods have been much less frequently used since the development of DNA-based markers, although applications are still published in the primary literature (e.g. Reyes-Zepeda et al. 2012).

DNA-based markers

DNA-based markers are based upon the premise that individuals can be discriminated through a comparison of regions of their genomes. If these DNA fragments differ, then the two genomes differ. This method does not rely on expression of the gene so that selectively neutral markers can be found so long as non-genic regions are targeted. Such an approach has an obvious advantage over the use of isozymes in which differences in the product of gene expression (enzymes) are measured. Higher levels of polymorphism may be identified using DNA-based markers compared with isozymes since variations in DNA do not necessarily change the amino acid composition or the charge of proteins (Dubreuil and Charcosset 1998). DNA does not change throughout the organism's lifespan, and could thus theoretically be harvested at any time. The most commonly used methods that are suitable for population ecologists compare fragments of nuclear DNA. However, comparisons of cytoplasmic DNA (chloroplast DNA and mitochondrial DNA) are also applicable. A detailed source of methodology and references is the PCR Jump Station web site at <http://www.horizonpress.com/pcr/>.

Restriction fragment length polymorphism (RFLP) *analysis* was the earliest DNA-based method. In this procedure, DNA is digested with restriction enzymes (typically four to six base pair cutters) and the resulting population of fragments is separated by gel electrophoresis and blotted onto a filter. Labelled probes are then hybridized to the bound DNA using the *Southern blot* procedure to allow discrimination of target fragments homologous to the probe. Fragments that have the same restriction sites (i.e. a similar DNA sequence variation) will migrate to the same location on the electrophoretic gel. Essentially, the genomic DNA from a mass of cells is cleaved into pieces and electrophoresis is used to see if the pieces are the same from one individual to the next. This method has the advantage that highly reproducible patterns are generated and heterozygotes are distinguishable because the markers are co-dominant. Disadvantages include: (a) a good supply of probes is needed and may need to be developed for each taxon, (c) the blotting and hybridization steps are time-consuming, and (c) quite large samples of DNA are needed, relative to other techniques (10 µg per digestion) to visualize the hybridization.

The *polymerase chain reaction* (PCR) allows the automated replication of exponential quantities of target DNA through successive cycles of amplification. Short oligonucleotide primers bind with single strands (templates) of denatured target DNA and heat-stable DNA-dependent DNA polymerase from thermophilic bacteria to catalyse the extension of the primers, using the DNA strand as a template. The primers are constructed to bind with known nucleotide sequences of the DNA template despite the unknown length and composition of the strand in between. Primers are constructed in pairs to target complementary DNA strand fragments. Thus, the amplification product is a large quantity of identical DNA fragments. The composition of these fragments depends, amongst other criteria, upon the composition of the primers used.

Modern molecular methods are based upon PCR [e.g. randomly amplified polymorphic DNA (RAPD), amplified fragment length polymorphism (AFLP), variable number of tandem repeats (VNTR), SNPs]. Like RFLP, PCR-based methods allow much clearer resolution of genetic differences than isozymes and allozymes (Cruzan 1998). As described below, PCR-based methods differ depending upon how the DNA is fragmented prior to amplification. An obvious advantage over RFLP analysis is that only a small amount of original DNA is needed for amplification (10 ng per reaction). Disadvantages include: (a) coverage of the genome is highly restricted, (b) few primer pairs for nuclear genes vary enough to allow detection of polymorphic differences below the species level, and (c) technical problems of contamination by DNA of other organisms (Karp et al. 1996). A number of the PCR-based methods described below have the limitation that they investigate variation in arbitrarily amplified DNA markers [AADs; including, for example, RAPD, AFLP, intersimple sequence repeat (ISSRs)]. Recent developments in gene-targeted and functional markers (GTMs and FMs, respectively) are outside the scope of this discussion, but are reviewed by Poczai et al (2013).

Randomly amplified polymorphic DNA. RAPD analysis is the simplest PCR-based method and works on the amplification of arbitrarily derived DNA segments. There is no requirement for a priori knowledge of the target sequences of the primers (hence their arbitrary nature or so-called randomness). PCR-amplified products are separated on agarose gels and visualized using an appropriate dye or fluorescent compound. Thus, the user does not know the base pair sequence of the sequences that are being amplified, only that co-migrating bands are assumed to represent identical genome segments for that particular primer pair. Loci can be clearly identified, but dominant expression of banding phenotypes is a problem (i.e. the markers are dominant and heterozygous individuals cannot be discriminated). Nevertheless, presence or absence of bands can be scored and the data converted into similarity matrices for the calculation of genetic distances. Other advantages of the method are that the procedure is quick, simple, automatable, and very small amounts of DNA can be used (10 ng per reaction). Despite the nature of the primers, correlations between RAPD markers and ecophysiological traits are possible for certain primers (McRoberts et al. 1999). Difficulties can include the following:

1. Reproducibility of banding profiles between different labs may be problematic due to amplification artefacts.
2. The presence of co-migrating non-homologous markers (most users assume that this is not a problem; it can be overcome with SCARs—sequence-characterized amplified regions).
3. Different sets of primers can provide different estimates of molecular diversity.

RAPD analysis is not now usually a method of choice.

Amplified fragment length polymorphism. AFLP analysis is essentially intermediate between RFLP and PCR. Genomic DNA is digested, followed by selective rounds of amplification of the cleaved fragments. AFLP is identical to RFLP in terms of its DNA target; however, the amplification process allows for much smaller quantities of DNA from each individual sampled. The amplified products are radioactively or fluorescently labelled before separation on sequencing gels. The actual analysis requires more DNA than RAPD (i.e. 1 μg per reaction), but provides a larger genome coverage than RAPD—usually 100 bands per gel compared with 20 per gel for RAPD. AFLP is considered a particularly efficient system because several bands can be revealed in a single amplification (Pejic et al. 1998). AFLP analysis is good for mapping and fingerprinting. Genetic distances can be calculated between genotypes. The limitations are similar to those of RAPD analysis regarding homologies and identities. AFLP banks may cluster on genetic maps around centromeres giving some selectivity to the portion of the genome analysed.

Variable number of tandem repeats. VNTR is also referred to as ISSR, SSR, EST-SSR, short tandem repeats (STRs), simple sequence length polymorphism (SSLP), microsatellites or minisatellites. VNTR is a hypervariable DNA-based marker that targets regions of the genome comprising tandemly repeated simple sequences (tandem arrays). Most of these are in the promoter regions of genes that control expression. The term microsatellite is applied when the basic repeat unit is around two to eight base pairs long; when it is longer the VNTR is referred to as a minisatellite. These sequence repeats are identified using specific primers. They are abundant in the eukaryotic genome and are highly polymorphic due to variability in the number of tandem repeats. Compared with RAPDs and RFLP/AFLPs, microsatellites have a much higher mutation rate due to their structure and origin via replication slippage. Length variant alleles are inherited in a Mendelian fashion and are considered to be selectively neutral. However, not all microsatellites are selectively neutral and some functional microsatellites can affect individual fitness. Hassl and Payseur (2012) provide simulation algorithms that can be used to test the neutrality of microsatellites and other VNTRs.

VNTR reveals high levels of allelic variation with co-dominant expression. This method can be used to distinguish varieties or even individuals, and reveal parentage identity. Of these approaches, the use of microsatellite markers is the most widespread in population ecology studies (e.g. Esselman et al. 1999; Reusch et al. 1999; Dixon et al. 2013) and along with the use of SNPs (see below) represent the most popular approach for identifying population-level molecular variation. The methodology for isolating microsatellites is discussed by Zane et al. (2002).

SNPs (or point mutations). This is a method for quantifying DNA sequence variation generated by restriction site-associated DNA sequencing (RAD-Seq; Davey et al. 2013) in which a single nucleotide (i.e. A, T, C, or G) differs between two DNA fragments at a location on a chromosome. When this occurs, the two differing DNA fragments are considered to be two alleles. SNPs generally occur in non-coding regions, but in coding regions they can affect gene expression and functional and/or structural changes in the encoded protein (i.e. in this case they are referred to as an expression SNP). The density of SNPs can expressed as frequency within a population. SNPs form the basis for DNA fingerprinting and, in humans, single base mutations that give rise to a SNP are associated with some diseases such as Alzheimer's. SNPs can be associated with known genes, allowing molecular variation associated with gene function to be quantified. For example Olson et al (2013) identified SNPs for 10 genes associated with bud flush or bud set in populations of the tree *Populus balsimifera*. Methodology is advancing rapidly to increase the efficiency of discovery and genotyping of large populations, particularly in the absence of a reference genotype (Baird et al. 2008; Lu et al. 2013) using approaches that are appropriate for molecular ecology of plant populations (Davey et al. 2013). For example, *genotyping-by-sequencing* (GBS) allows the discovery of SNPs and other genetic variants through next-generation sequencing of a subset of the genome (a reduced representation library) (Elshire et al. 2011) (see also 'RNA-seq', below). An advantage of GBS is that large numbers of individuals from species with large genomes or with little previously known genomic information can be screened (Narum et al. 2013). GBS was used to assess genetic diversity among populations of the highly heterogeneous polyploid perennial grass *Panicum virgatum* (Lu et al. 2013). In this study, GBS allowed discovery of 1.2 million putative SNPs.

Cytoplasmic DNA analysis. This is based on the chloroplast (cpDNA) or mitochondrial (mtDNA) genomes and allows uniparental (often maternal) modes of inheritance to be examined, including the contribution of seed dispersal to gene flow. Chloroplast DNA markers can be developed by conducting restriction enzyme digests of whole-organelle genomes, or by amplifying specific regions of organelle DNA using PCR and then using restriction enzymes to detect sequence variation. Low levels of intraspecific sequence variation in chloroplast genomes allow only limited use of cpDNA markers for the study of population processes. Chloroplast DNA analysis is best suited for assessing variation across large geographic areas (Cruzan 1998). Whereas cpDNA is highly abundant in leaves, mtDNA is less abundant and difficult to extract. Variation at the intraspecific and population levels can be detected using mtDNA analysis because of the high rates of structural rearrangements. A good example of the use of this method was the application of cpDNA to identify divergence of ancient lineages of a tropical dipterocarp species, *Shorea leprosula*, in conjunction with expressed sequence tag-based simple sequence repeats to distinguish geographically distant current populations (Ohtani et al. 2013).

Nuclear ribosomal DNA (rDNA). Nuclear ribosomal DNA analysis is widely used for diversity studies at the intergeneric level, but it is likely to be difficult to detect sequence variation below the species level (Karp et al. 1996).

Quantitative trait locus (QTL) analysis

Through use of one or more of the molecular methods outlined above, plant population ecologists are asking an increasingly wide range of fundamental questions. A promising area of development is the assessment of ecophysiological trait relationships using *quantitative trait locus* (QTL) analysis (Bachmann 1994; McRoberts et al. 1999). QTL analysis is a widely used genetic methodology that allows co-segregation between DNA markers and traits to be established in controlled crosses. This common technique is used to produce genetic linkage maps in cultivated plants; for example to detect genes for pathogen resistance, fruit types, or drought tolerance (Grandillo et al. 1999; Heusden et al. 1999). In plant breeding studies, QTLs allow for marker-assisted selection in breeding trials, including the identification of traits from native relatives of cultivated crops. The approach is to screen individuals in a population for a number of molecular markers (e.g. microsatellites) and ecophysiological traits. SNPs are the preferred molecular markers for developing QTLs as they are co-dominant markers allowing homozygotes and heterozygotes to be distinguished. Standard statistical methods (ANOVA, regression; see Chapter 7 and Knapp 1994) along with graphical plots [receiver operating characteristic (ROC) curves or 'log of the odds ratio' (LOD) scores; Young 1994; Murtaugh 1996] are used to illustrate the co-occurrence of the marker and trait when the value of the latter is greater than a chosen threshold. The general approach is reviewed in Paterson (1996). The R package qtl is available for mapping QTLs including the production of LOD plots (Broman et al. 2003).

In plant population ecology, QTL analysis allows the presence of genes for specific traits to be identified (using specific flanking molecular markers) in individuals within a population. For example, individuals possessing a QTL for early flowering (which might increase individual fitness) can be identified. Difficulties in establishing parent–offspring relationships in field populations limit the use of QTL analysis for population ecologists, but there is great potential for increased utility. Two examples illustrate the applicability to plant population ecology:

1. Recognition of convergent evolution for weediness: QTLs underlying weediness traits (heading date, plant height, growth rate, seed shattering, hull colour, and awns) were identified using 188 SSR, 5 SNP, and 3 insertion–deletion markers in two distinct F_2 mapping populations of rice (*Oryza sativa*) created from crossing two individuals with an *indica* cultivar individual (Thurber et al. 2013). However, although QTLs were identified on 9 of the 12 rice chromosomes, most did not overlap in location, suggesting different genetic mechanisms for adaption to the environment for the two populations.
2. QTLs for flowering in *Arabidopsis thaliana*: Mitchell-Olds (1996) used RFLP markers and 96 recombinant inbred lines from a cross between two *A. thaliana* ecotypes to identify QTLs influencing components of fitness. Specifically, QTLs near two markers had large and measurable effects on flowering date and flowering weight, but not growth rate (Fig. 5.11). These QTLs are thought to represent flowering genes present on chromosomes I and II, respectively. Overall, this study allowed identification of the early flowering and larger growing individuals in a population that will be favoured by natural selection. In an annual plant such as *A. thaliana* such traits are advantageous in environments subject to disturbance. A similar study used SSRs to identify three QTLs for flowering without vernalization for winter annual populations of *A. thaliana* (Grillo et al. 2013).

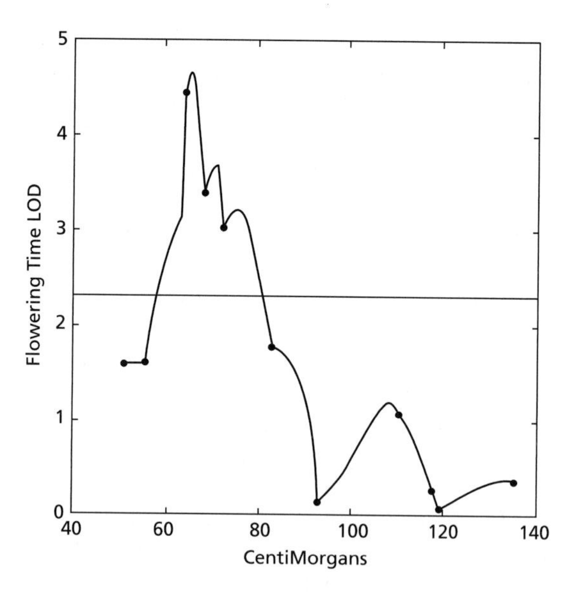

Figure 5.11 Log of the odds ratio (LOD) plot of flowering-time quantitative trait locus (QTL) on chromosome I of *Arabidopsis thaliana*. The uneven line shows the LOD score, which tests for odds of a flowering-time QTL at each point along the chromosome. Locations of restriction fragment length polymorphism (RFLP) markers are indicated by filled dots. The horizontal line indicates a 5% significance threshold. From Mitchell-Olds (1996). Reproduced with permission from John Wiley & Sons.

Gene expression profiling

Gene expression profiling allows quantification of the extent to which genes are being expressed at the time of tissue sampling based on isolated messenger RNA (mRNA). The abundance of mRNA of a specific gene is thus a continuous, quantitative trait. While probing the expression of individual target genes can be accomplished to a high degree of resolution using electrophoresis-based methods such as quantitative PCR, modern technologies can simultaneously quantify the expression of large numbers of genes. There are many applications of gene expression profiling, and for plant population ecologists this approach means that questions about within and among population-level genetic variability, relationships of phenotypic and genotypic variability with respect to environmental conditions, and speciation can be addressed. There are two approaches to gene expression profiling; microarrays and RNA-seq.

Microarrays. Microarrays simultaneously monitor the expression levels of thousands of genes. mRNA is extracted from a cell line or tissue (a total of 25–100 μg RNA is required per sample, although this can be down to 1 μg RNA following amplification). The mRNA is converted to fluorescently labelled complementary DNA (cDNA) which is used to generate a labelled sample ('the probe') that is hybridized in parallel to cDNA or oligonucleotide microarrays arranged on a glass slide or nylon membrane as 100–300 μm spots spaced at defined locations. The hybridized samples are detected by phosphoimaging or fluorescence scanning. Use of two fluorescent dyes (the fluorophores Cy5, Cy3) allows mRNA from two cell lines to be mixed and hybridized per microarray. Competitive binding of these mRNAs affects the relative intensity of gene expression to be determined for each gene on the array. Microarray studies generate gigabytes of raw data as expression levels for all genes in the genome are generated simultaneously. Microarray can be oligo-based if the whole genome has been sequenced, as it has for about 20 plant species (e.g. *A. thaliana*, *Helianthus annuus*, *Populus tremuloides*, *Zea mays*) allowing construction of a microarray chip. cDNA-based microarrays can be generated for species without sequenced genomes following cross-species hybridization against oligo-based microarrays of sequenced species with the assumption of comparable gene function between species. For example, variation in gene expression of the prairie grass *Andropogon gerardii* was determined through hybridization of cDNA onto *Z. mays*-spotted microarrays (Travers et al. 2010). In plant population ecology, microarray studies have provided insight into differential gene expression in intraspecific competitive interactions in laboratory-based settings (Geisler et al. 2012; see Fig. 5.12) and in response to variation in environmental conditions in the field (Travers et al. 2010). Having identified candidate genes involved in particular species interactions, methods such as quantitative PCR (qPCR) can be used to more closely target the expression of specific genes. The recent development of RNA-seq (see below) has meant that microarray studies are already becoming less frequent. At the time of writing, the cost of RNA-seq was equivalent to that of microarrays, and their greater sensitivity and accuracy may severely limit microarray utility.

RNA-seq (whole transcriptome shotgun sequencing, massively parallel cDNA). This is a high-throughput sequencing [next-generation DNA sequencing (NGS); high-throughput profiling (HTP)] technology used to generate millions of short reads from an organism's RNA providing transcriptome sequence information and a digital measure of gene expression (Ozsolak and Milos 2010). RNA-seq is more accurate than microarrays, allows a broader range of expression levels, and is not restricted to model organisms or closely related species; it thus provides a powerful, accurate alternative to microarrays. As with microarrays, differential gene expression among phenotypes or between experimental conditions can be quantified using RNA-seq. SNPs of particular interest associated with specific biological processes can be sequenced and expression profiled using this approach (although SNPs can be analysed with just normal (dideoxy) sequencing not HTP). An example of DNA-seq is the study of the progenitor diploid genomes of *Tragopogon dubius* and *Tragopogon pratensis*, and the allopolyploid derivative species *Tragopogon miscellus* (Asteraceae) by Buggs and colleagues (Buggs et al. 2010, 2012). In this comparison, 22% of 2989 SNPs in a single *T. miscellus* leaf showed apparent differential expression among homologous genes derived from both progenitor parents.

The promise of DNA-seq and other applications of NGS are just starting to be realized with applications in phylogenetics, genotype 'barcoding', polyploid genetics, and accession tracking in gene bank collections (Egan et al. 2012), such as GenBank (<http://www.ncbi.nlm.nih.gov/genbank/>) the genetic sequence database. Beyond DNA-seq, as advances in NGS continue through new technologies (e.g. Illumina, Ion Torrent, sequencing by ligation), NGS is facilitating expansion of the fields of ecological genomics and epigenetics (Ekbolm and Galindo 2011). Some of the challenges associated with obtaining and dealing with the massive numbers of data that can be generated are summarized in the somewhat satirical but sensible 'A field guide to genomics research' by Bild et al. (2014).

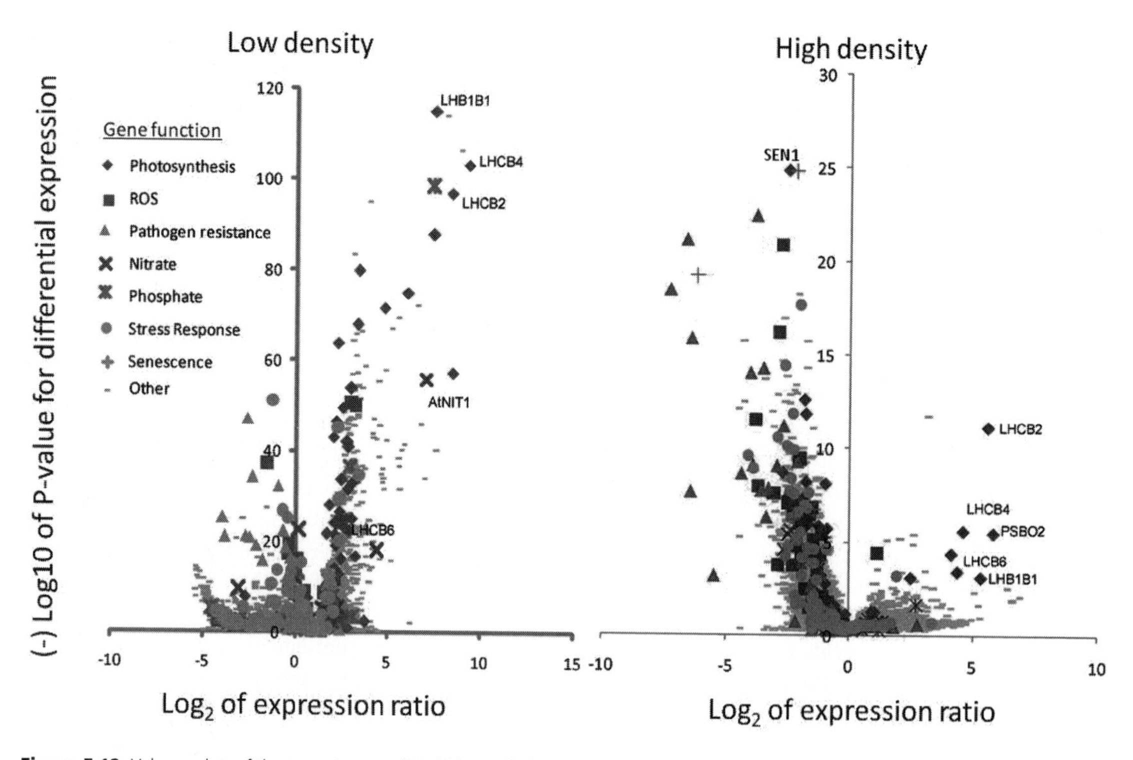

Figure 5.12 Volcano plots of the transcriptome of *Arabidopsis thaliana* plants showing gene expression from microarray analysis (from Geisler et al. 2012) of low (9 plants per pot) and high (100 plants per pot) intraspecific competition. Each point represents a single gene which is suppressed (left of *y*-axis) or induced (right of *y*-axis). Genes involved in known biological processes are indicated in the legend. The expression ratio (*x*-axis) was calculated as the expression value in competing plants over that in isolated plants (one plant per pot). The *P*-value for differential expression significance was calculated from a one tailed *t*-test of the pixel brightness for 80 pixels scored on each microarray corresponding to that gene. Key genes discussed in Geisler et al (2012) are LHC (light harvesting complex), PSB (photosystem subunit), AtNIT (nitrilase), SEN1 (senescence associated gene 1). Reproduced with kind permission from Springer Science+Business Media B.V.

Epigenetics

Epigenetics refers to meiotically or mitotically heritable trait variation that is not due to changes in DNA sequencing but rather is caused by other genetic modifications, especially DNA methylation and post-translational modifications (Ekbolm and Galindo 2011). Occurring independently of genetic variation, these epigenetic marks ('epigenes') affect the regulation of gene expression, and hence phenotypic plasticity. Most importantly for plant population ecologists, epigenetic variation can be environmentally induced and may vary among populations or among genotypes within populations. Epigenetic variation is a source of natural variation in ecologically relevant traits among individuals that may be subject to selection. Advances in NGS technologies are opening up opportunities for large-scale epigenetic surveys, allowing comparisons of epigenetic variation within and among populations, and their relationships to phenotypic variation and ecological interactions (Bossdorf et al. 2008; Richards et al. 2010; Bräutigam et al. 2013). A wide range of molecular techniques are used depending upon the epigenetic phenomena being investigated. For example, methylation-sensitive AFLP markers (MS-AFLP) were used to screen genome-wide methylation alterations triggered by herbivore and pathogen stress treatments and to assess the heritability of induced changes in genetically identical apomictic dandelion (*Taraxacum officinale*) plants (Verhoeven et al. 2009). A similar approach using MS-AFLPs was used by Herrera and Bazaga (2011) to show that epigenetic variation was involved in population differentiation and related to herbivore tolerance in *Viola cazorlensis*.

Functional metagenomics

The relatively new field of functional metagenomics allows functional-trait-based quantification of microbial community diversity (Orsini et al. 2013). High-throughout NGS can be used to determine genomic diversity in a sample (e.g. soil, root rhizosphere, leaf surface pathogens). DNA-seq gene expression profiling is subsequently used to characterize microbial diversity based upon genes of known function. For example, expression of genes for catabolic and anabolic functions (e.g. DNA metabolism) can be profiled or genes coding for specific pathways (e.g. photosynthetic pathway enzymes). Functional metagenomic diversity of the microbial community can then be compared with functional traits of the plant community (Powell et al. 2013). Using this approach, Zancarini et al. (2013) found that ecophysiological profiles of different genotypes of the legume *Medicago truncatula* were highly related to the functional diversity of associated rhizosphere bacterial communities. The functional metagenomic approach allows characterization of niche exclusion among plant species, and genotypes within species, on the basis of functional gene expression of associated microbial communities. The extent to which functional genomics can be related to plant population dynamics is poorly understood and relatively unexplored.

5.7 Indices for expressing plant growth and plant interactions

The growth of plants or whole populations is a vital rate (see the Preamble). There are several commonly used numerical metrics that allow plant growth to be expressed so that its relationship to the external environment can be explored. Indices for expressing plant growth described here include: relative growth rate (RGR), unit leaf rate (ULR), leaf area ratio (LAR), specific leaf area (SLA), leaf mass ratio (LMR), leaf area index (LAI), and tiller appearance rate (TAR) (Table 5.9). Full details of the derivation and relationships to each other of RGR, ULR, LAR, SLA, and LMR can be found in Evans (1972), Salisbury (1996), and Poorter and Garnier (2007).

Table 5.9 Indices for expressing plant growth. See text for nomenclature

Index	Equation	Typical range of values (Poorter and Garnier 2007) except where stated
Relative growth rate	$\mathrm{RGR} = \dfrac{\log_e M_2 - \log_e M_1}{t_2 - t_1}$	40–400 g increase kg^{-1} plant day^{-1}
Unit leaf rate assuming linear increase	$\mathrm{ULR} = \dfrac{M_2 - M_1}{[(L_{1A} + L_{2A})/2](t_2 - t_1)}$	2–20 g increase m^{-2} leaf day^{-1}
Unit leaf rate assuming exponential increase	$\mathrm{ULR} = \dfrac{M_2 - M_1}{t_2 - t_1} \dfrac{\log_e L_{2A} - \log_e L_{1A}}{L_{2A} - L_{1A}}$	
Leaf area ratio	$\mathrm{LAR} = \dfrac{L_{2A} - L_{1A}}{M_2 - M_1} \dfrac{\log_e M_2 - \log_e M_1}{\log_e L_{2A} - \log_e L_{1A}}$	5–50 m^2 leaf kg^{-1} plant
Specific leaf area	$\mathrm{SLA} = [(L_{1A}/L_{1M}) + (L_{2A}/L_{2M})]/2$	8–80 m^2 leaf kg^{-1} leaf
Leaf mass ratio	$\mathrm{LMR} = [(L_{1M}/M_1) + (L_{2M}/M_2)]/2$	0.15–0.60 g leaf g^{-1} plant
Net tiller appearance rate	$\mathrm{NTAR} = \dfrac{N_2 - N_1}{t_2 - t_1}$	−2.5 to 1.3 tillers day^{-1} for *Arrhenatherum elatius* (Gibson 2009)
Proportional net tiller appearance rate	$\mathrm{PTAR} = \dfrac{1}{N_1} \dfrac{N_2 - N_1}{t_2 - t_1}$	−1.5 to 3.3 tillers day^{-1} for *Arrhenatherum elatius* (Gibson 2009)

Eight indices for assessing plant interactions are described: i.e. relative yield (RY), relative yield total (RYT), relative yield of mixture, relative resource total (RTT), change in contribution (CC), competitive importance (C_{imp}), and relative efficiency index (Table 5.10). There are over a 50 indices of competition that have been proposed, and recent critiques include Williams and McCarthy (2001), Weigelt and Jolliffe (2003), and Bedoussac and Justes (2011). The full debate on how and what to measure to assess competitive interactions among plants is unresolved, and beyond the scope of this book; see Aarssen and Keogh (2002) and Trinder et al. (2013) for some background. There is an ongoing discussion about how to define and measure competitive importance as a process in habitats; see papers and references therein by Kikvidze et al. (2011) and Rees et al. (2012). As noted in Chapter 2, caution has to be exercised in using any of these derived variables.

Expressing plant growth

Relative growth rate. RGR [or specific absorption rate; g kg^{-1} day^{-1} (or week^{-1})] or the rate of relative growth in dry mass (M) allows plant growth to be expressed on a unit time (t) basis as

$$\frac{dM}{dt}\frac{1}{M}.$$

Reflecting change in dry mass of the plant through time, RGR is an overall measure, summation, or integration of the processes that increase plant dry mass. Plants do not grow at a constant rate, hence RGR is not a constant and must be expressed in terms of the time interval over which it is measured. RGR is calculated as

$$\text{RGR} = \frac{\log_e M_2 - \log_e M_1}{t_2 - t_1}$$

Table 5.10 Plant interaction indices. See text for nomenclature

Index	Abbreviation	Equation	Type of index (sensu Weigelt and Jolliffe 2003)	Reference
Relative yield	RY	$\text{RY}_A^D = \frac{Y_{AB}^D}{P_A Y_A^D}$	Effect	de Wit (1960)
Relative yield total	RYT	$\text{RYT}^D = \text{RY}_A^D + \text{RY}_B^D$	Effect	Connolly (1986)
Relative yield of mixture	RYM	$\text{RYM}^D = (Y_{AB}^D + Y_{BA}^D)/(P_A Y_A^D + P_B Y_B^D)$	Effect	Wilson (1988)
Relative resource total	RRT	$\text{RRT} = \frac{d_1}{d_{10}} + \frac{d_2}{d_{20}}$	Effect	Connolly (1987)
Change in contribution	CC[1]	$\text{CC}_A^D = [Y_{AB}^D/(Y_{AB}^D + Y_{BA}^D)]/[(P_A Y_A^D)/(P_A Y_A^D + P_B Y_B^D)] - 1$	Effect	Williams and McCarthy (2001)
Competitive intensity	CI	$\text{CI} = \frac{P_{NN} - P_{WN}}{P_{NN}}$	Intensity	Keddy (2001)
Competitive importance	C_{imp}	$C_{imp} = \{[w_m - w(N)]/w_m\} \times \{W_m/[W_{max} - W(N)]\}$	Intensity	Kikvidze and Brooker (2010)
Relative efficiency index	REI	$\text{REI}_{AB} = \text{RGR}_{Amix} - \text{RGR}_{Bmix}$	Outcome	Connolly (1987)

[1]Note: RY and CC are shown calculated for species A, but can also be calculated for species B.

where M_1 and M_2 are the total plant dry mass at the beginning (t_1) and the end (t_2) of the period, respectively. If weekly harvests are used then $t_2 - t_1$ = 1 week and RGR is simply obtained by subtracting the natural logarithms of the dry mass. RGR is formally equivalent to the intrinsic rate of natural increase of a population (r) (see Section 1.3) when a plant is growing in the absence of competition or self-limitation (e.g. self-shading). The value of RGR is unaffected by the units of mass, although the units usually used are g kg^{-1} day^{-1} (or $week^{-1}$). The unit g g^{-1} day^{-1} is commonly used, although it is technically incorrect because the gram is not an SI base unit and should not therefore be used in the denominator (Salisbury 1996). Typical values are in the range of 40–400 g kg^{-1} day^{-1}, with values of 31–151 g kg^{-1} day^{-1} for 15 tree species, 66–314 g kg^{-1} day^{-1} for 93 perennials, 120–299 g kg^{-1} day^{-1} for 22 annuals, and 113–365 g kg^{-1} day^{-1} for 24 herbaceous species (Lambers and Poorter 1992). In addition to biomass, RGR can be readily calculated based upon other measures of plant performance, including plant height or leaf, branch, or node number (e.g. Hastwell and Facelli 2003).

A comparison of RGR of three crop plants is provided in Fig. 5.13. Note how the RGR varies among the three crops and through time. *Zea mays*, for example showed maximum RGR 45 days after germination, whereas the other two species peaked much earlier before declining. Surveys have shown that RGR values display about a 10-fold range between species, whereas between-population, between-genet, and between-ramet variation is approximately two-fold (Hunt 1984). RGR decreases with plant age with increasing allocation to support tissues. RGR tends to decrease with increasing plant size, and non-linear models have been recommended to account for varying growth rates (Paine et al. 2012) and can be implemented in R using the lme4 and nlme packages (see the Appendix).

Relative growth rate$_{max}$ (RGR$_{max}$). This is a value measured on small, young isolated plants under standardized laboratory or greenhouse conditions over a short time interval (10–20 days) allowing comparisons among species. Recommended standard conditions are those of the ISP with PAR = 125 μmol m^{-2} s^{-1}, 16 hour daylength, and temperatures of 22 °C (day) and 15 °C (night) (Hunt 1993). Species grown under uniform conditions differ in RGR_{max}, with herbaceous species ranging from 100 to 400 g kg^{-1} day^{-1} and woody species from 10 to 150 g kg^{-1} day^{-1} (Poorter and Garnier 2007). Values of RGR_{max} in the laboratory or greenhouse are generally less than values obtained from field-grown plants. However, the relative ranking of species based on RGR_{max} values in both settings is generally maintained, indicating that the physiological, morphological, and anatomical components of RGR are under selective pressure.

Unit leaf rate (ULR). The ULR or net assimilation rate [g m^{-2} day^{-1} (or $week^{-1}$)] is a physiological index describing the rate of change of leaf area per unit dry mass per unit time. ULR is a measure of the net efficiency of a whole plant or population. Evans (1972) considered ULR to approximate to

ULR = daily photosynthetic rate
− respiration rate over 24 hours
+ daily mineral uptake
+ overall metabolic balance,

all expressed as change in dry mass per unit leaf area. Instantaneously, ULR is

$$\text{ULR} = \frac{dM}{dt}\frac{1}{L_A}$$

where L_A is leaf area. When dealing with seedlings, it is important to specify whether ULR, and any of the other indices which include leaf area, is based on true leaves only or true leaves plus cotyledons. This distinction is especially important if the cotyledons are thin, leaf-like, and photosynthetically active (i.e. paracotyledons). ULR can be calculated over a time period by assuming the simplest case of a linear increase in leaf area with plant age as

$$\text{ULR} = \frac{M_2 - M_1}{[(L_{1A} + L_{2A})/2](t_2 - t_1)}$$

where L_{1A} and L_{2A} are leaf area at times t_1 and t_2, respectively. If, however, leaf area increases exponentially then

$$\text{ULR} = \frac{M_2 - M_1}{t_2 - t_1}\frac{\log_e L_{2A} - \log_e L_{1A}}{L_{2A} - L_{1A}}.$$

ULR is unlikely to be constant under natural conditions, although it may fluctuate around a mean value with typical values of 2–20 g m^{-2} day^{-1} (Poorter and Garnier 2007).

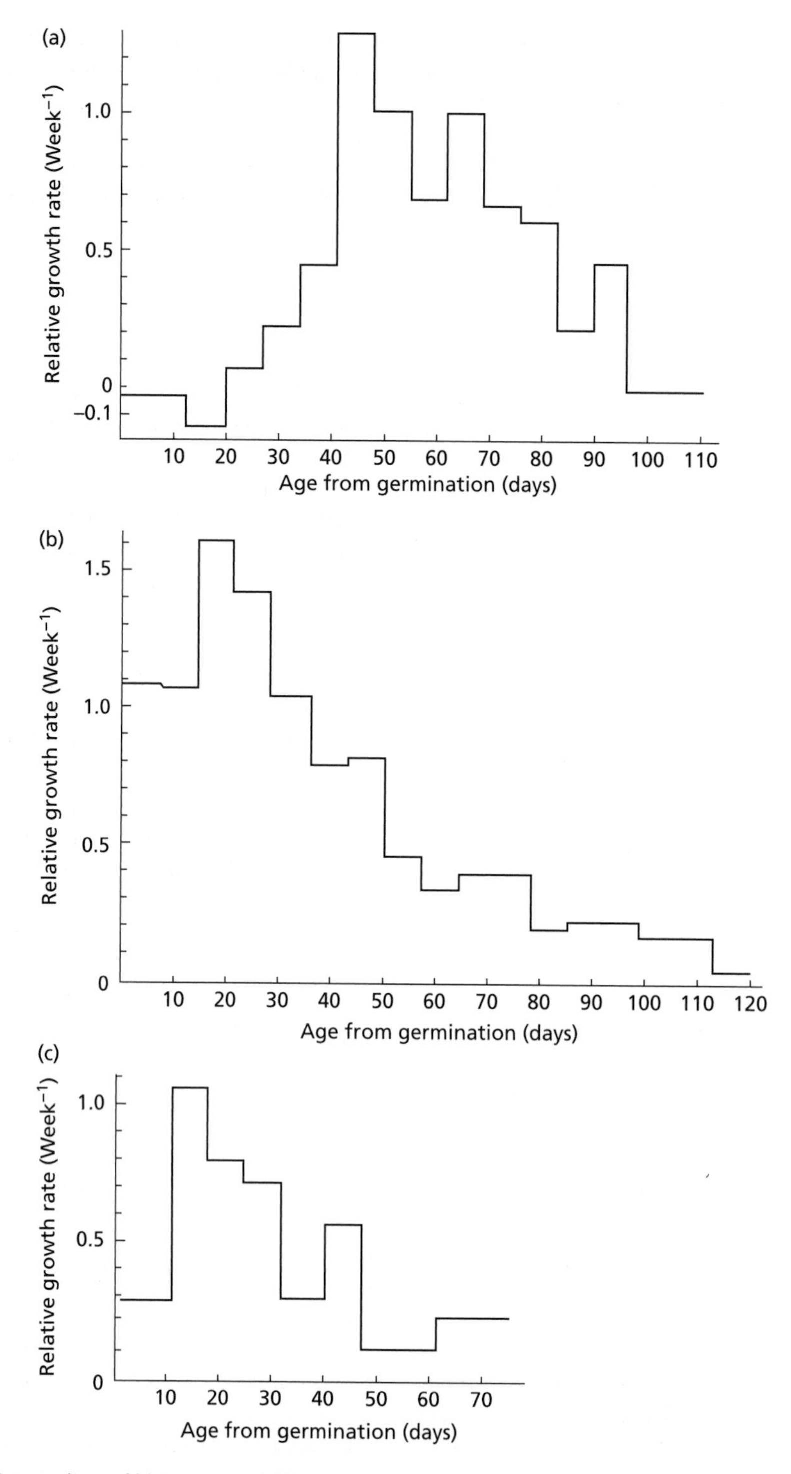

Figure 5.13 Relative growth rate of (a) *Zea mays*, (b) *Helianthus annuus*, and (c) *Gossypium arboreum*. From Evans (1972) with permission of the author.

Leaf area ratio (LAR). The LAR [m^2 kg^{-1} (sometimes mm^2 mg^{-1})] is a morphological index of plant form expressing leaf area per unit dry mass of the whole plant. Instantaneously, LAR = L_A/M and is calculated as

$$\text{LAR} = \frac{L_{2A} - L_{1A}}{M_2 - M_1} \frac{\log_e M_2 - \log_e M_1}{\log_e L_{2A} - \log_e L_{1A}}$$

with typical values of 5–50 m^2 kg^{-1} (Poorter and Garnier 2007), 0.1–4.5 m^2 kg^{-1} for 35 tropical trees, and 13–36 m^2 kg^{-1} for 24 herbaceous species (Lambers and Poorter 1992).

Specific leaf area (SLA). SLA (m^2 kg^{-1}) represents the average leaf expansion in area per unit dry mass of leaves (L_M). Instantaneously, SLA = L_A/L_M and is calculated as

$$\text{SLA} = [(L_{1A} / L_{1M}) + (L_{2A} / L_{2M})] / 2.$$

with typical values of 5–30 m^2 kg^{-1}, 6–37 m^2 kg^{-1} for 35 tropical trees, and 25–56 m^2 kg^{-1} for 24 herbaceous species (Lambers and Poorter 1992). SLA measures the extension of leaf dry matter in space.

Leaf mass ratio (LMR). LMR (g leaf g^{-1} plant) is the average fraction of the total dry mass represented by leaves. Leaf weight ratio (LWR) (also called leaf mass fraction, LMF) is the more commonly used term, but is incorrect because g and kg are units of mass not weight. Instantaneously LMR is

$$\text{LMR} = L_M / M$$

and is calculated over time as

$$\text{LMR} = [(L_{1M} / M_1) + (L_{2M} / M_2)] / 2.$$

Values of LMR range from 0.02 to 0.34 g g^{-1} for 35 tropical trees and 0.43 to 0.64 g g^{-1} for 24 herbaceous species (Lambers and Poorter 1992).

Components of RGR. The quantities RGR, ULR, and LAR, defined above, are related such that instantaneously

$$\frac{dM}{dt}\frac{1}{M} = \frac{dM}{dt}\frac{1}{L_A} \times \frac{L_A}{M}$$

that is, RGR = ULR × LAR. This relationship allows RGR to be split into two components related to dry weight increase (ULR based on rates of carbon assimilation) and leaf area development (LAR based on the organs most associated with carbon assimilation). This relationship is only valid for instantaneous quantities. It is only an approximation if used for time-interval formulae, particularly when area and mass are growing at unequal rates. Similarly, since the translocation of assimilates for new leaf production represents the integration of many physiological processes the dry weight and leaf expansion components of LAR can be partitioned as

$$\frac{L_A}{M} = \frac{L_A}{L_M}\frac{L_M}{M}$$

that is, LAR = SLA × LMR.

A simple example of the use of RGR, LAR, and ULR is provided in Fig. 5.14. Note how LAR is initially low, keeping RGR low. RGR increases as assimilates are used for new leaf production, allowing LAR to reach its maximum at 20 days after germination. ULR shows considerable fluctuation throughout, but declines overall as assimilates are diverted from leaf production, decreasing LAR and hence RGR. Figure 5.15 shows how 10 woody species differed in SLA under different light and nutrient regimes. Note how the increase in SLA in shade was more marked in high-nutrient soil. The effect differed among species and was most marked for *Rosa canina*.

Leaf area index (LAI). LAI is the one-sided area of leaf tissue per unit ground surface area and is a dimensionless ecophysiological characterization of the ecosystem canopy affecting within- and below-canopy microclimate, canopy water extraction, radiation extinction, and water and carbon exchange. LAI is an important component of many ecosystem plant canopy models. LAI can be measured directly through measurement of total leaf surface area above an area of ground surface and can be expressed in m^2/m^2. For direct measurements, leaves can be sampled by harvesting or collected in leaf traps placed below the canopy during leaf fall. Direct measurements of LAI can thus be time-consuming and are destructive. A number of instruments allow non-destructive, rapid, indirect estimates of LAI to be made based upon either (a) 'gap-fraction methods' using transmission of radiation through the canopy and making a number of assumptions about leaf-angle distributions or (b) 'radiation measurement methods' in which a comparison of incident (outside the canopy) and below-canopy measurements is made according to Beer–Lambert light extinction laws. LAI can also be

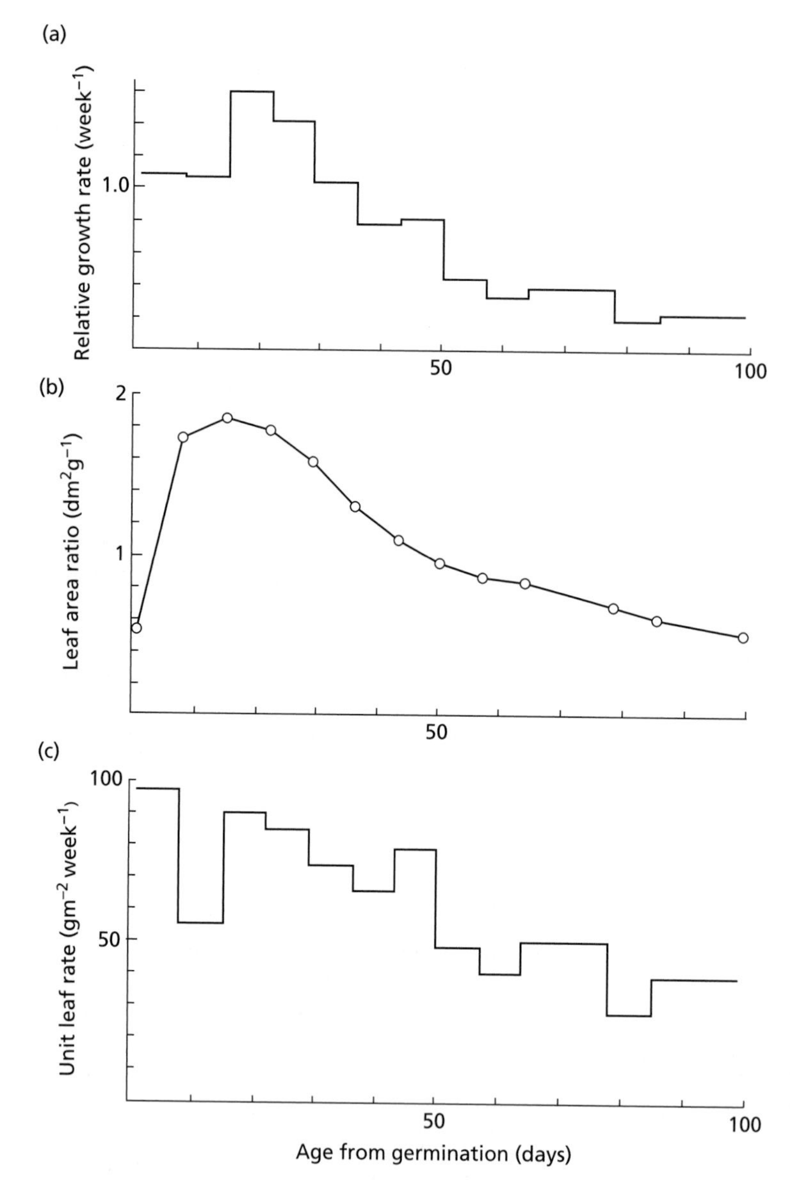

Figure 5.14 (a) Relative growth rate ($week^{-1}$), (b) leaf area ratio ($dm^2\ g^{-1}$), and (c) unit leaf rate ($g\ m^{-2}\ week^{-1}$) for *Helianthus annuus*. From Evans (1972) with permission of the author.

estimated indirectly using a point-intercept quadrat frame approach (Finzel et al. 2012). Indirect estimates of LAI are influenced by light zenith angles and the amount of non-leaf (i.e. stem) tissue in the canopy, and underestimate LAI. Hence, indirect measurements of LAI have been referred to as the plant area index. At the landscape scale, LAI can be estimated using remote sensing methodology, although again there are significant limitations to this approach (Gray and Song 2012). A full discussion of the methodological issues around the measurement of LAI is provided by Bréda (2003).

Tiller appearance rate (TAR). The axillary shoots of grasses consisting of a culm and associated leaves are referred to as tillers. Relative tiller appearance rate can be calculated on a unit time basis by taking repeated observations of the population of tillers in a sward or on a single individual and provides a measure of the dynamics of grass growth (Gibson 2009; Thomas 1980). Most frequently, TAR is

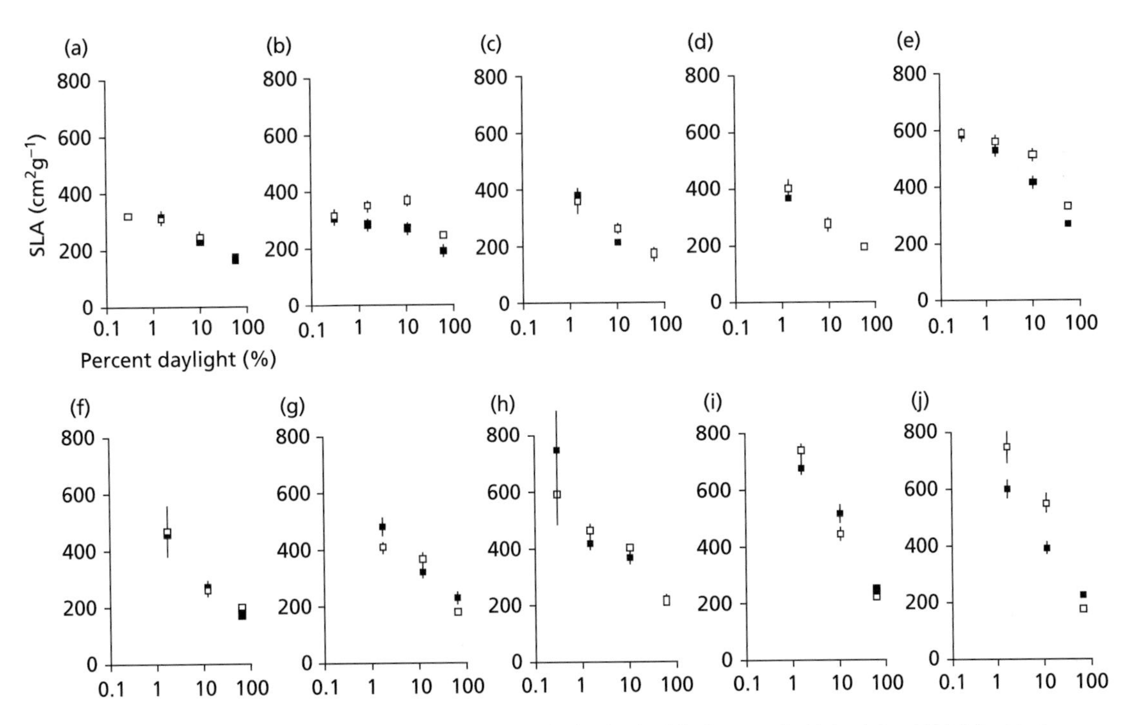

Figure 5.15 Specific leaf area (SLA) of seedlings of woody plants under four levels of shading on soil with low (■) and high (□) nutrients: (a) *Euonymus europaeus*, (b) *Viburnum lantana*, (c) *Juniperus communis*, (d) *Fagus sylvatica*, (e) *Viburnum opulus*, (f) *Crataegus monogyna*, (g) *Ligustrum vulgare*, (h) *Rhamnus catharticus*, (i) *Cornus sanguinea*, (j) *Rosa canina*. From Grubb et al. (1996). Reproduced with permission from John Wiley & Sons.

calculated as either net absolute TAR (NTAR) or proportional TAR (PTAR):

$$\text{NTAR} = \frac{N_2 - N_1}{t_2 - t_1}$$

where N_1 is the number of live tillers (per plant, pot, unit area) at time t_1, and N_2 is the number of live tillers at time t_2. If $N_2 = N_1$ + (number of new tillers between t_1 and t_2), then

$$\text{PTAR} = \frac{1}{N_1}\frac{N_2 - N_1}{t_2 - t_1}.$$

Negative values of NTAR or PTAR indicate tiller loss per time period. Monitoring the TAR of the perennial caespitose grass *Arrhenatherum elatius* over the course of a year quantified annual expansion and contraction of a grass clump in relation to the production of flowering tillers (Fig. 3.5; Gibson 2009). Alternatively, TAR can be standardized and expressed as the number of new tillers per 100 adult tillers per day (Scheneiter and Améndola 2012).

Plant interactions

Plant interaction indices can measure competitive effect, intensity, or outcome (Weigelt and Jolliffe 2003). Beyond these three categories, the indices emphasize and contrast different aspects of plant performance in mixture. It is important to choose an index that best suits the objectives of your experiment. The indices described here and in Table 5.10 were chosen because they are relatively straightforward and unambiguous to calculate and interpret. Extensive lists of other indices are provided by Williams and McCarthy (2001), Weigelt and Jolliffe (2003), and Reynolds and Rajaniemi (2007).

Following Williams and McCarthy (2001), plant interaction indices quantifying competitive effects (Table 5.10) are defined according to the following nomenclature when two species, A and B, are grown in monoculture or a mixture:

Y_A = yield of species A in monoculture

Y_B = yield of species B in monoculture

Y_{AB} = yield of species A in the presence of species B

Y_{BA} = yield of species B in the presence of species A.

To denote the density at which the plants are grown, a superscript is included after the Y:

Y_A^D = yield of species A grown at an overall density of D.

Proportional sowing densities are indicated as

P_A = proportion which species A was sown

and

P_B = proportion at which species B was sown,

such that $P_A + P_B = 1$. With more than two species, the addition of all proportions in a pot always sums to 1.0. Yield is generally measured in terms of aboveground biomass, but can also be belowground (root) biomass, or total (above + belowground) biomass, or some other measure of plant performance (e.g. Cahill 2002). To obtain measures on a per individual basis, divide by Y by D per experimental unit (e.g. per pot).

Relative yield (RY). RY is calculated as yield in a mixture divided by yield in monoculture accounting for species proportions. $RY_A = 1.0$ when species A grows as well in a mixture with species B as it does in monoculture, i.e. intra- and interspecific competition is equal. $RY_A > 1.0$ means that intraspecific competition > interspecific competition, and vice versa.

Relative yield of mixture (RYM). RYM provides the proportion of yield attained in a mixture compared with potential yield in monoculture. RYM > 1 indicates facilitation or avoidance of competition, whereas RYM < 1 indicates less biomass than expected, reflecting antagonism between competitors.

Relative yield total (RYT). RYT is generally the most popular index, and as it is related to the joint capture and use of resources it provides a single value describing niche relationships for a mixture (Connolly 1986). That is, RYT is the sum of the relative yields per species. RYT values equal to 1 indicate that two species perform equally in mixture. RYT > 1 implies a yield advantage in the mixture, whereas RYT < 1 implies antagonism or reduced effectiveness of resource use in a mixture. However, the RYT index provides a valid measure only if yields of the species when grown in monoculture are invariant across the range of densities used in the experiment.

Relative resource total (RRT). RRT was proposed to address the problems with RYT (Connolly 1987). RRT requires an examination of species responses (i.e. yield) in monoculture and mixtures across a range of densities. RRT is thus defined as the total pure stand area required to produce the same output as the unit area of the mixture. For two species, d_1 and d_2,

$$RRT = \frac{d_1}{d_{10}} + \frac{d_2}{d_{20}}$$

where d_i and d_{i0} are the yields per unit area in mixture and yield in the pure stand that produces that same output per individual as the mixture, respectively, for the ith species. Values of RRT range around unity in the same way as those of RYT, but it is a better measure of niche separation when values are greater than one. Caution needs to be used in interpreting RRT values when the estimated pure stand densities lie outside the range of the experimental densities (Connolly et al. 1990).

Change in contribution (CC). CC is a single species measure that takes into account the overall biomass contributed by each species in a binary mixture. CC_A indicates the proportional change of yield in mixture of species A compared with in monoculture (increase if positive, decrease if negative). Bedoussac and Justes (2011) found CC to be useful in quantifying the effects of competition in intercropped durum wheat and winter pea.

Competitive intensity (CI). CI is the combined (negative) effects of neighbours on the performance of an individual or population (Keddy 2001). It is measured by comparing the growth of a target plant or population in the absence of competition (P_{NN}) with the growth when surrounded by neighbours (P_{WN}). Thus

$$CI = \frac{P_{NN} - P_{WN}}{P_{NN}}.$$

Absolute measures of CI can be obtained by excluding the quotient terms, but may yield different

Plate 1 *Das große Rasenstück.* Watercolour by Dürer, 1503. Albertina, Vienna. (See Figure 1.1)

Plate 2 *Arabidopsis thaliana.* (Photo Rodney Mauricio.) (See Figure 3.2a)

Plate 16 Dune plants being damaged by flooding during Hurricane Andrew on Perdido Key, Florida, United States, in 1992. Both adults and seedlings of several species showed high mortality due to inundation of seawater (Gibson et al. 1995). Recording of events such as this are often best made by field observations as soon as possible after the event. (Photo Paul Looney.) (See Figure 6.5)

Plate 17 View looking to the south from the northern edge of the population of *Ceratiola ericoides* mapped in Fig. 8.1. (Photo David Gibson.) (See Figure 8.2)

results depending upon productivity in different communities. Relative CI values range from −1 to 1 where, CI > 0 reflects a negative effect of neighbours and CI < 0, or even < −1, reflects a positive effect of neighbours. Belcher et al. (1995) provided an example of the measurement of CI and showed that root and shoot competition intensity did not vary significantly along a soil depth gradient. Reader et al. (1994) measured CI in neighbour removal experiments with seedlings of the grass *Poa pratensis* in 12 plant communities on three continents. They noted that the magnitude of response varied among communities, but was unrelated to neighbour biomass in a consistent fashion.

Competitive importance (C_{imp}). This is 'the relative degree to which competition contributes to the overall decrease in growth rate, metabolism, fecundity, survival, or fitness of that organism below its optimal condition' (Weldon and Slauson 1986). C_{imp} is quantified based upon w_m = plant performance in the absence of neighbours (e.g. following experimental removal), $w(N)$ = plant performance in the presence of neighbours, and w_{max} = maximum performance of an isolated plant along the environmental productivity gradient being assessed (Table 5.10). Essentially, C_{imp} is the product of two ratios, with the first being an index of competitive intensity and the second being a ratio that scales the first by the impact of competitors relative to the overall negative impacts of the environment (Kikvidze et al. 2011) along a productivity gradient. Also see Seifan et al. (2010) for a modified index for quantifying interactions that shift from competition to facilitation.

Relative efficiency index (REI). The REI measures the outcome of competition over time and is based upon the RGRs of species A and B in mixture (RGR_{Amix}, RGR_{Bmix}, respectively) (Table 5.10). For a two-species mixture REI addresses the question 'Which species gains (wins) over time?' (Connolly 1987). REI indicates the efficiency of one species in a mixture relative to others over time. REI is independent of size with no bias comparing species of different sizes.

5.8 Fitness, fecundity, and reproductive effort

The terms fitness, fecundity, and reproductive effort are vital rates and are often confused, but they have fundamentally different meanings.

Fitness

In Chapter 1, fitness was defined as the 'chance of leaving descendants' (Harper 1977, p. 410). More formally, this can be quantified for an individual as the number of its offspring that survive to reproduce in future generations (Hutchings 1997). Thus, fitness can be measured as the intrinsic rate of increase (r) of the phenotype (p) in the environment (E) such that:

$$1 = \int e^{-rt} l_t(p,E) b_t(p,E) dt$$

where t is the age of the plant in years, $l_t(p,E)$ is the fraction of plants which survive from seed dispersal to age t, and $b_t(p,E)$ is the fecundity of phenotype p at age t in environment E (Crawley 1997a). Since this is an impractical measure to make in most field studies, and certainly in greenhouse or field trials, most investigators use a direct count of seed production of individual plants (i.e. per capita seed production or fecundity). However, seed production as a measure of fitness should be used with caution because 'it does not follow that an organism that produces a large number of progeny will also leave a large number of descendents' (Harper 1977). Natural selection, of course, acts to maximize individual fitness. At the individual level (the level at which natural selection acts), a plant might increase the chance of its progeny being more successful if increased vegetative vigour decreased its neighbours' chance of leaving descendants. At the population level, the finite rate of increase (λ; see Chapters 1 and 8) for each population has been described as a measure of its Darwinian fitness (W). For populations with discrete generations, λ can be estimated as the number of seeds produced on average for each seed sown one generation previously (Davy et al. 1988). Silvertown and Charlesworth (2001) noted that when comparing different genotypes fitness is a measure of the relative evolutionary advantage of one genotype over another under particular conditions. The genotype with the highest fitness has a value of $W = 1$, with the fitness of the other genotypes being given as a proportion of this.

Fecundity is total seed production per year (m_x). Fecundity is sometimes used as a surrogate measure of fitness, and the terms are often used synonymously. However, as it is clear from the discussion above, fecundity is a component of r (intrinsic rate of natural increase) which is itself a component of fitness. Fecundity is often a roughly linear function of plant size, although care has to be taken if the latter is used in its estimation.

In Case Study 4 in Chapter 1, Mauricio et al. (1993) considered three measures of plant fitness for the annual *Raphanus sativus*: the number of flowers produced, the reproductive biomass, and the total biomass of individuals. The effects of leaf damage treatments were similar for all three measures. The authors report a close relationship between the total number of flowers and seed production, pollen yield, and male reproductive success. However, in the absence of knowledge about how these measures relate to the reproductive success of the seed generation, it would be more appropriate to describe them as estimates of fecundity. Conversely, the use of seed number per adult as a measure of fecundity by McEvoy et al. (1993) in Case Study 3 in Chapter 1 is appropriate.

Reproductive effort and reproductive allocation

Reproductive effort (RE) is the investment of a resource in reproduction resulting in its diversion from vegetative activity (Bazzaz and Ackerly 1992). This is formulated as

$$RE = \frac{(R_r + R_v + S_r + A_r) - P_r}{T_r + S_v + A_v - P_r}$$

where R_r represents the reproductive pool, R_v is the vegetative biomass attributable to reproduction, S_r represents structural losses from reproductive organs, A_r represents atmospheric losses from reproductive organs, P_r is the enhancement of total resource supply due to reproduction, T_r is the total standing pool, S_v represents structural losses from vegetative organs, and A_v represents atmospheric losses from vegetative organs (Bazzaz 1997).

The value of RE is that it provides information about the allocation of resources that a plant is making in response to environmental constraints. RE obviously has implications for fitness. The choice of parameters required to measure RE is open to some interpretation, and this issue has been controversial (e.g. Solbrig 1980b; Thompson and Stewart 1981; Bazzaz and Reekie 1985; Bazzaz 1997). At issue is (a) the choice of 'currency' to measure and (b) which plant parts to measure. The choices of currency include energy (measured calorimetrically), biomass, carbon, or specific mineral nutrients (e.g. nitrogen, phosphorus). Of these, carbon is the most appropriate common currency for measuring RE because, in the form of sugars, it is used to 'purchase' other resources. Energy is correlated with carbon allocation but biomass may not be, and the allocation pattern of individual mineral nutrients may differ from each other and from carbon making it difficult to choose the most appropriate mineral nutrient. The issue of plant parts refers to the question of whether to measure only the allocation to actual flowers or seed, or to all supporting structures (flowering stems and associated cauline leaves, peduncles, etc.) produced in flowering individuals. Bazzaz and Ackerly's (1992) definition offered above suggests that all resources should be measured, as well as losses over the growth period or life of the plant. In many studies this is not practical, and many investigators actually measure *reproductive allocation* (RA)—the proportion of the total resource supply devoted to reproductive structures (Bazzaz and Ackerly 1992). RA can be expressed as

$$RA = \frac{R_r}{T_r} \quad (5.1)$$

$$RA = \frac{R_r + R_v}{T_r} \quad (5.2)$$

$$RA = \frac{R_r + R_v + S_r + A_v}{T_r + S_v + A_v}. \quad (5.3)$$

It is important to maintain the distinction between RE and RA; the former refers to changes in the resource supply due to reproduction, whereas RA refers to the proportion of resources devoted to reproduction. Measures of RA based upon biomass allocation to seeds and fruits (using Equations 5.1 and 5.2) should be interpreted with caution as they provide only a comparative measure of reproductive output (Reekie and Bazzaz 1987).

The *relative somatic cost of reproduction* (RSC) provides a measure for quantifying immediate resource costs (i.e. resource losses) in somatic (non-reproductive)

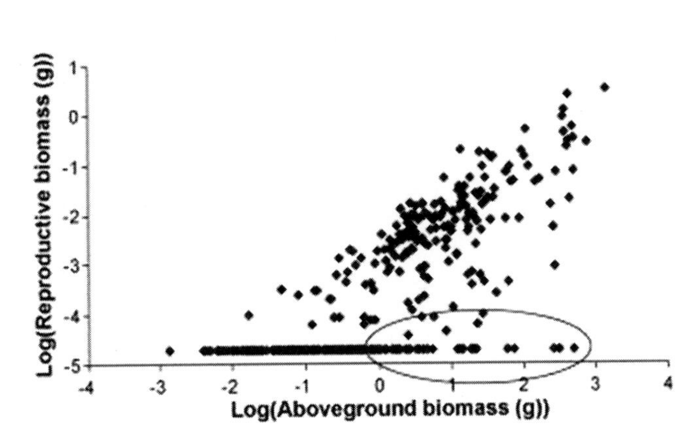

Figure 5.16 The logarithm of reproductive biomass versus the logarithm of aboveground biomass of individual plants of *Sinapis arvensis*. Approximately 49% of plants did not reproduce and the logarithm of their reproductive biomass is represented by −4.71. The plot suggests a linear allometric relationship between log(reproductive biomass) and log(aboveground biomass); however, the circled values do not support this relationship, indicating the need to use a mixture-model method to appropriately quantify the reproductive biomass relationship with aboveground biomass. From Brophy et al. (2007). Reproduced with permission from Elsevier.

tissues of reproducing individuals (Hemborg and Karlsson 1998) and can be estimated as

$$RSC = (I_{sn} - I_{sr}) / I_{sn}$$

where the components are the somatic investment in reproducing (I_{sr}) and non-reproducing (I_{sn}) individuals.

Allometric considerations

There is generally a strong positive allometric relationship between plant size and reproductive output, i.e. small plants have low RA and larger plants have a correspondingly larger RA. The relationship between reproductive biomass (R) on the y-axis and vegetative biomass (V) on the x-axis measured at senescence generally takes one of three forms:

short-lived herbaceous or clonal plants exhibit either i) a simple, linear relationship passing through the origin or ii) a positive x-intercept, or iii) larger, long-lived plants exhibit a log–log allometric relationship with slope <1 (Weiner et al. 2009).

The allometric relationships described above can be more complicated when there are large numbers of non-reproductive plants with zero RA, especially if some of these are large plants falling outside the log–log allometric relationship (Fig. 5.16). Excluding the plants with zero RA from calculations can lead to biased results, as can their inclusion in truncated regression models. A solution to this problem is to use a mixture-model method that specifically deals with these outlier values. A decision tree for identifying the correct analytical procedure along with SAS code is provided by Brophy et al. (2007). An application of this procedure that investigates the effects on RA of CO_2 level and plant size within size-structured populations is provided in Brophy et al. (2008).

5.9 Follow-up exercises

1. Make a list of the assumptions necessary when using surrogate measures, for example flower number, for estimating fitness and fecundity.
2. Design a field study to estimate fitness for (a) an annual herb and (b) a perennial shrub species in your local flora. What are the practical difficulties? Are you forced to use estimates of fecundity?
3. How would you relate molecular measurements and ecophysiological measurements to estimates of (a) individual plant fitness and (b) λ for a population of a named plant that you are familiar with?

CHAPTER 6

Abiotic and biotic measurements of a plant's environment

. . . the full elucidation of the life-history of any plant whatsoever in relation to its surroundings is a serious undertaking demanding of the investigator greater sacrifices than he has yet been accustomed to make. We need to cultivate closer touch with the individual plants in a way that may be compared to the unstinted devotion of the gardener.

Oliver (1913) in the first article in the first issue of the *Journal of Ecology*)

- **The biotic environment: herbivory, pathogens, allelochemicals**
- **The abiotic environment: the soil, radiation, atmospheric water, and disturbances**
- **Assessing the importance of spatial variation in the environment on plant performance within populations**

Preamble

Organisms have been viewed as existing in an '*n*-dimensional hypervolume' (Hutchinson 1957). This phrase was used to define the niche of an organism. More generally, the term niche is used to describe the role of an organism in its community; that is, as a producer, consumer, or detritivore of specific materials and resources. For plant populations, describing the microhabitat of individual plants quantifies the realized niche, that is, the actual relationship between the plant and its habitat. The microhabitat comprises many biotic and abiotic components, and their importance varies in both space and time. Temporal variation occurs not just as a function of seasonal or year-to-year variation in the environment, but also reciprocally in the response of a plant and its requirements as a function of its ontogeny. For example, the plant regeneration niche describes those components of the environment that are important during establishment and growth (Grubb 1977).

In this chapter, methods for measuring the components of a plant's environment are described. Quantifying the effect of the environment on a plant requires measurement of both the plant and the environmental factor of interest. For example, to determine the reduction in fitness caused by a herbivore requires quantification of the reduction in the number of viable seeds per plant as well as measurements of herbivore abundance.

In Section 6.1 components of the biotic environment are described, and in Section 6.2 the components of the abiotic environment. Many of the methods follow standard protocols, and for these reference is made to appropriate published procedures. In Section 6.3 methodology for relating variation in individual plant performance to variation in the environment is described.

6.1 The biotic environment

The biotic environment comprises the living component(s) of a plant's habitat. This environment ranges from neighbouring plants of the same species within a population (see Section 5.7) to organisms at different trophic levels. Interactions with

Methods in Comparative Plant Population Ecology. Second Edition. David J. Gibson.

other plants, herbivores, pathogens, and allelochemicals are considered in this section.

Herbivory

The establishment of experimental treatments to assess herbivory was described in Chapter 4. Here we discuss the most appropriate measurements on plants and herbivores to assess the effects of herbivory on plant performance.

There are no universal methods for measuring the effects of herbivory. Feeding by a herbivore on one part of a plant may lead to indirect changes in another part. For example, aboveground herbivory may lead to an increase in the root/shoot ratio due to removal of aboveground structures and stimulation of root growth. Immediate effects of herbivory have to be measured by considering the plant parts that are directly or indirectly affected by the herbivore. To assess the evolutionary consequences, however, measures of changes in plant fitness have to be made.

Reproductive characters (number of flowers, fruit, seeds), leaf number, plant height, plant biomass, survivorship, etc. have all been used to measure the effects of herbivory. Generally, the effects of herbivory are assessed by directly measuring the damage to the plant parts affected by the herbivore. Alternatively, changes in growth rate or yield (Chapter 5) can provide indirect measurements of the effect of a herbivore. Tissue damage can also induce changes in the plant's biochemistry, including the production of secondary compounds (Hartley and Jones 1997). Measuring changes in herbivore-induced plant secondary compounds may provide an important assessment of the effect of a herbivore. Gange et al. (1989) argue that leaf number is the best non-destructive measure of the effect of herbivory on plant performance. Plant damage due to herbivory can be expressed simply as the proportion of damaged leaves per plant (Brown et al. 1987). A finer-scale index of damage considers the product of the proportion of damaged leaves on a plant and the mean damage to each leaf. Thus, on a sample of 20–100 leaves per plant, Brown et al. (1987) and Gange et al. (1989) recorded individual leaf damage on a scale from 0 to 6 as follows:

Damage rating	Estimated leaf area removed (%)
0	0
1	1–5
2	6–25
3	26–50
4	51–75
5	76–99
6	Total removal (only petiole remaining)

with mean damage being estimated as (see Fig. 6.1)

$$\text{proportion of leaves damaged} \times \text{mean damage score.}$$

In addition to allometric changes, plant shape can also be altered by herbivory through the pruning of shoots and alterations to patterns of apical dominance. This damage can lead to the development of new branches (tillers in grasses), and can produce highly branched, bushy plants. Herbivory, bacteria, viruses, fungi, or somatic bud mutations can cause 'witches' broom' in trees, i.e. bushy shoot growth.

The integrated screening programme procedure for assessing the response of a plant to defoliation involves a standardized removal of 75% of the photosynthetic material from 45-day-old greenhouse-grown seedlings (Bossard and Hillier 1993). Plants are allowed to recover before non-destructive recording (shoot or tiller number, number of flowering shoots and flowers, and canopy height), with a final harvest of clipped and non-clipped control plants.

Granivory is a special case of herbivory, involving the removal and consumption of seeds by animals. Pre-dispersal granivory can be assessed by estimating seed loss from reproductive structures (e.g. reductions in seed mass or number of affected and unaffected seed heads). Post-dispersal granivory can be assessed by observing the loss of seed placed out in specific microsites, such as marked caches. Vilà and Lloret (2000) used two approaches to address this issue: (a) seeds were individually glued to pieces of fishing line and tied to a wire stake placed in the ground; (b) ant predation was studied by placing seeds in glass tubes laid horizontally on the ground. In

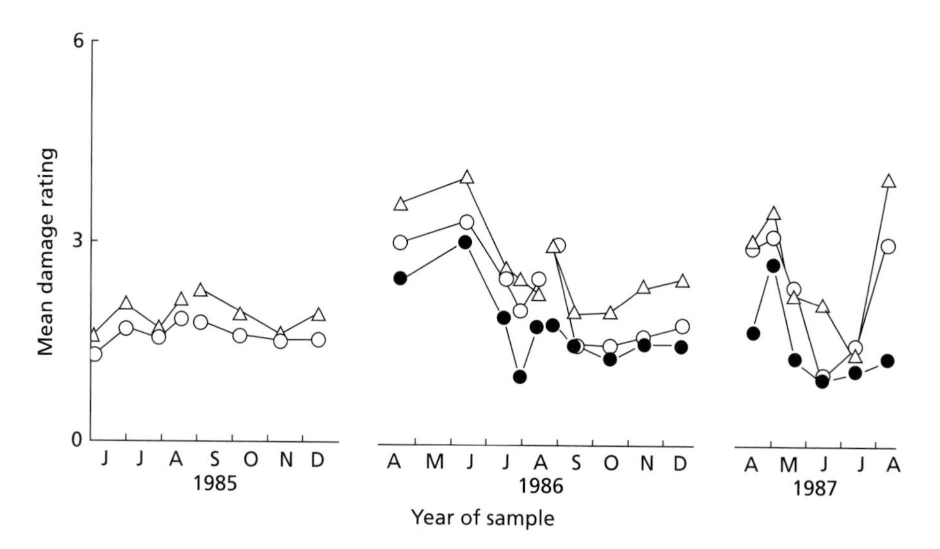

Figure 6.1 Seasonal variation in natural levels of insect herbivory on individual leaf damage on *Trifolium pratense* plants in three sites at Silwood Park, Berkshire. The sites were established on bare soil in 1985 (•), 1984 (O), and 1983 (Δ). From Gange et al. (1989). Reproduced with permission from John Wiley & Sons.

both cases, the seeds were checked for loss at regular intervals.

It is important to measure the abundance and temporal persistence of the herbivore (Chapter 4). With manipulative experiments, specific herbivore loads may be determined by the nature of the treatment. For example, in an experiment to test the effects of small mammal herbivory on tallgrass prairie, a high-density treatment consisted of plots with prairie voles maintained at a biomass of 250–400 g ha^{-1} over three growing seasons (Gibson et al. 1990). In other cases, especially mensurative experiments, herbivore abundance has to be measured. For example, in Case Study 3 in Chapter 1 insect densities were measured in exposed and protected plots to ensure the efficacy of the treatments. The unplanned intrusion of seed head fly was evaluated by counting the number of infested *Senecio jacobaea* seed heads. Sutherland (1996) provides methods for determining the abundance of animals, and Pedigo and Buntin (1994) detail sampling methods for arthropods.

Pathogens

The presence of pathogens in plants can be determined by noting standard visual symptoms of changes in leaf colour, mottling, burn, stem and shoot damage, etc., depending upon the particular pathogen. There are a number of texts and guidebooks available for the visual identification of pathogens (e.g. Gram and Bovien 1969; Fahy and Persely 1983). Quantification of the effects on the plant involves measurement of the relevant parameters described in Chapter 5, in a similar manner to that described for herbivory.

Accurate methods for determining the presence and abundance of a pathogen in plant tissues or the soil in which an infected plant is growing include the use of microscopic, histochemical, immunological, and molecular methods. Methods that are most likely to be useful for plant population ecologists are discussed here. General texts providing detailed sampling methods for pathogens include Hampton et al. (1990), Weaver et al. (1994), Dhingra and Sinclair (1995), and (Reid 2006).

Immunological techniques using monoclonal antibodies can provide a measure of the abundance of a pathogen in plant tissues (for review see Perotto et al. 1994). Monoclonal antibodies are produced in vertebrates (usually laboratory rabbits, mice, or rats) as an immunological response to infection by cell surface protein extracts (antigens) from pathogens. The antibodies are specific to the antigen contained in serum extracted from the animal and can be used to identify specific molecules originating, in this case, from pathogens. Specifically, dot blot or enzyme-linked immunosorbent assay (ELISA) techniques are used for determining the abundance of fungal mycorrhizae (Perotto et al. 1994) and endophytes (Welty et al. 1986), viruses (Converse and Martin 1990; Henry and Francki 1992), and *Rhizobium* bacteria (Perotto et al. 1994).

Nitrogen-fixing bacteria. Nodule number per plant root system provides an appropriate measure of the importance of nitrogen-fixing bacteria to the plant (e.g. Parker 1995). An estimation of the actual amount of nitrogen fixed by the bacteria can be obtained by assaying for nitrogenase enzyme activity using acetylene reduction, or by measuring the incorporation of $^{15}N_2$ into nodules and plant tissues (Caldwell and Virginia 1989). Results obtained using the acetylene reduction assay have to be viewed with caution because of the large diurnal and seasonal variation in nodule activity (Halvorson et al. 1992). In addition, a known non-nitrogen fixer of similar growth form to the nitrogen-fixer is needed for comparison, and there is then the assumption that there is similar fractionation in nitrogen uptake between fixer and non-fixer.

Fungal mycorrhizae and endophytes. Methods for determining the abundance of fungal mycorrhizae in plant roots are detailed in Schenck (1982), Norris et al. (1994), Vierheilig et al., (2005), and Utobo et al. (2011). For vesicular arbuscular (VA) mycorrhizae, the most commonly used procedure is determination of the extent of root colonization using the gridline-intersect method on cleared and mounted root samples (Rajapakse and Miller 1994; Widden 2001; Utobo et al. 2011). This method allows the percentage of colonized root length to be estimated. Production, standing biomass, and turnover of the extramatrical mycelium (EMM) of mycorrhizal fungi can be measured using a variety of techniques similar to those used for measuring plant roots in the soil (see Section 5.3) including minirhizotrons, ingrowth mesh bags, and cores. Biomass can also be estimated using a variety of biomarkers including chitin, ergosterol, and phospholipid fatty acids (PLFAs), and molecular methods such as quantitative polymerase chain reaction (qPCR; reviewed in Wallander et al. 2013).

For fungal endophytes, the presence of the fungus and density of the fungal hyphae is determined by microscopic examination of cleared and stained leaf sheath tissue (Bacon and White 1994). The density of fungal hyphae varies with plant age and position in the plant (Fig. 6.2) and can affect plant performance (Mack and Rudgers 2008). The polyclonal-based tissue print immunoassay (TPIA) (Gwinn et al. 1991; Hiatt et al. 1999; Hahn et al. 2003; Koh et al. 2006) is a fairly rapid and reliable method of detecting the presence of *Neotyphodium* or *Epichloë* fungal endophytes in plant tissues, including tillers and inflorescences, and is simpler than ELISA methods. A disadvantage of TPIA over ELISA is that false positives can arise from the presence of powdery mildews or rust fungi because of cross-reactivity with the antiserum and fungal hyphae. However, the localization of endophytes within rather than outside plant tissues makes this unlikely. TPIA kits can be obtained commercially (e.g. Agrinostics kits available from Agrinostics Co., Watkinsville, Georgia, United States; <http://www.agrinostics.com>). PCR-based methods can be used with primers designed to amplify specific genes on the fungal genome and detect the presence of fungal endophytes in plant seeds (Dombrowski et al. 2006).

Estimates of fungal biomass in plant tissues or soils can be obtained by assaying for fungal-specific compounds such as chitin (a cell wall component) (Jakobsen 1994) or ergosterol (a membrane constituent) (Nyland and Wallander 1994). DNA amplification using PCR can be used for quantifying the amounts of fungi in plant tissues (Ward et al. 1998) and for detecting and confirming the presence of particular fungi of interest (e.g. Mirlohi et al. 2006). Pitet et al. (2009) describe a technique that allows both clearing and staining of tissues for microscopic examination and subsequent extraction of ribosomal DNA from the same root samples. Fourier transform infrared (FTIR) spectroscopy followed by chemometrical data treatment can be used for rapid detection of fungal infections in dried and ground plant tissues (Brandl 2013).

Estimates of the deposition of fungal spores on plants can be made following the procedure used by Alexander (1990). Flowers of *Silene alba* were agitated in water and the concentration of spores from anther smut fungus (*Ustilago violacea*) in the resulting solution was determined using a haemocytometer.

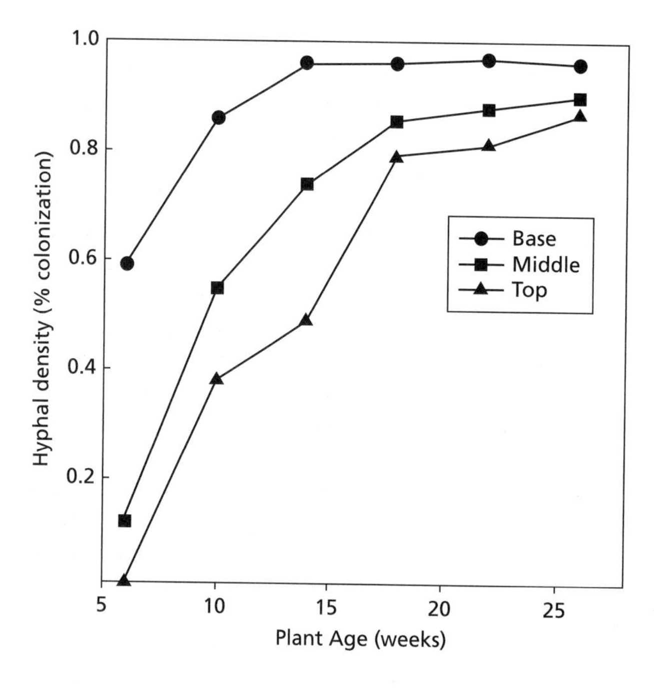

Figure 6.2 Endophyte hyphal density at three positions in the leaf sheaths of *Festuca arundinacea*. Drawn from data in Hickam (1998).

Spore counts per chamber were converted to spore number per millilitre, which is equivalent to spore number per flower. The effects of fungal disease following spore deposition can be assessed by experimentally infecting target plants with fungal spores collected from infected plants. For example, to test the effects of anther smut fungus on pollen tube growth in flowers of *Gypsophila repens*, flowers were experimentally pollinated with all combinations of dead (autoclaved) or living fungal spores and pollen (López-Villavicencio et al. 2005).

The soil microbial community

Total microbial biomass (bacteria, saprophytic fungi, and EMM of endomycorrhizal fungi) in soil can be estimated by measuring the difference in the flush of CO_2–C from chloroform-fumigated and unfumigated incubated soil samples (CFI) (Jenkinson and Powlson 1976). Alternative methods include using a respirometer to measure CO_2 efflux (Nordgren 1988; Setälä 1995), chloroform fumigation extraction (CFE), measuring glucose substrate-induced respiration (SIR) using near-infrared reflectance (NIR) (Palmborg and Nordgren 1993), and adenosine triphosphate (ATP) analysis (Horwath and Paul 1994). CFI is the oldest of these procedures and perhaps the most widely used as it allows estimation of microbial carbon as CO_2 and nitrogen as NH_4. All methods require calibration and standardization to allow precise and accurate results.

PLFA analysis is commonly used to estimate soil microbial biomass and profile microbial community structure under the premise that (some) PLFAs are specific to certain organisms. Although widely used, PLFA analysis has its limitations. Marker PLFAs are not necessarily indicators of particular groups as is sometimes assumed, and PLFAs should not be considered equivalent to microbial 'species' and used to calculate diversity indices (Frostegård et al. 2011). The related Microbial InDentification Inc. (MIDI) method should be used with great caution because MIDI fatty acid profiles can be related to plant-derived fatty acids as well as microbially derived fatty acids (Fernandes et al. 2013).

Allelochemicals

The nature of allelochemicals (see below) means that they can affect a large number of physiological processes in plants. The direct mode of action of allelochemicals include the following classes (Rizvi et al. 1992):

- cytology and ultrastructure
- phytohormones and their balance
- membranes and their permeability
- germination of pollen and spores
- mineral uptake
- stomatal movement, pigment synthesis, and photosynthesis
- respiration
- protein synthesis
- leghaemoglobin synthesis and nitrogen fixation
- enzyme activity
- conducting tissue
- water relations
- genetic material.

The plant population ecologist is unlikely to consider all of the above, but should measure the morphological and physiological (Chapter 5) variables that are most appropriate for any particular study. However, interpretation of any resultant effects upon plant performance should consider the physiological modes of action listed above.

The difficulties in establishing experimental treatments for allelopathy were discussed in Chapter 4. These difficulties extend to the isolation and identification of potential allelopathic compounds. Whatever method of isolation is used, the question remains as to whether these molecules are normally present in the environment at ecologically relevant concentrations for a long enough period of time to affect non-host plants in the manner that may be demonstrated in laboratory bioassays. The following issues should be considered in sampling the allelochemicals whether derived as soil-based chemicals from root exudates or leaf leachates, or as volatiles released from leaves (Cheng 1995). Laboratory sampling of volatile allelochemicals can be achieved by headspace sampling. In this procedure, a carrier gas (usually purified air) is passed over the plant sample (e.g. a tree branch with leaves) while it is confined in an entrainment chamber. Volatiles picked up in the carrier gas are collected in a porous organic polymer or activated charcoal solid trap before desorption for analysis by gas chromatography. Ruther and Hilker (1998) recovered and identified 12 volatile organic compounds (VOCs) from an elm (*Ulmus carpinifolia*) twig using this method. VOCs obtained from headspace sampling of artificially damaged and undamaged of *Artemisia frigida* were found to affect seed germination and seedling growth of common grasses in Mongolian pastures (Zhang et al. 2012). It would seem reasonable that this procedure could be extended to the field.

For root exudates, samples have to be representative of the allelochemicals within the plant rhizosphere without including contaminating root samples. For leachates, soil samples have to be taken from soils under the leaf litter or drip zone of the suspected allelopathic plants. Soil samples should be air-dried and stored in dark, non-humid conditions to avoid microbial degradation. Water is usually used for extracting allelochemicals from soils, but may not extract all water-soluble compounds because of retention on soil-surface particles. Chemical isolation of allelochemicals can be achieved through either gas or high-pressure liquid chromatography. Chemical characterization and quantification can be done by mass spectroscopy, FTIR spectroscopy, or nuclear magnetic resonance (Millar and Haynes 1998).

Allelochemicals are mainly secondary plant metabolites in the following major categories (Rice 1984):

- simple water soluble organic acids, straight chain alcohols, aliphatic aldehydes, and ketones
- simple unsaturated lactones
- long-chain fatty acids and polyacetylenes
- naphthoquinones, anthroquinones, and complex quinones
- simple phenols, benzoic acid, and derivatives
- cinnamic acid and derivatives
- flavonoids
- tannins
- terpenoids and steroids
- amino acids and polypeptides
- alkaloids and cyanohydrins
- sulphides and glucosides
- purines and nucleotides.

6.2 The abiotic environment (environmental variables)

Measurements of the abiotic environment around individual plants, or within populations, allow for comparisons among plants and populations as well as determination of the relationship between plant growth and the environment. Essentially, these are measurements of a plant's microclimate. Decisions about which components of the environment to measure will depend largely upon your experimental design and objectives. Manipulative experiments require the measurement of the treatments themselves and any treatments that involve structure. For example, if shade treatments are imposed then you have to measure photosynthetically active radiation (PAR; 400–700 nm). But since the shading itself is likely to affect temperature, humidity, and soil moisture it would be useful to measure these too. With mensurative experiments the choice of variables may also be obvious. For example, soil moisture should be measured frequently if the performance of plants along a putative soil moisture gradient is being assessed. But, again, all environmental factors that are likely to change along such a gradient should be considered. In some studies the choice of which environmental parameters to measure will be less obvious. For example, Totland and Nyléhn (1998) measured 15 environmental variables in their study of the perennial geophyte *Bistorta vivipara*. Of these, soil pH, elevation, and season length were the best predictors of performance. Clearly you have to make an informed decision at the beginning of your study, based upon prior knowledge of the plant and its habitat, about which environmental variables you suspect might be worth measuring. Abiotic measures can often be used as covariates in an analysis of treatment effects (Chapter 7).

Characterization of several aspects of the abiotic environment or microclimate can be readily made using the instrumentation available in micrometeorological stations. These measurements may include light availability, wind speed, soil and air temperature, and humidity. The measurements can be recorded using a data logger which can be programmed to take readings at set intervals, store the data, or allow remote wireless access. A set of standard guidelines for measuring and reporting the abiotic environment is provided in Table 6.1. Methods for measuring radiation, atmospheric moisture, and soil parameters are provided here. A concise discussion on measuring these and some other environmental variables is provided in Jones et al. (2006) with a quantitative account in Jones (1992). Selected examples that illustrate the relationship between abiotic environmental variables and plant populations are provided in Table 6.2. In all cases these examples are from studies in which the investigators were seeking to identify the abiotic environmental variables most closely related to plant performance. Bear in mind that correlation cannot be used to infer causation but it does provide the way forward for the generation of new hypotheses that may be worth experimental manipulation.

All components of the abiotic environment may show extreme variation in seasonal, daily, or even minute-to-minute values. This is particularly true for PAR, temperature, and humidity (e.g. Zipperlen and Press 1997; Horton and Neufeld 1998).

The soil

The soil is one of the most complex parts of the environment affecting plant populations, forming complicated plant–soil feedback interactions (van der Putten et al. 2013). Commonly measured abiotic components include soil moisture, texture, pH, nutrients, salinity, redox potential, and cation exchange capacity (Table 6.1). Good general descriptions of methodologies for describing and analysing the soil that are pertinent for plant studies are provided in Allen et al. (1986), Ball (1986), Binkley and Vitousek (1989), Westerman (1994), and Robertson et al. (1999). A discussion of instrumentation is provided in Smith (1983). Highly detailed descriptions of all aspects of soil sampling are described in the Soil Science Society of America's five-part *Methods of soil analysis* series (Klute 1986; Weaver et al. 1994; Sparks 1996; Dane and Topp 2002; Ulery and Drees 2008). Issues that are relevant for plant population ecology are discussed below.

Soil bulk density. Soil bulk density is the ratio of the mass of oven-dried soil to its bulk volume (Arshad et al. 1996). A cylindrical metal sampler is used to extract an intact soil core. A measurement of soil bulk density is needed to allow soil nutrient levels to be converted from a per gram basis to a volumetric basis before reporting. Bulk density varies with soil texture. Upper thresholds for bulk density that will restrict plant root growth are in the range

Table 6.1 Guidelines for measuring and reporting the abiotic environment for plant studies (from Rorison and Robinson 1986; Salisbury 1996; Style Manual Committee Council of Science Editors 2006)

Parameter	Units	Measurements		
		Where to take	When to take	What to report
Radiation				
Photosynthetically active radiation (PAR) (or radiation at specific wavelengths such as UV-A, UV-B or R:FR)				
(a) Photosynthetic photon (quantum) flux density (PFD): 400–700 nm with cosine correction (most ecophysiologists just use PFD with the designation that the waveband is 400–700 nm)	mol m^{-2} s^{-1}	Top of canopy. Obtain average over plant growing area	Start and finish of each study and biweekly if studies extend beyond 14 days	Average over containers or plants at start of study. Decrease or fluctuation from average over course of the study
Or				
(b) Photosynthetic irradiance (PI): 400–700 nm with cosine correction	W m^{-2}	Same as PFD	Same as PFD	Same as PFD
Total irradiance 280–2800 nm with cosine correction (see comments in text—probably best to use a pyranometer which will measure total shortwave irradiance)	W m^{-2}	Same as PFD	Start of each study	Average over containers or plants at start of study
Spectral irradiance (*S*): 250–850 nm in < 20 nm bandwidths with cosine correction (typically measured every nm with a spectroradiometer—especially important when comparing light quality for field and growth chamber grown plants)	W m^{-2} nm^{-1}	Top of canopy in centre of growing area	Start of each study	Same as Total irradiance
Daylength	Hours			Length of photoperiod in 24-hour cycle
Temperature				
Air				
Shielded and aspirated (>3 m s^{-1}) device	°C	Top of plant canopy. Obtain average over plant growing area	Hourly over the period of the study (continuous measurement advisable, although daily minimum and maximum values may suffice)	Average of hourly average values for the light and dark periods of the study with range of variation over the growing area
Leaf temperature	°C	Surface of representative leaves	Same as Air	Same as Air
Soil and liquid	°C	In centre of representative container	Same as Air	Average of hourly average values for the light and dark periods for the first day or over entire period of the study if taken

continued

Table 6.1 *Continued*

Parameter	Units	Measurements		
Atmospheric moisture				
Shielded and aspirated (>5 m s^{-1}) psychrometer, dewpoint sensor or infrared analyser	Relative humidity (%RH) or vapour pressure deficit (kPa)	Top of canopy in centre of plant growing area	Once during each light and dark period, taken at least 1 hour after light changes (hourly measurements over the course of the study advisable)	Average of once daily readings for both light and dark periods with range of diurnal variation over the period of the study (or average of hourly values if taken)
Air velocity	m s^{-1}	Top of canopy. Obtain maximum and minimum readings over plant growing area	At start and end of the study. Take 10 successive readings at each location and average	Average and range of readings over containers at start and end of the study
Carbon dioxide	m mol m^{-3} or µl CO_2 L^{-1} air	Top of plant canopy	Hourly over the period of the study	Average of hourly average readings and range of daily average readings over the period of the study
Watering	ml	At times of additions	Frequency of watering. Amount of water added per day and/or range in soil moisture content between waterings	
Substrate				Type of soil and amendments. Components of soilless substrate
Nutrition	Soil media: kg m^{-3}. Liquid culture: micro nutrients, µmol L^{-1}		At times of nutrient additions	Nutrients added to solid media. Concentration of nutrients in liquid additions and solution culture
	Macronutrients, µmol L^{-1}	Amount and frequency of solution addition and renewal		
pH	pH units	In liquid slurry for soil and in solution of liquid culture	Start and end of studies in solid media. Daily in liquid culture and before each pH adjustment	Mode and range during study
Conductivity	dS m^{-1} (decisiemens per metre)	In liquid slurry for soil and in solution of liquid culture	Start and end of studies in solid media. Daily in liquid culture	Average and range during study
Soil nutrients	µg g^{-1} or mg L^{-1} (p.p.m. is archaic but equivalent)	Soil extraction	At least once during the course of study, more often for mobile nutrients (e.g. NO_3^-, NH_4^+)	(a) Average and range during study
(b) Soil bulk density	(c) g cm^{-3}	(d) Cylindrical soil core extraction	(e) At least once (three or more replicates) during the course of the study	(f) Average dry weight of soil divided by its volume
(g) Soil moisture (volumetric/ gravimetric)	(h) kg water/kg dry soil	(i) Soil extraction	(j) Multiple times during the course of the study	(k) Average and range per unit time (e.g. per day or month)
(1) Soil texture	(m) % sand, silt, clay	(n) On air-dried and sieved (2-mm sieve to remove gravel) samples	(o) At least once during the course of study on replicate (often pooled) samples	(p) Soil texture class from soil texture triangle (<http://soils.usda.gov/technical/aids/investigations/texture/>)

Table 6.2 Studies showing a relationship between abiotic environmental variables and individual plant performance in naturally grown field populations

Life form/species	Abiotic variable(s)	Measure(s) of plant performance that were related to the abiotic variable(s)	Authors
Tree/*Acer mono*	PFD	Seedling biomass	Seiwa (1998)
Tree/*Pinus echinata*	PAR	Seedling growth (height, leaf number)	Gibson and Good (1987)
Tree/*Rhizophora mangle*	Water depth and sedimentation rate	Aerial root and shoot lengths	Ellison and Farnsworth (1996)
Annuals/*Polygonum* spp.	PAR, soil temperature, structure, moisture, and nutrients	Species presence	Sultan et al. (1998)
Annual/*Melampyrum lineare*	PAR	Fruit production, leaf number, height, number of branches, flowers, and leaves	Gibson and Good (1987)
Winter annual/*Amphianthus pusillus*	Soil depth	Seed production	Hilton and Boyd (1996)
Perennial herb/*Verbena officinalis*	Temperature	Survivorship	Woodward (1997)
Forest understorey perennial herb/*Viola blanda*	Soil P and Mg, tree canopy openness	Plant presence	Griffith (1996)
Perennial geophytes/*Bistorta vivipara*	Soil pH, elevation, season length	Plant biomass, flower number, bulbil number, weight	Totland and Nyléhn (1998)
Perennial grass/*Calamagrostis porteri* subsp. *insperata*	PFD, vapour pressure deficit, soil temperature	Leaf area, tiller number	Bittner and Gibson (1998)
Perennial grasses/*Agropyron cristatum* and *Pseudoroegneria spicata*	Soil colour and temperature	Seedling performance (tillers per plant and leaf area)	Boyd and Davies (2012)
Perennial vine/*Lonicera japonica*	18 environmental and biotic variables	Cover, vine density, number of internodes and internode length per vine, specific leaf area	West et al., (2010)
Cactus/*Opuntia polyacantha*	Rainfall	Biomass	Dougherty et al. (1996)
Aquatic mosses/*Sphagnum subsecundum*, *Drepanocladus exannulatus*	PFD, water temperature, CO_2 concentration in the water	Biomass, linear growth rates of side branches	Riis and Sand-Jensen (1997)

PFD, photon flux density; PAR, photosynthetically active radiation.

of 1.8 g cm^{-3} for sandy soils, < 1.55 g cm^{-3} for silty soils, and < 1.4 g cm^{-3} for clayey soils. Higher values of bulk density than these indicate a degree of compaction.

Soil nutrients. The procedure usually recommended for measuring soil variables is to use a soil corer (2–5 cm in diameter) to collect samples from the A horizon from across the sample area. These samples may then be composited into a single sample that is subsampled for laboratory analysis. Soil nutrient levels can be determined in samples as 'total' or 'available' amounts. Generally, available forms of nutrients are a more relevant measure for plant studies than total amounts in the soil, except when an assessment of the organically bound amounts of carbon, nitrogen, and phosphorus are required. Available nutrient levels are determined by chemical extraction with the goal of providing an assessment of the ecologically relevant amount of a nutrient in the soil. The methods for extracting available nutrients assume that the amount of nutrient released to the leaching solution or extractant (strong or weak solutions of various acids such as ammonium acetate, acetic, hydrochloric or sulphuric acid, potassium chloride, or EDTA) is equivalent to what is available in the soil to a plant root. In practice, this model may be simplistic because the empirical relationships between plant species and processes of nutrient solubilization and resorption of ions from solutions are only known for agricultural crops. To obtain ecologically meaningful data, an extractant that minimizes these processes should be used. Recommendations for particular extractants and guidelines for making informed decisions are provided in Allen et al. (1986).

Nitrogen is generally the most common limiting plant nutrient in the soil and is one of the most difficult of nutrients to measure. Total or available levels of soil nitrogen (ammonium or nitrate) based upon standard extractions have to be viewed with caution. This caveat is necessary because the amount of soil nitrogen available for uptake by plants is highly dependent upon microbial activity. Measuring nitrogen release following the stimulation of microbial activity allows assessment of the mineralized soil organic nitrogen fraction. In this procedure, soils are incubated in enclosed containers either in the laboratory for 14–28 days or in the field for 1–2 months. Mineralizable soil nitrogen is assessed as the difference in levels of extracted inorganic ammonium and nitrate nitrogen before and after incubation. The results obtained from field incubations are preferable to those obtained in the laboratory because the former integrate on-site temperature regimes. Field incubations should be carried out over the whole season to best measure plant-available nitrogen. Procedures and discussion about these and other methods of assessing plant-available nitrogen are provided in Hart et al. (1994).

Chemical analysis of composited soil samples removed from the field may not be appropriate for assessing the soil environment within plant populations, and particularly around individual plants. Spatial variation in soils can be extensive, ranging from resource depletion zones around the rhizosphere of individual roots to landscape-scale patterning reflecting the local topography or underlying bedrock. Individual plants will respond to this variation when individual soil factors are below limiting levels. Alternatively, plants may cause soil heterogeneity, such as the accumulation of nutrients under clumps of some caespitose grasses (Gibson 1988; Vinton and Burke 1995). Depending upon the goals of your study, you need to take this heterogeneity into account when sampling the soil. When dealing with individual plants it will be necessary to ensure that samples are taken from within the rooting zone of the target plant.

Soils are variable in time, especially with respect to moisture levels and some of the more mobile nutrients such as ammonium, nitrate, and potassium, which are readily taken up by plants as they become available and readily lost through volatilization (ammonium) or leaching (nitrate, potassium) when they are not taken up. Meaningful assessment of these nutrients may require repeated sampling. Non-destructive methods allowing repeated measures from the same location in the soil may be necessary for studies of individual plants; the more conventional analysis of collecting and removing samples to the laboratory for analysis is likely to be unsuitable. Ion-exchange resins in bags or spikes placed into the soil can be useful in these situations (Cain et al. 1999). Ion-exchange resins are synthetic polymers which bind covalently with ions in solution in exchange for weakly held anions or cations (depending on the type of resin) on the resin. Ion-exchange resins are available as beads which can be placed into small bags to be put in the soil (Gibson 1986; Gibson et al. 1985; Sibbesen 1977) or ~2 cm diameter mixed bed 'ion exchange resin capsules' (mesh balls) which can be placed directly in the soil (Jones et al. 2012; Jones et al. 2013). Alternatively, the resins are available already impregnated onto membranes which can be bonded onto tubes to make resin spikes (Cain et al. 1999). After removing the resin-filled bags, capsules, or resin spikes from the soil, the ions are washed off the resins for analysis in solution. A spatial resolution of soil nutrients in the soil down to 10 cm has been achieved using ion-exchange resins. The uptake of ions depends upon the movement of the soil solution. Ion-exchange resins provide a relatively simple, repeatable method for assessing the variability of soil nutrients.

Soil solution. Narrow-tension lysimeters allow repeated samples of the soil solution to be taken for analysis. A lysimeter consists of a plastic pipe placed into the soil. An unglazed ceramic cup at the base of the tube allows soil water to enter and be drawn up the tube when it is placed under a vacuum. Soil water can be sampled and removed in this way for chemical analysis when it is present at tensions less than the suction applied to the lysimeter. Lysimeters offer a method for repeatable analysis of the nutrient content of the soil solution. However, they can be problematic to install as they involve disturbance to the soil. Lysimeters do not work well in xeric soils as contact between the ceramic cup and the soil can be lost, making extraction of what little soil water is present difficult and uncertain. Further details about the use of lysimeters can be found in Chapman (1986).

Soil moisture content and soil water potential. Soil moisture shows high levels of spatial and temporal heterogeneity. A one-time measure cannot be used to represent the soil moisture environment around a plant. It is important to distinguish between soil moisture content and plant-available soil moisture (measured as soil water potential, see below), with the latter representing the difference in moisture content between field capacity (water content retained at 10 or 33 kPa after a water-saturated soil has been allowed to drain under gravity) and the permanent wilting point (water content retained at 1.5 MPa beyond which plant roots can no longer extract water) (see Bannister 1986). A general review of methodology is provided in chapters in Carter and Gregorich (2006) and by Bittelli (2011). Methods pertinent for local-scale (~0.01 m^2) measurements of soil moisture content are summarized here.

The simplest and most accurate measurement of soil moisture content is *thermogravimetric soil moisture*, in which a soil sample is removed, weighed, oven-dried to remove the water (105 °C for 24 hours), and weighed again. The mass of water is reported per unit mass of dry soil (e.g. kg of water/kg of dry soil). The disturbance necessary for sampling the soil means that gravimetric samples are inappropriate for repeated measurements of the soil moisture status around individual plants (except perhaps for trees or large shrubs). Gravimetric soil moisture (w) can be converted to volumetric soil moisture as $\theta = w\rho_b/\rho_l$ where ρ_b is soil bulk density and ρ_l is water density.

Indirect, but repeatable, surrogate measures of local soil moisture status can be obtained using in situ tensiometers, dielectric measurements, psychrometers, or soil moisture blocks. These methods measure another variable and relate changes in it to soil moisture status through calibration curves. Perhaps more importantly, these methods determine soil water potential rather than soil moisture content and are a better indication of the soil water available to plant roots than the amount of water in the soil, which is greatly influenced by soil texture.

Tensiometers measure the change in soil water tension when a water-filled porous cup buried in the soil is connected to a manometer and operate in the range -0.08 MPa $\leq \psi \leq 0$. Air-free soil is necessary to ensure continuity in soil tension, meaning that this method is limited to applications in mesic soils. Tensiometers also require maintenance to refill the ceramic cup after dry periods, and are slow to react to changing soil moisture conditions. Dielectric measurements are based upon the empirical relationship between the relative dielectric constant of soil and soil water content. Some sensors use time domain reflectometry (TDR) to measure the dielectric constant from the length of time taken for a voltage pulse to pass from metal rods placed a known distance apart (horizontally or vertically) in the soil (Topp and Reynolds 1998). Capacitance sensors employ a pair of electrodes (forming a capacitor) and use an oscillator to generate an AC signal and measure changes in operating frequency (i.e. changes in the dielectric properties of the soil) which are then related to soil moisture content. Recent advances in technology allow small-scale measurements to be taken, rocky soils to be assessed, and volumetric soil water content, soil solution electrical conductivity and temperature, and soil nitrate concentrations to be measured throughout the growing season using a variety of commercially available probes (sometimes the same probe). Disadvantages of dielectric measurements include the start-up cost of purchasing probes and data loggers, and some requirement for calibration.

Soil moisture blocks work on the principle that changes in the electrical resistance of porous materials buried in the soil reflect changes in soil moisture because of the equilibrium that is attained between the block and the soil. Gypsum blocks are normally used, but the blocks can be of nylon or fibreglass. The blocks are placed and left in the soil, and are subsequently attached to a suitable meter by wires to take readings. Calibration of the soil moisture blocks is necessary. Small mammals will often chew the wires unless they are well screened, for example with aluminium foil or encased in plastic piping. Further details about the use of this method is available in Bannister (1986) and Rundel and Jarrell (1989).

Psychrometers provide a direct measure of water potential through determination of the relative humidity inside a chamber within a porous cup (about 1 cm in diameter and 1 cm long), equilibrated with the soil, containing a small thermocouple. The principle is that the water potential of a porous material (ceramic, brass, or steel in this case) is directly proportional to the equilibrium water vapour pressure of the air surrounding the porous medium. Psychrometers have high sensitivity, especially in dry soils operating in the water potential range $\psi \leq -0.1$ MPa. However, psychrometers are slow to equilibrate with the soil, they are inaccurate in wet soils, have a small sensing volume, and exhibit poor accuracy in shallow soils because of their susceptibility to thermal gradients. Full details are available in Livingston and Topp (2006).

Radiation

Electromagnetic radiation has the properties of both a wave and discrete particles. Hence, measurements can be made of both photons (packets of electromagnetic radiation) and quanta (the amount of energy in a photon). The solar radiation utilized by plants has dual effects on them, i.e. photochemical (photosynthesis, photomorphogenesis) and thermal (energy and temperature balance), which determine how to characterize the light environment. For photosynthetic plants we are interested in recording the PAR. Photosynthetic photon flux density (PFD) at 400–700 nm is the incident photon flux density of the PAR and measures the number of photons incident per unit time on a unit surface. Photosynthetic irradiance (PI) is the radiant energy component of PAR incident per unit time on a unit surface. Measurements of PFD are preferable to those of PI for reporting in ecological studies, although solar (shortwave) irradiance (~300–3000 nm) is relevant when one is concerned about the thermal effects of sunlight on plants (e.g. energy balance, evapotranspiration rates). Measurements of illumination using photometric units (e.g. candela, lumen, and lux) are inappropriate for plant studies since they are based upon the spectral response of the human eye and do not correspond to the absorption spectra of plant pigments (Jones 1992). Table 6.1 shows the appropriate units for reporting PFD. The methods described below can all be adapted to record the intensity of a specific wavelength of interest, including UV-A and UV-B, through the use of wavelength-specific filters placed between the sensor and the light source, or through the use of wavelength-specific sensors (e.g. see discussion of red/far-red sensors below).

Repeated measurements of PFD (at least monthly throughout the growing season) incident upon study plants are necessary for characterizing the seasonal light environment. Single or infrequent measurements of PFD are likely to be of limited use, and may be misleading. For forest understorey plants, early season PFD or sunflecks may constitute the major source of light energy. For these, and other plants existing under a canopy, late-season measurements may underestimate the light environment. PFD is frequently reported relative to values obtained above or outside the forest canopy when a general assessment of the plant's light environment is required. In these situations, measurements are best taken between 10 a.m. and 2 p.m. on cloud-free days, or on days where continuous cloud cover affords an even, diffuse light.

A simple, cheap way to estimate the maximum intensity of solar radiation received in a location over a period of time is by using booklets of 10–20 sheets of light-sensitive diazo paper. Light intensity is estimated from the number of sheets of diazo paper that are bleached (following development with ammonia vapour) after exposure in the field. Although caution has to be taken in interpreting estimates of light intensity derived using this technique (Bardon et al. 1995), it can allow simultaneous measurements to be made at multiple locations at small scales. Antos and Allen (1999), for example, used this technique to relate estimated light intensity to reproductive effort of individual male and female shrubs of *Oemleria cerasiformis*.

There are several commercially available radiation sensors based upon the use of thermoelectric or photoelectric devices. Those with single, small photoelectric cells are particularly useful for measuring the PFD incident on individual plants.

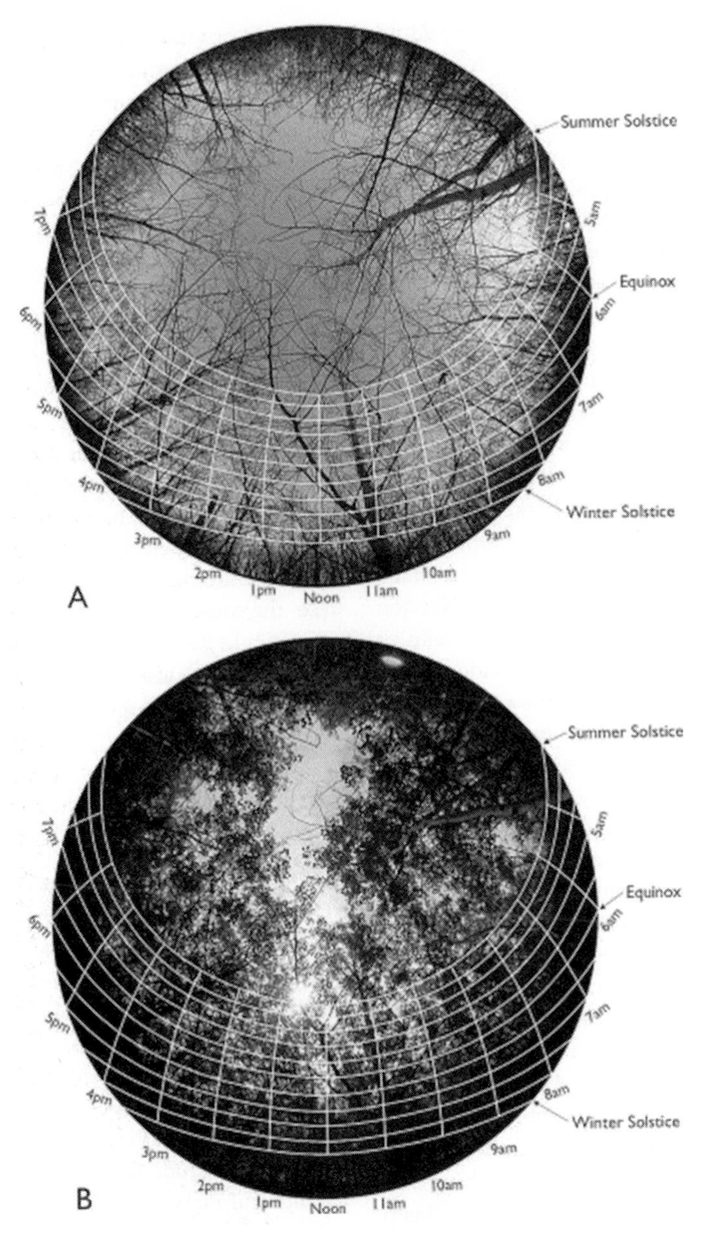

Figure 6.3 Hemispherical 'fisheye' photograph under an aspen (*Populus tremuloides*) canopy showing canopy structure in mid April before leaf-out (A) and near maximum leaf area in late July (B). The superimposed solar path indicates how the angle of the sun changes with time of day during the year. From Archibold and Ripley (2004). Reproduced with permission from Elsevier.

However, these single-cell sensors are very sensitive to fluctuations in radiation and need to be kept level while measurements are taken. Ceptometers (line quantum sensors) are linear arrays of radiation sensors that allow an integrated measurement of PFD to be taken over 1 m. These measurements may be particularly appropriate for estimates of radiation flux within populations. Specialized sensors are available to measure specific wavelengths. For example, the Skye SKR 110 red/far-red sensor

(Skye Instruments, Llandrindod Wells, UK) simultaneously measures red wavelengths (centred at 660 nm) and far-red wavelengths (centred at 730 nm) to provide an indication of the stimulation of the plant's photoreceptor phytochrome pigment. The red/far-red ratio (about 1–2 in unfiltered daylight decreasing to 0.1–0.2 under leaf canopies and leaf litter) can characterize much of the microsite spatial variation in forest understorey light levels (Capers and Chazdon 2004) and can significantly affect seed germination in light-dependent species (Jankowska-Blaszczuk and Daws 2007).

The spatial nature of the light environment incident upon plants can be described using either digital photographic images or though ocular estimation with a densiometer. Images taken with wide-angle hemispherical lenses (~8 mm, approaching 180°) attached to a digital camera or commercially available instrumentation (Bréda 2003; Newton 2007) can provide measurements of the relative and instantaneous light environment incident upon individual plants (Rich 1989). Analysis of these images (Fig. 6.3) can provide estimates of the spatial distribution and daily time course of direct and diffuse PAR (Chazon and Field 1987). The light environment of individual plants growing under forest canopies can be efficiently characterized using this approach (e.g. Cunningham 1997; Valverde and Silvertown 1997). Alternatively, the density of an overhead tree canopy in forests can be estimated using a spherical canopy densiometer, a simple handheld instrument (Lemmon 1956; Strickler 1959). A densiometer consists of a convex or concave mirror surface marked out with a grid. When held level, the overhead canopy cover is proportional to the number of grid squares in which the reflected tree canopy is observed. Densiometers are robust and easy to use, although care has to be taken to keep the instrument level while taking measurements. A detailed discussion of radiation and these and other sensors appropriate for ecological measurements is provided by Pearcy (1989b) and Newton (2007).

Atmospheric water

Measurements of the potential for the atmosphere to provide moisture to plants are frequently expressed as relative humidity (RH). RH is the amount of water in the air (actual vapour pressure or absolute humidity, e) as a proportion of that at full saturation (maximum vapour pressure, e_m) and at the same temperature and pressure. RH is conveniently measured with a hygrometer or sling psychrometer. A more meaningful measure is the atmospheric saturation deficit (sometimes referred to as the vapour pressure deficit) which is the difference between the actual vapour pressure and the saturation vapour pressure (this is equivalent to $e - e_m$). Tables of e_m for ecologically relevant temperatures are available in Cox (1990). Saturation vapour pressure approximately doubles for each 10 °C increase in temperature at a constant RH (Fig. 6.4). Saturation deficits provide a measure of the potential evaporation of moisture from leaves, soil, free water surfaces, etc. RH does not provide such a useful measure because the potential for evaporation (i.e. the saturation deficit) increases with temperature while RH remains invariant. Rundel and Jarrell (1989) provide a useful discussion on ecological measurements of atmospheric water.

Disturbance

In Section 4.4 we noted three main categories of natural abiotic disturbance that can affect plant populations: (a) fire, (b) weather (wind, ice, temperature, and precipitation), and (c) soil disturbance (erosion, deposition, flooding). By their very nature, disturbances may be unpredictable events and difficult to plan for. Wherever possible, it is important to make note of when or if these disturbances occur and affect plant populations. The most important characteristics of a natural disturbance that should be measured include the following (from Sousa 1984; White and Pickett 1985):

1. Areal extent—the size of the disturbed area.
2. Magnitude
   (i) intensity—the strength of the disturbance (e.g. fire temperature, wind speed) and
   (ii) severity—the amount of damage that occurred to the plants (e.g. stems killed).
3. Frequency—the number of disturbances per unit time.
4. Predictability—the variance in mean time between disturbances.

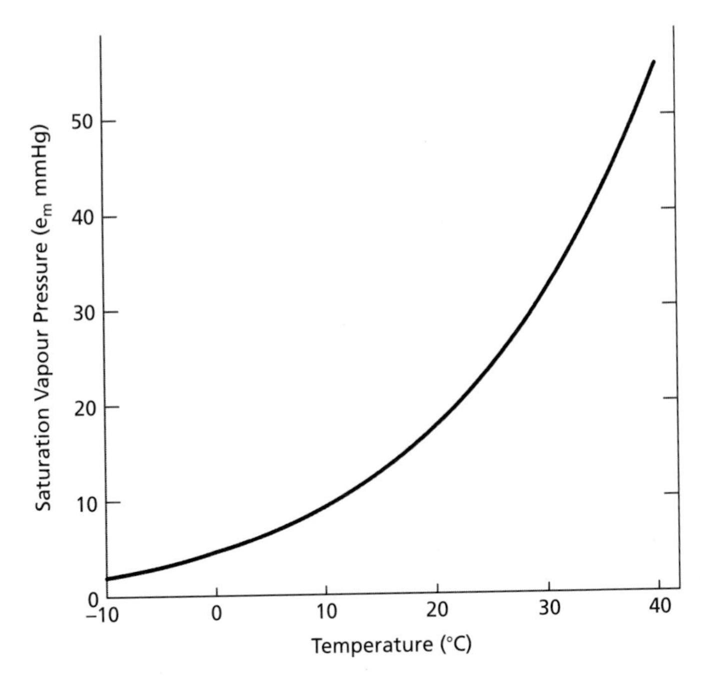

Figure 6.4 Relationship between saturation vapour pressure and temperature. Plotted from Cox (1990). Reproduced with permission from John Wiley & Sons.

5. Turnover rate or rotation period—mean time to disturb the entire plant population.
6. Return interval, cycle, or turnover time—mean time between disturbances.
7. Synergism—interactions among disturbances (e.g. drought increases fire intensity).

For each of these descriptors, measures of both central tendency and dispersion as well as frequency distributions should be made and reported.

Continuous recording devices connected to electronic data loggers can allow routine measurement of parameters such as wind, air and soil temperature, humidity, sunlight, and precipitation. Many of these electronic devices, such as those for temperature recording, are small enough to be connected to individual plants if necessary. The occurrence and magnitude of extreme weather events, such as an ice storm, may be more difficult to assess, especially at the level of an individual plant in the absence of first-hand observations in the field. For this reason it worth routinely checking marked plants, particularly following a suspected disturbance (Fig. 6.5 and Plate 16).

Fire. If fire is a planned experimental disturbance then the resulting fire temperature can be measured at the local scale of plant populations using thermocolour pyrometers, thermocouples, calorimeters, or estimates of fuel consumption (see review in Kennard et al. 2005). The intensity of unplanned wildfires is difficult to assess, although estimates can be made based on the presence of fire scars on trees and the depth or cover of ash on the soil surface. Henig-Sever et al. (2001) proposed that measurements of ash pH and the soil microarthropod community can provide an estimate of wildfire intensity.

Thermocolour pyrometers can be inexpensively constructed by painting spots of temperature-indicating paint onto the unglazed side of ceramic tiles. The tiles are placed in the field in advance of the burn and collected afterwards. The minimum temperature attained in the fire around each tile is determined as the highest-temperature paint spot that melted in the fire (Hobbs et al. 1984). This approach was used to assess the relationship between fire temperature and subsoil seed production in the grass *Amphicarpon purshii* (Cheplick and Quinn 1988). Similarly, maximum fire temperature can be estimated by placing out Tempil tablets designed to melt at different temperatures (Grace and Platt 1995). The tablets should be wrapped in aluminium foil before placing them out in the field (so

Figure 6.5 Dune plants being damaged by flooding during Hurricane Andrew on Perdido Key, Florida, United States, in 1992. Both adults and seedlings of several species showed high mortality due to inundation of seawater (Gibson et al. 1995). Recording of events such as this are often best made by field observations as soon as possible after the event. (Photo Paul Looney.) (See also Plate 16.)

that their remains can be found again) with their melting points corrected by calibration in the laboratory. Calorimeters work on the basis that fire temperature and intensity can be gauged from the evaporation of a liquid (water) from an open container. Basically, a calorimeter is a weighed and sealed container filled with water and having a hole in the top covered by a flammable material (cloth, paper, or tape) to stop spillage during transport and setup. The calorimeter is left in the path of the fire. The cover burns off during the fire allowing water to be lost as steam. After the fire, the container and any remaining water is weighed again to calculate loss (Pérez and Moreno 1998). Thermocouples measure temperature by monitoring the electromotive force gradient (voltage) produced when the temperature differs between two dissimilar metals that are brought together. The temperature of the sensing thermocouple is determined relative to the known temperature of a reference thermocouple. Thermocouples are inexpensive and allow relatively fast changes in temperature to be recorded. For example, the Type K 0.51-mm wire gauge glass-insulated chromel–alumel thermocouple (Omega Technologies, Stamford, Connecticut, United States) operates in a temperature range of −250 to 1260 °C, has a response time of 3 s, and a sensitivity of ±1 °C below 500 °C. Rates of fire spread and temperature profiles for various heights above ground can be recorded if a series of thermocouples are strategically placed out in the field (Jacoby et al. 1992). A disadvantage is that each thermocouple has to be wired directly to an electronic data logger. However, the advantage of thermocouples over thermistors and other devices is that each one does not require calibration. Ehleringer (1989) provides a good discussion on the use of thermocouples, albeit for physiological measurements.

6.3 Assessing the importance of spatial variation in the environment on plant performance within populations

It was noted at the start of Section 6.2 that an assessment of the ecological relevance of particular environmental variables upon plant performance in the early phases of an investigation may require some intuitive guesses. The examples in Table 6.2 illustrate this point. At a fairly large scale it is not too difficult to make observations about plant performance (including fitness) in relation to the environment. For example, it might be noted that individuals of a wetland species are limited in their distribution and grow better along streamsides than in other microhabitats. However, within the habitat of a species and within the spatial bounds of a population there also exists spatial heterogeneity in environmental resources (both biotic and abiotic). Individuals may be located preferentially in specific microsites, and within these microsites may show enhanced performance on certain types of microsite. Seedlings may have more specific or different microhabitat requirements than adults of the same species (cf. the regeneration niche; Grubb 1977). In this section a simple approach for assessing microsite preference for individuals in relation to environmental heterogeneity is described.

Characteristics of the microsites in which individuals of a species are located can be determined by comparing the value of environmental variables at sites containing the species with values obtained from random points where the species is absent. Griffith (1996) used this approach to assess within-habitat variation in relation to the distribution of *Viola blanda*, a forest understorey herb. She measured soil chemistry (pH, total N, P, K, Ca, Mg) and canopy openness at 100 random points and the nearest 100 points with *V. blanda* individuals along a 600-m transect. Subsequent statistical analysis (paired *t*-tests) showed that soil phosphorus and tree canopy openness were significantly higher and magnesium levels were significantly lower at points with *V. blanda* compared with random points where the plant was absent. The data collected with this method can also be usefully summarized using multivariate statistics (see Chapter 7), such as principal components analysis (Fig. 6.6). The use of this analysis shows that species are located in microsites constituting a subset of the total environment. The mechanisms responsible remain a matter for speculation without further investigation. Moreover, presence versus absence of a species from certain microsites could be due to unsuitability of the microsite (i.e. niche limitation) or it could be because of dispersal limitation. Direct seeding or planting of individuals into unoccupied microsites can help address this issue (Grace 1999; Gibson et al. 2011).

The procedure outlined above can be taken a step further if plant performance, rather than simply presence/absence within each microsite, is considered. Irrespective of data from random points, individual plant performance (as the dependent variable) can be regressed against the environmental variables in a stepwise procedure. For example, fitness of an annual forest herb (*Melampyrum lineare*) measured as the number of fruits was significantly

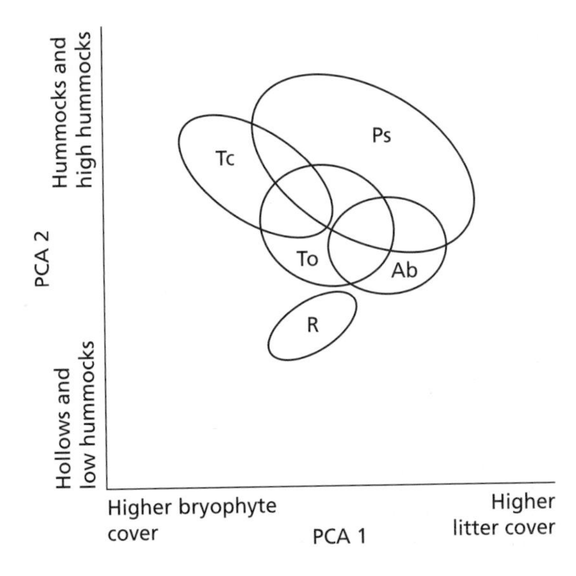

Figure 6.6 Principal components analysis (PCA) of bryophyte, litter, and muck cover among seedlings and random points for four tree seedlings: Ab, *Abies balsamea*; Ps, *Pinus strobus*; Tc, *Tsuga canadensis*; To, *Thuja occidentalis*; and R, random points. Ellipses represent 95% confidence ellipses. From St Hilaire and Leopold (1995). © 2008 Canadian Science Publishing or its licensors. Reproduced with permission.

related to canopy cover, PAR, and distance to the nearest shrub (Gibson and Good 1987). This observation followed the finding that the presence of individuals of this species was significantly related to high levels of moss cover, low leaf litter, and high PAR compared with random points within the forest habitat.

The size of the area to sample around plants and random points is an important consideration in using this approach. In both cases it should be the same. Some variables do not require a decision regarding scale to be made, for example distance to the nearest shrub or PAR incident upon a plant. Other parameters require a definition of the spatial extent of a plant neighbourhood, forcing a number of assumptions to be made. Most investigators use a circular quadrat of fixed radius with the plant at its centre for measuring cover of co-occurring plants, litter, etc., despite the knowledge that plant neighbourhoods are unlikely to be circumscribed so exactly. The quadrat should encompass only the ecologically relevant radius of the presumed microsite. This is a 'chicken-and-egg' type of task since it is not known in advance which parameters to include in such an assessment, and if we did then we would not need to carry out the investigation. In practice, investigators use quadrats with a radius approximately equivalent to or just less than the average plant height of the species concerned. For example, Gibson and Good (1987) used plots with a radius of 7 cm for *Melampyrum lineare* (13 cm tall) and *Pinus echinata* seedlings (7 cm tall), whereas Menges et al. (1999) used plots of radius 27 cm for *Dicerandra christmanii* shrubs (20–30 cm tall).

Plant neighbourhood sizes can be estimated and may provide guidance in determining the optimum quadrat size for the procedure outlined above. Non-linear hyperbolic regression equations can be fitted for a range of radii using target plant biomass as the dependent variable and distance, number, angular dispersion, and biomass of neighbours as independent variables. A maximum likelihood estimator allows the 'best' radius to be chosen to represent the plant neighbourhood. Details of how to apply this procedure are provided in Pacala and Silander (1990). Alternatively, Thiessen polygons (Chapter 8) can be used to determine the size of individual plant neighbourhoods which can then be related to plant performance. This analysis requires that all individuals in a population are mapped, or at least all individuals in a representative area. For example, reproduction in the annual kelp *Postelsia palmaeformis* was positively related to Thiessen polygon area (Barner et al. 2012). The correspondence between mapped point locations of plants and corresponding environmental parameters can also be analysed using kriging, a geostatistical interpolation method (Legendre and Legendre 2012), or spatial autocorrelation (Chapter 8).

A quite different probabilistic approach to inferring the importance of plant niches on plant performance was taken by Diez et al. (2014). Integral projection models (IPMs; see Chapter 8) were used to predict population growth rates of an orchid in response to environmental variables at microsite (2 m × 2 m cells), population (480-m^2 grids), and landscape (a deciduous forest watershed) scales. In this case, the orchid was frequently absent from microsites of high predicted suitability (consistent with dispersal limitation) and present in sites with low predicted suitability (consistent with ideas of source–sink dynamics). At larger scales, orchid abundance was positively correlated with predicted demographic suitability. The value of this approach is that by scaling up from microsite to landscape scales, the importance of niche characteristics on plant performance can be recognized.

6.4 Follow-up exercises

1. Consider one of the species that you study, or are interested in. What size of quadrat do you think would be appropriate for quantifying the ecological niche of the species? What criteria do you consider in making this decision? Would quadrat size vary with life-history stage?
2. Again, for a particular named species: which environmental variables do you think are limiting its performance? Do the same variables limit this species as a seedling or adult plant, or its fecundity?
3. How different is the methodology for measuring the importance of PAR in the following habitats: forest understoreys, grasslands, shallow water aquatic systems, desert, alpine tundra?

PART 4

Analysis

CHAPTER 7

Planning, choosing, and using statistics

Statistics are the triumph of the quantitative method and the quantitative method is the triumph of sterility and death.
(Attributed to Hilaire Belloc)

Some people hate the very name of statistics, but I find them full of beauty and interest. Whenever they are not brutalised, but delicately handled by the higher methods, and are warily interpreted, their power of dealing with complicated phenomena is extraordinary. They are the only tools by which an opening can be cut through the formidable thicket of difficulties that bars the path of those who pursue the Science of Man.
(Sir Francis Galton cited in Pearson (1930) *The life, letters and labours of Francis Galton*)

- **Choosing the right statistical test**
- **Different kinds of distributions**
- **Goodness of fit tests, contingency tables, checking, and correcting for normality**
- **Correlation and regression; relations between data sets**
- **Looking for differences between samples**
- **Multivariate techniques for ordering and classifying samples**
- **Bayesian analysis**
- **Reporting statistics**
- **Case studies**

Preamble

Many ecologists view data analysis with trepidation. Many of us feel that we are better at collecting data than analysing them, and the range of techniques available can be very daunting. Nevertheless, there is wide recognition of the importance of statistics in both undergraduate and graduate training. Statistics is not a bolt-on extra, rather it forms an integral part of our science that can help to inform the whole process from a study's design to its conclusion.

The data collected by plant population ecologists are almost always highly variable, and statistical techniques are necessary in order to make decisions about whether groups of data differ from each other by an amount which is due to more than the natural variation among replicates. In plant ecology, variation can come from a number of sources (e.g. the genotype of the plants being studied, the location in which the plants are growing, variation in the weather) in addition to the independent variables (Chapter 3) of specific interest. Statistics can help to unravel the contributions of each of these factors to variation in the data. Understanding the contribution of non-treatment sources of variation to differences between plants can itself be revealing (e.g. Rees et al. 2000).

While it is best to consider analysis from the outset of a study, not every study proceeds in this logical and sensible way. However, the majority of the best and most convincing studies are designed with analysis in mind. This advance planning ensures that the data are collected in an appropriate way to answer the questions of interest, and that the study progresses rapidly from data collection to analysis

Methods in Comparative Plant Population Ecology. Second Edition. David J. Gibson.

without a motivation-sapping pause while statistical advice is sought to rescue it.

Statistics text books are often large, door-stopping volumes which can seem quite daunting. However, there are a few smaller, more user-friendly texts which will help you to get started. For instance, Dytham (2011) uses non-technical language to describe a wide range of tests and how to perform them using three common statistical packages. His book also includes a useful annotated bibliography and a key to guide you to the most appropriate test for your question and data. Other useful introductory texts include Bailey (1995), Fowler et al. (1998), Heath (1995), Wheater and Cook (2000), and Gotelli and Ellison (2013). However good these smaller texts are, you will eventually get to the point where you will actually *want* to refer to the door-stoppers (really). Some of the more comprehensive texts commonly used by ecologists include Snedecor and Cochran (1989), Steel et al. (1996), Zar (2010), Sokal and Rohlf (2011), and Legendre and Legendre (2012). Scheiner and Gurevitch (2001) discuss statistical problems specifically encountered by ecologists and Underwood (1997) investigates the important connections between experimental design and analysis of variance (ANOVA). Waite (2000) describes some of the techniques needed by field ecologists.

Computer statistics packages are routinely used for statistical analysis. The advantages of these programs are immense and obvious in terms of time saving and accuracy of calculation. However, they do have some pitfalls. Paradoxically, one of the major pitfalls is the accessibility of complex analyses. Very little statistical knowledge is needed to run the programs, and it is easy to do inappropriate or poorly understood analyses. Sometimes the manual or help section gives some guidance about the assumptions behind a test, but this information is usually not comprehensive, nor intended to fully explain the purposes of the technique. Packages commonly used by ecologists include SAS (<http://www.sas.com/>), SPSS (<http://www-01.ibm.com/software/analytics/spss/>), MINITAB (<http://www.minitab.com/en-US/products/minitab/>), SYSTAT (<http://www.systat.com/>), and GENSTAT (<http://www.vsni.co.uk/software/genstat>). Many of these packages will also produce good-quality figures. In the Technological Tools section of its *Bulletin*, the Ecological Society of America occasionally reviews and compares statistical packages from the point of view of ecologists, and you might find one of these reviews helpful if you are trying to decide which package is best for you. However, most of these packages provide a similar range of statistics, and it is generally only the less commonly used tests that are not universally available. A major factor in deciding which package to use, especially if you are fairly inexperienced, should probably be to find out what your colleagues are using. A lot more informal support and advice will be available if you use the same program as your colleagues.

Increasingly, ecologists are using the free, downloadable R package to run their statistics (<http://cran.r-project.org/web/pacakges/>). Good introductions to R for ecologists, including sample code, are Gardener (2011), Beckerman and Petchey (2012), and Crawley (2012). A list of R packages for many statistical applications used in plant population ecology is provided in the Appendix. If you are transitioning to R from SAS or SPSS, then Muenchen's (2011) book is excellent.

The aim of this chapter is not to supply methods for performing statistical tests but to remind you of the range of statistical techniques available and to give some guidance as to direction. The assumption is that most readers of this book are familiar with (or at least have a vague receding memory of) basic statistical concepts. However, learning about statistics should be an ongoing process that continues throughout your scientific career. Tests included in this chapter are based upon personal experience and knowledge of current statistical practices in plant population ecology. The aim is to fill in the gap between the basic statistical advice available in introductory courses and some of the rather sophisticated techniques regularly used by plant ecologists and now available in many statistics packages.

The most basic text in which you will find useful information relevant to a particular test is referenced here, giving more advanced references where necessary. Many of the tests mentioned here are covered in most statistics text books. With the exception of some of the advanced methods covered in Chapter 8, the basic statistics used by plant population ecologists are not very different from those used in many other branches of ecology. The statistical tests used in the case studies referred to in Chapter 1 are detailed at the end of the chapter.

What statistics can do for you: P-values, Type I and Type II error, and power

The end point of any statistical analysis is a test statistic which is associated with a probability. This is the probability that your results could have been obtained by chance alone, given that your null hypothesis (Chapter 2) is actually correct.

With very few exceptions (which are quite hard to think of!) the aim of any statistical test should be to test a hypothesis. The clearer you are about the hypothesis you are testing, the easier it will be to choose your statistical analysis. It is conventional to test a null hypothesis against an alternative hypothesis (Chapter 2). If your null hypothesis is rejected, then your alternative hypothesis must be accepted. We perform statistics in order to find out the probability that the null hypothesis is true. Before we begin the analysis we set a level of probability that will allow us to confidently reject the null hypothesis in favour of our alternative hypothesis. This probability, termed α (alpha) is usually set at 0.05, or 5%, or 1 in 20. We consider that a 1 in 20 probability of an event happening by chance is sufficiently improbable that we can reject it as a probable explanation for our observed result. So if a test tells us that a result has a probability of 0.05 or less, we usually consider it a significant result and reject the null hypothesis. However, there may be times when you wish to choose a lower or higher level of significance. If it is very important that you do not reject a correct *null* hypothesis (for instance when doing so would lead to inappropriate management of a rare plant population); if so then you might choose to set a lower probability level as your cut off for significance. Conversely, if you are concerned not to falsely accept a null hypothesis, then you could choose a higher significance level. The important thing is to decide what significance level you are going to accept *before* you do the test and for the kinds of reasons outlined above, and not *after* the test when you are disappointed not to be able to reject your null hypothesis at $P = 0.05$. If you get a result that is very close to your pre-determined significance level (for instance $P = 0.052$ or $P = 0.049$) you should be extra cautious in your interpretation of the results. However, no statistical test can tell you whether your results are biologically significant. You, the investigator, must interpret the results.

Because statistical tests are based on probability, by the laws of chance there will be occasions when you reject a true null hypothesis or fail to reject a false null hypothesis. These two possibilities are termed Type I and Type II errors, respectively, and have been discussed in Chapter 3. The probability of rejecting a false null hypothesis (i.e. correctly identifying a significant result) is called power, and can be calculated for each analysis. Power depends on the test you are using, your sample size, the variance of the population being studied, the level of α that you have set, and the size of the effect you are looking for. It is also discussed in Chapter 3. You should consider the potential power of your tests at an early stage in your investigation. Prospective power analysis using variances from a pilot study or a previous similar investigation can be very useful in deciding how many replicates you need in a particular study (Toft and Shea 1983; Cohen 1988; Underwood 1997; Zar 2010). While no one wants to waste time inefficiently collecting unnecessary data, it can be very frustrating to realize retrospectively that if only you had taken a few more replicates your results might have been significant.

Pseudoreplication

We cautioned against pseudoreplication in Section 3.3. Pseudoreplication means that the data points are not independent of each other (Hurlbert 1984) as required by many statistical tests. It is common in large-scale studies such as grazing studies, where many measurements may be made within one large exclosure. Measurements made within a single experimental unit such as an exclosure are not truly independent. If you realize retrospectively that you have committed this statistical sin then you could investigate whether there is any possibility of analysing your data as a nested or split-plot design (Chapter 3), but this will lead to a reduction in the number of replicates and will sometimes leave you with no replication.

Descriptive statistics

After collecting data, it is often a good idea to do some simple preliminary data inspection before beginning the analysis. You can then reduce a huge mass of numbers to something you can start to get to grips with. By calculating some simple descriptive statistics that summarize your data you will get a good idea of problems that you are likely to encounter during analysis. Simple frequency plots and scatter plots will give you a good idea whether your data are normal or not. The appropriate descriptive statistics can tell you a lot about the magnitude and dispersion of your data. Occasionally, particularly if you are not testing a particular hypothesis, descriptive statistics are all that is required to answer your question.

When presenting your descriptive statistics it is important to make sure they are appropriate for the type of data that you have. For instance, if the data are not normally distributed, then presenting mean ± standard deviation might not be as appropriate as median and interquartile ranges. Similarly, bar charts and error bars may not be as informative as a box and whisker plot, or some other plot which shows the central tendency and dispersion of the data. Tufte (1983) and Ellison (2001) present a wide range of imaginative ideas for presenting data.

7.1 Choosing the right statistical test

Your choice of test should depend on the question you want to answer. If you are very clear about the null and alternative hypotheses in your study, and if these are stated as specifically as possible, then you will have gone much of the way towards identifying the appropriate test for your data. Many statistics texts have a key or table to help you choose between the many tests that are available (e.g. Dytham 2011; Sokal and Rohlf 2011). In some cases there may be more than one appropriate test. It is sometimes quite tempting to use a technique that you are familiar with rather than to learn a new one that is more appropriate for a new question. The design of your experiment or sampling regime should dictate your choice of statistical test, not vice versa (Potvin 2001).

It is also important to establish whether your data fit the assumptions of the test. Most tests have some built-in assumptions that should be met by your data in order for the test to be reliable. These may include the number of samples or replicates, whether the data are interval, ordinal, or categorical (Chapter 2), and whether the data follow a particular distribution.

Parametric tests assume that the data are normally distributed and are less reliable for data that are not normally distributed. However, if your data meet their assumptions they are often very powerful tests (i.e. they have a good chance of rejecting a false null hypothesis). If your data do not fit a normal distribution, yet a parametric test seems to be the most appropriate, you may be able to transform your data so that they become normally distributed (see below). However, there are many commonly used tests that do not make specific assumptions about the data being normal. These are non-parametric tests. They are often less powerful than their parametric equivalent but are 'safer' (i.e. less likely to lead you to falsely ascribe significance to a result and reject a true null hypothesis) if the normality of the data is in doubt. Some non-parametric tests are the most appropriate tests to use even if your data are normally distributed. For instance, the well-known chi-square test is a non-parametric test of the independence of data arranged in a contingency table. The text by Conover (1999) is devoted to non-parametric statistics and Potvin and Roff (1993) provide a useful review of the ecological uses of a number of non-parametric or 'distribution-free' tests. However, as Underwood (1997) points out, many non-parametric tests do make quite demanding assumptions about the data. Although they do not assume that the data are normal, they do assume that the samples being compared have similar distributions to each other. Most parametric tests have a non-parametric equivalent or can be conducted using rank transformation of the data (Conover 1999).

What are the risks of using an inappropriate test? If you seriously violate the assumptions of your test, you risk your analysis being completely meaningless or arriving at the wrong answer. At the very least you risk a loss of power, that is, you reduce your chances of finding a true significant result (Chapter 3).

Should your test be *one-tailed* or *two-tailed*? We are often interested in looking for any differences between treatments or groups, and a two-tailed test is most suitable for this purpose. Less commonly, we are only interested in whether one treatment or group is specifically larger (or smaller) than another, and then a one-tailed test is appropriate. If you already know that your sample is larger or smaller than the mean against which it is being compared, it might be tempting to use a one-tailed test, as it can increase the chances of rejecting the null hypothesis. However, one-tailed tests should only be used in those situations where only one result is relevant to the hypothesis under test (Heath 1995). As an example, you might hypothesize that a conservation management regime increases the size of a population of a rare plant compared with a regime without management. As you would only implement the management regime if it led to an increase in the size of the population, then a one-tailed test is justified. Choosing between one- and two-tailed tests is another situation where a very clear statement of the null and alternative hypotheses will help you to decide on the appropriate statistical test.

7.2 Different kinds of distributions

In Chapter 2, different kinds of ecological variables were described (see Table 2.3). In this section, the four main types of distribution that data may take are described.

There are a number of theoretical distributions that should be familiar to all biologists because many biological phenomena are reflected by them. Most of these are described early in any statistics course, and are covered in any basic text. The descriptions in Fowler et al. (1998) are particularly helpful. The aim here is merely to provide quick reminders. The use of these distributions for detecting spatial patterns in plant populations is discussed in more detail in Chapter 8.

The Poisson distribution

The Poisson distribution describes the number of times something occurs in time or space. It is appropriate when the observations consist of counts, and the counts are made within defined objects (e.g. a quadrat) or periods of time. An example would be the number of lichen thalli per individual tile on a roof. The assumptions of the Poisson distribution are that: (a) the observations or events are quite rare in space or time and the mean number is small relative to the maximum possible (e.g. there are a lot fewer lichen thalli per tile than a tile could hold); (b) occurrences are random and independent of each other (e.g. each lichen thallus is the result of an independent and random colonization event by a lichen propagule).

The mean and variance are equal in the Poisson distribution. Departures from this rule can provide useful information about the spatial pattern of the population. If the variance is greater than the mean, the population is underdispersed (or clumped) and if the variance is less than the mean, then the population is overdispersed (or regular). In fact, true randomness is not all that common in most population ecology studies (Chapter 8).

If it is found that the distribution of numbers of lichen thalli per tile matches that expected from the Poisson distribution then there is support for the hypothesis that the lichen distribution pattern is as a result of random processes. However, Dale (1999) has pointed out that the spatial arrangement of the sampling units must also be taken into consideration before concluding that a distribution which matches the Poisson one is necessarily random. For instance, although the numbers of thalli per tile might match a Poisson distribution, if the tiles which contain more thalli are all to be found in one area of the roof, then stochastic processes cannot be invoked for the pattern of lichen distribution.

The binomial distribution

The binomial distribution can be used to describe the pattern of distribution of events when there are only two possible outcomes from each event and the probability of each is constant. Although most text books illustrate the binomial distribution using the numbers of males and females that emerge from animal broods of various sizes, it is, in fact, appropriate for any situation where observations can be put into one of two possible categories, such as flowering or not flowering, dead or alive, *Festuca ovina* or *Festuca rubra*. If we know the probability of one of these events occurring, then as there are only two possible outcomes we must know the probability of the other. From that it is possible to calculate the probability of finding samples with different numbers of plants in each possible state. It is also possible to use the binomial to examine the distribution of objects when the variance of the count data is less than the mean and the objects are distributed regularly. A good fit of the lichen data described above to the binomial distribution would suggest that some process was preventing lichens from colonizing roof tiles in the vicinity of another lichen.

The negative binomial distribution

The negative binomial distribution can be used when, like the Poisson distribution, counts are from defined objects such as quadrats. However, unlike the Poisson distribution, the variance is larger than the mean and the counts are not randomly distributed but clumped, so that there are many quadrats (or other sampling units) with very few or very many counts. Only two parameters are needed to completely describe the negative binomial: the mean and k, a measure of clumping. The smaller k is, the more highly clumped the distribution (Waite 2000). The negative binomial reflects a large number of biological phenomena. If the lichens on roof tiles were found to follow a negative binomial distribution that would suggest that some underlying process is causing clumping, perhaps during colonization or subsequent establishment.

The normal distribution

The normal distribution is the most frequently encountered distribution for continuous data. It is important to be familiar with it, as many statistical tests assume that the data follow a normal distribution. Its shape is a familiar one, a symmetrical bell shape. Only two parameters are needed to completely describe a normal curve: the mean and the standard deviation. If the data are normally distributed 95% of the data points will lie between the mean and 1.96 standard deviations on either side of the mean. Large samples of other distributions such as the Poisson and the binomial may also converge on the normal distribution. Large samples of means of samples taken from other distributions should themselves follow a normal distribution (central limit theorem).

A data set can differ from normality in one of two ways, known as *skewness* and *kurtosis*. Skewness describes the position of the mean, which in a skewed distribution is nearer to one of the tails of the distribution rather than being at the centre. Distributions can be skewed to the right or the left, and these deviations are referred to as positive or negative skewness, respectively. A distribution which is symmetrical about the mean can also deviate from the normal in its degree of flatness, or kurtosis. A distribution which has more data points close to the mean and in the tails is leptokurtic, while one which is flatter and has more data points in the shoulders is described as platykurtic. If your data differ significantly (see 'Goodness of fit tests' in Section 7.3) from normal, then you may be able to transform them to become normal so that statistical tests which assume normality are still valid (see 'Transformations' in Section 7.3).

7.3 Goodness of fit tests, contingency tables, checking, and correcting for normality

Sometimes you need to find out whether or not your data fit a particular pattern or distribution. There are two main reasons for wanting to do this. The first is to test a hypothesis (about plants) that hinges on your data following a particular distribution. For example, you may want to test whether the number of seedlings per quadrat follows the Poisson distribution, indicating that their colonization pattern had resulted from random processes. The second reason is to investigate whether your data follow a normal distribution so that you can legitimately use parametric statistics.

Goodness of fit tests

The chi-square test is well known. It can compare a known (observed) distribution (divided into categories if it is continuous) with an expected distribution. The expected distribution is the pattern your data would follow if they conformed to the hypothesis being tested. The calculation of the expected distribution depends on the hypothesis you are testing, and should be thought about carefully. It should also be clearly explained when your analysis is reported. An advantage of the chi-square test is that the mathematics involved is very simple. As a quick reminder

$$\chi^2 = \sum (o_i - e_i)^2 / e_i \qquad (7.1)$$

where o_i is the number of observations in category i and e_i is the expected number of observations in category i. It can easily be calculated by hand. The test statistic (χ^2) is then compared with the chi-square distribution. Although it was thought that this test became unreliable if any of the expected values was less than 5, more recent advice (Zar 2010) is that the average expected value should be greater than 6, which is less restrictive.

An alternative to the chi-square test which is favoured by some authors for its theoretical and computational advantages is the *G*-test (Fowler et al. 1998; Zar 2010; Sokal and Rohlf 2011). This is also known as the log-likelihood ratio and can be simpler to carry out than the chi-square test for complicated designs. It also compares observed and expected values, and can be used in all cases suitable for the chi-square test. The *G*-test should be used in preference to the chi-square test when $|(o_i - e_i)| < e_i$ in one or more cells of the contingency table (Williams 19767). *G* can be calculated easily by hand, and the test statistic follows the χ^2 distribution. Its formula is

$$G = 2\sum o_i \ln(o_i / e_i) \qquad (7.2)$$

where o_i and e_i are as in Equation (7.1) and ln is the natural logarithm.

The Kolmogorov–Smirnov test can be used to test whether data from a large sample of continuous data fit a known distribution such as the normal distribution (for instance, before performing a parametric analysis). It can also be used to examine whether two samples follow the same distribution. The Shapiro–Wilk test is considered to be more powerful for detecting departures from normality for small samples ($n < 50$ measurements). Additional options are the Anderson–Darling test (Waite 2000; Dytham 2011) which determines whether sample data come from a specified distribution (usually normal, but others can be specified), and the D'Agostino–Pearson test (D'Agostino et al. 1990; Zar 2010) which can be used to find out whether data deviate from a normal distribution due to skewness and/or kurtosis.

Remember that in most tests for normality the null hypothesis is that the data are normal, so a significant result (i.e. $P < 0.05$) is indicative of a need to transform the data.

Contingency tables

When you are interested in whether or not two (or more) categorical variables are independently distributed the data can be displayed as a contingency table. For instance, if you are interested in testing the null hypothesis that flower colour is independent of soil type then you would record the number of flowers of each colour in each of several soil types and the data could be arranged in a contingency table with the number of rows and columns corresponding to the number of flower colours and soil types, respectively. The expected values can be calculated by holding the row and column totals the same and calculating the value that would be expected if individuals were allocated to categories based only on their proportional row and column abundance [(row total × column total)/grand total], or they can be calculated based on some other hypothesis. While the chi-square test or *G*-test are often used to analyse contingency tables, Zar (2010) and other authors recommend the Fisher exact test for the special case of 2 × 2 contingency tables. Sokal and Rohlf (2011) give an extensive discussion of the advantages and disadvantages of each of these three tests and the situations in which each is most appropriate.

Testing for homogeneity of variance

Another assumption of some parametric tests such as ANOVA is that the variances of groups of data that are to be compared are homogeneous or homoscedastic. That is, the variances of the groups aren't very different. If the variances are too different (heterogeneous or heteroscedastic) then some tests, such as ANOVA, become unreliable (Underwood (1997) has a useful discussion of this). As it is very unlikely that each of your groups of data will have identical variances you need to have a way of assessing how much variation is too much, although sometimes just examining the variances of the groups will make it clear that you have a problem. There are a number of tests available to help you decide whether your variances are sufficiently homogeneous. Fowler et al. (1998) recommend examining the ratio of the largest to the smallest group variance in the F_{max} test. Other commonly used tests are Bartlett's and Levene's tests. Underwood (1997) does not recommend either of these tests, on the grounds that Bartlett's test is too severe (ANOVA can easily cope with variances more heterogeneous than Bartlett's test would accept) and that Levene's test uses circular reasoning. Underwood (1997) recommends Cochran's C to determine if an estimated variance is significantly larger than a group of variances (or standard deviation). Cochran's C is equal to the largest variance divided by the sum of all the group variances. Critical values of C have been tabulated for different numbers of groups ('t Lam 2010). If C is significant, then the data are probably not suitable for ANOVA. However, all these tests are currently in use by plant ecologists, and Dytham (2011) recommends Levene's test as being readily available in computer statistics packages.

Transformations

As mentioned above, parametric analyses are generally more powerful than their non-parametric equivalents, as long as the data are normally distributed and homoscedastic (i.e. have homogeneous variances). However, if your data do not conform to these assumptions you may be able to transform them so that they do. A transformation is a mathematical function applied to the whole data set, which can, in certain circumstances, normalize the data and help to homogenize the variances. Both problems can often be solved with the same transformation. Most peoples' first reaction to hearing about data transformations is that they must be some sort of statistical massage (or be 'dodgy' at best), but they are perfectly legitimate as long as you transform the data for the reasons outlined above *before* analysis and not in order to try and change a result that you don't like!

There are three transformations that are commonly used to try to normalize data (Durrett and Levin 1994; Dytham 2011; Foster and Gross 1998):

1. Square root transformation. If the group variances are equal to (or similar to) the means, so that bigger means have proportionally bigger variances, then the square root transformation may help. Simply take the square root of each datum. However, if there are zero values in the data set, add 0.5 to each datum before transformation.
2. Arcsine (or angular) transformation. Data sets composed of proportions or percentages are unlikely to be normally distributed because their distributions are inevitably truncated at each end of the scale (by being limited to lie between 0 and 1 or 0 and 100). The arcsine transformation, in effect, stretches out both ends of the distribution. In this transformation the arcsin of the square root of each datum is taken. If you were doing this by hand you would first take the square root, then look up the angle whose sine takes that value. Warton and Hui (2011) argue that arcsine transformations are inappropriate for both binomial and non-binomial response variables. Instead, they suggest logistic regression for binomial data (e.g. yes/no, dead/alive variables; see Section 7.4) and a logit transformation for non-binomial data. The logit transformation is typically applied to linearize sigmoid distributions of proportions as the natural log $\ln[p/(1-p)]$ where p is a proportion.
3. Logarithmic transformation. If your data are skewed to the right or you have group variances which are positively correlated with their means, then the logarithmic transformation may normalize the data and homogenize the means. Simply, the logarithm of each datum is taken; if there are zeros in the data a small number (often 0.5 or 1.0) is added to each datum before transformation. Logarithms of any base of can be used, although base 10 is common. Figure 7.1 shows an example of a logarithmic transformation. O'Hara

and Kotze (2010) caution against using the log-transformation when dispersion of the data is small and mean counts are large. In these cases, they recommend quasi-Poisson or negative-binomial models (Ver Hoef and Boveng 2007) and the use of generalized linear mixed models (GLMMs) and their derivatives.

Other transformations are described in Waite (2000), Zar (2010), and Legendre and Legendre (2012).

After transformation you will need to check again that your data now conform to the required distribution before conducting any tests. It is unlikely that any transformation will normalize data containing a large number of zero values, in which case you may have to consider using a non-parametric analysis or a GLMM. Following analysis, it is also necessary to back-transform or use untransformed data (including appropriate measures of variation) in presenting the data. Transformed data values make little sense to a reader.

Rank transformation

For many non-parametric tests your data will need transforming into ranks, that is each datum is allocated a number based on its ranking within the group. Observations with the same value are usually given a shared rank value based on the mean of their ranks. In many statistics packages, ranking will be done as part of the program, if appropriate, and the transformation will be invisible to the user.

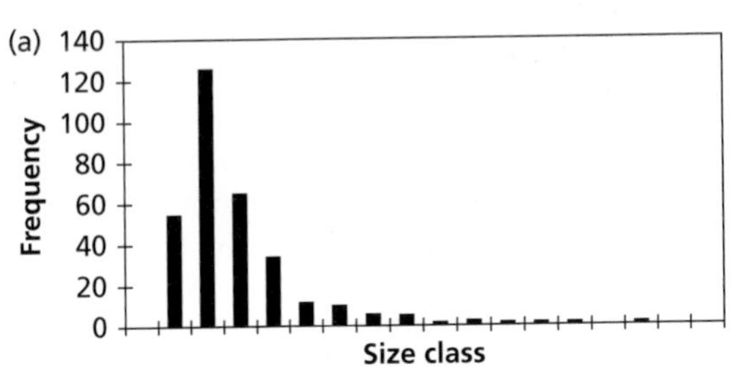

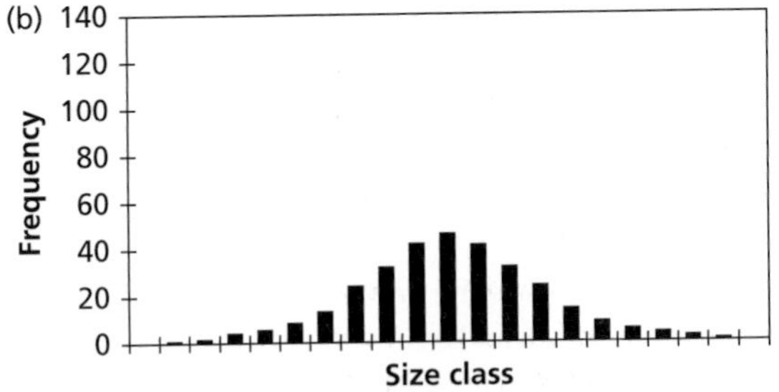

Figure 7.1 Example of logarithmic transformation. Frequency data are plotted (a) before and (b) after transformation. The data do not show a normal distribution before transformation, but are approximately normal afterwards.

7.4 Correlation and regression; relations between data sets

Ecologists often collect data on two variables (e.g. rosette diameter and number of flowers per plant) and want to know what the relationship between them is. Correlation and regression are two familiar sets of techniques for examining the relationship between variables. These tend to be considered together as alternative ways of examining data, and although the mathematics behind them is similar they address different questions. Correlation addresses the question of whether two variables vary together, whereas regression specifically addresses the question of whether there is evidence that one variable depends on the other, i.e. whether there is evidence for a causal relationship. If there is no a priori reason for thinking that one variable depends on the other then correlation is the more appropriate technique. As Dytham (2011) says, if you are unsure which of the two variables is the causal one then regression is almost certainly inappropriate.

Measures of correlation

The most commonly used measure of correlation is the *Pearson product–moment correlation*. This is a parametric test that requires both data sets to be normally distributed (Bailey 1995; Dytham 2011) and both to have been selected randomly from the wider population of values of which they are samples. If one or both of the variables have been controlled or selected to represent only particular values then a normal distribution is unlikely. Another requirement is that the relationship between the two variables is approximately linear. You may be able to transform one or both of your data sets (as appropriate) to meet the assumptions of normality and linearity (Fowler et al. 1998).

The correlation coefficient, r (i.e. the degree to which variation in the two variables is interdependent), ranges from −1 (perfect negative correlation) to +1 (perfect positive correlation). Details of how to calculate r can be found in Fowler et al. (1998). The significance of a particular value of r depends on the number of measurements used in the analysis. When reporting correlation, it is important to mention the sample size and the level of significance as well as the value of r.

If your data are not normally distributed, and cannot be transformed to normality, there are two non-parametric correlation coefficients that should be considered. *Spearman's rank correlation* (Spearman's rho, ρ) is similar to the Pearson correlation but is based purely on the ranks of the data. Each variable is transformed into ranks, and the only requirement of the data are that they can be sensibly ranked. The Spearman test produces the statistic r_s or ρ, which, like r, varies from −1 to 1 (indicating perfect negative to perfect positive correlation), although r and r_s will not be the same for a given data set. Although the Spearman test is less powerful than the Pearson one, it is much more likely to be appropriate for many data sets that do not meet the assumptions of the Pearson test (Dytham 2011). Kendall's tau (τ) is a less frequently used non-parametric measure of correlation which is appropriate in the same situations as the Spearman test. Its advantages and disadvantages over Spearman's ρ are discussed in Conover (1999).

There are many ecological situations where more than two variables are of interest and may covary. *Partial correlation* allows the investigation of the relationship between two variables when all other variables are held constant, so that the effect of other variables on the relationship of the two variables of interest is eliminated (Zar 2010; Legendre and Legendre 2012).

It is very important to bear in mind that finding that two variables are correlated does not mean that there is a causal relationship, i.e. that either variable is causing the change in the other one. For example, they both could be controlled by a third, unanalysed, variable.

Regression

Regression is used to test the dependence of one variable on another, i.e. whether a change in value of the putative controlling (independent) variable produces a predictable change in the dependent variable. However, as for correlation, even after a significant regression has been performed, deductions about cause and effect should be made very cautiously unless you have specifically and deliberately manipulated the value of the controlling variable. Regression fits a line to the data that best describes the relationship between the two variables. The line allows you to make a prediction about the most likely value of the dependent variable (the one on the y-axis) from a given value of the independent variable (the one on the x-axis, which is under your control).

Simple linear regression

Simple linear regression (Model I regression) is the most basic form of regression analysis. The analysis finds the straight line which best fits the data and minimizes the sum of the deviations of the values of the dependent variable from the line. Because the squared deviations (of each point from the line) are minimized in the analysis it is sometimes called the least squares method (Fig. 7.2). Regression lines are unlikely to be a perfect fit in plant ecology so there is usually some residual variation in values of the dependent variable either side of the line which is not accounted for by the regression model. Fowler et al. (1998) introduce the nuts and bolts of regression. More comprehensive treatments can be found in Zar (2010) and Sokal and Rohlf (2011). The output of regression analysis includes the parameters needed to describe the best fit straight line of the form

$$Y = a + bX \quad (7.3)$$

where Y and X are the values of the dependent and independent variables, respectively, b is the slope of the line (or the amount by which Y increases or decreases for each unit increase in X), and a is the value of Y when X is zero. In addition to the form of the straight line, it is usual to report r^2 after a regression. This is a measure of the proportion of the

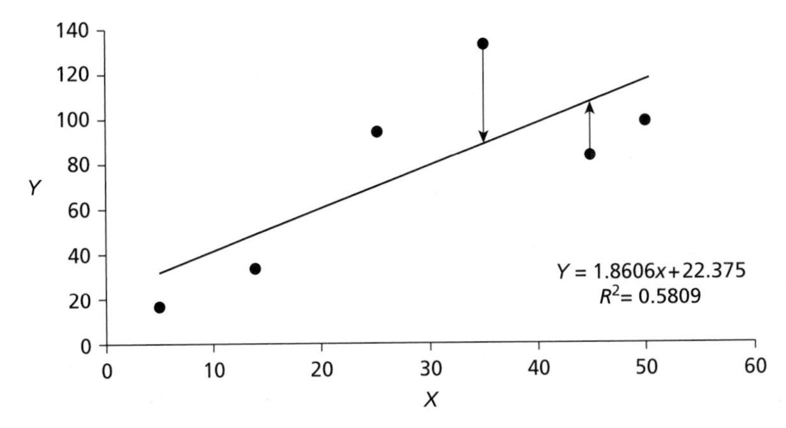

Figure 7.2 Graphical demonstration of the least squares regression method. The regression line is the one that minimizes the squared vertical deviations of the Y values from the fitted line. Unless the data all lie on the line (very unlikely in ecological applications) there will remain some variation that is not explained by the regression. This is indicated for two data points by the arrows. The equation at the bottom right describes the line of best fit (see text) and the r^2 for this line is also given.

variation in the values of the dependent variable that is explained by the regression line. If r^2 is 0.64 then 64% of the variation in the y values is explained by the regression model, with $1 - r^2$ being the unexplained proportion. r^2 is the square of the Pearson correlation coefficient that would be obtained with the same data.

As the magnitudes of b and r^2 are not in themselves measures of the statistical significance of the regression you will usually need to test its significance, either by using a t-test to test whether the slope, b, is significantly different from 0, or by using ANOVA (Fowler et al. 1998). ANOVA examines whether the variation explained by the fitted line is significantly greater than the residual (unexplained) variation. ANOVA may be done automatically by your statistics package when you ask for a regression. With very large sample sizes it is possible to have a significant relationship that only explains a small proportion of the variation in the independent variable. It may also be of interest to test whether two regression lines differ significantly in slope (steepness of the line) or elevation (height above the x-axis). Zar (2010) has a chapter devoted to this topic.

The assumptions underlying Model I linear regression are that there is a linear relationship between the two variables, that the independent variable is under the control of the investigator (and thus can be measured with no, or negligible, error), and that the variances of the x values associated with the various y values are all equal. If the relationship between the two variables is not a linear one, then a transformation may correct the situation (Fowler et al. 1998).

Model II regression is appropriate when the independent variable cannot be measured without error, as is likely to be the case in many ecological situations. Sokal and Rohlf (2011) and Legendre and Legendre (2012) discuss Model II regression.

In the biological interpretation of the results of a regression analysis it is important not to extrapolate beyond the boundaries of the data set, unless there is a very good reason to think that the relationship will not change with the magnitude of the variables.

Multiple regression

Multiple regression may be appropriate when you suspect that the dependent variable has a number of factors (independent variables) which are all affecting it. Multiple regression calculates the linear best fit of a number of independent variables to a single dependent variable (Sokal and Rohlf 2011), according to the model

$$Y = a + b_1X_1 + b_2X_2 + b_3X_3 + \cdots + b_mX_m \qquad (7.4)$$

where b_1 is the effect due to independent variable 1, b_2 is the effect due to independent variable 2, and so on to the final (mth) independent variable. The regression coefficient for each factor can be considered as the effect of that factor if everything else is held constant. The size of the different coefficients indicates the strength of the effect of each variable. The multiple regression can be tested for significance in much the same way as the simple linear regression.

Multiple regression is quite commonly used in plant ecology as we often suspect that several factors are simultaneously affecting the dependent variable. It allows us to develop models to examine how a number of factors contribute to an effect. For example, Escudero et al. (1999) used multiple regression to investigate which of several possible factors (features of soil, microhabitat, or surrounding plant cover) were affecting emergence and survival rates of *Helianthemum squamatum*, an endemic gypsophile of semiarid Spain. They concluded from the multiple regression that a complex set of abiotic and biotic factors was controlling emergence. Section 7.3 gives a discussion on the use of multiple regression for assessing the importance of environmental variation on plant performance.

The same assumptions must be met for multiple regression as for simple linear regression, with the additional assumption that each of the possible controlling variables must be independent of the others. Philippi (1993) discusses the use of multiple regression for building models in plant ecology.

A problem in multiple regression might be to identify which of the independent factors make an important contribution to the model describing the dependent variable. Stepwise regression is a form of multiple regression in which the independent variables are either sequentially added (forward selection or step-up procedure) or removed (backward selection or step-down procedure) from the model. At each stage the model is examined to see whether its explanatory power has increased or decreased by some threshold amount (often 5%), and a decision is made as to whether to retain or remove the independent variable in question (Zar 2010). The backwards procedure is recommended by Zar (2010).

Polynomial regression

If the line that best fits your data is not a straight one, and transformation does not seem to help, then you may be able to fit a curved line that gives a tighter relationship between your dependent and independent variables. The equation for a polynomial regression contains more terms than that for a simple linear regression:

$$Y_i = a + b_1X_1 + b_2X_i^2 + b_3X_i^3 + \cdots + b_mX_i^m. \quad (7.5)$$

Although theoretically m could be quite large, the most common form of polynomial regression is quadratic regression (including only the first three terms of Equation 7.5), which will describe a parabola. For each extra term in the analysis a degree of freedom (d.f.) (if this term is mysterious to you see Underwood (1997) or Dytham (2011)) is lost, meaning that for small data sets the chances of getting a significant result are rapidly reduced. Also it is very unlikely that complex curves can be readily explained in a biologically meaningful way. Once again, a typical rule of thumb is that each extra polynomial term should increase the r^2 by at least 5% to merit inclusion in the final model.

Logistic (logit) regression

Logistic regression can be used to predict the probabilistic relationship between a binary dependent variable (Y, coded '0' or '1') and an explanatory (X, predictor, independent) variable (continuous or categorical). When there are more than two predictor variables, logistic regression can be extended as multiple logistic regression. For example, logistic regression can be used to predict the relationship between the probability of individuals flowering or not flowering along an environmental gradient. Logistic regression calculates the probability, $\pi(x)$ that the predictor variable (y) equals 1 for a given value of x according to the logistic function (see below). Because of the binary nature of the predictor variable, the error terms are not normally distributed but have a binomial distribution. Hence, maximum likelihood estimation rather than ordinary least squares estimates of model parameters is necessary. The Wald test and the G^2 test are used assess the fit of logistic regression models. A detailed explanation of the application of logistic regression is provided in Quinn and Keough (2002) with an evaluation of its use in wildlife habitat selection studies in Keating and Cherry (2004). As an example, the individual survival probability (y: alive = 1 and dead = 0) of the annual herb *Amaranthus palmeri* to the herbicide glyphosate as a function of *EPSPS* gene copy numbers was modelled for two levels of glyphosate treatment using the logistic function

$$y = \frac{e^{\alpha+\beta x}}{1+e^{\alpha+\beta x}}$$

where x is the *EPSPS* gene copy number and α and β are unknown parameters to be estimated (Vila-Aiub et al. 2014). The *EPSPS* gene codes for 5-enolpyruvylshikimate-3-phosphate synthase in the shikimic acid pathway and is inhibited by glyphosate but amplified as the resistance mechanism in *A. palmeri*. The logistic regressions in this study indicated that there was a 95% survival probability of *A. palmeri* individuals with an *EPSPS* gene copy number of 21 at the lower glyphosate treatment level, increasing to an *EPSPS* gene copy number of 53 at the high dose.

Path analysis and structural equation modelling

Path analysis (PA) and structural equation modelling (SEM) can be used to untangle the relationships between a number of factors (i.e. independent variables, referred to in PA as *measured variables*) that are thought to be acting sequentially to affect the variable of interest. PA has similarities to multiple regression, except that the independent variables may have indirect effects on the dependent variable via their effects on each other. From previous knowledge, the investigator determines a likely causal order for events in a sequence, such that factor 1 may affect factor 2, which may in turn affect factor 3, but factor 3 cannot affect factor 2, and so on (Legendre and Legendre 2012). A path diagram showing the likely interactions is prepared (not all potential interactions are sensible and are not included) and regression is used to calculate path coefficients for each interaction in the diagram (see Figs. 5.10 and 7.3). For instance, Purrington and Schmidt (1998) used path analysis to try to untangle the relative effects of seedling emergence time and flowering time on the number of plants of different sex produced in *Silene latifolia* populations. They showed that emergence time affects flower number directly, but also indirectly via its effect on flowering time. Mitchell (1993) discusses the potential uses of path analysis for plant ecologists. He gives the example of factors contributing to pollination success in *Ipomopsis aggregata*, where number of flowers, nectar production

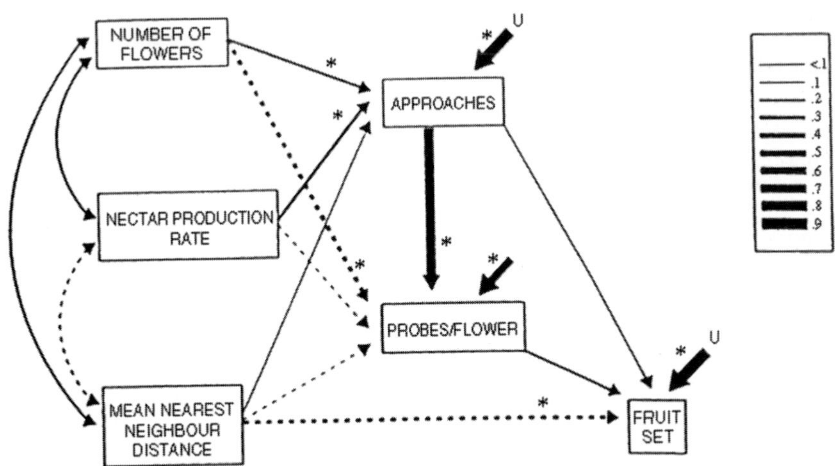

Figure 7.3 Example of a solved path diagram from Mitchell (1993). Mitchell (1993) was interested in factors affecting fruit set in *Ipomopsis aggregata*. He considered number of flowers, nectar production rate, and mean nearest neighbour distance as factors that could all potentially affect the number of approaches made by hummingbirds and the number of probes per flower. These factors in turn, along with mean nearest neighbour distance, were predicted to affect fruit set directly. The arrows link putative causal effects. Note that in a path diagram some factors, such as nectar production rate in this example, only have an indirect effect on the factor of interest. In this solved path diagram solid lines denote positive effects and dashed lines negative effects. The width of each line is proportional to the strength of the relationship found, and significant relationships are denoted with an asterisk. Unknown factors not included in the model are shown with a U, and can be seen to have a large impact on fruit set. Reproduced with permission from Oxford University Press.

rate, distance to nearest neighbour, approaches by pollinating hummingbirds, and number of probes by birds per flower all potentially affect fruit set (Fig. 7.3). Kingsolver and Schemske (1991) discuss the applicability of path analysis to understanding the contribution of many different factors in natural selection. The workings of path analysis are clearly explained by Sokal and Rohlf (2011).

SEM is a more sophisticated form of PA that takes latent variables into account as well as the measured variables in testing the proposed interaction model. Latent variables are unseen constructs (i.e. they are not observed directly) which are inferred from the covariance among two or more measured variables. PA and SEM can be used to address a number of problems in plant population ecology and their application is likely to increase in the near future. The use of SEM in ecology is discussed in Grace et al., (2010). PA and SEM can be conducted in a number of software packages such as the TCALIS and CALIS procedures in SAS/STAT (SAS Institute Inc. 2008), or the sem and OpenMx packages in R (see the Appendix).

7.5 Looking for differences between samples

In order to test a hypothesis in ecology we very often need to decide whether groups of measurements differ by a significant amount, that is, whether two or more groups can be said to represent samples from different populations.

Tests for differences between two groups

If the data are normally distributed and the variances of the two samples are sufficiently similar, then a series of familiar parametric tests are available. However, there is also a suite of parallel non-parametric tests.

Tests for unpaired data

In many situations when we want to compare two groups of data, the data are not paired (see below) and each sample can be considered independent of the others. If the data are normally distributed, measured on a continuous scale, and both groups have similar variances, the *t-test* is most appropriate, or for large samples (containing more than 30 data points per group) the *z-test*. These tests examine the hypothesis that the samples in the two groups actually come from the same population with the same true mean. They both examine the ratio of the difference between the means divided by the standard error of the difference (Fowler et al. 1998). Although Zar (2010) reports that the *t*-test is fairly robust to deviations from the assumptions outlined above, if you are unsure, or if the deviations are large, then the *Mann–Whitney U-test* is more appropriate. This, like most non-parametric tests, is based on ranked data, and is slightly less powerful than the equivalent parametric test.

Tests for paired data

Measurements are sometimes made in pairs. For example, you might have a series of sites, at each of which you have two permanent quadrats, one inside and one outside a herbivore exclosure. At each site you measure the height of your study plant in both quadrats. Thus you have a series of paired measurements, one pair from each site. If the data are continuous and normally distributed and the variances equal, the *paired t-test* would be appropriate (Dytham 2011; Fowler et al. 1998). If these assumptions are not met then the non-parametric equivalent is *Wilcoxon's signed ranks test*. The assumptions of this test are that the data are continuous and there are at least six pairs. Another non-parametric test, the *sign test*, makes no assumptions about the data but is of low power.

Problems with simultaneous comparison of more than two groups

Caution has to be exercised in making a series (family) of separate pairwise comparisons. The probability of the family-wise Type I error goes up as more tests are performed. By 'family' we mean a collection of simultaneous tests based upon a single data set arising from a single experiment or survey. The family-wise Type I error for a set of independent (orthogonal) tests can be calculated as

$$1-(1-\alpha)^c \qquad (7.6)$$

where α is the significance limit (e.g. 0.05) and c is the number of tests (Quinn and Keough 2002). So, for example, the family-wise probability of at least one Type I error while conducting a set of 40 independent t-tests is 90%.

In this sort of situation it is usually more appropriate to use ANOVA or its non-parametric equivalents, the Kruskal–Wallis and Friedman tests. If, however, you feel that ANOVA is not suitable for your data, and that pairwise comparisons are necessary, then you will need to find a method of protecting yourself against an elevated Type I error by using a *Bonferroni procedure* or one of the many available multiple comparisons procedures (Day and Quinn 1989; Zar 2010). These are used after an ANOVA has shown a significant difference between the means of groups (see below). The Bonferroni procedure adjusts the significance level used in multiple testing situations so that each comparison is tested at α/c (e.g. $0.05/10 = 0.005$ for a set of 10 simultaneous t-tests). This is a very conservative approach, although it is easy to use and has broad applicability. A less conservative procedure is the sequential Bonferroni. In this case, the P-values from c-test statistics (F, t, etc.) are ranked from largest to smallest. The smallest P-value is tested at α/c, the next at $\alpha/(c-1)$, the next at $\alpha(c-2)$, etc., until a non-significant result occurs at which point the testing is halted as all other comparisons are then deemed non-significant (Quinn and Keough 2002). West et al. (2009) used sequential Bonferroni to adjust significance levels following multiple t-test comparisons of environmental variables measured at exotic species locations and random locations in a shale glade.

Tests for differences between more than two groups: mostly about ANOVA

ANOVA allows you to investigate whether there are differences between the means of many (i.e. two or more) groups, as long as the data meet certain assumptions. There are good reasons why it is the statistical technique most used by plant ecologists. It is a very powerful and flexible technique, and it is important for all plant ecologists to be comfortable with it.

If you have done an experiment or a study with three or more levels of a test factor (e.g. several intensities of herbivory as in Mauricio et al. (1993), or three or more locations between which you are comparing population parameters), or in which a factorial design (Chapter 3) has been used, then you will need to consider ANOVA. The simplest kind of ANOVA is a *one-way ANOVA*, which can examine the effect of a single factor at any number of levels (e.g. the effects of three different levels of nutrient supply on the size of a plant population).

Factorial ANOVA can examine the effects of two or more factors simultaneously, each at two or more levels. For instance, you might be interested in the effects of two factors, nutrient supply and vertebrate herbivory, on the size of a plant population. In fully factorial experiments, each level of each factor must be combined with each level of each other factor. So, if there were three levels of nutrient supply and two levels of herbivory (e.g. vertebrate herbivores present or absent—see Chapter 4) in an experiment, there would have to be $3 \times 2 = 6$ treatments in total. (This would then have to be multiplied by the number of replicates per treatment to get an idea of the amount of work involved, which can rapidly become huge in factorial experiments.) In this case the appropriate *two-way ANOVA* would tell you whether there was an effect due to each of the main factors (nutrient supply and herbivory) and, in addition, whether there was a significant interaction between the two factors. An interaction is a non-additive effect of the two factors together. For example, Fraser and Grime (1999) investigated the effects of fertilizer and invertebrate herbivory on the aboveground biomass of plants grown in outdoor microcosms. They found that aboveground biomass was affected by the addition of fertilizer, by herbivory, and by the interaction between fertilizer and herbivory (Table 7.1). Examination of the data presented in Fig. 7.4 shows that the interaction between fertilizer and herbivory results in a greater impact of herbivores on community biomass at high levels of fertilizer addition, that is, the interaction between herbivores and fertilizer could not have been predicted from their individual effects.

A more complex experiment could include a third factor. For instance, in addition to fertilizer and vertebrate herbivory, you might also be interested in investigating the impact of water availability

Table 7.1 Example of two-way factorial ANOVA output (data from Table 2 of Fraser and Grime 1999). There were two factors in their experiment, soil fertility level (FERT), with three levels, and herbivory (HERB), with two levels. The table reports the effects of these factors on aboveground biomass of an experimental plant community. Significant effects of both main factors and a significant interaction effect are shown. The direction of the effect of the interaction is shown in Fig. 7.4. (See 'Examining and interpreting ANOVA output' in Section 7.5 for more information on how to interpret this table)

Source of variation	d.f.	Mean square	F-ratio	P
FERT	2	18,621	963.0	< 0.001
HERB	1	3997	206.7	< 0.001
FERT × HERB	2	2320	120.0	< 0.001
ERROR	30	19		

d.f., degrees of freedom.

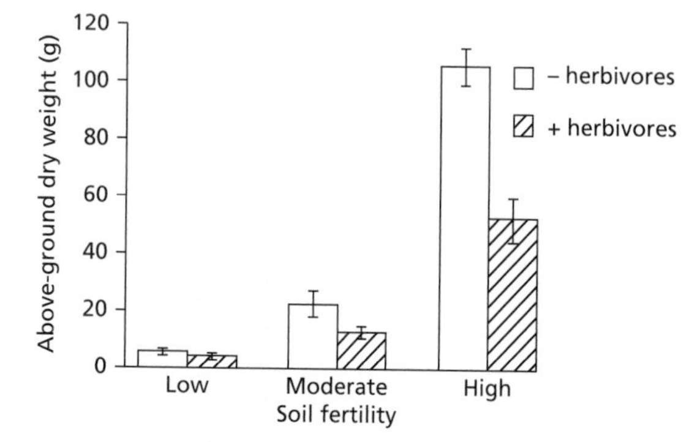

Figure 7.4 Example of an interaction effect (redrawn from figure 2 of Fraser and Grime 1999). In the absence of invertebrate herbivores the effect of fertility is much higher than would be predicted. Open bars are the herbivore absent treatment, hatched bars are the herbivores present treatment. Reproduced with permission from John Wiley & Sons.

on the plant population (say at two levels; water supply supplemented or not supplemented). There would now be three factors (fertilizer, herbivory, and water), which when combined factorially will provide 3 × 2 × 2 = 12 treatments (see Chapter 3 if you are unclear about this). The appropriate *three-way ANOVA* will now analyse whether each of the three main factors has had an effect and whether there are significant interactions between each possible pair of factors. It will also investigate whether there is a significant three-way interaction (nutrient supply × vertebrate herbivory × water supply). Such interactions can be very difficult to interpret from a biological point of view. Technically, there is no reason why you could not have four-way or 'more-way' ANOVA, but the results will become more and more difficult to interpret. McEvoy et al. (1993) provide an example of a *four-way ANOVA* table (Table 7.2; and see case studies in Section 7.9). Luckily (from the point of view of interpretation), the higher-order interactions in this study were not significant.

Factorial experiments and their associated ANOVAs are very efficient techniques for analysing the

Table 7.2 Example of four-way factorial ANOVA output (data from Table 2 of McEvoy et al. 1993). In this experiment there were four main factors: time, beetles, and moths with two levels each, and competition with three levels. For more information on this study see Case Study 3 in Chapter 1. The effects of each of these factors and their interactions on the number of emerging ragwort seedlings is shown. The effect of competition, with three levels, is decomposed to two orthogonal contrasts, each contributing one degree of freedom, as described in Section 7.5 ('Planned and unplanned comparisons'). The block effect is also shown. Of the main factors, only time and the contrast between unaltered and clipped vegetation had any significant effect on the number of emerging seedlings (i.e. are associated with a probability of less than 0.05). There were no significant interactions.

Source of variation	d.f.	Mean square	*F*-ratio	*P*
Block	3	2883	6.25	0.0008
Time (T)	1	3553	7.70	0.0071
Beetle (B)	1	301	0.65	0.4221
Competition (C)				
Removed versus other	1	4	0.01	0.9227
Unaltered versus clipped	1	2174	4.71	0.0334
Moth (M)	1	210	0.46	0.5022
B × T	1	726	1.57	0.2140
B × M	1	876	1.90	0.1728
M × T	1	561	1.21	0.2743
C × B × M	2	291	0.63	0.5354
C × T	2	165	0.36	0.7007
C × B	2	154	0.33	0.7173
C × M	2	801	1.73	0.1841
C × T × B	2	116	0.25	0.7790
C × T × M	2	376	0.81	0.4473
T × B × M	1	67	0.14	0.7051
C × B × T × M	2	263	0.57	0.5688
Error	69	462		

effects of two or more factors simultaneously. You might be wondering whether it would be better to conduct a number of separate experiments on each of the factors of interest. This is inefficient, for two reasons. First, it would mean that you would not be able to find out about the interactions which are probably very important in the natural environment. Second, from a practical point of view, some of the replicates for a particular level of each factor are effectively shared with levels of another factor (Bailey 1995). Thus, factorial experiments are usually very cost-effective. However, as indicated above, complex multifactorial ANOVAs can be difficult to analyse and interpret.

Why is a test that is used to examine differences between groups called analysis of variance? In essence, it examines whether the variance, caused by the different levels of the factors under consideration, is greater than the variation within each level of the factors (termed residuals or error). ANOVA partitions the variation among the individual replicates according to the various possible causes of variation. For instance, the variation caused by the treatments can be separated out from that caused by the blocks (or locations of the treatments; Chapter 3), and there will also be some residual (error) variation that cannot be accounted for by any aspect of the experimental design (due to inherent variation among replicates). The variation is partitioned by calculating the sums of squares for each putative cause of variation (sums of squares are the sum of the squared deviations of individual values from the appropriate mean, and their calculation is described in almost any statistics text). The test statistic for ANOVA is the *F*-ratio. In essence the *F*-ratio numerator is a measure of the variance due to the treatment and 'other things' (Newman et al. 1997). The denominator (divisor) is the variance due to the 'other things'. When the *F*-ratio is high, the variance due to the factor of interest (the treatment) is large relative to the variance due to 'other things', and the test is likely to be significant. There is a helpful discussion of the basic principles of ANOVA and the use of sums of squares in Chapter 7 of Underwood (1997).

What ANOVA tells you is whether there is a significant effect of a particular factor, not which treatments differ from each other (unless you happen to have only two levels of a factor, in which case they must be different if there is a significant effect). To find out which means differ from each other you need to go on to perform previously decided comparisons (a priori, or planned comparisons) between means or unplanned (a posteriori or post hoc) tests, which are discussed below.

As with all parametric tests, ANOVA requires data to be normally distributed with homogeneous variances between treatments. Statisticians are of the opinion that the assumption of normality is not as critical as that of homoscedasticity, but as with all tests if you are aware of violating any of the assumptions of the test then any significant results which are close to $P = 0.05$ should be interpreted with caution. (See above for tests to help you decide whether your data meet these assumptions.)

In statistics packages, or in the literature, you will often see the term general linear model, or GLM, associated with ANOVA. That is because ANOVA comes from a family of statistical techniques which are built around a simple linear model of the relationship between the data and the treatments.

The basic one-way ANOVA model can be represented by the equation

$$x_{ij} = \mu + A_i + e_{ij} \qquad (7.7)$$

which simply says that each data point (x_{ij}; replicate *j* of treatment level *i* of factor *A*) is equal to the overall mean (μ) plus an effect due to the treatment you have applied (A_i; the amount by which level *i* of the factor *A* increases or decreases the value of each replicate) plus the error associated with that data point (e_{ij}). The error referred to in ANOVA is not necessarily error in the sense that you may have measured something inaccurately (although this could contribute to the error term), but indicates variation due to the fact that plants differ from each other in the real world for a variety of reasons apart from the treatments that might be applied to them in experiments. The term residual is also used, as this is the variation that is left after other causes of variation have been partitioned out.

You can build up this model, depending on the type of experiment or study you are examining. For instance, in the case of a simple two-way factorial design you would have to add to the equation the effects of the second factor (B) and the effects due to an interaction between the two factors (AB):

$$X_{ijk} = \mu + A_i + B_j + AB_{ij} + e_{ijk} \quad (7.8)$$

where x_{ijk} (the value of the kth replicate in the ith level of factor A and the jth level of factor B) is equal to the overall mean, plus the effect due to level i of factor A (A_i), plus the effect due to level j of factor B (B_j), plus the effect due to the interaction between level i of factor A and level j of factor B (AB_{ij}), plus the error associated with that particular replicate. You can perhaps begin to see how you could adjust this model to add in terms to account for variation caused by, for instance, block effects or plots in a split-plot design (Chapter 3). This is one of the great strengths of ANOVA; it is incredibly flexible and can analyse a wide range of experimental designs. However, it is not a goal of this chapter to provide a long list of alternative models. You can read more in Underwood (1997), Scheiner and Gurevitch (2001), Zar (2010), and other texts. Be aware, though, that it is important to specify exactly what you want to test when you are using a statistics package to run your analysis. Obviously, the computer will not know what your model was unless you tell it, although it will probably run a default analysis.

Unbalanced designs

A term that you will often see associated with ANOVA is balance. In balanced ANOVA there are the same number of replicates for each treatment and factorial designs are fully factorial (i.e. every level of every factor is combined with every level of every other factor; see above). Balanced designs in ANOVA are the most statistically robust, but in the real world of ecology there are many reasons why a particular treatment might not be feasible, or some replicates might be lost before measurement. Many statistics packages can cope with unbalanced designs, but you may be risking a greater Type I or II error. Shaw and Mitchell-Olds (1993) explain how the method used to calculate the sums of squares becomes critical in an unbalanced ANOVA and can have a dramatic impact on the conclusions of your analysis. The default in a statistics package may not be the most appropriate in an unbalanced situation (Goldberg and Scheiner 2001). Milliken and Johnson (1984) and Underwood (1997) suggest a number of remedies to make up for lost or missing data, but most authors agree that it is best to avoid unbalanced designs wherever possible.

Blocks, split-plots, and nested designs

As mentioned above, ANOVA can deal with a very wide range of experimental designs. Blocks, split-plots, and nested designs (Chapter 3) are all very commonly encountered in the ecological literature. In each case, ANOVA partitions the variance according to the model prescribed by your experimental design. For instance, if you have conducted a blocked experiment then you should analyse the experiment using a model that contains blocks. Not doing so could have serious consequences for the interpretation of your ANOVA (Potvin 2001).

Fixed and random effects

One aspect of any ANOVA model which needs thought is whether the model includes fixed or random effects. Sometimes the levels of a factor are chosen deliberately (e.g. grams of fertilizer added per plot or species included in a comparative experiment because they have specific characteristics relevant to your hypotheses). The hypothesis under test is that the chosen levels are affecting the dependent variables, and conclusions can only strictly be applied to those levels. These are fixed effects. On the other hand, there will be cases where the levels of the factor have not been specifically chosen but are representative of a wider range of levels, from which the levels in your study are just a selection (e.g. a selection of sites chosen simply to represent different levels of natural fertility rather than because they represent particular levels of fertility, or species included in a comparative experiment chosen because they were available in a seed catalogue rather than because they had specific features relevant to the hypothesis). These are random effects. The hypothesis being tested is a more general one about global effects of the factor, of which your levels are just a selection (places with different fertility or the responses of different species). Newman et al. (1997), suggest that in order to decide whether effects are fixed or random, consider whether another scientist could exactly replicate the levels of the treatment. If that would be possible, then you are probably dealing with fixed effects; otherwise you have random effects. Underwood (1997) also provides help on deciding whether factors should be fixed or random. However, as Newman et al. (1997) pointed out, in ecology it is not always easy to decide whether a factor is fixed or random. They use the example of blocks in blocked experiments, which some authors of statistical door-stoppers treat as fixed effects and others as random. So if you're feeling that this concept is a bit difficult to clarify, you are in good company. Of course, in most cases it should follow clearly from your hypothesis whether you need to use a fixed or random effect in your analysis, and your experiment should be designed accordingly.

ANOVAs which have only fixed effects are sometimes called Model I ANOVAs, while those with only random effects are termed Model II. Fixed and random factors are often both present, and these mixed models may be referred to as Model III. This issue is of more than theoretical interest. For single-factor ANOVA there is, in fact, no difference in the calculation of Model I and Model II ANOVA, but once two or more factors are involved then the calculation of the *F*-ratio depends on whether each factor is fixed or random, so it is something that you do need to think about. For instance, Newman et al. (1997) make it very clear that the decision of whether or not to include blocks as random or fixed effects in an analysis can affect whether or not one of the main treatment factors is found to be significant.

Underwood (1997) and Zar (2010) both tabulate the calculation of *F*-ratios for various random effects and mixed model ANOVAs. An example of a mixed model ANOVA can be found in Larson and Barrett (1999). They investigated whether pollen supply limits female fertility in populations of *Rhexia virginica* in Ontario, Canada. They performed an experiment in which plants in two populations either received a supplementary pollen treatment or were left as controls. This was done over a period of 7 days. Their three-way mixed model analysis of variance had pollination treatment as a fixed effect and population and day of treatment as random effects. Another example is supplied by Krupnick and Weis (2000). They investigated the effect of herbivory on the export of pollen from flowers of *Isomeris arborea*. They experimentally protected some plants from herbivory and compared the export of pollen from protected and unprotected plants on a number of dates. They analysed the data using a two-way-mixed model ANOVA with plant treatment as the fixed effect and date of trial as the random effect. One situation in which there will always be at least one random factor is the nested experimental design (Chapter 3). In the nested, or hierarchical, design one factor is not independent of the other and is nested within it. For instance, samples taken from individual plants growing within a treatment plot are nested within the plot. Nested factors are always random and are often (but not always) nested within a fixed factor, so often contribute to a mixed model ANOVA.

Examining and interpreting ANOVA output

The output you get from your computer package will probably be similar to that presented in Table 7.3, which is an example of output from SYSTAT.

The first column usually names the source of the variation (the factor or interaction under consideration). In another column are the sums of squares; a measure of the amount of variation between the levels of each factor. A further column will include

Table 7.3 Example of a summary ANOVA output. There were two main factors in the experiment being analysed here: plant association (ASSOC) and nutrient treatment (TREAT). These, their interaction, and the error are the only sources of variation considered in this analysis, which was a fixed-factor ANOVA. Column 2 shows the sum of squares associated with each of these sources of variation, and column 3 the relevant d.f. There were four levels of ASSOC [hence (4 − 1 =) 3 d.f.] and seven levels of TREAT (6 d.f.). Thus the interaction has 3 × 6 = 18 d.f. The error d.f. is 337 because there were 364 observations in the data set (13 replicates of each combination of ASSOC and TREAT). The mean square was calculated by dividing the sum of squares by the relevant degrees of freedom. The *F*-ratio is calculated by dividing the mean square for each factor of interest by the error mean square, and the *P*-value associated with that *F*-ratio is reported.

Source	Sum of squares	d.f.	Mean square	*F*-ratio	*P*
ASSOC	165.503	3	55.168	4.740	0.003
TREAT	62.745	6	10.457	0.898	0.496
ASSOC × TREAT	128.232	18	7.124	0.612	0.890
ERROR	3922.319	337	11.639		

the d.f. for that factor. This should be the number of treatments (or levels) of the factor minus 1. For interaction terms the d.f. should be equal to the product of the d.f. of each factor included in the interaction. A subsequent column will list the mean square for each factor or interaction. This is the sum of squares divided by the relevant d.f. Below (or sometimes above) the rows of the table reporting on the factors and their interactions there should be a row which reports the sum of squares, d.f., and mean square for the error (or residual). The error d.f. should be equal to the total number of replicates in your experiment or study, less the number of d.f. due to the elements in your model (factors and their interactions for instance), less 1. The penultimate column of the table is usually the *F*-ratio for each element in your model. For fixed effect ANOVAs the *F*-ratio is the mean square for that factor (or interaction) divided by the error mean square. Thus, the bigger the error mean square the less likely your *F*-ratio is to be significant. The calculation of the *F*-ratio may be different if random factors are present (Underwood 1997; Zar 2010). The final column of the table is the *P*-value associated with the *F*-ratio for the appropriate d.f.

In addition to the above, some computer packages will report a line for the sum of squares, d.f., mean square, *F*-ratio, and *P*-value for the whole ANOVA model (e.g. see Equation (7.7) for an example of the simplest kind of model).

You should check the output carefully to see if everything matches your expectations. The d.f. can provide a very useful check as to whether your whole data set was analysed, for example. If your d.f. adds up to the number of plants in the analysis, rather than the number of replicate plots, then perhaps this is an indication that you have inadvertently pseudoreplicated.

A temptation is to ignore the significant interactions and concentrate on the main (individual factor) effects. Interactions can be very hard to interpret, especially when there are no significant main effects. They indicate a non-additive effect of the two (or more) factors together, and can complicate interpretation of the results. However, assuming that you have designed a reasonable experiment, they are telling you something about real complications that are important to plants and the way they respond to their environment. It could be true, for example, that plants are only able to respond positively to the addition of nutrients when herbivory is prevented, which could be a very important result. It is often helpful to plot the means of each of the treatment groups when trying to interpret interactions. When you have found out which means differ significantly then the effect of the interaction will become clearer to you (e.g. see Fig. 7.4).

Reporting ANOVA

You will find ANOVA results reported in many different ways, even in published papers. Sometimes the complete ANOVA table (sums of squares, d.f., mean square, *F*-ratio, and *P*-value) is reported. In other cases only the *P*-value is supplied. The *F*-ratio, its d.f., and the *P*-value comprise the absolute minimum amount of information necessary in a published paper [i.e. $F_{\text{between groups d.f.,within groups (error) d.f.}}$ = value, P = P-value; e.g. for the plant association (ASSOC) factor in Table 7.3, $F_{3,337} = 4.74$, $P = 0.003$]. The mean square or sum of squares are also very helpful. If you are writing up a study for your thesis, for your advisory committee, or for the first stages of submission to a journal, then more information will be very helpful to those assessing your work. Presenting the complete ANOVA table makes it clear what was done in the analysis, and the caption to the table should make the details clear.

Repeated measures ANOVA

Sometimes in population ecology our hypotheses require us to return to a particular plant or population and make a series of measurements on it. For such a design to be necessary there would normally be some interest in the treatment × time interaction (i.e. whether the effect of the treatment changes with time). An example might be a study of how the application of fertilizer at different rates affects the number of plants in a population over several years. Replicate populations would be fertilized at different rates and the size of each population measured annually.

Repeated measures taken from the same plant or population are clearly not independent of each other and are likely to be strongly correlated. It might be tempting to analyse these data as a two-way ANOVA with treatment as one factor and time as another, but this would not be appropriate because of the lack of independence between the samples. Performing a series of one-way ANOVAs for each time interval would be similarly risky and lead to a high level of Type I error.

However, there are two commonly used methods of tackling this problem (see the helpful discussion in von Ende 2001). One is to use a univariate repeated measures ANOVA which treats the experiment as a split-plot design, in which the individual populations (or individual plants, etc.) are considered as if they were plots. Time is then treated as if it were a treatment allocated within the plots (as in a split-plot design; see Chapter 3). In statistical terminology, the populations are subjects, the fertilizer application is the between-subject treatment, and time is the within-subject treatment. However, this application of univariate repeated measures ANOVA makes some fairly severe assumptions about the degree of covariance between the elements of the analysis (von Ende 2001), some of which are unlikely to be met in many ecological examples of repeated measures ANOVA. As an example, one of these assumptions is that the variance–covariance matrix should be 'spherical' (don't worry if you don't know what a variance–covariance matrix is, but if you want to find out more look in von Ende (2001) or Manly (2004)). One of the criteria for this to be met, is that measurements made on the same subject at consecutive time intervals should not be any more highly correlated than measurements made at any other intervals (i.e. the degree of correlation of two within-subject measurements should be independent of the time interval between them). This would often be violated in ecological work. von Ende (2001) discusses the consequences of such violations, and the adjustments that can be made to make the test more conservative (i.e. less likely to result in Type I error). The Huynh–Feldt epsilon is a multiplier that reduces the degrees of freedom for the *F*-ratio and thus makes the test more conservative (Crowder and Hand 1990; von Ende 2001). An even more conservative adjustment is provided by the Greenhouse–Geisser epsilon, and the results of both adjustments may be reported by your computer package. An alternative is to use a multivariate repeated measures ANOVA (see below), which makes less rigorous assumptions about the variance–covariance matrix. However, multivariate ANOVA (MANOVA) has its own limitations (see below).

Gerdol et al. (2000) used a repeated measures ANOVA to examine the effects of experimental nutrient addition and neighbour removal on sub-alpine dwarf shrubs over 3 years. Their data did not meet the assumption of sphericity, so they used the adjustment provided by the Huynh–Feldt epsilon.

As both univariate and multivariate repeated measures ANOVA are commonly used nowadays, it is important to be clear about which one you have chosen to report and why. It may not be immediately obvious which method your statistics package uses when you ask for repeated measures ANOVA.

Multivariate ANOVA

In addition to repeated measures on the same plant or plot, other situations in which data could not be presumed to be independent of each other would arise when several variables from the same plant have been measured at one point in time. For example, root, shoot, and flower weights from the same plant are likely to be correlated with each other, and could not be regarded as independent measures. Similarly, the abundances of different species in a community are unlikely to be independent of each other. In such situations MANOVA should be used. MANOVA is a complex and sophisticated analysis, but is now available in many computer packages and is frequently used in published studies (e.g. Case Study 4 in Section 7.9; Mauricio et al. 1993).

Multivariate statistics have many uses, but there are always a number of dependent variables for each independent variable (i.e. several measurements have been made on each sample). In this section we will only consider MANOVA, but there are many other multivariate techniques commonly used by plant population ecologists. Some of these will be discussed briefly in Section 7.6. Although now in common use, multivariate techniques are usually dealt with in specialist texts. Two of the most helpful are Manly (2004) and Tabachnick and Fidell (2012). The latter is aimed primarily at psychologists (who are also prodigious users of multivariate analyses) but is written very sympathetically for those with only limited mathematical or statistical knowledge, and the examples are surprisingly easy to consider in an ecological context. Johnson and Wichern (2007) is a slightly more technical treatment of multivariate statistics. MANOVA can tell you whether there is an overall significant effect of the treatments on all the variables considered together. For instance, whether there is a significant overall change in root, shoot, and flower weight. There are several statistics used to report overall significance of a MANOVA, but the most widely reported is Wilks' lambda (λ) (Manly 2004; Tabacknick and Fidell 2012). Wilks' λ is essentially a ratio dividing overall effects due to error by effects due to error and treatments, so that large values of Wilks' λ (close to 1) are less likely to be significant than smaller values (Tabacknick and Fidell 2012). Others you will come across include Pillai's trace, Hotelling's T, and Roy's largest root. These statistics can all seem rather mysterious, but are introduced, albeit briefly, in the most recent edition of Zar's text book (Zar 2010). All these statistics can be converted to an F-statistic, and you will need to check whether your computer program has done this.

If Wilks' λ is reported as significant, meaning that there is an overall effect of the treatments, it is considered legitimate to go on to test each of the dependent variables using univariate ANOVA (i.e. ANOVA on each dependent variable separately). For instance, having shown that there is an overall effect of the factor of interest on root, shoot, and flower weights together, you can then go on to look at whether each of roots, shoots, and flowers is being affected by the factor. Just because there is an overall significant effect of a factor does not mean that each of the dependent variables will have responded significantly. For instance, in Mauricio et al. (1993) two separate MANOVAs were performed on various growth and reproductive measures and both produced significant values for Wilks' λ. Mauricio et al. (1993, Table 1) then went on to perform univariate ANOVAs on each attribute separately. Of 15 univariate ANOVAs performed in total, only four were significant.

One of the limitations of MANOVA is that the power of the test depends on the ratio of the number of independent samples to the number of dependent variables (von Ende 2001). The inequality $N - M > k$ must be met for MANOVA, where N is the number of subjects (in all treatments), M is the number of between-subject treatments and k is the number of dependent (within subject) variables measured. However, if $N - M$ is close to k then the power of the analysis is very low, leading to the recommendation that it is best to keep the number of within-subject measurements relatively low, or to have as many subjects (replicates) as possible.

ANCOVA

In an experiment you might feel that some factor not manipulated by you (and possibly not specifically of interest to you) could be affecting your dependent variable. For instance, you might be interested in investigating whether the number of flowers produced by a rare orchid is affected by a factor of interest such as the conservation management regime. However, the plants you are studying vary in rosette size, and you already know that larger plants produce more flowers, so this is not of interest to you. Although you would try to design an experiment that ensured that rosette size did not vary systematically with the treatment, you might not be able to completely achieve this. You might then choose to try and account for the effect of rosette size by using it as a covariate in analysis of covariance (ANCOVA). ANCOVA, in effect, performs regression as well as an analysis of variance. In the case above, ANCOVA would perform a regression of the effects of rosette size on flower number and a one-way ANOVA on the effects of the management regime on flower number. The effects of management can then be effectively separated out from the effects of the covariate, as long as certain (quite demanding) assumptions of the ANCOVA are met. The relationship between the covariate and the dependent variable should be a linear one and there should be no interaction between the independent variable and the covariate. In our simple example, rosette size must affect flower number independently of the treatment regime, otherwise effects are confounded. However, as Underwood (1997) points out, differences in regression slopes between treatments are probably of great interest and should not be ignored. As an example, in their study of the effects of floral herbivory on pollinator service to *Isomeris arborea*, Krupnick and Weis (1999) wanted to control the effects of plant size in their experiment. To do this they used ANCOVA with the number of branches per plant (a surrogate measure of plant size) as the covariate.

Another example of the use of ANCOVA is the study by Feldman et al. (1999) of the costs to pine trees of living in multigenet clusters (caused by seed caching by the corvid Clark's nutcracker). Their paper is a very helpful one in the current context, because their ANCOVA models are explicitly set out and can be used to illustrate the principles behind ANCOVA. For instance, their first ANCOVA model describes the interaction between tree height (the dependent variable), tree age (the covariate), stand location, and growth form:

$$H_{ijk} = \beta_0 + \beta_1 A_{ijk} + \alpha_i + \gamma_j + \varepsilon_{ijk}. \qquad (7.9)$$

In this model H_{ijk} is the height of the kth tree in the jth growth form in the ith stand, β_0 is the overall intercept with the y-axis, β_1 is the slope of the height–age relationship, A_{ijk} is the age of the tree, α_i is the change in the intercept due to the ith stand (i.e. the effect on height of being in stand i), γ_j is the effect on height of the jth growth form, and ε_{ijk} is the error. Note the similarities between this model (Equation 7.9), the regression equation (Equation 7.3) and the ANOVA model equation (Equation 7.7). In other words, if height were to be plotted against age, the slope of the line would be β_1, but the height of the line above the x-axis would depend on the stand and growth form variables. Feldman et al. (1999) report that the model was a good fit to the data. Age was a significant predictor of tree height, stand had a significant effect on the intercept of the regression, but growth form had no significant effect. The slope of the regression line between age and height was not affected by stand or growth form, suggesting that the relationship between height and age is independent of stand and growth form.

ANCOVA is a widely used technique, so it is worth becoming conversant with it, even though it is not covered in all statistics texts. Underwood's (1997) chapter covering ANCOVA is particularly helpful, and the details of how to do it are covered in Sokal and Rohlf (2011).

Non-parametric alternatives to ANOVA

If your data are not normally distributed, or have very heterogeneous variances, and transformations (see above) are unable to resolve these problems, then there are few non-parametric tests that are available as alternatives. These will be less powerful than ANOVA but less likely to lead to Type I error if the data do not meet the ANOVA assumptions. The Kruskal–Wallis test (Dytham 2011) can replace the one-way ANOVA. Friedman's test can be used for randomized block designs (Zar 2010). Dytham (2011) also reports that the Scheirer–Ray–Hare test can replace two-way ANOVA, but it is not widely available. However, Zar (2010) reports poor performance for this test. Conover (1999) suggests that any experimental design can be analysed by first transforming the data to ranks.

Planned and unplanned comparisons

If one or more of the factors in your ANOVA have been shown to have a significant effect you will want to go on to find out which treatment means differ from each other. At this point, many researchers take an exploratory approach to their data and perform an unplanned multiple comparison procedure (Day and Quinn 1989). This means that, for each significant factor, the mean of each level of that factor is compared with the mean of each other level. At first sight it might seem as though these comparisons could be made with a series of t-tests comparing all possible pairs of means. However, Day and Quinn (1989) explain why this would be unacceptable. They point out that if you had four treatments with equal population means (i.e. no 'real' differences between them) then the Type I error rate (EER) for each comparison would be 5%. However, because many comparisons are being made using the same data, the experiment-wise EER for all possible comparisons would be much greater than this—about 26%. That is, you would have an unacceptably high chance of rejecting a true null hypothesis (or finding a false significant result). However, there is a wide range of tests designed specifically to get around this problem by comparing means while offering 'protection' against enhanced Type I error. You will have come across the names of many of them: the Student–Newman–Keuls (SNK) test, Duncan's multiple range test, Tukey's test (also called the HSD, honestly significant different, test), the least significant difference (LSD) test, Scheffe's test, the Bonferroni adjustment, and the Dunn–Sidak test are all commonly used tests. As Day and Quinn (1989) pointed out, most statistical texts written at that time only tended to consider two or three of these many tests, and more recent texts (e.g. Fowler et al. 1998; Zar 2010; Dytham 2011) continue this tradition. However, Day and Quinn (1989) provide an extensive comparison of a very wide range of tests suitable for multiple unplanned comparisons and planned comparisons (see below). They recommend a test not listed above (Ryan's *Q*) as providing maximum power and minimizing the EER for pairwise tests where the groups have equal variances. While Day and Quinn's advice doesn't seem to have been taken on board by the majority of plant ecologists [although it clearly was by Mauricio et al. (1993) (Case Study 4 of Section 7.9), who used it for their a posteriori (post hoc) comparisons], their paper is well worth looking at when you are deciding which multiple comparison test to use. It makes clear the assumptions that need to be met for each test to operate efficiently, and also gives the formula for each test. However, your decision about which test to use will have to be based on a number of factors, such as the suitability of the test for your data and the availability of particular tests in the computer program you are using, as well as what your most trusted statistical text has to say on the matter. If the ANOVA effect is not a very strong one it can happen that the multiple comparison test does not find a significant difference between any of the means. This can be very frustrating, but can happen because ANOVA is more powerful than the multiple comparison test (Zar 2010).

The above paragraph discusses what are known as post hoc (after the fact) tests or unplanned comparisons, performed after the ANOVA has found significant effects. The mean of every group is compared with every other mean in a search for differences. Once you find the differences you then have to explain them, in terms of a rather vague null hypothesis such as 'the treatments will have no effect'. However, many statisticians and ecologists recommend the use of planned or a priori comparisons (e.g. Day and Quinn 1989; Underwood 1997; Sokal and Rohlf 2011). This approach involves testing predetermined hypotheses involving differences between particular means being tested. It means that your null and alternative hypotheses have to be thought through carefully before you perform the experiment, and forces you to consider your experimental design very carefully. Planned comparisons can be orthogonal (independent of each other) or not (Underwood 1997; Sokal and Rohlf 2011). If they are not orthogonal then there is no independence between any of the comparisons, and levels of α must be adjusted accordingly so as to protect against Type I error. For instance, Lehtilä and Strauss (1999) used non-orthogonal planned comparisons to examine the effects of different kinds of herbivory on various floral traits. They compared the results of both of two different sources of herbivory against the results from the same undamaged controls. Thus, the comparisons were not independent. To protect themselves from enhanced Type I error they used Dunn–Sidak adjustment of probabilities. However, orthogonal contrasts are recommended where possible, as their independence increases the power of the analysis and the level of α does not need to be adjusted. In a factorial design, it may be possible to consider the effect of a factor as a number of orthogonal contrasts. For instance, in one of the case studies (McEvoy et al. 1993) one factor, competition, with three levels (vegetation removed, vegetation clipped, or vegetation left unaltered) was decomposed to two 1-d.f. orthogonal contrasts. The first contrast was to compare plots with no vegetation with plots with vegetation present (i.e. the mean of the plots with vegetation removed was compared with the mean result for both kinds of plots with vegetation). The second contrast examined the effects of different heights of vegetation, so that the mean of plots with clipped vegetation was compared with the mean of plots with unaltered vegetation (Table 7.2). The total d.f. for the factor, competition, was 2 (the number of treatments minus 1) and the d.f. for the two contrasts add up to 2. The sums of squares for the contrasts should also sum to the sums of squares for the overall factor if the contrasts are orthogonal. Note that in the McEvoy et al. study, no individual mean was used in more than one comparison. If one of the means (say vegetation removed) had been compared with each of the other two means in turn then the contrasts would not have been orthogonal. Sokal and Rohlf (2011) explain very clearly how to perform and analyse planned comparisons, and there is also a slightly more technical treatment in Steel et al. (1996).

Generalized linear mixed models (GLMMs)

Many population data sets are non-normal or include predominant random effects (e.g. experimental blocks replicated across time or space, or measurements of the variation among individuals or genotypes spanning multiple time periods) that are either ignored or treated as fixed effects. These data can be difficult to appropriately handle using conventional parametric ANOVA-based methods, even after attempts to transform data to normality and homogeneity of variance. GLMMs combine linear mixed models with random effects and generalized linear models (which use link functions and exponential family distributions to handle non-normal data). The main things to decide about in using a GLMM are to construct a full (the most complex) model, and to specify the distribution, link function, and structure of the random effects. Of course, there is more to it that this, and making these decisions is not easy. A full description of GLMM model fitting, methods of best practice, and a decision tree for GLMM fitting and inference is provided in Bolker et al. (2009). The GLIMMIX and NLMIXED SAS procedures and the lme4 R package (see the Appendix) are suitable for running GLMMs. For example, in a study of the competitiveness of wild type, cultivars, and wild type × cultivar hybrids of *Plantago lanceolata*, plant samples were included as random factors and the three plant types previously listed were specified as fixed effects (Schröeder and Prasse 2013). In this study, GLMMs were fitted with maximum likelihood using Bayesian methods (see Section 7.7) and Akaike's information criterion (AIC) as a measure of fit.

7.6 Multivariate techniques for ordering and classifying samples

Plant ecologists often have to interpret multivariate data, where each observation is associated with a large number of measurements. For instance, an investigator may have measured a number of attributes of floral morphology on many individual flowers collected from different parts of the species' range, and be interested in how these attributes vary within and between populations. In some cases, such data may not have been collected with a specific hypothesis in mind but rather as the initial stage in a study of variation in floral morphology. There is a range of multivariate techniques available for the ordering and classification of such data, including MANOVA described in Section 7.5. The technique that is chosen will depend on the nature of the data and on the question being asked. Most of the techniques mentioned below do not automatically result in a significance test (although one may be available) and may be more appropriate for hypothesis generation than hypothesis testing. Ordination and classification techniques are widely used in plant community ecology as well as plant population ecology. In all cases the data take the form of a matrix, with each individual sample (whether it be plant or quadrat) associated with a number of measurements (morphological variables or species abundances, for example). Useful texts that describe the techniques mentioned below (and many others) in more detail include Waite (2000), Manly (2004), and Coker (2011). Annotated bibliographies of ordination and classification methods are available in Gibson (2012) and Wildi (2012).

Ordination

Ordination techniques are used to order or arrange complex multivariate observations in an easily visualized manner. They can be considered as a way of reducing multidimensional data to a few easily visualized dimensions. Ordinations are commonly used by community ecologists to summarize species-by-sites data sets and their relationship to environmental factors. Plant population ecologists can use ordination analyses to analyse data sets of plant growth or performance metrics among individuals of a single species. A brief summary of ordination methods is provided below; a more detailed introduction is provided by Legendre and Legendre (2012) and Greenacre and Primicerio (2014). Ordination analysis can be conducted in most standard statistical packages, although there are a number of specialized software programs that have been developed for ecologists (summarized in Gibson 2012). The ade4 and VEGAN packages in R can run several ordination methods (see the Appendix).

Principal components analysis (PCA). PCA is one of the earliest and best known ordination techniques (Pielou 1977; Waite 2000; Manly 2004; Coker 2011). Imagine a scatter plot in which each observation is represented by a point, the location of which is determined by the score for that observation along axes representing each of the measured variables (or traits). If there were only two variables this would be a simple two-dimensional graph, three variables would produce a three-dimensional graph, and more than that would produce a multidimensional scatter plot. Four or more dimensions are hard to visualize. PCA uses basic matrix manipulation techniques to find the line that passes along the longest axis of the scatter of points present, whether the number of variables is 2 or 20. That line is the first principal component and is the line along which there is most variation between the observations. The two most dissimilar observations should lie at opposite ends of the line. As with regression there will be some portion of the variation in the data which is unexplained by the first principal component, unless the points all lie on a completely straight line. The next principal component is at right angles to the first, and explains less variation than the first component but more than the third, and so on. Note that this requirement of orthogonality (being at right angles) means that all the principal components are completely independent of each other and uncorrelated. The number of principal components can (in some cases) be as large as the number of variables (traits) measured, but in most analyses only the first two or three reveal anything interesting about the data, or explain an appreciable amount of the variation among the observations. Each variable or attribute (e.g. floral trait) is associated with an eigenvector for each component. The eigenvector is the weighting for that variable on the component, and can vary from −1 to 1. An eigenvector close to −1 or 1 indicates that the variable is an important determinant of the component (and thus of variation between observations), whereas eigenvectors close to 0 indicate otherwise. Each principal component has an eigenvalue that is a measure of its explanatory power (i.e. the proportion of the variation in the data set that it explains). After running a PCA it is usual to plot the data using their scores on each of the first and second (or higher) axes (if significant; see Franklin et al. 1995). In such a plot, it is easy to visualize which the most dissimilar observations are.

Reich et al. (1999) used PCA to analyse the relationships between five leaf traits from a variety of species from six biomes (alpine tundra to tropical rainforest). They showed that the first principal axis (explaining the largest portion of the variation between species traits) was correlated with high values of nitrogen, photosynthetic capacity, and specific leaf area, and short leaf life spans (Fig. 7.5 and legend). In another context, Suding and Goldberg (1999) used PCA to create a composite index of productivity from three other variables that were thought to be related to productivity.

One of the assumptions behind PCA is that the traits are linearly related to the principal components. In community ecology (where quadrats are the observations and species the traits) this restricts its use to fairly short sections of environmental gradients, over which the assumption of linearity might hold for most species. However, over longer environmental gradients a unimodal response is more likely, with a species having an optimum abundance somewhere along the gradient with a lower abundance at either end. In this case, correspondence analysis is more suitable.

Correspondence analysis (CA). In CA a process of iterated reciprocal averaging (RA) is used to arrive at species scores for each axis. The algorithm begins by allocating arbitrary trait scores and using these to calculate observation scores, then use the observation scores to recalculate trait scores. This processes is reiterated until the scores stop changing (it may sound incredible, but worked examples in Waite (2000) and Coker (2011) will convince you that it works). CA thus allows simultaneous

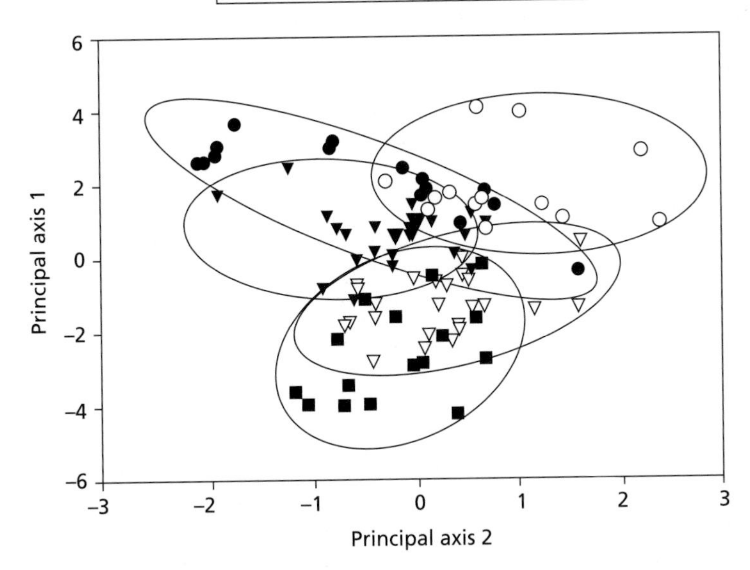

Figure 7.5 Application of principal components analysis (PCA). The PCA is of species from six biomes, ordinated according to five leaf traits (leaf nitrogen, photosynthetic capacity, life span, specific leaf area, and leaf diffusive conductance). The main contributors to high values of the first axis were increasing nitrogen, photosynthetic capacity, specific leaf area, and decreasing leaf life span. Leaf diffusive conductance was the main contributor to the second axis. Redrawn with permission of the Ecological Society of America from Figure 6 of Reich et al. (1999).

analysis of observations and traits. Prentice et al. (2000), in a study of the effects of a field experiment on allozyme frequencies in *Festuca ovina*, used a CA to examine the way in which the allele frequencies differed between treatments (Fig. 7.6). The CA showed that the plots that had received different treatments could be separated out on the first and second axes of the CA based on the allele frequencies of the plants within them.

Non-metric dimensional scaling (NMDS). NMDS operates on a ranking of distances between samples (i.e. individual sites or plants) obtained from a resemblance matrix. This ranking procedure provides the advantage that NMDS does not have the assumptions inherent in other metric scaling methods, and as a result NMDS is regarded by many as the most robust method of ordination (Minchin 1987). NMDS can be used to summarize genetic and population data. For example, Spooner et al. (2004) used NMDS to analyse population stem-class frequency data to identify relationships among roadside populations of three species of the shrub *Acacia* (*A. pycnantha*, *A. montana*, and *A. decora*).

Classification

Ordination techniques arrange multivariate data along a continuum while classification techniques place the data into groups, the members of which have more in common with each other than they do with those in other groups. Any group of observations (quadrats, plants, etc.) can be classified *agglomeratively* or *divisively*. In the agglomerative approach, quadrats start as different entities and are put into progressively larger groups based on their similarities. The divisive approach starts with all the quadrats together and separates off distinct groups based on their differences from each other. Agglomerative techniques are often called cluster analyses. They pull groups of observations together into clusters. A similarity coefficient, such as Sørensen's index of similarity, is used to investigate similarity of all observations to each other, and the two most similar quadrats are then fused together into a single group. The index is then recalculated. The investigator has to decide where to stop the process, otherwise all groups would eventually be

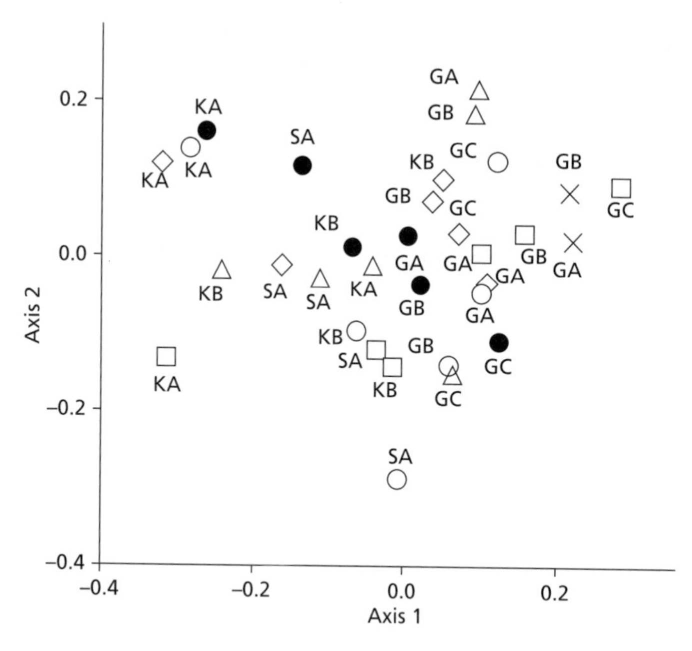

Figure 7.6 Demonstration of the use of correspondence analysis (CA) in plant population ecology from Prentice et al. (2000). Allozyme differentiation in *Festuca ovina* was examined under different experimental regimes at three sites. The figure shows the first two axes of a correspondence analysis based on counts of the 15 most common alleles (at four loci) within each of their replicate plots. Each point on the graph represents one of the 32 plots used in the experiment. The alphanumeric labels differentiate replicates and the symbols differentiate treatments. Reproduced with permission from John Wiley & Sons.

fused together. Keinänen et al. (1999) used a cluster analysis in their examination of the chemical responses of silver birch clones to defoliation and herbivory. The cluster analysis showed that the chemical profiles of the birch trees depended more on the clone from which they came than the experimental treatments.

Discriminant function analysis

Discriminant function analysis (DFA) is a way of placing individual observations into predetermined groups based on whether or not their characteristics match those of the group. Discriminant function analyses could be used to place individuals with measured morphological characteristics into species, for instance, or vegetation observations into community types. DFA assumes that the data are multivariate normal. A discriminant function is calculated for each predetermined group, and is used to predict which group an individual or observation is most likely to belong to (Fowler et al. 1998). The discriminant function is composed of linear combinations of the measured variables. Manly (2004) describes the mathematical principles behind the alternative discriminant function methods, and how statistical tests can be performed on the results. Menges and Dolan (1998; see Case Study 2 in Section 7.9) used discriminant analysis to predict the extinction risk of particular populations. They achieved an 80% success rate in classifying populations to groups based on extinction risk. Region, genetic variation, and fire were the most important predictors.

7.7 Bayesian analysis: an alternative to classical hypothesis testing

Bayesian analysis is an alternative approach to classical hypothesis testing which is becoming increasingly popular in ecology and environmental science, including plant population ecology (Hilborn and Mangel 1997; Clark 2005). This approach operates on the assumption that competing hypotheses about a process or situation have an estimable probability of being true. The goal of this approach is to determine which hypothesis is most likely to be true given a model and the probabilities derived from a set of data or prior assumptions.

Given two models, M_1 and M_2, representing hypotheses H_o and H_a, likelihood analysis determines the likelihood that a model is an appropriate description given the data at hand. For model M_1, the likelihood (L) is denoted by $L_1\{M_1|\,\text{data}\}$. The likelihoods of the two models are now compared.

If $L_1\{M_1|\text{data}\} \gg L_2\{M_2|\text{data}\}$, then model M_1, and hence H_o, is accepted as being the best representation of the true situation; while the converse is true if $L_1\{M_1|\text{data}\} \ll L_2\{M_2|\text{data}\}$. If $L_1\{M_1|\text{data}\} \sim L_2\{M_2|\text{data}\}$ then the data do not allow a distinction between the models and neither hypothesis can be accepted as being more likely than the other. The statistical details of determining the likelihoods and the basis for determining >>, <<, and ~ need not concern us here, but can be found in many textbooks including Edwards (1992), Hilborn and Mangel (1997), and Gotelli and Ellison (2013).

Bayesian analysis can be viewed as an extension of the maximum likelihood approach where a priori (previous) knowledge about a situation (e.g. model M_1) is incorporated to estimate the parameters of a statistical model. This initial knowledge is summarized as the 'prior probability' that model M_1 is true, denoted by P_1. With two models then $P_1 + P_2 = 1$. The a priori knowledge used in calculating the prior probabilities can be investigator estimates. Upon provision of some additional data, the prior probabilities can be updated to provide 'posterior probabilities' that model M_1 is true, given the data. Thus, probabilistic statements of confidence can be made comparing hypotheses and models, based upon the posterior probabilities (Hilborn and Mangel 1997). In the context of comparing hypotheses, the posterior probability for the null hypothesis conditional on the data is

$$P(H_o \mid x_1, \ldots, x_n) = \frac{P(x_1, \ldots, x_n \mid H_o)P(H_o)}{P(x_1, \ldots, x_n \mid H_o)P(H_o) + P(x_1, \ldots, x_n \mid H_a)P(H_a)} \quad (7.10)$$

where H_a is an alternative hypothesis under consideration, $P(H_o)$ is the prior probability associated with the null hypothesis, and $P(H_a)$ is the prior probability associated with the alternative hypothesis (Reckhow 1990). The more is known about the prior probability distributions (i.e. the lower variance they have) the more influence the priors have on the posterior probability distributions and the less influence the likelihood function has. This is both an advantage and a disadvantage of Bayesian analysis. The advantage is the explicit inclusion of prior information in the analysis; the disadvantage is the influence that this information has and hence a lack of objectivity. Different investigators may have different prior information and so can come up with different results.

Gotelli and Ellison (2013) specify six steps in Bayesian inference:

1. Specify the hypothesis: $P(H \mid \text{data})$.
2. Specify parameters as random variables (with known distributions).
3. Specify the prior probability distributions: $P(\text{hypothesis})$.
4. Calculate the likelihood:

   $$L, P(\text{data}_{\text{observed}} \mid \text{hypothesis}).$$

5. Calculate the posterior probability distribution: $P(\text{hypothesis} \mid \text{data})$.
6. Interpret the results.

Example 1: Acid rain damage on red spruce (Ellison 1996)

Ellison (1996) describes the hypothetical example where two ecologists are interested in assessing damage to the leaf canopy of red spruce (*Picea rubens*) by acid rain in a previously unstudied location in Vermont, United States. Based upon their knowledge of the plant, the effects of acid rain, and red spruce forests in Vermont, each ecologist can estimate the fraction of foliar area likely to be affected by a defined amount of acid precipitation (i.e. hypotheses, denoted β). The estimates of β by the two ecologists may be quite different, but nevertheless represent prior probability estimates for a single model of leaf loss due to acid rain. A pilot experiment (see Section 2.5) is conducted where seedlings of red spruce are grown in the laboratory and watered with different concentrations of nitric acid. The data from the pilot experiment can be combined with the ecologists' prior probabilities using Bayesian inference to calculate a posterior probability distribution of β. The posterior probability can be used as a new prior probability distribution in the development of new hypotheses, including a comparison with actual field data.

Example 2: Canopy–gap transitions and seedling recruitment in mixed-oak forest (Beckage et al. 2007)

Bayesian analysis was used to estimate the location of change-points (canopy–gap transitions) and seedling recruitment (λ) along transects in a mixed-oak forest. Seedling counts of *Acer rubrum* in quadrats along the transects were used to update estimates of the prior probability distribution of change-points. A vector corresponding to a uniform distribution of change-points represented the prior distribution. The posterior distributions provided estimates of seedling densities associated with gap and canopy conditions as well as the likely location of change-points. Results of this analysis were superior to a more conventional analysis where seedling estimates were compared in quadrats clearly identified as gap or canopy. In the conventional analysis, transition quadrats were eliminated from the analysis resulting in a loss of information and less powerful analysis.

Example 3: Estimating population viability of an epiphytic moss (Ruete et al. 2012)

Population growth rates may vary stochastically in response to environmental conditions. However, estimating these stochastic population growth rates, log λ_s, using conventional, frequentist (probabilistic) approaches does not account for knowledge of the parameters related to such stochasticity. Posterior distributions of log λ_s were estimated for the moss *Buxbaumia viridis* based on estimated posterior distributions of year-specific growth rates and a priori knowledge of 'effect-size' parameters (i.e. weather variables and measurement error). This approach allowed the incorporation of natural variability and sampling uncertainty across regional and local scales and is an advantage of previous frequentist methods of estimating log λ_s based upon point estimates or confidence intervals.

Discussion of Bayesian analysis examples

The examples described above each provide a confidence estimate for a single model (fraction of leaves damaged, distribution of change-points and seedling density, and population viability, respectively). However, multiple models (or hypotheses) could be evaluated through a comparison of posterior probabilities.

The statistical details of applying Bayesian analysis are beyond the scope of this book but can be found in several textbooks (Lee 1989; Box and Tiao 1992; Ghosh and Meeden 1997). Hilborn and Mangel (1997) argue strongly for the ecological use of Bayesian analysis, and a series of eight papers introducing the topic to ecologists appeared in the journal *Ecological Applications* (see introduction by Dixon and Ellison 1996). Bayesian analysis is increasingly being used for studying questions relating to plant populations and has been used to study yearly apple crop production and to examine alternative outcomes in competition experiments (Chen and Deely 1996; Damgaard 1998). Bayesian analysis is available in the most recent versions of SAS/STAT 9.2 and above using the BAYES statement in the GENMOD, LIFEREG, and PHREG procedures. Code for running Bayesian analysis in R is described in Kruschke (2011), and specifically for population ecology in King et al. (2010). This approach has an application for the analysis of extinction risks of rare plant species as demonstrated by a number of animal population studies (e.g. Ludwig 1996; Taylor et al. 1996). The potential advantages of Bayesian analysis over classical hypothesis testing include: (a) all the available information can be included in the analysis, (b) posterior probabilities allow the results of multiple experiments or data to be combined, (c) multiple hypotheses can be compared in a single analysis, (d) null and alternative hypotheses are explicitly parameterized, and (e) data are treated as fixed as opposed to being a random representative sample (Dixon and Ellison 1996; Ellison 1996; Clark and Lavine 2001).

7.8 Reporting statistics

Fowler (1990) bemoaned the fact that the statistics used by ecologists were often poorly described in manuscripts submitted to the journal *Ecology*. This is still the case for numerous papers submitted to a wide range of ecological journals. It is important to clearly describe exactly how the statistics were performed, so that another researcher who accesses your data could do exactly the same tests and come up with the same results. Fundamental information about the design of your study (such as number of replicates or samples and arrangement of plots) is essential for readers to judge the appropriateness of your statistics. You should make it clear whether and how you transformed your data to meet the assumptions of a particular test, and if there is any possibility of confusion you should make it very clear how a particular test was performed. This is particularly important for potentially complex methods such as ANOVA, where you must clearly state which model was used and how the *F*-ratios were calculated if there is any possibility of confusion.

It is also important to be unequivocal when reporting the outcome of statistical tests. At the very least, a *P*-value needs to be associated clearly with the name of the test, the d.f., and the test statistic. Often, in more complex analyses, a fairly detailed table should be supplied (see the section on ANOVA above).

Figures that present data should be clear and self-explanatory. The reader should be able to understand easily what the figure is showing. Three-dimensional graphs and graphs with any 'extra' information that does not explicitly contribute to illustrating the data should be avoided (Dytham 2011; Ellison 2001). As pointed out by Tufte (1983), the information to ink ratio of each figure should be maximized. Ellison (2001) provides examples of different ways to present the same data. Error bars should be clearly identified in the legend (as, for instance, 95% confidence intervals, standard deviations, or standard errors). If standard errors of the mean are used, then the sample size should be clearly stated. Consider alternative methods of presenting data. For example, box and whisker plots can be far more informative and appropriate than means and standard deviations (Figs. 7.7 and 7.8).

It is customary to indicate means that differ significantly from each other using letters adjacent to the data point or bar. Means sharing the same letter are not significantly different from each other. If you are new to this you may find the application of letters to means quite tricky. An easy way to do this is first to write down your means in rank order in a line. Draw lines under those groups of means which do not differ from each other. All means in the group containing the largest mean get the letter a, the next group gets the letter b, and so on.

One word that should be used with care in the results section of a report is 'significant'. Because this word has both a colloquial meaning and a formal statistical meaning, confusion can arise. This term should be reserved for describing effects that are actually statistically significant. If a difference

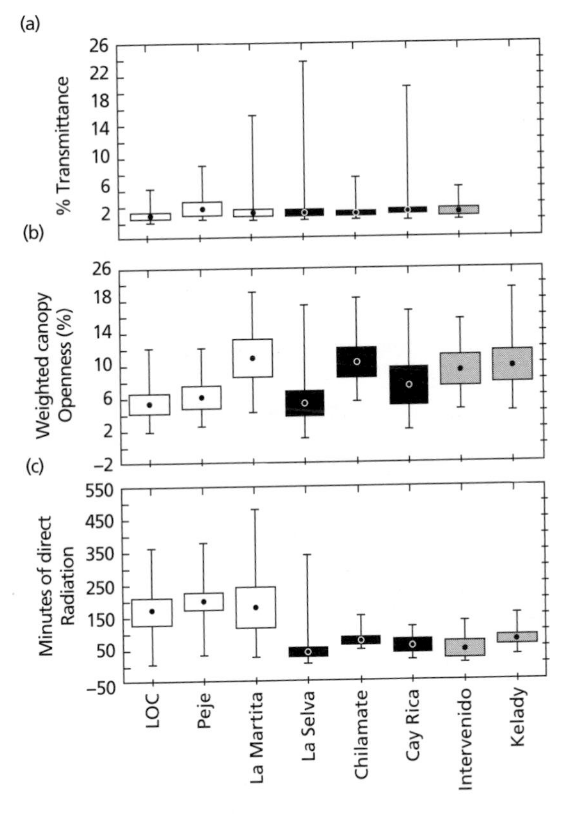

Figure 7.7 Effective presentation of data. Box and whisker plots are used to illustrate features of the forest floor environment experienced by seedlings at different locations. Dots, median; box, 25th and 75th percentiles; whiskers, minimum and maximum values. Measurements were made in three second-growth (unshaded boxes), three old-growth (darkly shaded boxes), and two selectively logged stands (lightly shaded boxes). Reproduced with permission of the Ecological Society of America from Nicotra et al. (1999).

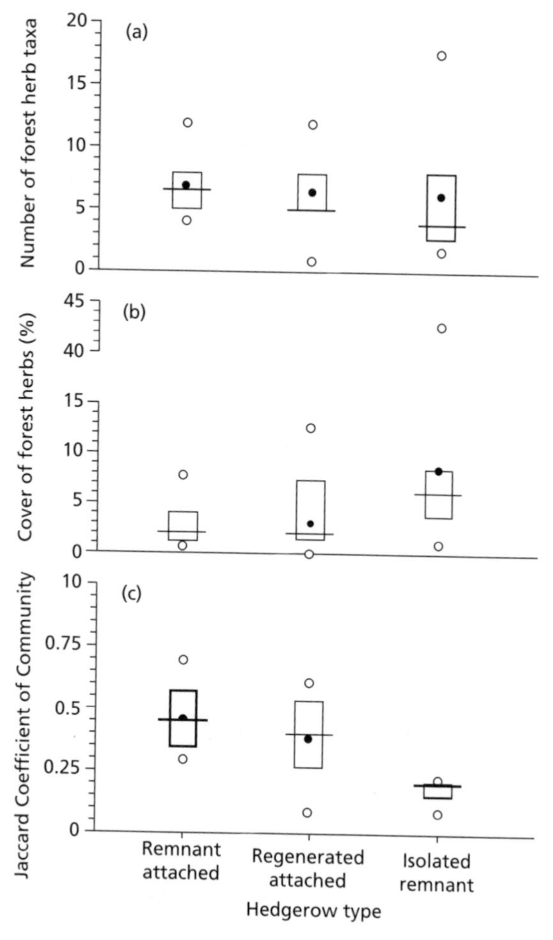

Figure 7.8 Effective presentation of data. Box plots are used to compare the forest herb flora of three different types of hedgerows: remnant attached ($n = 14$), regenerated attached ($n = 11$), and isolated remnants ($n = 7$). Horizontal bar, median; box, quartiles; open dots, range; filled dot, mean. (a) Richness, (b) abundance, (c) similarity to forest. Reproduced from Corbit et al. (1999) with permission from John Wiley & Sons.

is large or dramatic, but has not been shown to be statistically significant, then an alternative word should be used.

7.9 Case studies

The main statistical tests used in the four case studies described in Chapter 1 are outlined below.

Case Study 1 (Zuidema et al. 2010)

To test their hypothesis that sink subpopulations are maintained by recruitment subsidies from a larger source population, the authors needed to demonstrate recruitment from source to sink populations. First-order spatial analysis was used to compare dispersion of adult trees and juvenile recruits of *Scaphium borneense* in each of three habitats. Specifically, Hamill and Wright's test of dispersion of one type of group of plants to another based upon the cumulative distribution of distances of juveniles (recruits) to nearest and second-nearest adults was used (Hamill and Wright 1986; Section 8.1). Differences between the observed and expected distributions under a null model were tested using a Kolmogorov–Smirnov test. A *G*-test was used to compare recruitment rates between habitats, although only the *P*-value and not the test statistic is presented. Multistate matrix models based on size-based Lefkovitch matrices (Section 8.2) were used to simulate population dynamics within and between habitats. Submatrices along the leading diagonal of the transition matrix quantified recruitment dynamics within valley, slope, and ridge habitats, and off-diagonal submatrices quantified recruitment from one habitat to another. Elasticity analysis was used to investigate stasis, progression, and recruitment elements of asymptotic growth rates λ estimated for each habitat.

Case Study 2 (Menges and Dolan 1998)

The aims of this study of the demographic viability of populations of *Silene regia* under different management regimes were discussed in Chapter 1. The modelling methods used in the paper are discussed in Chapter 8. While a wide range of statistical approaches was used in this important and interesting paper it is not always clear why the authors used certain statistics in some situations and not others, and it is not always possible to be exactly sure how the statistical tests were performed. A principal components analysis was used to generate an index of genetic variation. Individual populations were then placed along this axis using their PCA scores. One-way ANOVA was used to investigate whether there were differences in elasticity in populations from different management regimes. When testing for differences between values

continued

Case Study 2 *Continued*

of the population growth rate λ, only the *F*-ratio and its associated *P*-value were reported for each ANOVA. However, when testing for differences between *G*-values, *F*, *P*, and one part of the d.f. are reported—it is recommend that more information is provided (see above). The ANOVA was followed by the Student–Newman–Keuls test for differences between groups. Fisher's exact test was used to test the significance of a contingency table examining whether or not seedlings were more likely to be found in sites managed by fire. Goodman and Kruskal's τ was also used to analyse contingency tables of the effects of management and fire on population viability. Spearman correlation coefficients were used to examine correlations between λ and genetic measures. Discriminant analysis was used to investigate whether extinction risk could be predicted from measured variables. It was successful in separating out the populations at most risk of extinction.

Case Study 3 (McEvoy et al. 1993)

This study of factors affecting the successful control of ragwort populations is discussed in more depth in Chapter 1. Linear regressions were used to predict plant weights. A four-way ANOVA (the four factors were time, beetle, competition, and moth, with block effects also being examined) was used to analyse a field experiment with a factorial design. The ANOVA table is very clear (their Table 2, our Table 7.2), so that although the authors do not explicitly state that it reports a Model I ANOVA it is clear that this is the case (error mean square used in all cases to calculate *F*). One factor, competition, with three levels (vegetation clipped, removed, or left unaltered) was decomposed to two, 1-d.f. orthogonal contrasts: removed versus control, unaltered versus clipped. This is also very

Case Study 3 *Continued*

unambiguously set out in the ANOVA table. The authors state clearly that they used Type III sums of squares in ANOVA (as recommended by Shaw and Mitchell-Olds 1993) where the design was not balanced.

Case Study 4 (Mauricio et al. 1993)

Chapter 1 gives a detailed description of the aims of this study of the effects of patterns of leaf damage on the fitness of *Raphanus sativus*. Prior to analysis, variables were tested for normality and homogeneity of variance, though the authors do not say how this was done. Proportional data (allocation data) were arcsine-transformed, as is generally recommended. A priori contrasts were used to identify groups which could be pooled for subsequent analysis. Because many of the measured variables could not be considered independent of each other, two separate MANOVAs were used for the reproductive measures and the growth measures, respectively. The d.f., Wilks' λ, and its associated probability were reported for each MANOVA. Both MANOVAs were significant, indicating an overall effect of the treatments on the measured growth parameters. Thus, it was appropriate to follow them by 15 univariate ANOVAs on each reproductive and growth measure. The ANOVAs were presumably one-way, with four treatment levels. For most of the individual ANOVAs only *F* and *P* are presented as output. For the four significant ANOVAs the d.f.s are also reported. Very few of the individual ANOVAs were significant. For those ANOVAs which produced significant results, a posteriori (post hoc) Ryan–Einot–Gabriel–Welsch tests (as recommended in Day and Quinn 1989) were used to determine which treatment levels differed significantly.

CHAPTER 8

Advanced statistical techniques: spatial patterns, life tables, modelling, and population viability analysis

Seek simplicity, but distrust it.

(G. E. Lagrange, mathematician (1736–1813) in Hutchinson (1978))

. . . without making these simplifying assumptions, this model, like any other model, would have been self-defeating because it would have been as difficult to interpret the model as it is to understand the real world.

(Silvertown et al. 1992)

- **Spatial pattern analysis**
- **Life tables and matrix models**
- **Cellular automata, individual-based models and integral projection models**
- **Population viability analysis**

Preamble

In Part 2, and then in Chapter 7, the importance of thorough advance planning for efficient field work was emphasized. This planning ensures that adequate data are collected and that appropriate statistical analyses can be used to address your research hypotheses. In this chapter we re-emphasize this along with discussion of the more advanced statistical and mathematical procedures used by plant population ecologists. The application of these methods for plant population ecology is stressed rather than their mathematical derivation.

A general goal of this chapter is to discuss the methodology used to understand the spatio-temporal dynamics of plant populations. In particular, neighbourhood (phenomenological, spatial) models designed for studying the population dynamics of sessile organisms are described. The assumptions implicit in these models are (Czárán and Bartha 1992) as follows:

1. Individuals in populations are sessile or with limited movement for most of their lifetime.
2. Population changes reflect local events on the spatial scale of an individual.
3. The fate of individuals can be followed.
4. Interactions are between individuals.

In large part, the differences among the spatio-temporal neighbourhood models reflect the definition of the neighbourhood that is used.

It is important to have a basic understanding of models and their purpose. Models are analogies that represent important features of a system (Ford 2000), with their primary purpose being to expose ideas rather than to necessarily provide an exact description of a biological system. By constructing

Methods in Comparative Plant Population Ecology. Second Edition. David J. Gibson.

a model we can: (a) make predictions, (b) better understand the system, and (c) study complexity (Thornley 1998). A model is not a complete synthesis of a system.

A simple classification recognizes two types of model (Pacala 1997). A *model of ideas* is a relatively simple test of 'if–then' statements rather than a direct empirical measurement of parameter values and functional forms. For example, Tilman's R* model of species coexistence makes qualitative predictions that can be tested through field observations and experiments (e.g. Wedin and Tilman 1993). By contrast, *models of natural systems* include estimates of parameters and functional forms allowing predictions to be made and tested from the value of parameter estimates. In either case, deviations from predictions provide clues as to which other factors not included in the model might be important. The most complex of the population models are dynamic and can be used to predict population changes through time. In practice, it is useful to recognize the following types of models (Jeffers 1978):

1. Deterministic models, for which the predicted values may be computed exactly.
2. Stochastic models, for which the predicted values depend on probability distributions.

In addition, models fall into two further categories:

1. Analytical models, for which explicit formulae are derived for predicted values or distributions. These include regression and multivariate models.
2. Simulation models, which can be specified by a routine of arithmetic operations, such as the repeated application of a transition matrix, use of random numbers, or the solution of differential equations.

Whatever model is constructed there is value in conducting a sensitivity analysis in which the model is run with a range of parameter estimates to see how robust the conclusions are. Also, analytical approximation must be developed in the construction and testing of these models. Simulation models, for example, are relatively straightforward to construct but difficult to analyse, it is also easy to make mistakes when coding. Therefore it is extremely useful to have analytical approximations (mathematical equations) which allow the simulation model to be understood and checked. An analytical approximation comprises a system of equations that ignore some aspect of the system, allowing simple analytical insights. These equations can then be analysed and the behaviour of the simulation model understood. For example, Rees and Paynter (1997) developed a complex simulation model for *Cytisus scoparius* incorporating age structure, local seed dispersal, and a seed bank. There were nine parameters that defined the model: exploring each of these in turn and in combination would have taken a prohibitively large amount of time and even then the huge amount of output generated would have been difficult to interpret. Mathematical approximations for the simulation model were developed in which the spatial structure was ignored. By comparing the results of the approximation and the simulation model the impact of spatial effects could be assessed. This approach showed that, under a wide range of conditions, the detailed spatial structure did not matter and that three parameters determined the abundance of *C. scoparius* (probabilities of disturbance, sites becoming available for colonization, and maximum plant longevity). Before developing the approximations, the authors had spent several weeks running the simulation model, but really had no idea what parameters were important or why. The analytical approximations also accurately described the simulation results, suggesting that no mistakes had been made in coding the model.

A complete survey of plant population models is not provided here, rather the focus is on models of natural systems that are immediately tractable and useful for general applications. The development of these models as an extension of descriptive statistics is emphasized. A general review of plant population models is provided by Silander and Pacala (1990) and Berger et al. (2008) and in a number of books (e.g. Tuljapurkar 1980; Heafner 1996; Hilborn and Mangel 1997; Roughgarden 1998; Ford 2000; Grimm and Railsback 2005). Examples of models that address specific, applied, or conceptual issues include Rees and Paynter (1997), Freckelton and Watkinson (1998), and Conlisk et al. (2013).

In Section 8.1 methods of spatial pattern analysis are discussed. The (usually) non-random distribution of individuals within a population has important ramifications for how neighbour interactions are viewed. Distance and tessellation neighbourhood models are described. In Section 8.2 the construction of life tables, survivorship curves, and the use of matrix models for estimating the finite rate of growth of a population and other parameters are presented. The use of cellular automata models, individual-based dynamic population models, and integral projection models as spatially explicit developments of matrix models is discussed in Section 8.3. One particular model, SORTIE, which is used to model forest dynamics is discussed in detail. The use and application of population viability analysis (PVA) is discussed in Section 8.4. Within each section, the use of simple approaches proceeds to those that are statistically more complex and sophisticated. While bearing in mind Lagrange's quote at the start of the chapter, remember the dictum of Ockham's razor, namely to make things no more complex than necessary. A selection of textbooks for more advanced treatments is provided in Table 8.1.

8.1 Spatial pattern analysis

Plants seldom occur at random across the landscape. The distribution of plants is rarely unrelated either to other plants, of the same or different species, or to environmental discontinuities. Given this observation (the demonstration of which is discussed below), as plant population ecologists we need to know how the distribution of individuals, and the processes that effect their distribution, affects the performance of individual plants. The objectives of this section are to present ways of characterizing individual plant distributions and to relate them to population structure and the environment.

Here we concentrate on the most useful methods for plant population ecologists: broader reviews of spatial pattern analysis in ecology are available elsewhere (Table 8.1). Our focus is on spatial point patterns in which the position of individual plants in a two-dimensional plane is investigated. Quadrat-variance transect methods are not described as these

Table 8.1 Some textbooks detailing topics covered in this chapter.

Topic				Text
Spatial pattern analysis, distance, and tessellation models	Matrix models	Cellular automata and individual-based models	Population viability analysis	
		✓		*Identification of cellular automata* (Adamatzky 1994)
✓				*Perspectives on spatial data analysis* (Anselin and Rey 2010)
				Population viability analysis (Beissinger and McCullough 2002)
✓				*Applied spatial data analysis with R* (Bivand et al. 2008)
	✓		✓	*Matrix population models* (Caswell 2001)
✓				*Spatial pattern analysis in plant ecology* (Dale 1999)
✓				*Statistical analysis of spatial point patterns* (Diggle 1983)
	✓			*Plant and animal populations: methods in demography* (Ebert 1999)
✓				*Quantitative plant ecology* (Greig-Smith 1983)
		✓		*Individual-based modeling and ecology* (Grimm and Railsback 2005)
	✓	✓		*Modeling biological systems* (Heafner 1996)
✓				*Numerical ecology* (Legendre and Legendre 2012)
✓				*Statistical ecology* (Ludwig and Reynolds 1988)
			✓	*Quantitative conservation biology* (Morris and Doak 2002)
✓	✓			*Primer of ecological theory* (Roughgarden 1998)
✓	✓			*Statistical ecology* (Young and Young 1998)

relate primarily to the description of community-based plant species patterns (Greig-Smith 1979). However, these methods do have value to the plant population ecologist for identifying the occurrence, scale, and intensity of pattern for groups (patches) of individuals. As the location and size of patches of individuals varies so will the number and size of individuals within patches.

Two types of methodology are described. *First-order analyses* use mean distances from one plant to another or to a random point and are often based on incompletely mapped data. By contrast, *second-order analyses* use the variance of the plant-to-plant distances from a completely mapped population (or sample of the population). Consequently, second-order analyses are more informative than first-order analyses. Nevertheless, the latter are frequently used because of the logistical constraints involved in complete mapping.

The arrangement of points (or plants in our case) in a plane is referred to as the spatial pattern or *dispersion* (Dale 1999). The null model for dispersion is complete spatial randomness (CSR) where all plants occur independently of each other. In this case, all regions of the same size have the same probability of containing a given number of points. These types of data follow a Poisson distribution in which the variance is equal to the mean (see Section 7.2). If plants are not randomly distributed then they can either be clumped (over-dispersed, aggregated, or contagious), like seedlings emerging from a seed cache, or regularly distributed (under-dispersed or uniform), like pine trees in a plantation or apple trees in an orchard. Departure from a Poisson distribution forms the basis of several tests for spatial pattern. In general, distributions are clumped if the variance is significantly greater than the mean or uniform if it is significantly less than the mean. Tests of this include distributions in accord with a negative binomial or positive binomial, respectively (for details see Ludwig and Reynolds 1988):

1. The Poisson distribution (variance = mean) for random patterns.
2. The negative binomial (variance > mean) for clumped patterns.
3. The positive binomial (variance < mean) for uniform patterns.

Data sets

The field data necessary to conduct the analyses below consist of mapped *x*–*y* coordinates of plants in the field (Fig. 8.1), referred to as an 'event' set. Data on plant performance or identity (referred to as 'marks', *m*; e.g. height, biomass, number of flower stems) are necessary to relate the pattern of an event set to performance. Obtaining the *x*–*y* coordinates of a population can be quite a challenge, especially when the plants are above head height and the underbrush is thick. In these situations the distance of each plant to the ends of a baseline can be measured and the coordinates recovered through triangulation (see Fig. 5.4). In some cases it is easier to divide the whole area into a set of smaller grid squares and then locate plants first within each grid square. When the canopy is open enough to obtain satellite fixes, a global positioning system can be used to accurately record plant locations. When the plants are small it may then be possible to record coordinates directly onto acetate sheets held in frames over the vegetation. The individual shrubs of *Ceratiola ericoides* shown in Fig. 8.1 were mapped by first marking a 30 m × 45 m plot into 5 m × 5 m grid cells in the field. The *x*–*y* coordinate of each plant was determined from the distances from the corners of the grid cell in which it occurred (Fig. 8.2). In addition to the location of each shrub, its sex (male, female, or juvenile), age (estimate from node counts), and size (height, maximum horizontal diameter, and maximum diameter perpendicular to the first diameter) were noted (Gibson and Menges 1994). These data are used to illustrate several of the techniques described below and allow the following general questions to be addressed:

1. Are individuals or different plant types (e.g. life-history stages, sexes) distributed at random?
2. Is there a relationship between the dispersion of one plant type and another (e.g. juveniles and females)?
3. At what scale does non-randomness occur?
4. Is performance related to the area around individuals in a non-randomly dispersed population?

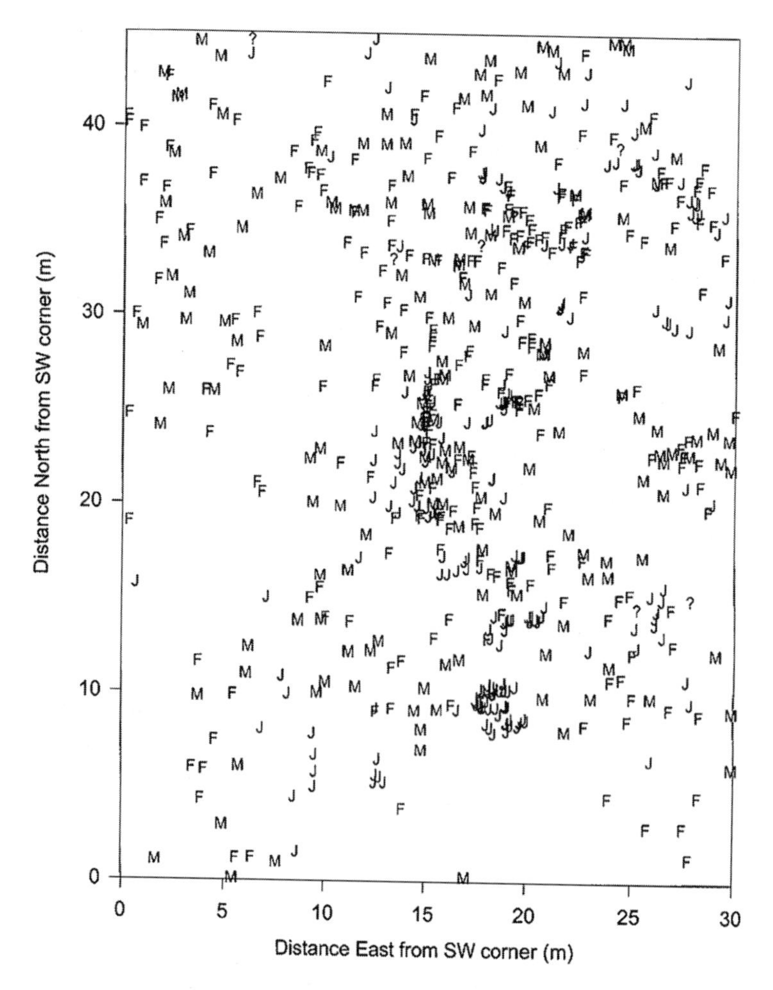

Figure 8.1 Mapped population of the dioecious shrub *Ceratiola ericoides* showing the location of male (M), female (F), and juvenile (J) individuals (Gibson, unpublished).

Figure 8.2 View looking to the south from the northern edge of the population of *Ceratiola ericoides* mapped in Fig. 8.1. (Photo David Gibson.) (See also Plate 17.)

Spatial analysis software

The analyses presented in this section were conducted using a combination of the GS+ program (<http://www.gammadesign.com/>) (spatial autocorrelation), the Links program in 'The R Package' (Legendre and Vaudor 1991) (Dirichlet tessellations), ADE-4 (Thioulouse et al. 1997) (Ripley's *K*-function), the TSQUARE BASIC program in Ludwig and Reynolds (1988) (T-square index), and a BASIC program in Hamill and Wright (1986) (Kolmogorov–Smirnov test). Spatial analyses are also available in the free PASSaGE 2 program (Rosenberg and Anderson 2011), the ade4 and deldir R packages (see the Appendix, Bivand et al (2008), and <http://cran.r-project.org/web/views/spatial.html> for complete listings), and many geographical information system (GIS) and mapping programs.

Are the individuals or different plant types distributed at random? First-order analysis

Methods for detecting departure from randomness for individual plants within a sample area are based upon measurements of nearest-neighbour distances or point-to-plant distances. Because these methods use mean distances they are known as first-order analyses. Two of the methods that have been proposed to address this issue (see Greig-Smith 1983; Dale 1999; Dixon 2006; Perry et al. 2006) are discussed.

T-square index. N sampling points (O) are selected randomly within the population. Two distances are measured: the distance (x) from the random point to the nearest plant (P), and the distance (y) from P to the nearest neighbouring plant (Q) beyond the plane line at P perpendicular to O–P. If the distribution of plants is random then the expected square of point-to-individual distances (x) will be approximately one-half the expected square of nearest-neighbour distances (y). From this an index of spatial pattern (C) is derived as a ratio of squared x and y distances as

$$C = \frac{\sum_{i=1}^{N}\left[x_i^2/(x_i^2 + \frac{1}{2}y_i^2)\right]}{N}.$$

Values of C of approximately 0.5 indicate random patterns, values significantly less than 0.5 indicate uniform patterns, and values significantly greater than 0.5 indicate clumped patterns. The significance of a departure of C from 0.5 is based upon calculating z as

$$z = \frac{C - 0.5}{\sqrt{1/12N}}.$$

At $P = 0.05, z = 1.96$, so calculated test values exceeding this critical value indicate a significant departure of C from 0.5. Other critical values of z can be obtained from a probability table for the standard normal distribution (Ludwig and Reynolds 1988).

The individuals of *C. ericoides* mapped in Fig. 8.1 were randomly distributed as a whole population but the juveniles were clumped, as shown by a C-statistic of 0.63 that exceeds 0.5 (Table 8.2). The z-statistic of 6.86 is larger than the critical value (1.96) indicating that the deviation from a random distribution was significant.

Hopkins' CA: Hopkins (1954) suggested a coefficient of aggregation (CA) for detecting non-randomness based upon the squared distance (x) from a random point to the nearest plant, and the squared distance (y) from a randomly chosen individual to its nearest neighbour where:

$$\mathrm{CA} = \frac{\sum x^2}{\sum y^2}.$$

Values of CA = 1 indicate a random distribution, >1 a clumped distribution, and <1 a regular (uniform) distribution. CA is dependent upon random

Table 8.2 Tests of spatial pattern on the mapped population of *Ceratiola ericoides* shown in Fig. 8.1. From Gibson and Menges (1994) and Gibson (unpublished).

	Male	Females	Juveniles	All plants
N	172	243	227	650[1]
T-square *C*	0.47	0.46	0.63	0.52
T-square *z*	−1.47	−2.26	6.86*	1.88
Coefficient of aggregation	0.71	1.19	3.05	1.33
X	0.41*	0.54	0.75*	0.57
Overall interpretation	Uniform	Random	Clumped	Random

[1]Eight plants were not coded as male, female, or juvenile but are included in the analysis of all plants.
*$P < 0.05$ indicating a departure from a random distribution.

samples and the mean CA from at least 20 iterations has been found to produce stable estimates (Briggs and Gibson 1992). The significance of the departure of CA values from 1 is assessed by calculating the parameter $x = CA/(1 + CA)$ which has a mean of 0.5 for random distributions. Greig-Smith (1983, p. 73) provides a plot of values of x for $P = 0.05$, 0.01, and 0.001 at $n = 20$–50 observations. If $n > 50$ then the expression $2(x - 0.5)\sqrt{(2n + 1)}$ should be checked against standard tables of the normal distribution.

Calculation of CA and x using the data for the shrub *C. ericoides* in Fig. 8.1 indicates that the male plants showed a uniform dispersion and the juveniles a clumped distribution (Table 8.2). The females and all the plants considered together were randomly dispersed. However, this simple analysis should not be used without some caution, as noted below.

There are limitations to the *T*-square index, Hopkins' CA, and other statistical indices based upon nearest-neighbour and random-point-to-plant distances. Based upon nearest-neighbour distances these indices do not consider the range of spatial scales in the data set, ranging in this case from 0.01 m (the smallest distances recorded in locating the shrubs) to 30 m (the length of the shortest axis of the sample plot). Furthermore, first-order tests such as the procedures described here include only nearest-neighbour distances, and second or third nearest neighbours are not considered. However, a plant-to-all-plants distance analysis, which uses the distances between all possible pairs of plants appears not to give clear results with simulated data and may not be worth the trouble of the extra calculations (Dale 1999). A second general problem with these methods is that the random points will tend to fall in the space between tightly packed clumps of individuals that may occur. This means that the random-point-to-plant distances will tend to most frequently sample plants on the edge of clumps as opposed to the plants within clumps. The result is that individuals at the edge of clumps will be indistinguishable from single individuals and indications of aggregation in the mapped dataset may be underestimated. Finally, these techniques ignore scale and do not provide a measure of the scale of spatial pattern, i.e. the size of clumps. An advantage of the *T*-square index and Hopkins' CA over some of the other indices is that the density of individuals in the study area does not have to be known or measured.

Is the dispersion of one species/class of individual related to that of another?

When investigating the distribution of plants in a population one may wish to incorporate the relative dispersion of one type of plant to another type of plant, i.e. bivariate data sets in which points in an event set have different marks (see 'Data sets', above). Two indices that are useful for addressing this issue are Pielou's coefficient of segregation and Hamill and Wright's test (see Dixon (2006) for a discussion of other methods).

Pielou's coefficient of segregation. A simple test is Pielou's (1977) coefficient of segregation, S, where

$$S = 1 - \frac{\text{observed number of unlike nearest-neighbour pairs}}{\text{expected number of unlike nearest-neighbour pairs}}.$$

The observed number is drawn from sampling at random individuals and their nearest neighbours from the mapped population. The expected number is calculated from the known proportion of individuals of each type in the population. For example, if a population had 50 males and 50 females then a random sample of 20 individuals would be expected to sample 10 unlike nearest-neighbour pairs (50% of the time you would expect the nearest neighbour to be of the other sex in the absence of spatial segregation). A sample from a population of 80 males and 20 females yields an expectation of 13 unlike nearest-neighbour pairs from a sample of 20 individuals. This can be simplified by using a 2 × 2 table like that below for a sample of N individuals of m males and n females:

Sex of base plant	Sex of neighbour		
	Male	Female	Σ
Male	a	b	m
Female	c	d	n
Σ	r	s	N

where a is the number of male–male nearest-neighbour pairs, b the number of males with female nearest neighbours, c the number of females with male nearest neighbours, and d the number of female–female nearest-neighbour pairs. Using this approach S can be calculated as

$$S = 1 - \frac{N(b+c)}{ms+nr}.$$

Values of S vary between −1 and + 1 with values greater than 0 indicating a positive spatial association of the sexes (male–male and female–female neighbours being predominant) and values less than 0 indicating spatial segregation of the sexes (e.g. male–female nearest neighbours being predominant). An example of the application of this index is given in Wheelwright and Bruneau (1992). Pielou's test is only appropriate for sparsely sampled data in which just a subsample of all individuals in the sample area has been sampled. Dixon (1994, 2006) provides formulae for calculating the appropriate test statistics from a nearest-neighbour contingency table for completely sampled data.

Hamill and Wright's test. Pielou's S is a scaleless measure and provides no information about how far away one plant type may be from another. For example, a measure of the tendency of juveniles to be located near or far from maternal plants provides important information on plant neighbourhoods. Hamill and Wright (1986) provide a test of the null hypothesis: are juveniles located at random with respect to adults? Their measure compares the observed cumulative distribution of distances from juveniles to the nearest adult. This distribution is then compared with the null hypothesis distribution using a two-sided Kolmogorov–Smirnov test. This test compares the maximum difference between the observed and null distributions with a critical value based upon the number of distances used. For descriptive purposes the observed and null distributions can be graphed.

In an x by y plot the cumulative distribution function of juvenile-to-nearest-adult distances under the random hypothesis is

$$F(s) = \frac{1}{A}\int_{x=0}^{x_{\max}}\int_{y=0}^{y_{\max}} I(s)\,dy\,dx$$

where A is the total area of the plot and $I(s)$ is the indicator function

$$I(s) = \begin{cases} 1, & \min\left\{\sqrt{(x-a_i)^2 + (y-b_i)^2} : i = 1,2,K,n\right\} \le s \\ 0, & \text{otherwise} \end{cases}$$

with n adults of coordinates a_i, b_i. $I(s) = 1$ for all (x, y) values in a plot that has a nearest adult distance $\le s$.

Comparing plots of the null versus observed cumulative distributions of juvenile-to-adult distances, six different cases of juvenile distribution can be identified (Fig. 8.3):

1. Randomness (R). The null case, and only likely if resource patches are randomly and independently distributed, too small to support more

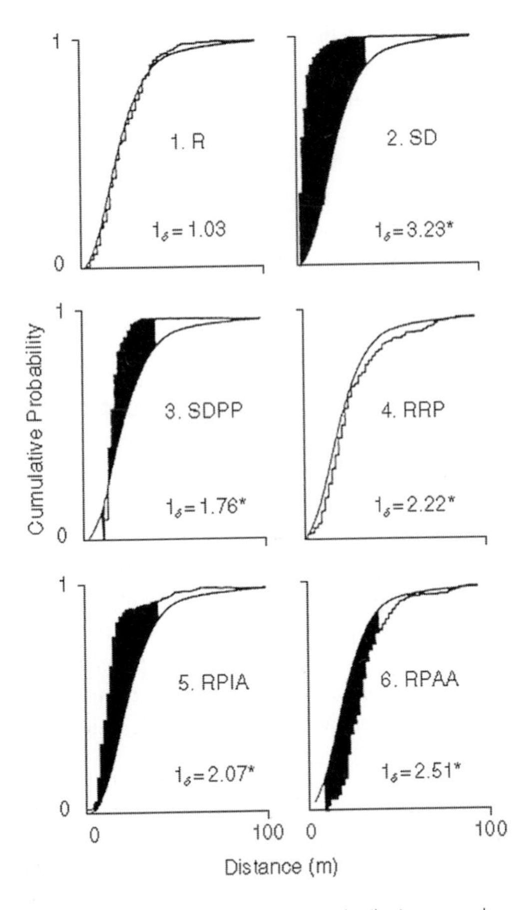

Figure 8.3 Six different cases of juvenile distributions around adult plants. Shaded areas indicated a significant departure of the observed from the expected distribution (the smooth line). From Hamill and Wright (1986) with permission of the Ecological Society of America.

than one juvenile, or too ephemeral for subsequent generations to influence each other.

2. Seed dispersal (SD). Mechanisms of SD are unlikely to lead to a random distribution. When seeds fall near to parents then clumping of juveniles will occur relative to both adults and juveniles.
3. Seed dispersal plus a seed or seedling predator (SDPP). This supports the hypothesis that seed/seedling predators are more likely to kill juveniles near adults.
4. Random resource patches (RRP). Juveniles are unlikely to survive unless in a particular resource patch, but the resource patches are independent of each other and the adults. Indistinguishable from R in this analysis.
5. Resource patches include adults (RPIA). If resource patches are permanent and associated with adults then juveniles will be clumped with respect to the adults. This case is like case (2), SD.
6. Resource patches avoid adults (RPAA). If the juveniles are clumped but in resource patches negatively associated with adult distributions (e.g. light gaps), then the observed cumulative distribution is significantly less than the null distribution at scales corresponding to the resource patch size.

Analysis of the *C. ericoides* data set (Fig. 8.1) showed significant aggregation of juveniles around female plants at 0.5–0.75 m (Fig. 8.4, shaded area between the two curves). This is consistent with case (2), SD. The median size of female plants in this population was 0.75 m × 0.60 m (maximum by perpendicular diameter), suggesting that juveniles were occurring in clumps under the canopy of female plants. This suggests that juveniles were originating from seed dropped directly just inside the edge and below the canopy of female mother plants.

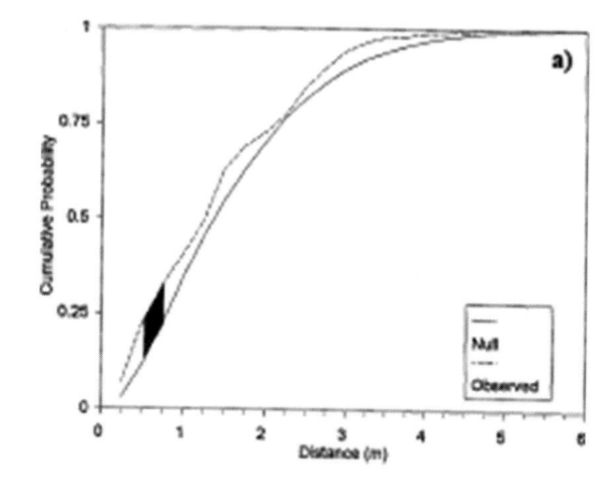

Figure 8.4 Dispersion of juveniles around females of *Ceratiola ericoides*. The shaded area indicates a significant ($P < 0.05$) departure of the observed distribution from the null model. From Gibson and Menges (1994). Reproduced with permission from John Wiley & Sons.

At what scales does non-randomness occur? Second-order analysis

The scales at which individual plants within a population exhibit departures from randomness can be investigated using second-order statistics (Franklin 2010). This means the variance of the plant-to-plant distances is of interest, not the mean of the distances in the first-order measures described above. The most widely used of the second-order techniques is based upon Ripley's *K*-function. Wiegand and Moloney (2004) review second-order analyses providing a set of recommendations for best practice to help determine the role of heterogeneous environmental factors in determining plant patterns.

Ripley's K-function. This method is based upon determining the number of plants within circles of radius *t* centred on a number of random points (Dale 1999; Haase 1995). Values of *t* are determined by the investigator and thereby provide the scales at which the analysis provides information. If the *n* plants are randomly (Poisson) distributed within an area *A* then the density $\lambda = n/A$ (do not confuse λ here with its use to designate the annual or finite rate of population increase in the rest of this book). The expected number of plants for any given radius around an arbitrary point is simple average density times area, i.e. $\lambda K(t)$, with an expected value of $K(t)$ equal to πt^2. The goal of this analysis is to obtain an estimate of the function *K* at each value of *t*, i.e. $K(t)$, the average number of individuals in a circle of radius *t*. An estimate of $K(t)$ as $\hat{K}(t)$ can be obtained as

$$\hat{K}(t) = n^{-2} A \sum_{i \neq j}^{n} \sum^{n} w_{ij}^{-1} I_t(u_{i,j})$$

where w is a weighting factor for reducing edge effects which can be set at 1 when the value of t describes a circle that remains within the study area (see Haase (1995) for details). Otherwise, w is the reciprocal of the proportion of the circle's circumference lying within the study area. I_t is a counter variable which takes the value 1 if the distance between i and j is less than t, otherwise it takes the value 0. Note that the term $I_t(u_i j)$ indicates that the interpoint (point-to-plant) distances are summed for each radius, t. This distance can be calculated as

$$u_{i,j} = \sqrt{(x_i - x_j)^2 + (y_i - y_j)^2}.$$

Each pair of points is considered twice, because plant j and plant i may have different spatial neighbourhoods or relationships to the plot boundary requiring a different weighting factor.

Because of the Poisson expectation mentioned above for randomly distributed plants in a study area, we assumed already that $K(t) = \pi t^2$; hence, a plot of $\sqrt{K(t)}$ versus t should be linear. Similarly, a plot of

$$\hat{L}(t) = \sqrt{\frac{\hat{K}(t)}{\pi}} - t$$

will have a value of zero at all values of t. $\hat{L}(t)$ will be negative when plants are uniformly dispersed and positive when clumped, depending on the scale, t. The advantage of plotting $\hat{L}(t)$ versus t is that the spatial pattern of the individual plants can be resolved at all scales tested within the study area. Monte Carlo simulations using 100–1000 simulations can be used to compute a confidence interval for the null hypothesis of a Poisson point pattern (Young and Young 1998).

$\hat{L}(t)$ was calculated for the mapped *C. ericoides* data (Fig. 8.1). $\hat{L}(t)$ values were positive and exceeded the upper 90% confidence limit for all plants, juveniles, males, and females at all spatial scales, except 0.5 m and below for male plants (Fig. 8.5). The interpretation of this analysis is that *C. ericoides* individuals displayed aggregation at all spatial scales. Notice too how much larger the $\hat{L}(t)$ values were for juveniles compared with that for males and females, indicating more intensive clustering of the former. This analysis suggests that the scaleless point pattern analyses presented in Table 8.2, in which only the juveniles which appeared to be clumped, were insensitive to the less intense clustering of males and females. The advantage of the more complicated analysis in calculating Ripley's K-function is that the potential for departures

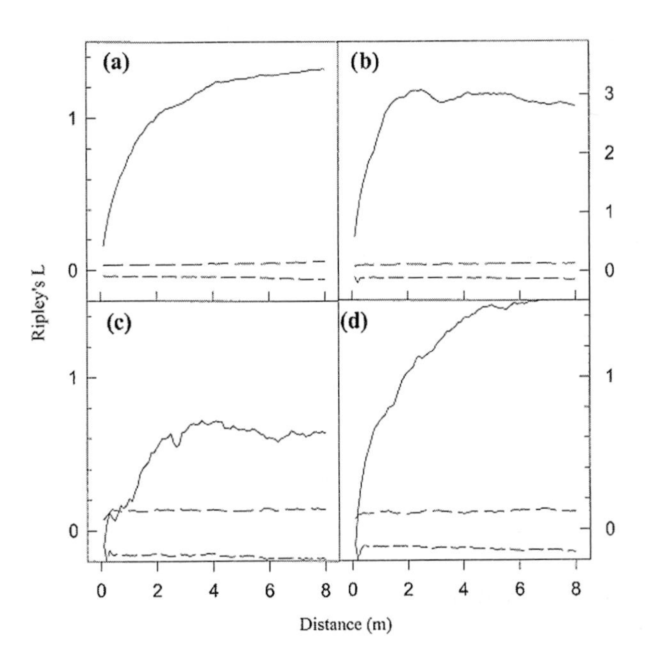

Figure 8.5 $\hat{L}(t)$ versus distance for *Ceratiola ericoides* (solid line; 90% confidence envelope shown by dashed lines): (a) all plants, (b) juveniles, (c) males, (d) females (Gibson, unpublished).

from a random distribution are considered at all spatial scales.

Is performance related to the area around individuals in a non-randomly dispersed population?

Spatial correlograms. Mapped point data in which a measure of plant performance has also been recorded (variable m, e.g. plant height) can be used to analyse the *spatial autocorrelation* of the variable as a function of distance around individual plants. Spatial autocorrelation refers to the dependence of z on values of the same variable at geographically adjoining locations whose spatial positions are known. In other words, spatial autocorrelation plots (*correlograms*) allow us to obtain a measure of our ability to predict values of, say, plant height, at other sampling locations within the mapped area. Positive spatial autocorrelation over short distances is commonly observed and reflects phenomena that are contagious (clumped) over short distances, such as limited seed dispersal, clonal growth, nutrient-rich patches, and positions on an environmental gradient. By contrast, negative spatial autocorrelation over short distances would suggest avoidance or competition among neighbours.

Two coefficients of spatial autocorrelation are commonly used: Moran's I and Geary's c are calculated at various distance classes, d. Both test the null hypothesis that there is no significant spatial autocorrelation among the mapped points with respect to the measured variable. Moran's I is given by

$$I(d) = \frac{(1/W)\sum_{h=1}^{n}\sum_{i=1}^{n} w_{hi}(y_h - \bar{y})(y_i - \bar{y})}{(1/n)\sum_{i=1}^{n}(y_i - \bar{y})^2} \quad \text{for } h \neq i$$

and Geary's c by

$$c(d) = \frac{[1/(2W)]\sum_{h=1}^{n}\sum_{i=1}^{n} w_{hi}(y_h - \bar{y})^2}{[1/(n-1)]\sum_{i=1}^{n}(y_i - \bar{y})^2} \quad \text{for } h \neq i$$

where y_h and y_i are the values of z observed for a plant at locations h and i, w_{hi} is a weighting coefficient taking the value of 1 when the sites are at distance d otherwise it equals 0, and W is the sum of weights for a given distance class (Legendre and Legendre 2012). As calculated above, Moran's I resembles a Pearson's product–moment correlation coefficient with values usually varying within the range −1 to +1. Positive values indicate positive autocorrelation. Geary's c is a distance-type coefficient and ranges from 0, indicating perfect positive autocorrelation, to 1 (or more), indicating an absence of spatial autocorrelation. The coefficient values for both indices are plotted against distance classes and take the form shown in Fig. 8.6. Details of how to calculate statistical tests of significance of Moran's I and Geary's c are provided in Legendre and Legendre (2012).

A non-standardized form of Geary's c, corresponding to the numerator in the equation, above is called the semivariance and is calculated at different distances (d). The semivariance is plotted in a variogram against different values of d (Fig. 8.7). The terminology in Fig. 8.7 reflects the original use and development of this analysis for geostatistical prospecting. The *range* is the distance at which the variance levels off, while the *sill* is the variance of the variable. When present, the sill reflects an important requirement for data that are to be used in this analysis, i.e. the data are second-order stationary. This means that the expected mean and spatial covariance of the data (numerator of Moran's I) are the same throughout the mapped area, and that the variance (denominator of Moran's I) is finite (Legendre and Legendre 2012). A non-zero intercept on the y-axis of the variogram is referred to as the *nugget effect* and represents the variance below the spatial resolution of the analysis (i.e. the smallest spatial scale). The shape of the line described by the variogram can be modelled (i.e. linear, Gaussian, exponential, or spherical) and is important for identifying the sill, range, and nugget, and also for making spatially interpolated maps of the z variable using kriging (see Legendre and Legendre (2012) for details).

As described above, the spatial autocorrelation is assumed to be equal in all directions across a plot (i.e. *isotropic*). When this case does not hold, for example if there is a gradient in plant size from one side of the plot to another, then the spatial autocorrelation is *anisotropic*, requiring the calculation of directional variograms (see Legendre and Legendre 2012).

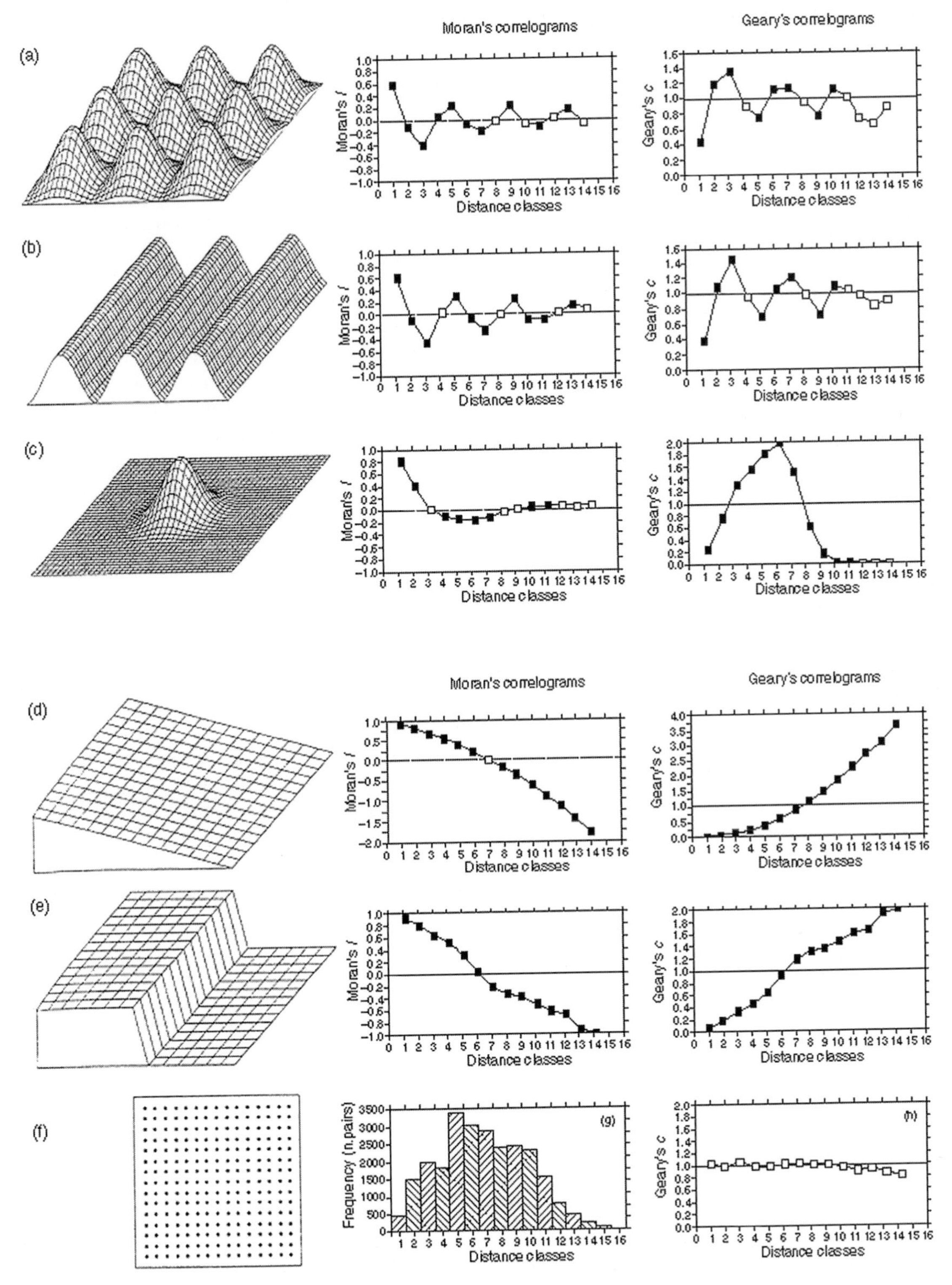

Figure 8.6 Spatial autocorrelation analysis of artificial data sets shown on the left: (a) nine bumps, (b) waves, (c) single bump, (d) gradient, (e) step, (f) sampling grid (15 × 15), (g) histogram, (h) random values: ■, significant autocorrelation at $\alpha = 0.05$ (Bonferroni correction); □, non-significant. Reprinted from Legendre and Legendre (2012) Reproduced with permission from Elsevier.

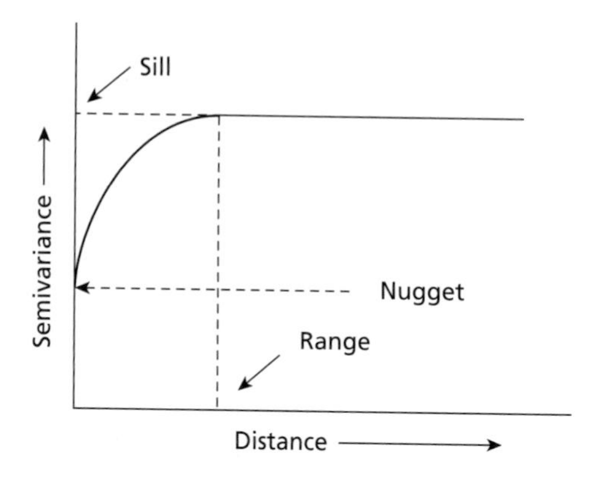

Figure 8.7 General form of a variogram.

The *C. ericoides* mapped data (Fig. 8.1) show little evidence for spatial autocorrelation of ages for all individual plants (Fig. 8.8a). Values of Moran's *I* were close to zero except for distances less than 5 m where *I* was weakly positive reflecting positive spatial autocorrelation. The isotropic variogram of these data was best fitted with a linear model in which a range and sill cannot be identified; that is, the spatial variance continued to increase even at the largest scale sampled (30 m). By contrast, the juveniles showed evidence of spatial autocorrelation of individual plant ages, again at the smallest scales (Fig. 8.8b). An exponential model best fitted the variogram with a range at 7.2 m, indicating that the age of juveniles was spatially autocorrelated at scales up to that distance. The interpretation of this analysis would be that juveniles occurred in similarly aged (not necessarily even-aged) patches up to approximately 7.2 m in diameter. The lack of spatial autocorrelation of all individuals (i.e. adults plus juveniles) may reflect density-dependent thinning and the loss of similarly aged patches as the plants age.

Tessellation models

The neighbourhood of individual mapped plants can be studied by analysing the partitioning of space among individual plants. One way of doing this is by dividing the mapped area into *Dirichlet* (*Voronozï* or *Thiessen*) *tessellations* or polygons. Each polygon describes the region of the mapped area closer to one (particular) plant than to any other plant (Fig. 8.9). Each polygon is convex and bounded by a vertex common to three polygons. The lines joining pairs of plants with contiguous polygons form a *Delaunay triangulation*. The frequency distribution of plant-to-plant distances (the lines) in a Delaunay triangulation is unimodal with low variance for uniform populations and bimodal with high variance in clumped populations (lots of short distances within clumps and many long distances between clumps).

A qualitatively similar approach to describing plant neighbourhoods was used by Aguilera and Lauenroth (1993). Their procedure started with a map of the outline of the basal area for 200 individuals of the grass *Bouteloua gracilis*. The outlines of each plant were expanded radially one raster cell

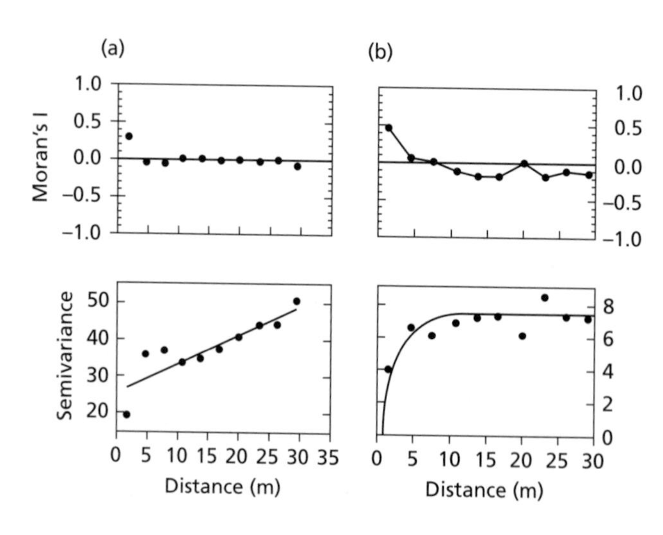

Figure 8.8 Spatial autocorrelation analysis of *Ceratiola ericoides* by plant age. Upper panels show Moran's *I*, lower panels show variogram of semivariance: (a) all individuals ($n = 470$ plants); (b) juveniles ($n = 179$ plants). The line shown on the variograms (bottom panels) indicates the best model (i.e. highest r^2 value) fitted to the semivariance (Gibson, unpublished).

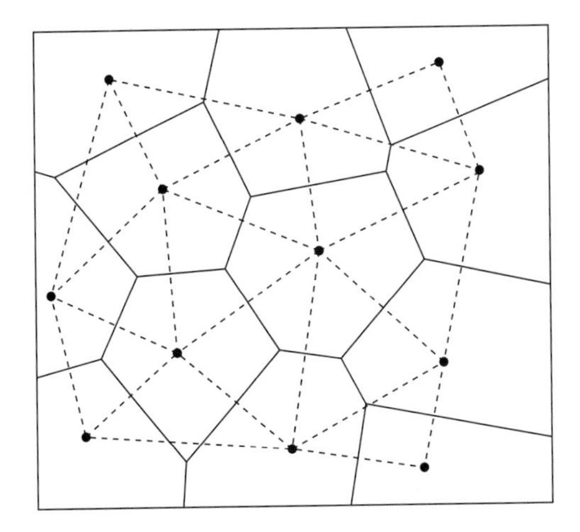

Figure 8.9 Construction of Dirichlet tessellations. The dotted lines show the Delaunay triangulation. From Diggle (1983). Reproduced with permission from Elsevier.

at a time using a geographic information system until they contacted the boundaries of adjacent expanding neighbourhoods. This approach produced a map of irregularly shaped plant neighbourhoods directly related in size to each plant and distance to its nearest neighbours.

Representations of plant neighbourhoods using either Dirichlet tessellations or the raster-based approach of Aguilera and Lauenroth (1993) can be used to test the relationships between neighbourhood size (area) and plant performance (e.g. Mithen et al. 1984; Matlack and Harper 1986; Wyszomirski and Weiner 2009). The assumption is that resources are most available to the nearest plant. Furthermore, they allow identification of the nearest, and potentially the most important, neighbours.

Polygons of *C. ericoides* drawn from the mapped data in Fig. 8.1 show a large variation in neighbourhood size (Fig. 8.10). Note that the largest neighbourhoods appear to be around the periphery of the sampled plot. This is misleading, and simply reflects the lack of known neighbouring plants around the edges. These polygons were excluded from subsequent analysis. Plant age and size were both positively related to polygon area (Fig. 8.11). Juveniles occurred in smaller polygons (mean of 1.45 ±1.44 m^2) than reproductive males and females (which themselves did not differ in polygon area; mean of 5.11 ± 1.04 m^2). The interpretation of this analysis is that the mature plants occur in relatively large neighbourhoods, probably through competitive exclusion as the plants get larger with age. Juveniles occur in tight clusters (smaller neighbourhoods) in an aggregated pattern. They are recruited into the pre-existing pattern imposed by the mature plants.

Distance models

The neighbourhood around an individual can be defined in two ways—as either the set of individuals within a given distance from a focal individual or as a circular zone defined by some measure of the pairwise distances between individuals. The first definition assumes that each of the individuals within the neighbourhood exerts an unweighted effect on the mortality and fecundity of the focal individual. The choice of neighbourhood distance is critical and should be based upon some biological criteria (Chapter 6). Some applications use a single, fixed-radius neighbourhood (Weiner 1984), whereas others test several neighbourhood sizes. As an example of the latter approach, Silander and Pacala (1990, and references therein) developed a series of neighbourhood population models in which the interspecific neighbourhood is defined as the number of individuals of species j that occur within a circle of radius r_{ij} from the focal individual of species i. A similar intraspecific neighbourhood is also defined. Thus, a series of concentric circles of radii r_{ij}, $j = 1, 2, \ldots, N$ are constructed in an N-species model. The integration of four submodels (germination, survivorship, fecundity, and dispersal) allows plant performance and the fate of an individual of species i to be forecast throughout its life cycle as well as population sizes and the spatial distribution of individuals. Neighbourhood simulations use initial seedling population size and spatial distribution to follow the fate of seedlings. Analytically tractable models include probability density functions that specify the spatial location of individuals in each generation and allow calibration and testing in the field. For a population of annuals comprising two species, Silander and Pacala (1990) were able to predict accurately the dynamic behaviour and equilibrium conditions. Paradoxically, the spatial distribution of individuals had little effect on the dynamics of the

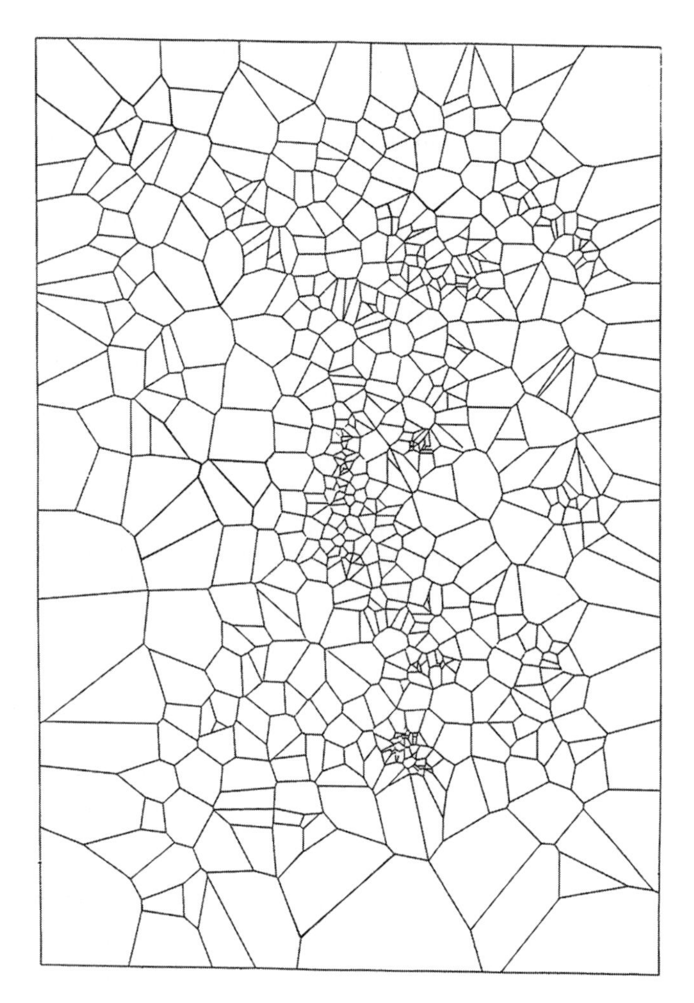

Figure 8.10 Dirichlet tessellations of the *Ceratiola ericoides* mapped data. Note the large polygons around the edge. These should be excluded from subsequent analysis (Gibson, unpublished).

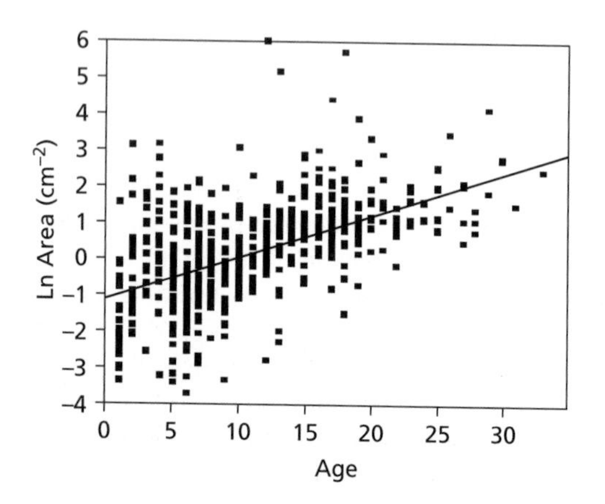

Figure 8.11 Relationship between Dirichlet tessellation polygon area and plant age for *Ceratiola ericoides* ($r^2 = 0.28$, $F_{1,603} = 237.6$, $P < 0.0001$) (Gibson, unpublished).

modelled system, although neighbourhood interactions remained an important and meaningful part of the model.

By contrast, pairwise-distance models assume that neighbouring plants influence each other as a function of distance. Ideally, these models incorporate the size, identity, and angular dispersion of neighbouring plants in assessing the influence of plants in the neighbourhood of the focal plant (Mack and Harper 1977; Hara 1988). Ramstad and Hestmark (2000) used this approach to predict lichen performance. A measure of local crowding was calculated from the mean distance to the n nearest neighbours of a focal individual, with values of n varying from one to eight neighbours. Predictive power was best with $n = 4$ or 5 neighbours, and was better than the results from an analysis using Dirichlet tessellations or neighbourhoods of a fixed radius.

8.2 Life tables and matrix models

The analyses described in this section are used to obtain quantitative information about the life history of populations. Several measures of population structure and dynamics can be obtained which can be useful in understanding the status of a population (whether it is increasing, decreasing, or stable). The life-history stages that are most limiting to population growth can be identified. These are potentially very useful in conservation management because they can have a direct bearing upon the short-term survival potential of a population as well as extinction threats. Conversely, knowledge of the critical life-history stages may assist in understanding the success of invasive species. Comparisons among different species can be made, allowing a comparison of life histories. The effects of environmental treatments on population parameters (*vital rates*, i.e. rates describing the movement of individuals through the life cycle; Caswell 2001) can be made using life table response experiments (LTRE; see below), and transition matrix models form the basis for PVA (Section 8.4).

Data sets

The data necessary to conduct the analyses described in this section are quite easily, but laboriously, obtained. Depending upon which comparisons among populations and treatments are planned, the minimum data set would comprise a permanently marked study plot in which the fate of marked individuals is followed over several census periods. For annual plants, individuals should be followed from seedling emergence until senescence. Following the fate of marked individuals may not be possible for long-lived perennial plants, in which case a complete life table (see below) cannot be calculated. Determining appropriate and justifiable interpretations that can be obtained for analysis of short-term surveys is an area of active research (Crone et al. 2013). Informative transition matrix models (see below) can be calculated by following individuals as they progress through different ontogenetic life-cycle stages. Ideally, the construction of life tables also requires data on seed production, seed viability and germination, seed dormancy, field germination, and seedling emergence. Although it is preferable to do so, these data need not be obtained from the same individuals that are marked and censused through time.

One of the first decisions to make is the choice of which life-cycle structure to investigate (Fig. 8.12). Age-structured life cycles are only possible when the plants in a population can be accurately aged, at least to within a number of discrete categories. Again, aging may be difficult or impossible for long-lived perennials. Age structures may also miss ecologically relevant information for species that are constrained developmentally by the environment (see the discussion on ontogeny in Section 4.2). Stage-structured life cycles are probably the most informative for population ecologists as they provide information on what individuals in a population are doing. Size-structured life cycles have value when interest lies in understanding the dynamics of individual growth. However, size-structured life cycles alone are likely to have only limited application. In many cases, investigators use a combined stage- and size-structured life-cycle approach where some of the life-cycle stages, such as vegetative stages, are also categorized into a number of size classes. Decisions about where to place the upper and lower bounds of size categories are quite important and are considered further in our discussion of matrix models.

Software

The main calculations described here and in Box 8.1 can be readily performed using a spreadsheet program such as Microsoft Excel. The PopTools add-in for 32-bit PC versions of Microsoft Excel runs matrix population models and stochastic simulations (<http://www.poptools.org/>). A number of BASIC language programs are available in Ebert (1999), and software containing simulations suitable for classroom teaching is available in *Populus* (Alstad 2001), downloadable from <http://www.cbs.umn.edu/populus>. The ULM package allows the user to run matrix models (Legendre and

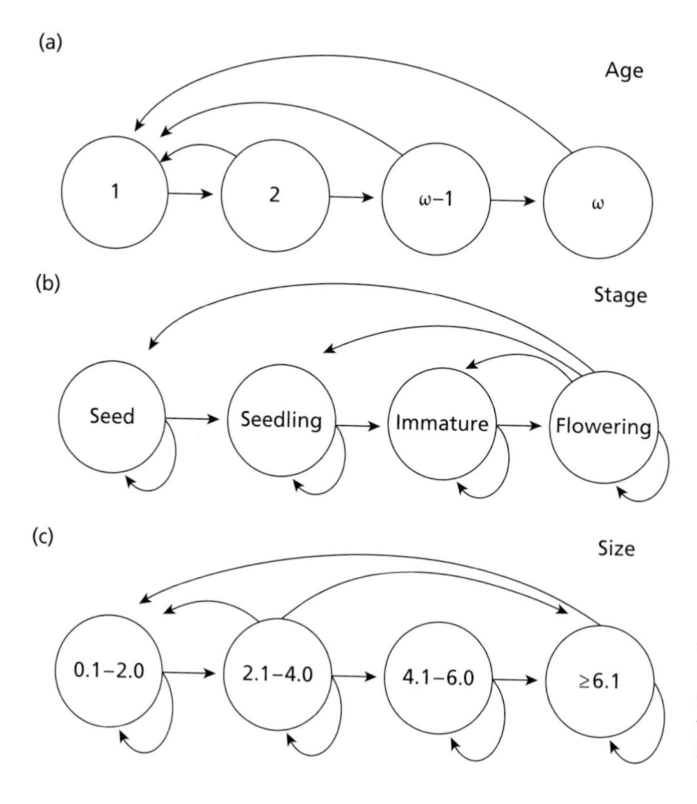

Figure 8.12 Life cycle structures. (a) Age-structured life cycle with transitions (survival probabilities) from one age to the next and return transitions to age 1, which are reproductive contributions; ω is the final age class and no individuals survive past this age. (b) Stage-structured life cycle; during a time period, individuals may remain in a stage or transfer to the next stage. In the absence of seed dormancy, the transition from flowering plants to seedlings can occur within a year. (c) Life cycle structured by size classes showing some of the possible transfers including shrinking to a smaller size and skipping a size class. Adapted from Ebert (1999) with permission from Elsevier.

Clobert 1995). popbio and Popdemo are R packages for matrix models whereas IPMpack provides tools for running Integral Projection Models (see the Appendix). R code to get started on the popbio package is provided by Stubben and Milligan (2007).

Population flux

It was noted in Chapter 1 that a central goal of plant population ecology is to understand changes in plant populations through time and across space. We also noted that the finite rate of increase, lambda (λ), provides a measure of population change. In this section, the components of λ as well as other parameters that are important in understanding the life history of a population are described.

Recall from Chapter 1 that the number of plants (or modules or genets) per unit area N_t at time t is related to the number N_{t-1} the previous year:

$$N_t = N_{t-1} + B - D + I - E$$

where B is the number of births, D is the number of deaths, and I and E are the number of immigrants and emigrants into and out of the population, respectively. Change in population size from one time to another is obtained by subtracting N_t from both sides of the equation giving

$$\Delta N = B - D + I - E.$$

If ΔN is negative then the population is in decline, if positive then it is increasing. In practice, this model is too simplistic since it does not account for density dependence, or, more importantly, the interaction between density-dependent and density-independent processes affecting B, D, I, and E.

Recall also that $N_{t+1}/N_t = \lambda$. When $B + I > D + E$ then the population will multiply each year in an exponential manner under unchanging conditions (Fig. 8.13):

$$N_{t+x} = N_t e^{rx}$$

where x is the population x years following time t, e is the base of natural logarithms, and r is the *intrinsic rate of natural increase* of the population. r is related to λ by the equation $\lambda = e^r$ which can also be written $r = \ln(\lambda)$.

Box 8.1 Worked example

There are numerous examples of the use of transition matrix models in the plant population ecology literature. The analytical procedure described in this chapter is presented in Case Study 2 in Chapter 1 using a transition matrix for one of the *Silene regia* populations. The programs MATRIX.BAS and SIMPOS.BAS in Ebert (1999) were used for the calculations, with confirmatory results obtained using the R package popbio (Stubben and Milligan 2007).

The R script using popbio is:

```
library(popbio)
stages <- c("seedlings", "vegetative", "smallflowering", "medflowering", "largeflowering", "aliveundefined")
A <- matrix(c(0, 0, 5.32, 12.74, 30.88, 0,
0.308, 0.111, 0, 0, 0, 0,
0, 0.566, 0.506, 0.137, 0.167, 0.367,
0, 0.111, 0.210, 0.608, 0.167, 0.300,
0, 0, 0.012, 0.039, 0.667, 0.167,
0, 0.222, 0.198, 0.196, 0, 0.133), nrow = 6, byrow =
TRUE, dimnames = list(stages, stages))
n <- c(1, 5, 9, 16, 34,9)
p <- pop.projection(A, n, 15)
p
stage.vector.plot(p$stage.vectors, col = 2:4)
eigA <- eigen.analysis(A)
eigA
```

The authors of Case Study 2 used a Lefkovitch stage/size-based approach and placed the plants that they censused into one of six categories (seedling, vegetative, small flowering, medium flowering, large flowering, and alive undefined). The algorithm from Moloney (1986) was used to determine the optimal division of plants into the categories based upon plant height. By following marked plants over 12 months the transition matrix *A* was calculated (transitions representing stasis in bold):

$$A = \begin{bmatrix} - & - & 5.32 & 12.74 & 30.88 & - \\ 0.308 & \mathbf{0.111} & 0 & 0 & 0 & 0 \\ 0 & 0.566 & \mathbf{0.506} & 0.137 & 0.167 & 0.367 \\ 0 & 0.111 & 0.210 & \mathbf{0.608} & 0.167 & 0.300 \\ 0 & 0 & 0.012 & 0.039 & \mathbf{0.667} & 0.167 \\ 0 & 0.222 & 0.198 & 0.196 & 0 & \mathbf{0.133} \end{bmatrix}.$$

Any set of values can be used in the initial column vector (n_{ij}), but for the sake of argument let us start with values that correspond to the eventual reproductive value of each stage/size category (derived below), i.e. [1, 5, 9, 16, 34, 9]. For the first time step, the transition matrix is multiplied by the column vector to give the age structure at time $t + 1$; that is, column vector n_2:

$$\mathbf{A} = \begin{bmatrix} - & - & 5.32 & 12.74 & 30.88 & - \\ 0.308 & \mathbf{0.111} & 0 & 0 & 0 & 0 \\ 0 & 0.566 & \mathbf{0.506} & 0.137 & 0.167 & 0.367 \\ 0 & 0.111 & 0.210 & \mathbf{0.608} & 0.167 & 0.300 \\ 0 & 0 & 0.012 & 0.039 & \mathbf{0.667} & 0.167 \\ 0 & 0.222 & 0.198 & 0.196 & 0 & \mathbf{0.133} \end{bmatrix}$$

$$\times \begin{bmatrix} 1 \\ 5 \\ 9 \\ 16 \\ 34 \\ 9 \end{bmatrix} = \begin{bmatrix} 1301.64 \\ 0.86 \\ 18.56 \\ 20.55 \\ 24.91 \\ 7.23 \end{bmatrix}.$$

Immediate changes in the numbers of individuals in each stage/size category can be seen. Further multiplication over 20 time steps (generations) give the values shown in Box Fig. 8.1. A levelling off in numbers in each category has not occurred after 20 generations. Systems such as this do not model local interactions accurately and so the quantitative projections of population numbers cannot be expected to be accurate. The unrealistically high numbers shown in Box Fig. 8.1 reflect the fact that models like this can only be used for predictions of the system's qualitative properties. In reality,

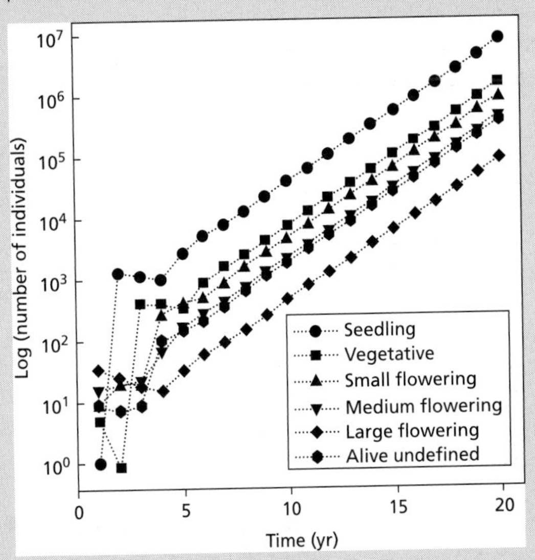

Box Figure 8.1 Simulation of population structure of *Silene regia* over 20 generations. Based on data in Case Study 2 and Menges and Dolan (1998).

continued

Box 8.1 *Continued*

demographic parameters vary and seedling recruitment is episodic. Stochastic approaches in which matrix elements vary according to a predetermined distribution provide a more realistic population projection.

The transition matrix yields a stable stage structure (w) after 11 generations with a value for λ of 1.70 (Box Fig. 8.2) reflecting the steep increase in numbers observed in Box Fig 8.1. The stable stage structure and the reproductive value (v) vectors (scaled to show the first category to 100) are:

$$w = \begin{bmatrix} 100 \\ 19 \\ 11 \\ 6 \\ 1 \\ 5 \end{bmatrix} \text{ and } v = \begin{bmatrix} 100 \\ 550 \\ 900 \\ 1560 \\ 3390 \\ 870 \end{bmatrix},$$

respectively.

Seedlings will dominate populations of *Silene regia* (numerically), and reproduction is affected most by large flowering individuals.

Using the formulae shown in Section 8.2, sensitivity values are obtained as

$$s_{ij} = \begin{bmatrix} - & - & 0.024 & 0.019 & 0.002 & - \\ 0.144 & \mathbf{0.222} & 0 & 0 & 0 & 0 \\ 0 & 0.362 & \mathbf{0.215} & 0.107 & 0.021 & 0.092 \\ 0 & 0.627 & 0.373 & \mathbf{0.185} & 0.037 & 0.159 \\ 0 & 0 & 0.812 & 0.403 & \mathbf{0.081} & 0.347 \\ 0 & 0.351 & 0.209 & 0.104 & 0 & \mathbf{0.089} \end{bmatrix}.$$

Similarly, elasticity values are:

$$e_{ij} = \begin{bmatrix} - & - & 0.075 & 0.089 & 0.043 & - \\ 0.207 & \mathbf{0.014} & 0 & 0 & 0 & 0 \\ 0 & 0.121 & \mathbf{0.064} & 0.009 & 0.002 & 0.020 \\ 0 & 0.041 & 0.046 & \mathbf{0.066} & 0.004 & 0.028 \\ 0 & 0 & 0.006 & 0.009 & \mathbf{0.032} & 0.034 \\ 0 & 0.046 & 0.024 & 0.012 & 0 & \mathbf{0.007} \end{bmatrix}.$$

Note how different the ranges of values for the sensitivity and elasticity values are. The latter sum to 1 across the matrix, whereas the former do not. The elasticity values show that λ for this population is most sensitive to transitions between seedlings and vegetative plants ($e_{1,2} = 0.207$). The transition from vegetative plants to small flowering plants is also important. Summing the elasticity values to obtain G, L, and F values, one obtains:

$$\begin{aligned} G &= E_3 + E_6 \\ &= (0.02 + 0.028 + 0.034) + (0.207 + 0.121 \\ &\quad + 0.041 + 0.046 + 0.006 + 0.024 + 0.009 \\ &\quad + 0.012) = 0.585 \end{aligned}$$

$$\begin{aligned} L &= E_4 + E_5 \\ &= (0.009 + 0.002 + 0.004) + (0.014 + 0.064 \\ &\quad + 0.066 + 0.032 + 0.007) = 0.198 \end{aligned}$$

$$F = E_1 + E_2 = (0.075 + 0.089 + 0.043) = 0.207$$

These values indicate that λ in *Silene regia* is most sensitive to demographic parameters related to growth (G) and places the species to the high-G side in G, L, F space with other iteroparous herbs of open habitats. The importance of growth for λ is consistent with its natural prairie habitat. Also, fire affects seedling recruitment and survival and is important for λ and extinction.

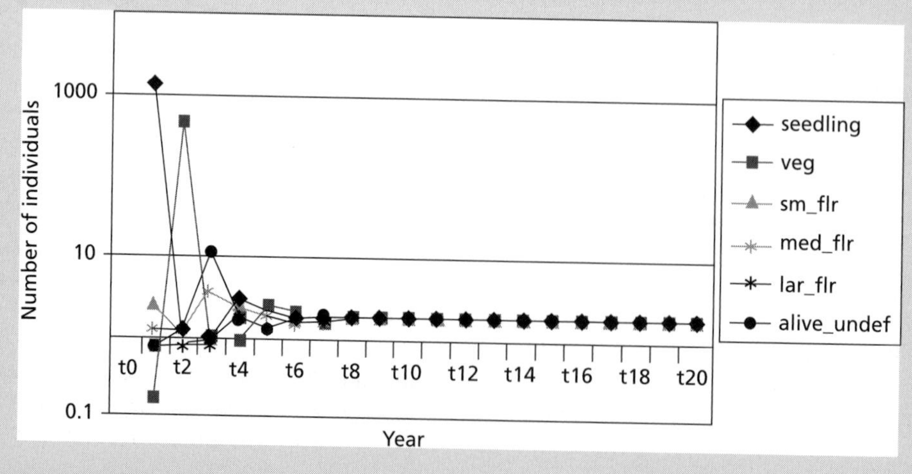

Box Figure 8.2 Stabilization of the stable stage structure of *Silene regia* over 20 generations to provide an approximation of λ. The y-axis is calculated as the number of individuals projected in a stage class at time $t + 1$ divided by the number of individuals in the same stage class at time t. Based on data in Case Study 2 and Menges and Dolan (1998).

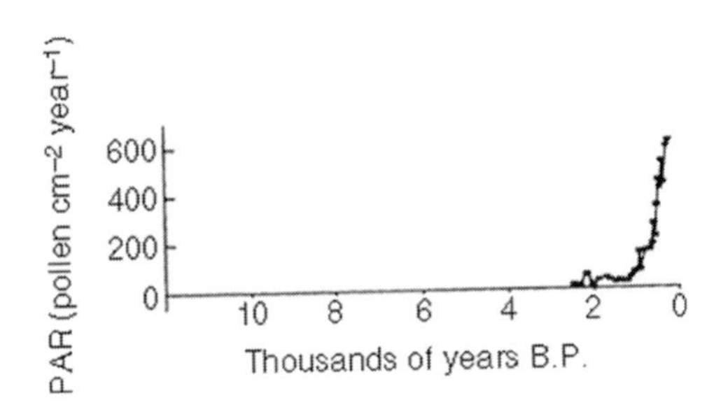

Figure 8.13 Exponential growth in a population of lodgepole pine (*Pinus contorta*) from Flamingo Lake, Yukon Territory, Canada over the last 3000 years (from MacDonald and Cwynar 1991). PAR, fossil pollen accumulation rate. Reproduced with permission from John Wiley & Sons.

r can be solved for in discrete time from the Euler or Lotka equation

$$1 = \sum_{x=0} e^{-rx} l_x m_x$$

where l_x is survivorship and m_x is fecundity. Solving for r in the equation above involves numerical iterations adjusting an initial value until the right-hand side is equal to 1 (Crawley 1997a, p. 79 provides a worked example). Because of the relationship between r and λ, r can also be approximated according to the equation

$$r = \ln R_0^{1/\tau}$$

where τ is average generation time (the time between a mother giving birth and her daughter doing the same; sometimes shown as T) and R_0 (R nought) is the *net reproductive rate* over τ years:

$$R_0 = N_{t+\tau} / N_t.$$

Important points to note from the above are: (a) a simple model of population growth can be obtained from basic demographic parameters, and (b) estimates of survivorship (l_x) and fecundity (m_x) are important for these models. However, these models are simplistic because they take no account of density dependence, plant life-history components (annual plants versus modular plants, presence of a seed bank, etc.), or stochasticity with changing demographic environments over time. Silvertown and Lovett Doust (1993), Watkinson (1997), and Ebert (1999) show how these limitations may be accounted for and provide more extensive treatments that consider the limitations. Stochasticity can be incorporated through calculation of the stochastic population growth rate, λ_s (Caswell 2001, 2010).

Life tables

Data on the number of individuals of a cohort that survive and (when it occurs) reproduce at intervals from birth can be summarized in a life table. A life table is a schedule of probabilities for individuals in a population. Cohort (or dynamic or horizontal) life tables are constructed to represent the fate of a particular population from initial recruitment until death of the last member. By contrast, static (time-specific or vertical) life tables are constructed to represent observations made from a population at a single (sometimes two) point in time. The advantage of a cohort life table is that it enables populations with non-overlapping generations to be summarized; however, it cannot be constructed when the species is too long-lived to follow a cohort from its recruitment to death. Conversely, static life tables do not allow for age-specific or year-to-year changes in birth and death rates.

To illustrate the procedure, consider the cohort life-table analysis for the small annual herb *Minuartia uniflora* (Sharitz and McCormick 1973) (Table 8.3). The following important life-cycle components can be calculated in a life table (some of the exact parameters are specific to the example, such as ages being measured in months). From l_x onwards the population is standardized to an initial cohort of 1000 individuals. The only field data necessary to complete the table are the unstandardized numbers of individuals surviving at each life cycle stage (l_x).

1. Life cycle duration, D_x, the number of months of each life cycle stage. This is very important since the rest of the analysis depends upon the partitioning made here. Some investigators will divide the established stage into a number of size classes (e.g. small, medium, and large established plants).
2. Age, A_x, the population age in months since, in this case, seed formation.

Table 8.3 Cohort life table for *Minuartia uniflora*.

Life-cycle stage, x	D_x	A_x		l_x	d_x	q_x	$\log_{10} l_x$	k_x	L_x	T_x	e_x
Seed produced	4	0–4	−100	1000	790	0.79	3.00	0.68	605	2632	2.6
Seed available for germination	1	4–5	−52	210	146	0.70	2.32	0.51	137	212	1.0
Germinated seedlings	1	5–6	+90	64	53	0.83	1.81	0.77	38	74	1.2
Established seedlings	2	6–8	+128	11	3	0.27	1.04	0.14	10	37	3.4
Rosettes	2	8–10	+204	8	3	0.38	0.90	0.20	6	18	2.3
Mature plants in flower	2	10–12	+280	5	5	1.00	0.70	0.70	2	5	1.0

See text for explanation of column headings.
After Sharitz and McCormick (1973), with permission of the Ecological Society of America.

3. Percentage age, A'_x, the age of the population at the beginning of each life-cycle stage as a percentage of the mean life span of the population.
4. Survivorship, l_x, the number of individuals surviving at the beginning of each life-cycle stage. Note how l_x is standardized from A_x to a starting density of 1000 seeds. The columns that follow are calculated from this standardized value.
5. Senescence, d_x, the standardized number of individuals that die during each life-cycle stage. Obtained by successively subtracting from l_x the number alive at the beginning of the next stage ($l_x + 1$). Note that d_x values can be summed to obtain the total number of individuals dying over a period of time; for example, the total number of seeds that died during the 'seed produced' stage ($d_x = 790$) + seeds that died during the 'seed available for germination' stage ($d_x = 146$) = the total number of seeds that had died by the end of the second life-cycle stage 'seed available for germination' (936).
6. Mortality rate ('chance of death'), $q_x = d_x/l_x$.
7. $\text{Log}_{10}\, l_x$, a step in the calculation of k.
8. Killing power, $k_x = \log_{10} l_x - \log_{10} l_{x+1}$. k_x reflects the intensity or rate of mortality. k_x values can be summed to obtain the total killing power over a time period; for example, killing power of k-value from seed production through germination = 0.68 + 0.51 + 0.77 = 1.96.
9. Stationary population, $L_x = (l_x + l_{x+1})/2$. The mean number of individuals that are alive during each life-cycle stage.
10. Residual population life span:

$$T_x = d_x(D_x/2) + d_{x+1}(D_x + D_{x+1}/2) + d_{x+2}(D_x + D_{x+1} + D_{x+2}/2) + \cdots,$$

which simplifies to $T_x = D_x L_x + T_{x+1}$. The total number of individuals times age units (plant-months) remaining to all members of the population currently living at the beginning of the life-cycle stage.

11. Life expectancy, $e_x = T_x/l_x$.

Note that the example given in Table 8.3 is for an annual species at different stages in its life cycle. An alternative approach is to represent the population in terms of the biological age of individuals. However, age is not the best measure of an individual's biological status. Many long-lived plants, for example, may stay in a vegetative stage for many years until the optimum conditions for reproduction occur. Thus, the ontogenic life-history stage (such as the stages shown in Fig. 8.12 for example) may be more ecologically appropriate; this also, incidentally, gets around the real, logistical problem of estimating plant age.

Two additional statistics that can be computed on reproductive plants in a population at demographic equilibrium are (Silvertown and Charlesworth 2001):

1. Current fecundity as seeds/plant, m_x.
2. Reproductive value, V_x, where

$$V_x = m_x + \sum_{i=1}^{i=\infty} \frac{l_{x+i}}{l_x} m_{x+i}.$$

In other words, reproductive value (V_x) provides a value for current (m_x) and future (m_{x+i}) reproduction weighted by the probability of survival ($l_x + i/l_x$) from now (age x) through future age classes. Reproductive value can be quite different at each plant age or life stage and represents the contribution that at average individual aged x makes to the next generation before it dies. The age at which V_x is maximized represents the optimum age at which individuals in the population should begin to reproduce and minimize the cost in terms of growth and increased probability of death.

Survivorship

The importance of assessing survivorship (l_x) and the component D in ΔN was noted above and is relevant because most seeds and seedlings in a population do not survive to maturity. As we will see later when we discuss matrix models, the contribution of seeds and seedlings to population processes may be quite small.

The survival of plants in populations can be analysed using either depletion curves or survivorship curves. For both, the log of the proportion of plants (or ramets or modules) surviving is plotted against an arithmetic time axis. A depletion curve involves plotting the survival of all plants present on a given census date through time. These plots are drawn when the population has an unknown age structure. Figure 8.14 shows depletion curves for populations of *Ranunculus repens*. Depletion curves are often remarkably linear and allow half-lives (the time taken for a population to decrease by 50%) to be calculated. For the populations in Fig. 8.14 the half-lives were 23–25 weeks. By contrast, a survivorship curve involves plotting the survival of a particular cohort of plants at each census date (i.e. against plant age). The latter are preferred because depletion curves do not reflect age-specific mortality risks for the whole population. There are three fundamental types of survivorship curve (Fig. 8.15):

1. Type I: mortality risk increases as the maximum life span is reached. Examples include many annual species.
2. Type II: constant, age-independent mortality risk throughout the life of the cohort. A constant pro-

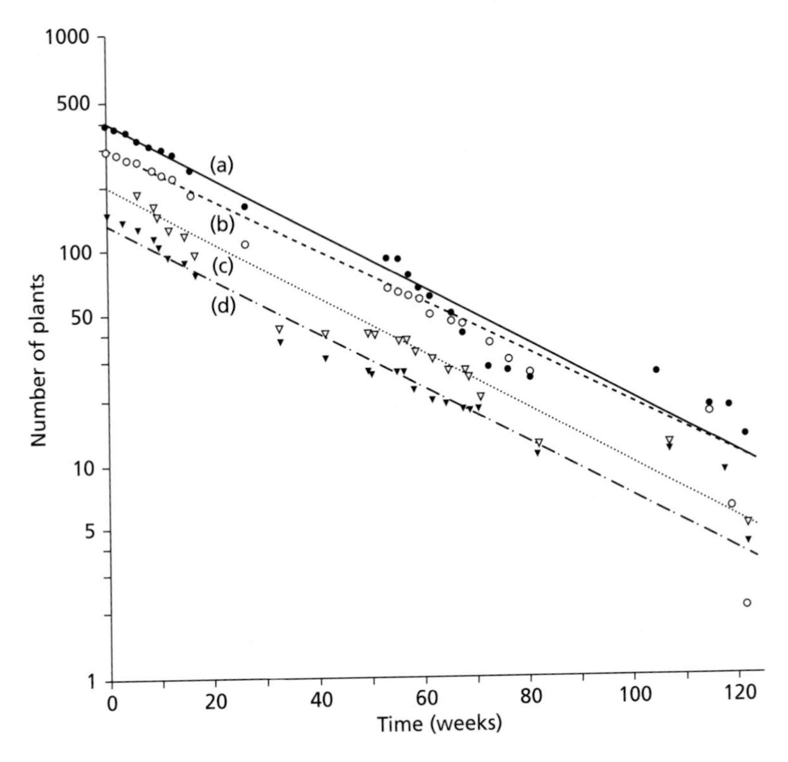

Figure 8.14 Depletion curves for four populations of *Ranunculus repens* in 1 m^2 permanent plots; time is weeks after first observation (April 1969). From Sarukhán and Harper (1973). Reproduced with permission from John Wiley & Sons.

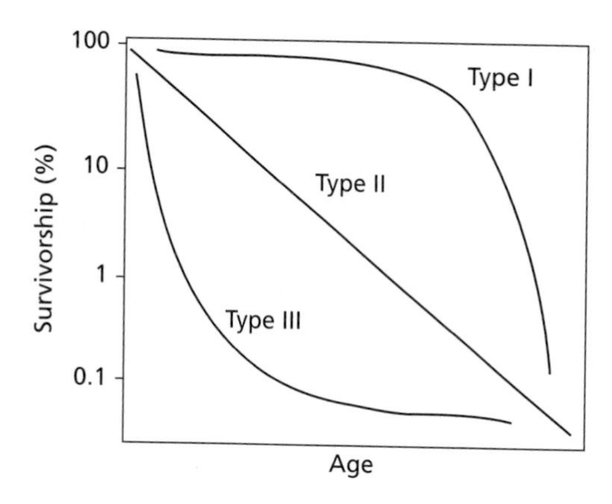

Figure 8.15 Three types of survivorship curve.

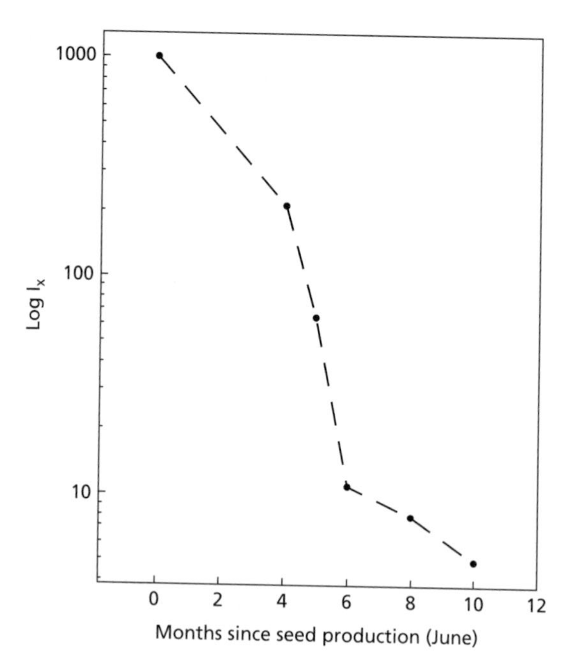

Figure 8.16 Survivorship curve for *Minuartia uniflora* drawn from data in Table 8.3.

portion of the surviving population dies in each time interval, although the decline in the number of survivors is exponential. Type II survivorship is common in many plant populations across a wide range of species, especially herbaceous perennials.

3. Type III: mortality risk is highest for young plants and declines with age. Only a few individuals reach the maximum life span. This type of survivorship curve is observed when large number of individuals perish early in life. Some trees have shown this pattern of survivorship.

Survivorship of *Minuartia uniflora*, the example provided in Table 8.3, can be obtained by a plot of l_x against time (months in Fig. 8.16). Mortality risk was fairly constant over the entire life span from seed production in June, and survivorship most closely resembled a Type II curve. From germination onwards (4–6 months), survivorship of *M. uniflora* most closely followed a Type III survivorship curve reflecting high rates of seedling mortality (between months 5 and 6). Survivorship curves were used in Case Study 3 in Chapter 1 to compare disturbance, insect, and competition treatments on *Senecio jacobaea* populations (Fig. 1.5). Depending upon the experimental treatment Type I or Type II curves were obtained for this species. Survivorship curves provide a direct visual assessment of the risks that a population faces at different times over its life span. Survivorship curves can be drawn for a population from seed production, seedling emergence (the usual manner), or for each life stage separately (from which it is often observed that the shapes differ). It is important that the populations are censused regularly, and both early on and late over the course of the plant's life span. Regular monitoring may be particularly critical for long-lived plants where much of the demographic change in populations may occur in the first few months after seedling emergence. Survivorship curves can be used in management to focus upon the portion of the life span that is most critical for ensuring species success and that might require active protection. Conversely, a survivorship curve might indicate which portion of the life span to focus on for controlling or limiting population growth of an invasive species.

The failure-time method is used to analyse survivorship curves with the goal of testing for equality of survival distributions and it determines whether survivorship curves between different populations differ significantly from each other (e.g. Gibson et al. 2011). Fox (2001) discusses forms of log-rank, Wilcoxon, and chi-square tests (Chapter 7) that have been used in these calculations and provides sample SAS code. These procedures can also be used to compare populations for other parameters derived from life tables, such as emergence time and time to flowering. The survival R package can be used to conduct a number of survival analysis tests (see the Appendix).

Life-cycle diagrams

The information from a life table can be usefully displayed in graphical form as a life-cycle diagram. Generalized life-cycle diagrams are shown in Fig. 8.12, with a detailed example for the monocarpic perennial herb *Dipsacus sylvestris* in Fig. 8.17 and the shrub *Verticodia fimbrilepis* in Fig 8.23. Each age, stage, or size class is represented by a node (displayed as a circle) and transitions between nodes by arrows. This format is sufficiently flexible to accommodate complex life cycles or perennial species with overlapping generations. It is important to draw up a life-cycle diagram early on in an investigation to ensure that data on the numbers required for the various steps are obtained. Further details on the construction of life-cycle diagrams are in Begon and Mortimer (1981).

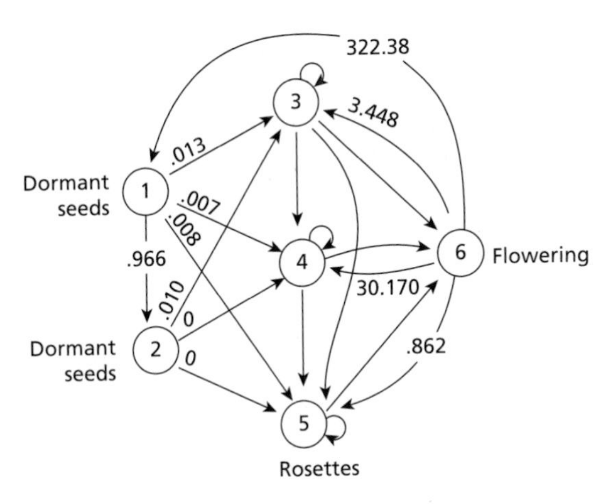

Figure 8.17 Life-cycle diagram for the perennial herb *Dipsacus sylvestris*. From Caswell (2001). Stages: 1, first-year dormant seeds; 2, second-year dormant seeds; 3, small rosettes; 4, medium rosettes; 5, rosettes; 6, flowering plants. Arrows and accompanying values represent transitions and probabilities from one stage to another. The transition 1 to 6 reflects fecundity (seed production).

Transition matrix models

Matrix models allow the dynamic structure of a population to be modelled mathematically. Future population states (number of individuals in each stage, size, or age class) can be predicted through projection and the contribution of different life-cycle components to λ calculated. One of the basic assumptions of these models is that populations exhibit internal structure that can be summarized by looking at the age, stage, or size structure (Fig. 8.12). The analyses are based upon the use of matrix algebra, but the operations are, for the most part, relatively simple and are illustrated using standard matrix algebra notation. The discussion here is based upon conventions used in Ebert (1999), Caswell (2001) and Silvertown and Charlesworth (2001). A worked example based upon Case Study 2 in Chapter 1 is provided in Box 8.1.

The basic model takes the form

$$\boldsymbol{n}_{t+1} = \boldsymbol{A}\boldsymbol{n}_t$$

where $\boldsymbol{n}_t$ is a column vector describing the age or stage structure of the population at time t and $\boldsymbol{A}$ is the matrix of associated demographic parameters.

The age or stage state of a population at time $\boldsymbol{n}_t$ can be described by a column vector:

$$n_t = \begin{bmatrix} n_1 \\ n_2 \\ n_3 \\ \vdots \\ n_i \end{bmatrix}$$

where $n_1, \ldots, n_i$ are the numbers of individuals in each age/stage class (the nodes in Fig. 8.12). A projection matrix (A) corresponds to the complete life-cycle graph (e.g. Fig. 8.12 or 8.17) with the elements representing the transitions (probabilities) from one age/stage to another. For example, for a plant with three stages, seed, rosette, and flowering, $\boldsymbol{A}$ can be represented as shown below with the arrows representing transitions from one state to another (Silvertown and Charlesworth 2001):

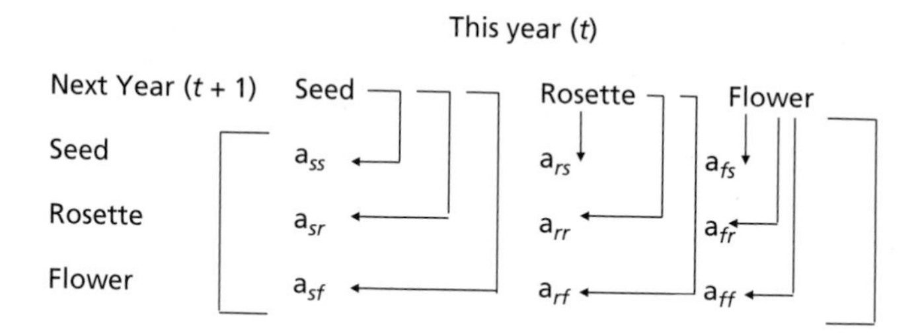

For example, the probability that a seed in the current year (t) remains a seed in the next year ($t + 1$) is given as a_{ss}.

There are two basic types of projection matrix that can be used. A *Leslie matrix* is used for age-classified models and a *Lefkovitch matrix* for stage/size-classified models.

A Leslie matrix takes the form

$$\mathbf{A} = \begin{bmatrix} f_1 & f_2 & f_3 & \cdots & f_{n-1} & f_n \\ p_1 & 0 & 0 & \cdots & 0 & 0 \\ 0 & p_2 & 0 & \cdots & 0 & 0 \\ 0 & 0 & p_3 & \cdots & 0 & 0 \\ \vdots & \vdots & \vdots & \cdots & \vdots & \vdots \\ 0 & 0 & 0 & \cdots & p_{n-1} & 0 \end{bmatrix}$$

where p_i is probability that an individual will survive (0 to 1 values) and f_i is the rate of reproduction (fecundity) for an individual in age class i (values of any magnitude). The elements on the leading diagonal are all zero (plants cannot remain the same age). The Leslie matrix has non-zero elements only in the subdiagonal (annual survival: p_i) and top row (births: f_i).

The Lefkovitch matrix takes the form

$$\mathbf{A} = \begin{bmatrix} s_1 & f_2 & f_3 & \cdots & f_n \\ g_1 & s_2 & 0 & \cdots & 0 \\ 0 & g_2 & s_3 & \cdots & 0 \\ \vdots & \vdots & \vdots & \vdots & \vdots \\ 0 & 0 & \cdots & g_{n-1} & s_n \end{bmatrix}$$

where s_i is the probability that an individual will survive and remain in the same stage class (0 to 1 values), g_i is the probability that an individual will survive and transfer to the next stage class (0 to 1 values), and f_i is the rate of reproduction (fecundity) for an individual in stage class i (values of any magnitude). Elements below the leading diagonal relate to the probability of growth to the next stage class, whereas elements above the leading diagonal relate to fecundity. In contrast to the Leslie matrix, elements in the leading diagonal of a Lefkovitch matrix (s_i) cannot be zero (there is always a chance that a plant will remain in the same life stage the next year). In addition, information about the reproductive contribution of each size class is included (fecundity: f_i). The exact form of the matrix depends upon the species, the nature of the life cycle categories, and the time intervals used (e.g. Eriksson 1988; Charron and Gagnon 1991; Økland 1995). Plants can skip ahead or regress from one stage to another. There can be special stages for clonal growth, dormant plants, the seed bank, etc., to allow the various different life histories to be incorporated. As an example, consider the Lefkovitch projection matrix below which corresponds to the six-stage life cycle graph for *Dipsacus sylvestris* shown in Fig. 8.17 (Caswell 2001):

$$\mathbf{A} = \begin{bmatrix} 0 & 0 & 0 & 0 & 0 & 322.38 \\ 0.966 & 0 & 0 & 0 & 0 & 0 \\ 0.013 & 0.01 & 0.125 & 0 & 0 & 3.448 \\ 0.007 & 0 & 0.125 & 0.238 & 0 & 30.170 \\ 0.001 & 0 & 0 & 0.245 & 0.167 & 0.862 \\ 0 & 0 & 0 & 0.023 & 0.750 & 0 \end{bmatrix}.$$

Note that fecundity is limited to the rightmost element on the top row that represents the seeds produced by flowering plants (Fig. 8.17).

It is important to choose the most appropriate stage or size categories for a Lefkovitch matrix in order to avoid inaccurate transition probabilities. Many investigators choose size classes based upon both biological relevance and sample size. One approach is to divide the population into a set of groups with an equal number of individuals per group. However, there are two types of error that can occur in making this choice. 'Distribution error' (DE) occurs when the categories are too large and the assumption that individuals falling within the same category have the same transition probability is violated. Conversely, 'sample error' (SE) occurs when the classification categories are too small and contain few individuals. Moloney (1986) proposed a generalized algorithm to minimize these two types of negatively correlated errors. However, use of Moloney's algorithm can be problematic when comparisons among populations of the same species are of interest. A fixed number of categories among all populations may lead to the absence of individuals in one or more size categories in some populations. Caswell (2001) shows that the algorithm fails to really provide an 'optimum' set of size classes in any real sense and does not recommend its use. For this reason, many investigators prefer to establish categories that subdivide their plants into a number of well-represented groups (e.g. small, medium, and large reproductive plants; Valverde and Silvertown 1998).

As noted above, population structure in the next time step is calculated as:

$$\boldsymbol{A} \times \boldsymbol{n}_t = \boldsymbol{n}_{t+1}.$$

Using the notation above for a simple Lefkovitch matrix, this is accomplished as:

$$\begin{bmatrix} s_1 & f_2 & f_3 & f_4 \\ g_1 & s_2 & 0 & 0 \\ 0 & g_2 & s_3 & 0 \\ 0 & 0 & g_3 & s_4 \end{bmatrix} \times \begin{bmatrix} n_{1,t} \\ n_{2,t} \\ n_{3,t} \\ n_{4,t} \end{bmatrix}$$

$$= \begin{bmatrix} s_1 n_{1,t} & +f_2 n_{2,t} & +f_3 n_{3,t} & +f_4 n_{4,t} \\ & g_1 n_{1,t} & +s_2 n_{2,t} & \\ & g_2 n_{2,t} & +s_3 n_{3,t} & \\ & g_3 n_{3,t} & +s_4 n_{4,t} & \end{bmatrix} = \begin{bmatrix} n_{1,t+1} \\ n_{2,t+1} \\ n_{3,t+1} \\ n_{4,t+1} \end{bmatrix},$$

i.e. projection matrix × column vector = new population age/stage structure. Repeated iterations are carried out to find $\boldsymbol{n}_{t+1}$, i.e. age/stage structure in subsequent time steps:

$$n_{t+1} = An_t,\ n_{t+2} = An_{t+1},\ n_{t+3} = An_{t+2},\ \text{etc. until } n_{t+i}.$$

The projection matrix $\boldsymbol{A}$ never changes in this procedure and thus represents unchanging demographic conditions.

Repeated iterations will reach the point where the *relative* values in the vector $\boldsymbol{n}_{t+1}$ cease to change with further multiplications by the projection matrix. This vector is known as the right eigenvector (w) and represents a stable age/stage structure, independent of values in the initial vector. This stable structure is a mathematical property of matrix algebra known as *ergodicity*, meaning that the eventual behaviour of the population is deterministic and independent of its initial state (Caswell 2001). At this point, λ can be calculated as the dominant eigenvalue of A:

$$\boldsymbol{A}\boldsymbol{n}_t = \lambda \boldsymbol{n}_t$$

or

$$\lambda = \text{size of } n_i / \text{size of } n_{i-1}.$$

In other words, λ is the factor by which each age class changes from one time interval to the next. Estimates of λ can be obtained by comparing the ratio of the numbers in any one age/size class with the numbers in the same age/stage class in the previous time interval; i.e. all age/stage classes grow at the same rate of λ. The closer the structure that is reached is to the stable state then the closer the value of λ obtained is to the true value. At equilibrium an exact estimate of λ is obtained.

λ can also be obtained more directly from A without projecting to the stable age/stage distribution (Ebert 1999):

$$\det(\boldsymbol{A} - \lambda \boldsymbol{I}) = 0$$

where 'det' is the determinant yielding the characteristic or polynomial of the matrix, and I is the identity matrix (a matrix with ones on the diagonal and zeros elsewhere).

So far, the stable age/stage distribution (the dominant or right eigenvector, w) and λ (the dominant eigenvalue of A) has been obtained. A reproductive value vector (referred to as the left eigenvector, v) can also be calculated. v provides a measure of the present and future contribution of individuals in an age/stage class to reproduction. The larger the value of v the greater the contribution. This is obtained as

$$(A' - \lambda I)n_t = 0$$

where A' is the transpose of A. Solving this equation yields the vector v as

$$v_t = \begin{bmatrix} v_1 \\ v_2 \\ v_3 \\ \vdots \\ v_i \end{bmatrix}$$

Note that each element in v corresponds to an age/stage class category.

A final set of calculations can be carried out to determine how sensitive λ is to the different elements in the projection matrix. '*Sensitivity*' (s_{ij}) for each matrix element (a_{ij}) of A can be calculated based upon the scalar products of the stable age/stage distribution vector (w) and the reproductive vector (v), respectively:

$$s_{ij} = \frac{v_i w_j}{\langle v, w \rangle}$$

where

$$\langle v, w \rangle = v_i w_i + v_2 w_2 + \cdots + v_n w_n$$

with the vectors adjusted so $v_1 = w_1 = 1$.

However, sensitivity values are often less than satisfactory. It is difficult to compare sensitivities calculated from probabilities versus those calculated from fecundity because of the different range of units concerned. *Elasticity* (e_{ij}) is a modified index of sensitivity that gets around this problem as elasticity values sum to 1 within a matrix, i.e. e_{ij} represents the proportion of λ due to the transition a_{ij}. Thus, elasticity provides a measure of the contribution of a matrix element to fitness (de Kroon et al. 1986):

$$e_{ij} = \frac{a_{ij}}{\lambda} s_{ij}.$$

Elasticity values within a matrix can be assigned to one of six categories representing a different part of the plant's life cycle (Fig. 8.18). These are (Silvertown et al. 1993):

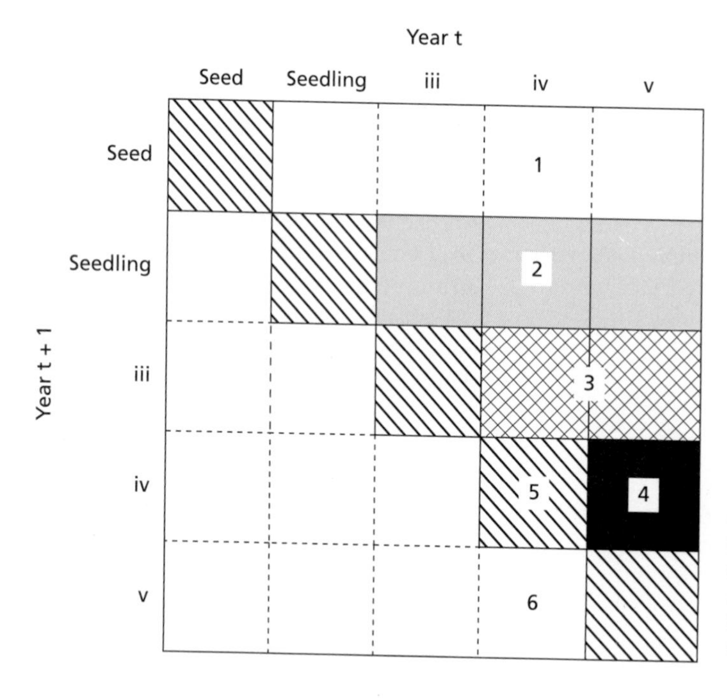

Figure 8.18 The six regions of the stage projection matrix: E_1, seed production; E_2, seedling recruitment; E_3, clonal growth; E_4, regression; E_5, stasis; and E_6, progression (from Silvertown et al. 1993). Reproduced with permission from John Wiley & Sons.

1. Recruitment of seeds to the seed bank.
2. Recruitment of seedlings or juveniles from current seed production.
3. Clonal growth.
4. Retrogression due to plants decreasing in size or reverting from a flowering state to a vegetative one, or becoming dormant.
5. Stasis, or survival from one year to the next in the same stage class.
6. Progression to later stage classes.

Elasticities can be summed within each of the six regions to give measures (E_1–E_6, respectively) (Fig 8.18) of the importance of each component of the life cycle to λ and fitness. However, despite the property of elasticities summing to one, negative relationships among elasticities should not be used to imply or detect trade-offs between traits (Shae et al. 1994).

Three major components of life cycle components can then be identified as:

G; growth = $E_3 + E_6$

L; survival = $E_4 + E_5$

F; fecundity $E_1 + E_2$.

Each species or population studied can then be placed within a 'demographic triangle' according to their G, L, and F scores (Fig. 8.19). The position of species within the demographic triangle allows a categorization of patterns of demography and life history. The *GLF* demographic triangle allows species to be compared (see examples in Silvertown et al. 1993). Intraspecific variation can also be investigated, as shown in Case Study 2 of Chapter 1 (see Box 8.1).

Limitations and shortcomings of the matrix approach for estimating and projecting population parameters can include insufficient data (e.g. too few plants or too few years of data to provide accurate transition probabilities) and the failure of the model to account for density-dependent vital rates. In addition, values of λ may sometimes be misleading as

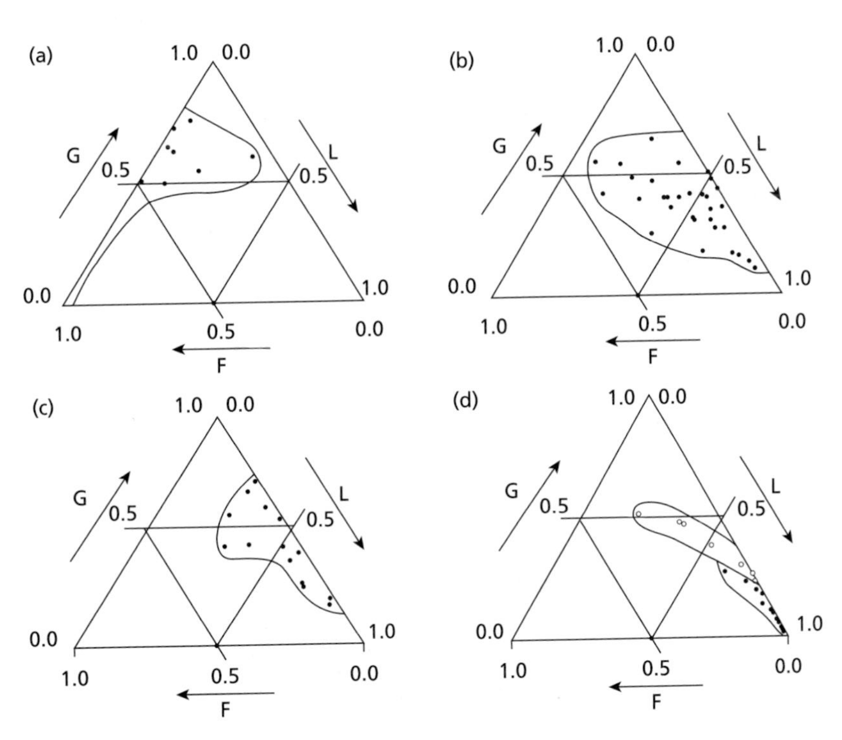

Figure 8.19 Variation in the location of different species groups in *G, L, F*, space: (a) semelparous herbs, (b) iteroparous herbs of open habitats, (c) iteroparous forest herbs, (d) woody plants (closed circles denote trees; open circles denote shrubs). From Franco and Silvertown (1997). Reprinted with permission of Cambridge University Press.

a measure of population's short-term prospects for survival (Bierzychudek 1999).

As described here, estimates of λ, projected population structure, and sensitivities are deterministic and based upon invariant elements of the matrix A. Thus, λ is a deterministic, asymptotic growth rate assuming indefinite maintenance of the vital rates in the matrix. In reality, demographic parameters vary due to the stochastic nature of environmental conditions. One approach for dealing with this is to calculate an average value for λ based upon transitions measured over several years or among seasons (e.g. Vavrek et al. 1997). Monte Carlo simulations and resampling methods are statistically more rigorous approaches based upon stochastic matrix modelling, making it possible to estimate demographic properties in fluctuating environments. Monte Carlo simulations assume known distributions, often normal, for vital rates. The mean and variance for λ is calculated following several runs of the analysis in which vital rates are drawn each time from the assumed distribution. By contrast, resampling methods do not require assumptions about the statistical distribution of vital rates. The original sample of individuals is resampled multiple times by random selection of individuals with replacement (bootstrapping) or random omission of individuals (jackknifing) to obtain λ pseudovalues. The pseudovalues are then used to obtain unbiased estimates of the mean and variance of λ (Alvarez-Buylla and Slatkin 1991; McPeek and Kalisz 1993). Applications of stochastic matrix modelling include Shea and Kelly (1998) and Bierzychudek (1999). Detailed statistical procedures are described in Caswell (2010).

The analytical procedures used with matrix models have advanced considerably in the last decade (Salguero-Gómez and De Kroon 2010). As already noted, population ecologists are increasingly modelling the stochastic population growth rate and incorporating environmental stochasticity (see 'Life table response experiments', below). Nevertheless, most population datasets are relatively short term, limiting the interpretation of model predictions (Crone et al. 2013). The COMPADRE Plant Matrix Database (<http://www.compadre-db.org/>) contains plant projection matrices for over 900 plants, allowing comparative evolutionary and ecological questions to be addressed.

Life table response experiments (LTREs)

A LTRE analysis uses a life table (set of vital rates) as a response variable in an experimental design in which the effects of different levels of two or more treatments (e.g. environmental factors such as fire or grazing) on the vital rates are tested. Demographic transition matrix models (described above) are used to decompose treatment effects at the population level on the response variable, usually expressed as the deterministic population growth rate λ, although other demographic summary statistics can be used such as net reproductive rate R_0. A full discussion of LTREs is provided in Caswell (2001, Chapter 10). Just like classical ANOVA, LTRE designs can be fixed (one-way or factorial), random, or regression designs (treatments are quantitative factors). LTRE analysis has been extended to analyse treatment effects on the stochastic population growth rate λ_s (Caswell 2010). Small noise approximation LTRE (SNA-LTRE) allows one to model the contributions of changes in mean vital rates to the difference in λ_s (Jacquemyn et al. 2012). LTRE analyses is widely used by plant demographers to address questions related to management, invasive species, pathogens, abiotic disturbance (e.g. fire, flooding, rainfall), and herbivory (see examples in Caswell 2010). For example, Li et al. (2011) conducted a fixed-design LTRE on vital rates to determine the effect on population growth rate of the shrub *Artemisia ordosica* at three stages of dune stabilization (semi-fixed dunes, fixed dunes, and fixed dunes colonized with microbiotic crust). LTRE showed that a decrease in λ during dune stabilization was caused mainly by reductions in fecundity.

8.3 Cellular automata, individual-based population models, and integral projection models

In this section the use of matrix-based population models is expanded to incorporate neighbour relationships between individuals. This expansion is important since in Section 8.1 it was shown that plants are often distributed non-randomly and that neighbour relationships can be important for understanding the performance of an individual. Here, we first consider cellular automata models before briefly discussing SORTIE, an individual-based dynamic model of forest dynamics, and integral projection models (IPMs). Even so this will only provide a brief overview of the much larger field of plant population modelling that underlies the topics in this chapter. Jeltsch et al. (2008) identified four research themes in plant population modelling: (a) demography and population dynamics of single species, (b) mechanisms of species interactions, (c) number and relative abundance of species in a community context, and (d) spatial population structure and dynamics. Transition matrix models (Section 8.2) addressed themes (a) and (d) (and see PVA in Section 8.4), while the models discussed in this section build upon the methodology for spatial pattern analysis (Section 8.1) and transition models (Sections 8.2) in addressing themes (b) and (c).

Cellular automata (CA; individual-based dynamic automaton models, tessellation automaton, grid-based models) are dynamic models that are discrete in time, space, and state (Balzter et al. 1998; Wang et al. 2003). In these models the transition matrix is given a spatial dimension by constraining the 'rules' (transition probabilities) through which a plant changes from one age/stage state to another by a second set of rules governing interactions between neighbours. Thus, in these models, plants are assigned a location in space (a cell) relative to neighbouring plants (neighbouring cells). By repeated iterations of such a model, we obtain simultaneously the discrete age/stage state of single individuals at each location (cell). Thus, a simple and crude population map and associated population statistics can be generated from this dynamic, deterministic simulation model. *Coupled map lattices* (CML) are similar to cellular automaton models, except that (a) continuous states (variables) are tracked, such as several individuals per cell, and (b) values are not restricted to integers and are generated deterministically (Hendry et al. 1996; Kaneko 1993; Rees and Paynter 1997; Yokozawa et al. 1999). Thus, CA models are individual-based, whereas CML are continuum systems.

The population is modelled as an $n \times n$ grid of cells. Each cell corresponds to an individual plant i in the lattice. The lattice can be one of two basic types: a *'von Neumann neighbourhood'* is a diamond/hexagonal lattice of the type below which shows the distances from a central focal plant in cell '0':

```
                4
              4 3 4
            4 3 2 3 4
          4 3 2 1 2 3 4
        4 3 2 1 0 1 2 3 4
          4 3 2 1 2 3 4
            4 3 2 3 4
              4 3 4
                4
```

where the numbers 1–4 (or more) represent neighbours (1 = nearest neighbours to the focal plant, 2 = second nearest neighbours, etc.). Note that the focal plant in cell '0' has four nearest neighbours.

By contrast, a *'Moore neighbourhood'* is square with more points less efficiently packed in space:

```
4 4 4 4 4 4 4 4 4
4 3 3 3 3 3 3 3 4
4 3 2 2 2 2 2 3 4
4 3 2 1 1 1 2 3 4
4 3 2 1 0 1 2 3 4
4 3 2 1 1 1 2 3 4
4 3 2 2 2 2 2 3 4
4 3 3 3 3 3 3 3 4
4 4 4 4 4 4 4 4 4
```

Note the focal plant at '0' has eight nearest neighbours. As a simple example, Inghe (1989) represented the probabilities of a clonal herb colonizing an empty site using a Moore neighbourhood as

```
                  0.1
       0.25       0.4       0.25
  0.1  0.4    empty site    0.4   0.1
       0.25       0.4       0.25
                  0.1
```

To avoid edge effects, both CA and CLM lattices are constrained as a torus with wrap-around margins, i.e. the top and bottom edges are joined as are the left and right boundaries.

Despite these differences in structure, Durrett and Levin (1994) suggest that the type of neighbourhood model has little effect upon the qualitative behaviour of the model.

In a similar manner to the transition models described in Section 8.2, analysis proceeds by multiplying each cell in the lattice by the probabilities of plants going from one state to another (this can be a transition matrix; see Section 8.2). The result of the multiplications at each iteration, however, is weighted by a set of constraints reflecting neighbour interactions. Thus, the current state of the plant, the transition probability, and the state of its neighbours determine the state of a plant in a cell at the next time interval. It is also important to realize that the changes to plants in each cell are all carried out at the same time (in parallel); thus the previous state of neighbours is used, not that resulting from the current changes. The constraints can be any parameters that you consider to be important in affecting the interaction between neighbours. They may reflect the identity or history of the plant in the cell, or surrounding cells (perhaps within a defined radius). Thus both density-dependent and density-independent constraints can be included in the model. For example, you may decide that a seedling can only replace a flowering plant in a cell if it will be located within four cell distances (four neighbours) of an existing flowering plant. This constraint would introduce seed dispersal and density dependence into the model, albeit in a simplistic manner. Another constraint might impose mortality on an individual within two cell distances of a mature plant that remains a seedling for two generations. Environmental constraints and disturbances can be included. For example, Hochberg et al. (1994) included the constraint that tree seedlings were susceptible to fire-induced mortality following burning, imposed at the start of each year, if not surrounded by a certain number of mature (protecting) trees. Obviously, the constraints should be grounded in features considered to be ecologically relevant for the species concerned. When using coupled map lattices where cells are occupied by more than one plant, systems of difference equations linked by migration terms may specify the rules governing the transition from one time period to the next (Hassell et al. 1991).

Required data. To conduct a cellular automaton analysis you need a data set that provides transition probabilities among plant life-cycle stages within permanent plots. The size of the permanent plots will determine and be equal to the cell size in the lattice that is established. As noted below, an alternative data set would provide transitions among species within plots. Information is also needed on the relevant processes that govern neighbour interactions in order to establish the set of constraints that will be imposed on the transitions. Examples are provided in the worked example below.

An example of a cellular automaton model: population dynamics of wild daffodil (*Narcissus pseudonarcissus*) (Barkham and Hance 1982). The authors collected 10 years of field data on the demography of individually located wild daffodils, a perennial woodland herb, from permanent plots. This data set allowed them to estimate the probabilities of (a) adult death (pm = 0.056 and 0.038 in shaded and open sites, respectively), (b) an adult arising from clonal growth (pv = 0.015 and 0.045), and (c) an adult arising from seed reproduction (ps = 0.00087 and 0.0049). A coppice management regime, typical of the wild daffodil habitat, was instituted by switching from shade to sun probability values in alternate 5-year periods. A 10 000 unit square Moore neighbourhood lattice of 1 cm^2 units was established into which 100 adult individuals were randomly distributed, each from a different genet. A model such as this contains built in constraints and assumptions. Constraints reflect limitations imposed on the operation of the model (i.e. the rules between neighbouring cells in the lattice), whereas assumptions are features of the plant's life history taken as being true for the purpose of model construction. The constraints imposed upon the model were the following:

1. The area occupied by each shoot module is a constant 1 cm^2.
2. A vegetative daughter must occupy randomly any one of the eight 1 cm^2 locations adjacent to the parent.
3. An offspring produced from seed must occupy randomly a 1 cm^2 location 15 cm distant from the parent. This distance approximates a bent over scape of an individual dispersing seed at senescence.
4. Offspring are only produced in cells that are unoccupied (density-dependent fertility).
5. Vegetative offspring out-compete seed offspring.

6. Offspring allocated a position outside the 1 m^2 plot are lost from the analysis. Note: an alternative approach which can avoid these edge effects and losses is for the modelled universe to have wrap-around margins (upper and bottom rows are neighbours, and left- and rightmost rows are neighbours; e.g. Inghe 1989).
7. Individuals within 1 cm of each other constitute a clump. Multiple genets may be represented in a single clump.

There were at least five main assumptions associated with the model in this form. These were:

1. All individuals are capable of reproduction. Juveniles and subadults were not considered.
2. Interspecific interactions were not included.
3. Population growth is two-dimensional, despite the knowledge that the volume of soil is important for this species.
4. The response to tree canopy management (i.e. coppicing) was instantaneous. Random climate variation was not included.
5. All processes important to the demography of wild daffodils occur between 1 cm^2 and 1 m^2. Processes operating at smaller or larger scales, such as animal disturbances, were not included.

The authors ran 1000-year simulations initially using the probabilities for *pm*, *pv*, and *ps* derived from their field data (given above), both with and without imposing coppice management cycles. They found that in open conditions the population levels became unrealistically high, whereas under shaded conditions the populations went extinct. It was necessary to adjust the probabilities to produce realistic and stable populations. This reflects the sensitivity of population sizes to small changes in reproduction and mortality. The spatial distribution of two genets over 900 years is shown in Fig. 8.20, revealing substantial redistribution of clones over the time period and clonal fragmentation. The changing number of individuals in each clone in relation to one another suggests intraspecific competition and density-dependent regulation.

Value of cellular automaton studies. As with most modelling studies, the value lies in producing a qualitative estimate of the processes that affect plant demography (Durrett and Levin 1994). This type of model is a spatially explicit method that can be used to identify the life-cycle parameters that are important for population regulation. Insight is gained into the processes that govern the spatial dynamics and the attainment of spatial heterogeneity of populations. The relative importance of the processes included in the model constraints are assessed; for example, abiotic or biotic disturbances. These allow the development of new and specific testable hypotheses regarding the population ecology of plant species. As noted earlier, coupled map lattices are a suite of models that are similar to cellular automaton models, except in the former the number in each cell is tracked.

The applications of single-species demographic models have been stressed here. However, cellular automaton models can also be developed using a matrix of replacement probabilities of one species with another. This allows an assessment of the importance of interspecific competitive effects, or bottom-up forces, in the assembly of plant communities (Balzter et al. 1998). A limitation of CA models is the spatial resolution (cell size) and the uniformity within grid cells (individual cells are homogeneous within themselves) (Berger et al. 2008).

Individual-based population models

Individual-based neighbourhood models (IBMs) have been used extensively to predict the behaviour of plant populations in forest ecosystems (Shugart 1984). They are not an extension of cellular automaton models but they are considered here because they incorporate spatially explicit behaviours. Four criteria have been proposed that help define IBM models (Grimm and Railsback 2005): (a) a complexity of individual life cycles; (b) explicit representation of the dynamics of resources used by individuals; (c) the use of real or integer numbers to represent the size of a population; and (d) consideration of variability among individuals of the same age are. The extent to which these criteria are considered and represented vary among IBMs. IBMs track the fate of all individuals throughout their life in the modelled universe (e.g. so many hectares of forest). The state of the population is determined by summing the condition and state of each individual. These models may follow several hundred

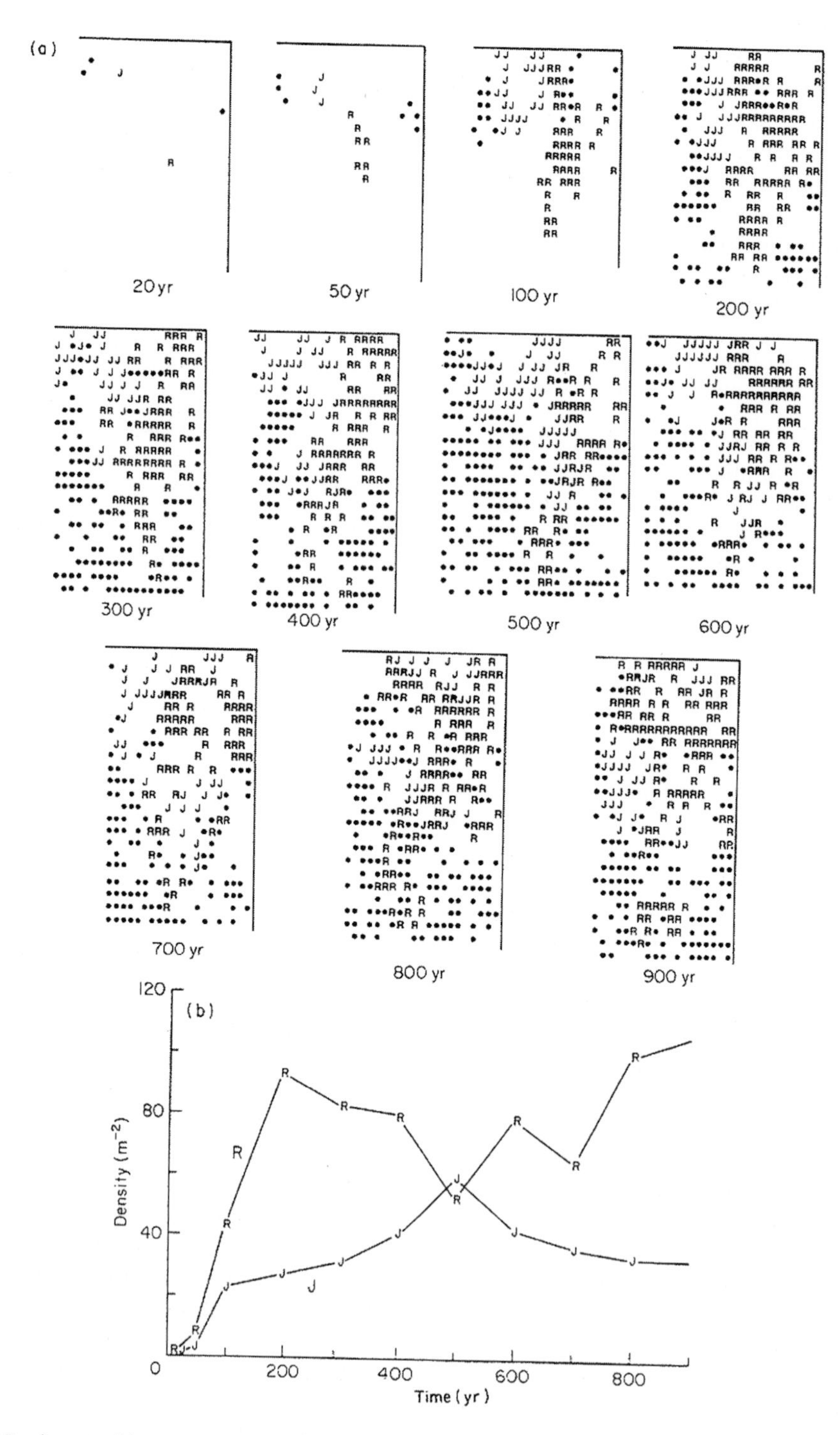

Figure 8.20 (a) Development of the spatial distribution of two clones (J and R) in a simulated population of wild daffodil. Filled symbols are bulbs belonging to other clones. (b) Changes in the number of individuals of clones J and R over 900 years. From Barkham and Hance (1982). Reproduced with permission from John Wiley & Sons.

or thousand individuals simultaneously. As they do not use population averages for the parameters that generate population dynamics, IBMs are especially useful in systems where individuals interact (i.e. most, if not all, systems).

IBMs are only discussed briefly here because the complexity of the computer programs necessary for running IBMs is outside the scope of this book. A detailed overview is provided elsewhere (Shugart 1984; Grimm and Railsback 2005). The IBM SORTIE is an appropriate example through which to illustrate the development and value of IBMs to plant population ecologists.

SORTIE. SORTIE is a spatially explicit individual-based model of forest dynamics (Pacala et al. 1996). In common with other IBMs, SORTIE is used to predict the fate of individual forest trees throughout their life cycle using submodels which predict growth, mortality, production of recruits, and local resource availability. Species-specific functions are parameters derived from field measurements of nine species in hardwood forests of the northeastern United States (Table 8.4). Light is the only resource considered in SORTIE because field evidence showed little evidence of limitation of seedling and sapling growth by water or nitrogen. The program allows the user to specify the plot size (9–100 ha) and initial condition consisting of location (x and y coordinates), species identity, and size of each individual at the start of a simulation run. The plot has wrap-around edges in the shape of a torus to avoid edge effects. A single iteration of the model is for 5 years and proceeds by first calculating the growth rate for each seedling, sapling, and adult plant based upon species-specific functions and light. Competition occurs when taller neighbours shade a plant. The probability of mortality is next calculated from the growth rate and the mortality submodels and a pseudorandom coin-toss determines which individuals die. Finally, the recruitment submodel determines the spatial location of new recruits. Repeated iterations then occur up to 2000 or more years.

As an example of the utility of the SORTIE model, and other IBMs like it, Ribbens et al. (1994) showed that final abundance in forest stands of beech and hemlock were very sensitive to the value of dispersal parameters in the model (parameter R_1 in Table 8.4). Fifteen-hundred-year runs of the model compared the effect of varying the mean dispersal distance of the two species from actual field-calibrated dispersal parameters to high-modified dispersal parameters for each of the two species in turn (Fig. 8.21). Thus, the abundance and dynamics of the two forest species were highly sensitive to recruitment when considered in a spatially explicit, individual-based population model. Deutschman et al (1997) used SORTIE to explore the effect of disturbance on forest dynamics, with the resulting simulations being captured in video animations (available at <http://www.sciencemag.org/site/feature/data/deutschman/index.htm>).

The value of IBMs to the plant population ecologist. IBMs allow the importance of population and individual-based parameters to community, and

Table 8.4 Definitions of parameters in the individual-based model SORTIE.

Submodel	Symbol	Description
Resource	H_1	Asymptotic tree height (m)
	H_2	Initial slope of height–diameter relationship (m/cm)
	C_1	Tree crown radius/stem diameter ratio (m/cm)
	C_2	Crown depth/tree height ratio (m/m)
	E_i	Light extinction coefficient
Growth	G_1	Asymptotic high-light growth rate of saplings (year^{-1})
	G_2	Slope of sapling growth rate at low light (year^{-1} global light index^{-1})
	G_3	Maximum allowed growth of adult stem area (cm^2)
Mortality	M_1	Probability of growth at zero growth (2.5 year^{-1})
	M_2	Decay of growth dependent mortality function (cm^{-1})
	–	Probability of growth due to random disturbance (year^{-1})
Recruitment	R_1	Distance decay of recruitment shadow (m^{-3})
	R_2	Number of recruits produced by a tree 100 cm in diameter
	–	Initial radius of new recruits

From Pacala et al. (1996) with permission of the Ecological Society of America.

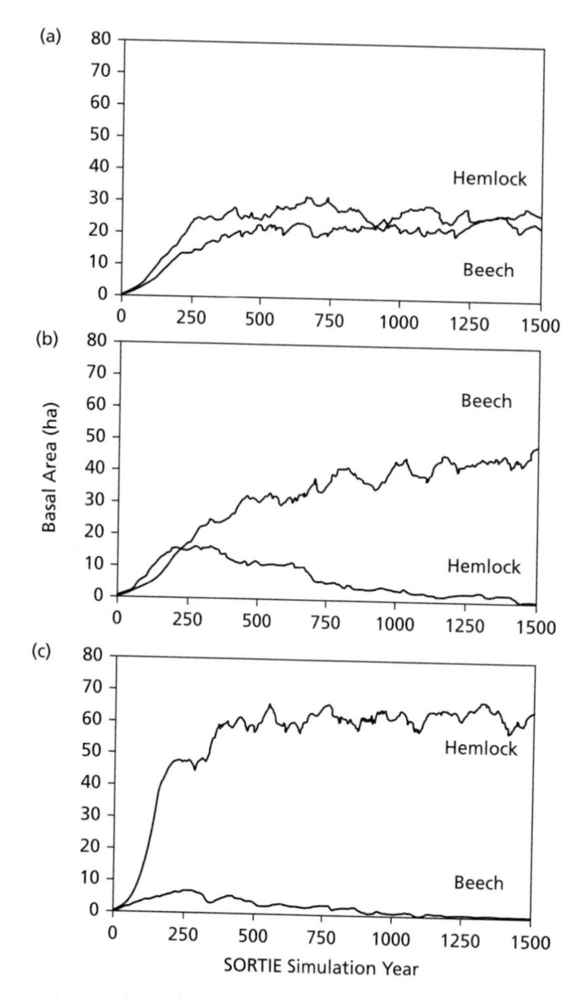

Figure 8.21 Results of SORTIE simulation of beech–hemlock forest dynamics: (a) actual dispersal, (b) high beech dispersal, (c) high hemlock dispersal. From Ribbens et al. (1994), reprinted with permission of the Ecological Society of America.

indeed ecosystem, processes to be quantified. Like the cellular automaton and neighbourhood models discussed above, they provide clear evidence of bottom-up forces in natural systems. However, their complexity means that the majority of plant population ecologists are unlikely to become involved with the direct application of these models. The enormity of the field-calibration and computer code in the development of an IBM can be daunting. Furthermore, they require extensive field calibration that makes them impractical for all but the most comprehensive studies. Recent developments of IBMs include models that consider competition in three dimensions through inclusion of aboveground plant architecture (e.g. tree height and branching), the field-of-neighbourhood (FON) approach, and the particle-in-cell (PIC) approach (Berger et al. 2008). FON models calculate the influence of all neighbours around all plants in reducing the maximum growth rate of individuals. PIC models consider the explicit location of individuals within grid cells and consider this spatial information in predicting growth. Belowground interactions and facilitation require additional consideration (e.g. Schiffers et al. 2011). For example, the spatially explicit IBM DINVEG was used to examine the coexistence of shrubs and grass in semiarid Patagonian steppe (Cipriotti et al. 2014). In this case, demographic bottlenecks and species interactions among life stages controlled the coexistence and relative abundance of growth forms.

Integral projection models

Integral projection models (IPMs) are developed from individual-based models and incorporate continuous functions of state variables (Easterling et al. 2000). An IPM uses the basic matrix model tools in an analytical context but offers a different understanding of how local demographic processes and individual variation affect population growth. IPMs were developed as a way to deal with the problem in transition matrix models of assigning individuals into discrete classes or size stages despite the continuously varying nature of size traits such as plant height or other measures of body size. Defining optimal stage boundaries is a common problem, and is sometimes somewhat arbitrary in transition matrix model studies, but the decisions can greatly affect model output and interpretation. IPMs can also produce smaller bias and variance for λ than matrix models when small data sets are being analysed, and so may be of particular value in assessing population performance for poorly known endangered or invasive species (Ramula et al. 2009).

The same data on the observed fate, size distribution, and reproduction of individuals that are used in discrete-state transition matrix models can be used for an IPM. Following the notation in Easterling et al. (2000) and Ellner and Rees (2006), a continuous individual-level size or state variable x is used (e.g. the number of individuals of known height). Compared with a transition matrix model, a distribution function $n(x,t)$ is equivalent to the population vector n_t where $n(x, t)\mathrm{d}x$ is the number of individuals with their state variable in the range $[x, x + \mathrm{d}x]$ (see Williams et al. (2012) for methodological issues related to range size). Accordingly, the projection matrix A is represented by a projection kernel (K) of two parts (survival/growth, and fecundity) where $K(y,x) = s(x)g(y,x) + f(y,x)$, where from state x to state y (or size distribution at time t to the size distribution at time $t + 1$ one time step or census interval later), s represents survival, g growth, and f fecundity. In estimating the projection kernel, $s(x)g(y,x)$ represents the probability that an individual survives the census interval and the probability of the size distribution of individuals at the second census interval, and $f(y, x)$ represents the number and size distribution of offspring produced by reproductive individuals during the census interval. Population dynamics can then be modelled as

$$n(y,t+1) = \int_L^U k(y,x)n(x,t)\mathrm{d}x,$$

where $[L,U]$ is the range of possible states. Through application of this approach, population growth metrics similar to transition matrix models can be obtained, including estimates of size- and age-dependent population growth rates, probabilities of flowering and mortality, and sensitivity/elasticity analyses. Population IPMs can be scaled up from the plot level to projection at broader geographic scales suitable for capturing the effects of climate change through dynamic modelling, including the introduction of stochastic effects through time (Gelfand et al. 2013).

The R package IPMpack can be used for building IPMs (Metcalf et al. 2013 and see the Appendix). A practical guide for developing IPMs including R code for a range of plant life histories is provided by Merow et al. (2014); additional guidance based primarily on animal populations that can be adapted for plants is in Coulson (2012) and Rees et al. (2014). Of particular importance in developing IPMs are the time step lengths (census intervals), structure and diagnostics of the kernel, and model selection for parameterization.

Example: an IPM for the iteroparous, long-lived Orchis purpurea Huds *(lady orchid).* Miller et al (2012) used IPMs to identify the flowering size that maximized fitness (using net reproductive rate, R_0 as a proxy) in light and shaded habitats. The IPMs for each of the two light environments incorporated plants that varied continuously in size ($N(x)$) and in three discrete stages (protocorms, tubers, and dormant plants). Individuals of *O. purpurea* have an estimated life span of 44–60 years, and IPMs identified flowering and fruit production as a non-lethal cost that lowered growth rates, especially for the smallest plants, regardless of habitat (Fig 8.22). A general interpretation from the IPMs in this study is that the cost of reproduction in *O. purpurea* has selected for the evolution of reproductive delay.

8.4 Population viability analysis

This section briefly explains what PVA is, its uses and limitations, and software than can be used to run PVA (Table 8.5). An example application is also presented (another is Case Study 2 in Chapter 1). A detailed methodology is beyond the scope of this book, although the basics of transition matrix models that form the statistical background for PVA have been described in Section 8.2. Full details are available in Morris and Doak (2002) and Beissinger and McCullough (2002).

PVA is a conservation biology tool that makes use of population models to estimate the probability that a species will go extinct in a given number of years. PVA was derived from the concept of minimum viable population (MVP) proposed by Shaffer in the 1970s (Shaffer 1971, cited in Shaffer 1981) while working on populations of grizzly bears. The MVP was defined as 'the smallest isolated population (of a given species in a given habitat) having a 99% chance of remaining in existence for 1,000 years despite the foreseeable effects of demographic, environmental, and genetic stochasticity, and natural catastrophes' (Shaffer 1981).

PVA extends the MVA concept to allow the assessment of population trends, the impacts of

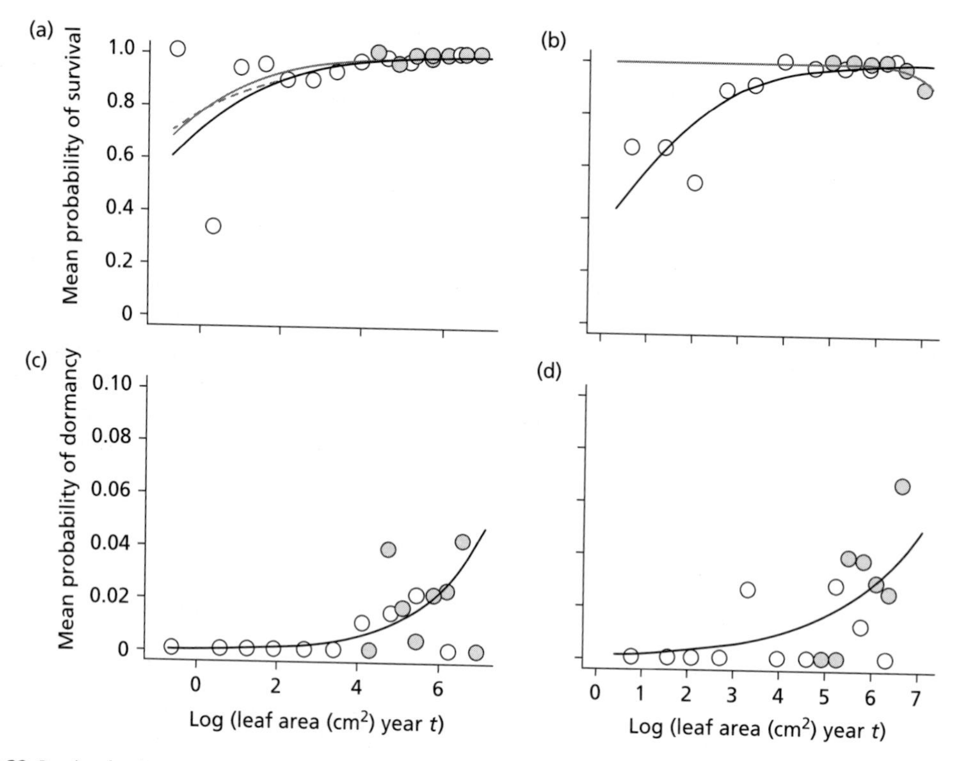

Figure 8.22 Results of an integrated projection model for *Orchis purpurea.* (a, b) Size-dependent survival and (c, d) dormancy in (a, c) light and (b, d) shady habitats. Points show binned proportions for plants that were vegetative (white circles) or flowering (grey circles) in year *t*. Thick black lines show the fitted null models. Thin grey lines in (a, b) show the best-fit model that incorporates differences between vegetative (dashed) and flowering (solid) plants (note that these differences are driven by the concentration of mortality at pre-reproductive sizes). From Miller et al. (2012) by permission of the Royal Society.

management and recovery plans, as well as extinction risks and population viability estimates for threatened species. Aspects of a species' life history can be identified, allowing research and management priorities to be determined. The vulnerability of particular populations to extinction can be identified. Finally, the impacts of human activities (e.g. the sensitivity of a population to anthropogenic disturbances such as logging or harvesting) and a ranking of management options can be produced. Morris and Doak (2002) identified eight potential uses of PVA:

Table 8.5 Software programs and packages available for population viability analysis (PVA) (adapted and updated from Newton 2007). Note that popbio and PVAClone are R packages (see the Appendix for additional information).

Program	Reference	Website
META-X	Frank et al. (2003), Grimm et al (2004)	–
RAMAMetapop and RAMAS GIS	Akçakaya and Root (2004)	<http://www.ramas.com>
ULM (unified life models)	Ferrière et al. (1996), Legendre and Clobert (1995)	<http://www.biologie.ens.fr/~legendre/ulm/ulm.html>
VORTEX	Lacy (1993), Lacy and Pollak (2013)	<http://vortex10.org/Vortex10.aspx>
popbio	Stubben and Milligan (2007)	<http://cran.r-project.org/web/packages/popbio/>
PVAClone	Nadeem and Lele (2012)	<http://cran.r-project.org/web/packages/PVAClone/index.html>

1. Assessing the risk of extinction of a single population.
2. Comparing relative risks of two or more populations.
3. Analysing and synthesizing monitoring data.
4. Identifying key life stages or demographic processes as management targets.
5. Determining how large a reserve needs to be to achieve a desired level of protection from extinction.
6. Determining how many individuals to release to establish a new population.
7. Setting limits on the harvest or 'take' from a population that are compatible with its continued existence.
8. Determining how many (and which) populations are needed to achieve a desired overall likelihood of species persistence.

PVA was developed for animal populations, and subsequent extension of PVAs to plants has brought additional challenges, including plant dormancy, seed banks, clonal growth, and periodic recruitment (Menges 2000). PVA in plants has also proven useful in planning re-introductions of threatened species (Knight 2012). These studies have shown that PVA is useful as a comparative tool that can incorporate and compare the influence of demographic aspects (e.g. vital rates, environmental drivers) to predict population growth rates (λ).

Table 8.6 The DAC-PVA protocol delineating the steps and elements of good model in general and in a useful population viability analysis (PVA) (from Pe'Er et al. 2013).

	Elements of a good model	Elements of a good PVA
During design	Includes stakeholders	Includes stakeholders in design, validation and interpretation
		Builds on (long-term) high-quality data
	Formulates objectives; justifies choice of model approach and complexity	Performs and justifies a careful model selection
During application	Careful parameterization	Includes relevant parameters based on knowledge of the system and the literature and in consideration of gaps
		Applies careful parameter estimation and parameterization
	Calibration, verification, validation	Performs calibration, verification, and validation or directs further monitoring and validation efforts
	Quantification of uncertainties	Performs sensitivity analyses and addresses uncertainty in a systematic and transparent way
	Applies multiple models	Compares the outcomes of alternative models where possible
		Differentiates among parameters affecting the model, the real world, and those that are relevant to management
		Ranks management scenarios to support decision making
During communication	Formulates assumptions	Communicates the entire modelling cycle and justifies decisions and assumptions along the way
	Effective documentation and transparency	Reports all inputs and outputs systematically to allow repeatability
		Uses carefully selected time horizon and viability measures and reports using consistent units to allow comparability
		Demonstrates that the PVA serves its purpose by, e.g., leading to on-the-ground actions
		Enhances collective learning and potential generalization
	Peer reviewed	Both the model (design, code, application) and the report are peer reviewed

An important component of PVA is the use of population models, most often incorporating transition matrix models (Section 8.2). Software for PVA is summarized in Table 8.5. The programs differ largely in how they incorporate density dependence, spatial structure (through linked transition matrices), and dispersal. Increasingly, the most sophisticated PVAs use Bayesian statistics (Section 7.7) to incorporate parameter uncertainty into the analysis (Wade 2002; Evans et al. 2010; Ruete et al. 2012). A comprehensive overview of running PVA, including MATLAB computer code, is provided in Morris and Doak (2002). As with the use of any 'canned' software, the user needs to take care in interpretation the output because the different packages can provide contrasting results (Mills et al. 1999; Brook et al. 2000). DAC-PVA is a protocol describing the three primary elements of design, application, and communication that characterize a useful PVA (Table 8.6).

While the goals of PVA are laudable, the usefulness and reliability of PVA has been questioned because of limitations, lack of precision and power, and even potential errors in the underlying statistical models, data, and parameter estimates (Coulson et al. 2001). The accuracy of PVA can be tested and used to identify the weakest aspects of the underlying models (McCarthy et al. 2001): see <http://www.ramas.com/mistakes.htm> for a list of concerns and advice on avoiding mistakes in PVA. Newton (2007) summarized these potential mistakes as:

- invalid model assumptions
- overly complex or overly simplistic models
- lack of internal consistency in the model
- bias in estimation of fecundity or survival, and uncertainty in estimates of these parameters
- too many (or too few) age classes or stages
- the wrong type of density dependence, or failing to consider density dependence

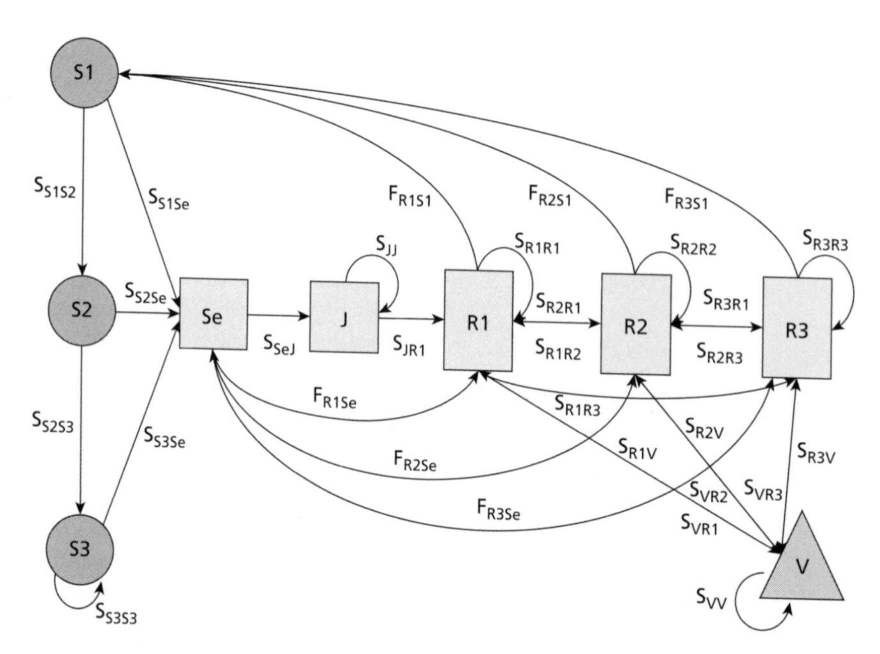

Figure 8.23 Conceptual model of the life cycle of *Verticordia fimbrilepis*. Each arrow represents the possible transitions between life stages. Arrows labelled with S represent survival and change of stage of plants. Arrows labelled with F represent fecundity. Stages identified in this model are 1-year-old seeds (S1); 2-year-old seeds (S2); > 3-year-old seeds (S3); first year seedling (Se); plants that have not reached reproductive maturity (J); three sizes of reproductive plants (R1, R2, R3) and vegetative senescing plants (V). From Yates and Ladd (2010). With kind permission from Springer Science and Business Media.

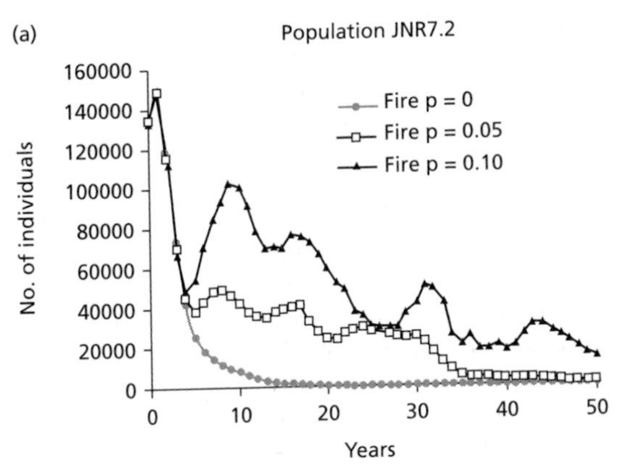

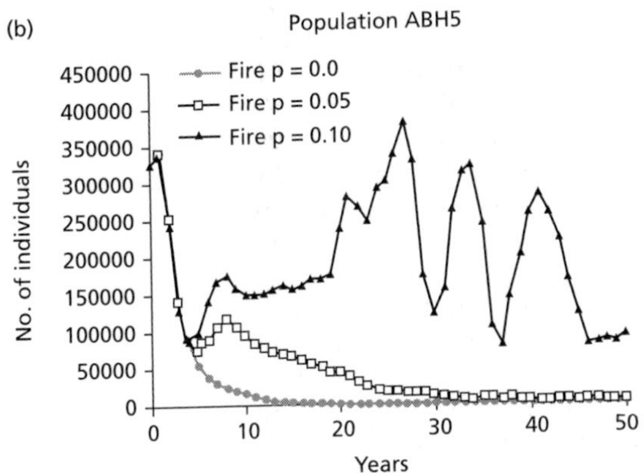

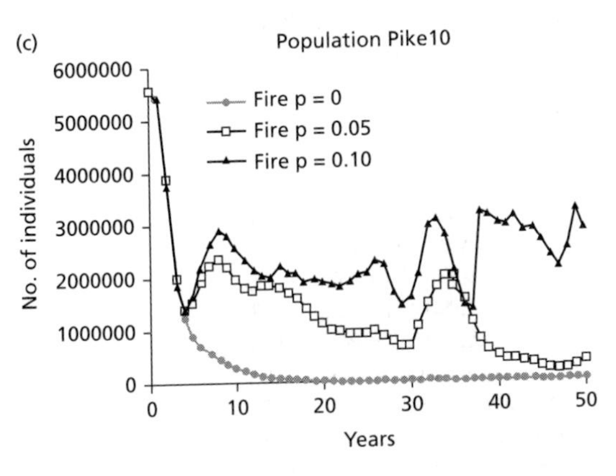

Figure 8.24 Mean population trajectories for three populations of *Verticordia fimbrilepis*, (a) JNR7.2, (b) ABH5, and (c) Pike10, under three fire regimes. All simulations were run for 50 1-year time steps and were replicated 1000 times. Counts include all seed and plant stages in the model. From Yates and Ladd (2010). With kind permission from Springer Science and Business Media.

- over- or under-estimation of the maximum growth rate of the population(s)
- not considering demographic stochasticity
- too long or too short a duration (simulation time horizon)
- ignoring spatial structure
- considering too many (or too few) populations
- incorrect or uncertain dispersal rates
- estimating risk of extinction rather than decline.

An example of PVA: the response to fire of Verticordia fimbrilepis *subsp.* fimbrilepis *(Myrtaceae) populations. Verticordia fimbrilepis* subsp. *fimbrilepis* (*Verticordia* hereafter) is an endangered shrub that grows in south-west Australia in fire-prone Mediterranean climate habitats. Thirteen disjunct populations are dispersed across fragmented remnant, fire-suppressed habitat in a matrix of agriculture. Transition models were used by Yates and Ladd (2010) as part of a PVA to evaluate burning as a management tool and test the hypothesis that prescribed fire will improve the probability of persistence of *Verticordia* in habitat fragments.

Demographic census data (life stage, plant height, size, and flowering of all individuals within 1 m × 1 m sample plots) were collected annually from three populations over 5 years and were used to establish a life-cycle model (Fig. 8.23) and develop transition matrix models. Population trends and viability without burning and with different probabilities of fire were modelled using the stochastic simulation program RAMAS (Table 8.5). The extinction risk (probability that a population falls below a threshold population size) over 50 years and expected minimum abundance were estimated.

PVA showed that *Verticordia* populations were highly susceptible to fire interval (Fig. 8.24). In the absence of fire, all populations were projected to decline or become locally extinct over 50 years. Fire reduced extinction risk depending on the population, with the highest fire frequency at one population decreasing extinction risk by 53% (Fig. 8.24). Seedling recruitment was also higher in wetter years, especially in unburned sites. Under projected drier future climates, fire will be necessary to promote population viability. Unfortunately, with increasing habitat fragmentation, the probability of landscape fires decreases. This study is a relatively simple example illustrating how PVA can help to identify management options (prescribed fire) to help conserve populations of a threatened species. More involved analyses, such as the inclusion of Bayesian models to estimate covariation among vital rates, can allow greater levels of demographic and environmental stochasticity to be incorporated in PVA (e.g. Evans et al. 2010).

8.5 Follow-up exercises

1. List some of the advantages and disadvantages of first-order and second-order methods of spatial pattern analysis.
2. Suppose that you are trying to manage an area to enhance the population of a rare plant species. How important is it to know which of the survivorship curves illustrated in Fig. 8.15 pertain to that plant? How would you use this knowledge in establishing a recovery plan? Can you find any examples where knowledge of survivorship curves has proved useful in this respect?
3. The worked example in Box 8.1 included seedling recruitment. How are the results different (i.e. λ, population growth, elasticity) in the absence of seedling recruitment?
4. Construct life-cycle graphs and then Leslie and Lefkovitch projection matrices (without values in the cells, unless you have the data) for an annual, perennial, vine, and tree species that you are familiar with. How do the matrices differ? What age, size, or life-history stage would you consider to be important?
5. List five or more rules that you would use to constrain population growth in a cellular automata model for *Silene regia*, the long-lived iteroparous prairie perennial that is the focus of Case Study 2 in Chapter 1 and Box 8.1.
6. Outline the data that you would need for a PVA of a rare plant that you are familiar with. What are the difficulties in obtaining the data that you would need. What assumptions about the life cycle of the plant will you have to make?
7. For a named plant of your choice representing a specified life history, contrast the vital rate data needed and the approach you would take to estimating population projections using transition matrix models versus IPMs.

APPENDIX

R Packages for Analysis of Plant Population Ecology Data

R (R Development Core Team 2011) is a freely available, open source statistical environment that is increasing in popularity and use. While many conventional statistical analyses can be accomplished with 'standard' R modules, a large number of packages have been developed for implementing specific analyses. Be warned, R packages are 'a bit of a moving feast' (Rob Freckleton, personal communication)! R packages that are useful for plant population ecology are provided here. Most are available for free download from the Comprehensive R Archive Network at <http://cran.r-project.org/web/packages/>. If you are not familiar with R then there is a cottage industry of books and online sources available to help including Bolker (2008), Beckerman and Petchey (2012), and Crawley (2012), and R manuals available at <http://www.R-project.org> (see Chapter 7 for more details).

Package name	Application	Reference and description and download site if not CRAN
Population dynamics		
demoniche	Simulates stochastic population dynamics in multiple populations of a species	Nenzén et al. (2012), <https://r-forge.r-project.org/projects/demoniche/>
IPMpack	Suite of demographic tools for running IPMs	Metcalf et al. (2013). See also the blog at <http://ipmpack.r-forge.r-project.org/> and R code in the appendix of Merow et al. (2014).
popbio	Basic analyses of demographic models, including matrix models and PVA	Stubben and Milligan (2007)
popdemo	Novel methods for analysis of population projection models	Stott et al. (2012)
PVAClone	PVA with data cloning	Nadeem and Lele (2012)
survival	Survival analyses and tests	Additional R packages are listed at <http://cran.r-project.org/web/views/survival.html>
Spatial patterns		
ade4	Multivariate methods including ordinations and spatial pattern analysis	Thioulouse et al. (1997), <http://pbil.univ-lyon1.fr/ADE-4/home.php?lang = eng>
betapart	Spatial patterns of beta diversity	Baselga and Orme (2012)
circular	Circular statistics	Pewsey et al. (2013)
deldir	Delaunay triangulation and Dirichlet or Voronoi tessellations	
Spatstat	Analysis of spatial point patterns	Baddeley and Turner (2005)
spdep	Spatial regression analysis including spatial correlograms based on Moran's *I* and Geary's *c*	

continued

Package name	Application	Reference and description and download site if not CRAN
Phylogenetics		
ape	Comparative phylogenetics, PICS	Paradis (2006), <http://ape.mpl.ird.fr/?>
CRAN Task View: phylogenetics, especially comparative methods	Listing and description of R packages for comparative phylogenetic methods	<http://cran.r-project.org/web/views/Phylogenetics.html>
caper	Comparative analyses of phylogenetics and evolution in R	Orme et al (2012)
phytools	Methods for comparative phylogenetic biology, including tree inference, phylogeny input/output, plotting, and manipulation	Revell (2012)
picante	Package for analysing the phylogenetic and trait diversity of ecological communities	Kembel et al. (2010)
treebase	Provides access to phylogenetic data in the TreeBASE repository	Boettiger and Lang (2012)
Mixed models and related Statistics		
abc	Approximate Bayesian computation for parameter inference and model selection under complex models	Csilléry et al. (2012)
coxme	Mixed effects Cox models containing Gaussian random effects (frailty models)	
lme4	Linear mixed effects models, GLMMs	Bolker et al. (2009)
MCMCglmm	Markov chain Monte Carlo GLMMs	
NCF	Spatial non-parametric covariance functions	
nlme	Non-linear mixed effects models	
odprism	Optimal design and performance of random intercept and slope models. Investigates accuracy, precision, and power of regression models used for quantifying reaction norms	van de Pol (2012)
pamm	Power analysis with mixed models. Measuring individual differences in reaction norms	Martin et al. (2011) (and corrections in Erratum)
Miscellaneous		
bipartite	Analysis of patterns in ecological webs, e.g. pollinator networks or seed disperser networks	Dormann et al. (2008)
OpenMx	Advanced structural equation modelling	Boker et al. (2011), <http://openmx.psyc.virginia.edu/>
pwr	Basic functions for power analysis	Champely (2009)
qtl	QTL mapping and LOD plots	Broman et al. (2003)
sem	Structural equation modelling	Fox (2006)
taxize	Taxonomic information and tasks including verification, hierarchies, and names	
VEGAN	Multivariate methods including ordinations	Dixon (2009), <http://vegan.r-forge.r-project.org/>

IPM, integral projection model; PVA, population viability analysis; PICS, phylogenetically independent contrasts; GLMMs, generalized linear mixed models; QTL, quantitative trait locus; LOD, log of the odds ratio.

References

Aarssen LW and Clauss MJ (1992). Genotypic variation in fecundity allocation in *Arabidopsis thaliana*. *Journal of Ecology* **80**: 109–114.

Aarssen LW and Epp GA (1990). Neighbor manipulations in natural vegetation: a review. *Journal of Vegetation Science* **1**: 13–30.

Aarssen L and Keogh T (2002). Conundrums of competitive ability in plants: what to measure? *Oikos* **96**: 531–542.

Abrahamson WG and Weis A (1997). *Evolutionary ecology across three trophic levels: goldenrods, gallmakers, and natural enemies*. Princeton University Press, Princeton, NJ.

Adamatzky A (1994). *Identification of cellular automata*. Taylor and Francis, London.

Adams DE, Perkins WE, and Estes JR (1981). Pollination systems in *Paspalum dilatatum* Poir. (Poaceae): an example of insect pollination in a temperate grass. *American Journal of Botany* **63**: 389–394.

Adler PB and HilleRisLambers J (2008). The influence of climate and species composition on the population dynamics of ten prairie forbs. *Ecology* **89**: 3049–3060.

Aerts R (1990). Nutrient use efficiency in evergreen and deciduous species from heathlands. *Oecologia* **84**: 391–397.

Aerts R, Berendse F, De Caluwe H, and Schmitz M (1990). Competition in heathland along an experimental gradient of nutrient availability. *Oikos* **57**: 310–318.

Affandi NAM, Kamali B, MZ R, Tamin NM, and Hashim R (2010). Early growth and survival of *Avicennia alba* seedlings under excessive sedimentation. *Scientific Research and Essays* **5**: 2801–2805.

Aguilera MO and Lauenroth WK (1993). Neighborhood interactions in a natural population of the perennial bunchgrass *Bouteloua gracilis*. *Oecologia* **94**: 595–602.

Akçakaya HR and Root W (2004). *RAMAS Metapop: viability analysis for stage-structured metapopulations*, version 4.0. Applied Biomathematics, Setauket, NY.

Albert CH, Grassein F, Schurr FM, Vieilledent G, and Violle C (2011). When and how should intraspecific variability be considered in trait-based plant ecology? *Perspectives in Plant Ecology, Evolution and Systematics* **13**: 217–225.

Alexander HM (1990). Epidemiology of anther-smut infection of *Silene alba* caused by *Ustilago violacea*: patterns of spore deposition and disease incidence. *Journal of Ecology* **78**: 166–179.

Alexander HM and Antonovics J (1988). Disease spread and population dynamics of anther-smut infection of *Silene alba* caused by the fungus *Ustilago violacea*. *Journal of Ecology* **76**: 91–104.

Ali MA, Ashraf M, and Athar HR (2009). Influence of nickel stress on growth and some important physiological/biochemical attributes in some diverse canola (*Brassica napus* L.) cultivars. *Journal of Hazardous Materials* **172**: 964–969.

Allen LH, Drake BG, Rogers HH, and Shinn JH (1992). Field techniques for exposure of plants and ecosystems to elevated CO_2 and other trace gases. *Critical Reviews in Plant Sciences* **11**: 85–119.

Allen SE, Grimshaw HM, and Rowland AP (1986). Chemical analysis. In PD Moore and SB Chapman (eds) *Methods in plant ecology*, pp. 285–344. Blackwell Scientific Publications, Oxford.

Alstad D (2001). *Basic populus models of ecology*. Prentice-Hall, Inc., Upper Saddle River, NJ.

Alvarez-Buylla ER and Slatkin M (1991). Finding confidence limits on population growth rates. *Trends in Ecology and Evolution* **6**: 221–224.

Anderson JE and McNaughton SJ (1973). Effects of low soil temperature on transpiration, photosynthesis, leaf relative water content, and growth among elevationally diverse plant populations. *Ecology* **54**: 1220–1233.

Anderson JT, Lee C-R, Rushworth CA, Colautti RI, and Mitchell-Olds T (2013). Genetic trade-offs and conditional neutrality contribute to local adaptation. *Molecular Ecology* **22**: 699–708.

Andres MW and Wilcoxson RD (1984). A device for uniform deposition of liquid-suspended urediospores on seedling and adult cereal plants. *Phytopathology* **74**: 550–552.

Anselin L and Rey SJ (eds) (2010). *Perspectives on spatial data analysis*. Springer-Verlag, Berlin.

Antonovics JA, Bradshaw AD, and Turner RG (1971). Heavy metal tolerance in plants. *Advances in Ecological Research* **7**: 1–85.

Antonovics J and Levin DA (1980). The ecological and genetic consequences of density-dependent regulation in plants. *Annual Review of Ecology and Systematics* **11**: 411–452.

Antonovics J and Primack RB (1982). Experimental ecological genetics in *Plantago*. VI. The demography of seedling transplants of *P. lanceolata*. *Journal of Ecology* **70**: 55–75.

Antonovics J, Ellstrand NC, and Brandon RN (1988). Genetic variation and environmental variation: expectations and experiments. In LD Gottlieb and SK Jain (eds) *Plant evolutionary biology*, pp. 275–303. Chapman and Hall, London.

Antos JA and Allen GA (1999). Patterns of reproductive effort in male and female shrubs of *Oemleria cerasiformis*: a 6-year study. *Journal of Ecology* **87**: 77–84.

Applebee T, Gibson DJ, and Newman JA (1999). Elevated atmospheric CO_2 alters the effects of allelochemicals produced by tall fescue on alfalfa seedlings. *Transactions of the Illinois Academy of Science* **92**: 23–31.

Archibold OW and Ripley EA (2004). Assessment of seasonal change in a young aspen (*Populus tremuloides* Michx.) canopy using digital imagery. *Applied Geography* **24**: 77–95.

Arena G, Symes CT, and Witkowski ETF (2013). The birds and the seeds: opportunistic avian nectarivores enhance reproduction in an endemic montane aloe. *Plant Ecology* **214**: 35–47.

Argyres AZ and Schmitt J (1992). Neighbor relatedness and competitive performance in *Impatiens capensis* (Balsaminaceae): a test of the resource partitioning hypothesis. *American Journal of Botany* **79**: 181–185.

Arntz AM, DeLucia EH, and Jordan N (1998). Contribution of photosynthetic rate to growth and reproduction in *Amaranthus hybridus*. *Oecologia* **117**: 323–330.

Arntz AM, DeLucia EH, and Jordan N (2000). From fluorescence to fitness: variation in photosynthetic rate effects fecundity and survivorship. *Ecology* **81**: 2567–2576.

Aronson J, Kigel J, and Shmida A (1993). Reproductive allocation strategies in desert and Mediterranean populations of annual plants grown with and without water stress. *Oecologia* **93**: 336–342.

Arshad MA, Lowery B, and Grossman B (1996). Physical tests for monitoring soil quality. In JW Doran and AJ Jones (eds) *Methods for assessing soil quality*, pp. 123–142. Soil Science Society of America, Madison, WI.

Ashenden TW, Baxter R, and Rafarel CR (1992). An inexpensive system for exposing plants in the field to elevated concentrations of CO_2. *Plant, Cell and Environment* **15**: 365–372.

Askew AP, Corker D, Hodkinson DJ, and Thompson K (1997). A new apparatus to measure the rate of fall of seeds. *Functional Ecology* **11**: 121–125.

Asseng S, Aylmore LAG, MacFall JS, Hopmans JW, and Gregory PJ (2000). Computer-assisted tomography and magnetic-resonance imaging. In AL Smit, AG Bengough, C Engels, et al. (eds) *Root methods: a handbook*, pp. 343–363. Springer-Verlag, Berlin.

Atkinson D (2000). Root characteristics: why and what to measure. In AL Smit, AG Bengough, C Engels, et al. (eds) *Root methods: a handbook*, pp. 1–32. Springer-Verlag, Berlin.

Auger S and Shipley B (2013). Inter-specific and intra-specific trait variation along short environmental gradients in an old-growth temperature forest. *Journal of Vegetation Science* **24**: 419–428.

Augspurger CK (1983). Seed dispersal of the tropical tree, *Platypodium elegans*, and the escape of its seedlings from fungal pathogens. *Journal of Ecology* **71**: 759–771.

Augustine DJ, Frelich LE, and Jordan PA (1998). Evidence for two alternate stable states in an ungulate grazing system. *Ecological Applications* **8**: 1260–1269.

Austin MP (1982). Use of a relative physiological performance value in the prediction of performance in multi-species mixtures from monoculture performance. *Journal of Ecology* **70**: 559–570.

Avinoam D and Orshan G (1990). The distribution of Raunkiaer life forms in Israel in relation to the environment. *Journal of Vegetation Science* **1**: 41–48.

Axmanová I, Zelený D, Li C-F, and Chytrý M (2011). Environmental factors influencing herb layer productivity in Central European oak forests: insights from soil and biomass analyses and a phytometer experiment. *Plant and Soil* **342**: 183–194.

Bachmann K (1994). Molecular markers in plant ecology. *New Phytologist* **126**: 403–418.

Bacon CW and Hill NS (eds) (1997). *Neotyphodium/grass interactions*. Plenum Press, New York.

Bacon CW and White JF (1994). Stains, media, and procedures for analyzing endophytes. In CW Bacon and JF White (eds) *Biotechnology of endophytic fungi of grasses*, pp. 47–55. CRC Press, Boca Raton, FL.

Baddeley A and Turner R (2005). spatstat: an R package for analyzing spatial point patterns. *Journal of Statistical Software* **12**: 1–42.

Baer SG, Collins SL, Blair JM, Knapp AK, and Fiedler AK (2005). Soil heterogeneity effects on tallgrass prairie community heterogeneity: an application of ecological theory to restoration ecology. *Restoration Ecology* **13**: 413–424.

Bailey NTJ (1995). *Statistical methods in biology*, 3rd edn. Cambridge University Press, Cambridge.

Baird NA, Etter PD, Atwood TS, Currey MC, Shiver AL, Lewis ZA, et al. (2008). Rapid SNP discovery and genetic mapping using sequenced RAD markers. *PLoS ONE* **3**: e3376.

Baker A (ed.) (2000). *Molecular methods in ecology*. Blackwell Science Ltd, Oxford.

Bakker JP, Bakker ES, Rosén E, Verweij GL, and Bekker RM (1996). Soil seed bank composition along a gradient from dry alvar grassland to *Juniperus* shrubland. *Journal of Vegetation Science* **7**: 165–176.

Baldwin AH, Platt WJ, Gathen KL, Lessmann JM, and Rauch TJ (1995). Hurricane damage and regeneration in fringe mangrove forests of southeast Florida, USA. *Journal of Coastal Research* **21**: 169–183.

Ball DF (1986). Site and soils. In PD Moore and SB Chapman (eds) *Methods in plant ecology*, pp. 215–284. Blackwell Scientific Publications, Oxford.

Balzter H, Braun PW, and Köhler W (1998). Cellular automata models for vegetation dynamics. *Ecological Modelling* **107**: 113–125.

Bannister P (1986). Water relation and stress. In PD Moore and SB Chapman (eds) *Methods in plant ecology*, pp. 73–143. Blackwell Scientific Publications, Oxford.

Barbour MG, Burk JH, and Pitts WD (1987). *Terrestrial plant ecology*, 2nd edn. Benjamin/Cummings Publishing Company, Inc., Menlo Park, CA.

Bardon RE, Countryman DW, and Hall RB (1995). A reassessment of using light-sensitive diazo paper for measuring integrated light in the field. *Ecology* **76**: 1013–1016.

Barker AV and Pilbeam DJ (eds) (2007). *Handbook of plant nutrition*. Taylor and Francis, Boca Raton, FL.

Barkham JP and Hance CE (1982). Population dynamics of the wild daffodil (*Narcissus pseudonarcissus*). III. Implications of a computer model of 1000 years of population change. *Journal of Ecology* **70**: 323–344.

Barner AK, Pfister CA, and Wootton JT (2012). The mixed mating system of the sea palm kelp *Postelsia palmaeformis*: few costs to selfing. *Proceedings of the Royal Society B: Biological Sciences* **278**: 1347–1355.

Barot S, Gignoux J, and Menaut J-C (1999). Seed shadows, survival and recruitment: how simple mechanisms lead to dynamics of population recruitment curves. *Oikos* **86**: 320–330.

Barrett JP and Silander JA (1992). Seedling recruitment limitation in white clover (*Trifolium repens*; Leguminosae). *American Journal of Botany* **79**: 643–649.

Barrs H and Weatherley P (1962). A re-examination of the relative turgidity technique for estimating water deficits in leaves. *Australian Journal of Biological Sciences* **15**: 413–428.

Barto K, Friese C, and Cipollini D (2010). Arbuscular mycorrhizal fungi protect a native plant from allelopathic effects of an invader. *Journal of Chemical Ecology* **36**: 351–360.

Baselga A and Orme CDL (2012). betapart: an R package for the study of beta diversity. *Methods in Ecology and Evolution* **3**: 808–812.

Baskin JM and Baskin CC (1972). Ecological life cycle and physiological ecology of seed germination of *Arabidopsis thaliana*. *Canadian Journal of Botany* **50**: 352–360.

Baskin JM and Baskin CC (1983). Seasonal changes in the germination responses of buried seeds of *Arabidopsis thaliana* and ecological interpretation. *Botanical Gazette* **144**: 540–543.

Baskin CC and Baskin JM (1998). *Seeds: ecology, biogeography, and evolution of dormancy and germination*. Academic Press, San Diego, CA.

Bazzaz FA (1996). *Plants in changing environments*. Cambridge University Press, Cambridge.

Bazzaz FA (1997). Allocation of resources in plants: state of the science and critical questions. In FA Bazzaz and J Grace (eds) *Plant resource allocation*, pp. 1–38. Academic Press, San Diego, CA.

Bazzaz FA and Ackerly DD (1992). Reproductive allocation and reproductive effort in plants. In M Fenner (ed.) *Seeds: the ecology of regeneration in plant communities*, pp. 1–26. CAB International, Wallingford, Oxon.

Bazzaz FA and Grace J (eds) (1997). *Plant resource allocation*. Academic Press, New York.

Bazzaz FA and Reekie EG (1985). The meaning and measurement of reproductive effort in plants. In J White (ed.) *Studies on plant demography: a festschrift for John L. Harper*, pp. 373–387. Academic Press, London.

Bazzaz FA and Stinson KA (1999). Genetic vs. environmental control of ecophysiological processes: some challenges for predicting community responses to global change. In MC Press, JD Scholes, MG Barker (eds) *Physiological plant ecology*, pp. 283–295. Blackwell Science, Oxford.

Beckage B, Joseph L, Belisle P, Wolfson DB, and Platt WJ (2007). Bayesian change-point analysis in ecology. *New Phytologist* **174**: 456–467.

Beckerman AP and Petchey OL (2012). *Getting started with R: an introduction for biologists*. Oxford University Press, Oxford.

Beddows AR (1967). Biological flora of the British Isles: *Lolium perenne*. *Journal of Ecology* **55**: 567–587.

Bedoussac L and Justes E (2011). A comparison of commonly used indices for evaluating species interactions and intercrop efficiency: application to durum wheat–winter pea intercrops. *Field Crops Research* **124**: 25–36.

Beerling DJ (1999). Long-term responses of boreal vegetation to global change: an experimental and modelling investigation. *Global Change Biology* **5**: 55–74.

Begon M and Mortimer M (1981). *Population ecology: a unified study of animals and plants*. Sinauer Associates, Sunderland, MA.

Beissinger SR and McCullough DR, eds. (2002). *Population viability analysis*. University of Chicago Press, Chicago, IL.

Belcher JW, Keddy PA, and Twolan-Strutt L (1995). Root and shoot competition intensity along a soil depth gradient. *Journal of Ecology* **83**: 673–682.

Bell G and Lechowicz MJ (1991). The ecology and genetics of fitness in forest plants. I. Environmental heterogeneity measured by explant trials. *Journal of Ecology* **79**: 663–685.

Benamar A, Pierart A, Baecker V, Avelange-Macherel M-H, Rolland A, Gaudichon S, et al. (2013). Simple system using natural mineral water for high-temperature phenotyping of *Arabidopsis thaliana* seedlings in liquid culture. *International Journal of High Throughput Screening* **2013**: 1–15.

Bender EA, Case TJ, and Gilpin ME (1984). Perturbation experiments in community ecology: theory and practice. *Ecology* **69**: 1–13.

Berendse F (1983). Interspecific competition and niche differentiation between *Plantago lanceolata* and *Anthoxanthum odoratum* in a natural hayfield. *Journal of Ecology* **71**: 379–390.

Berger U, Piou C, Schiffers K, and Grimm V (2008). Competition among plants: concepts, individual-based modelling approaches, and a proposal for a future research strategy. *Perspectives in Plant Ecology, Evolution and Systematics* **9**: 121–135.

Bertness MD and Shumway SW (1993). Competition and facilitation in marsh plants. *American Naturalist* **142**: 718–724.

Bever JD (1994). Feedback between plants and their soil communities in an old field community. *Ecology* **75**: 1965–1977.

Bever JD, Westover KM, and Antonovics J (1997). Incorporating the soil community into plant population dynamics: the utility of the feedback approach. *Journal of Ecology* **85**: 561–573.

Bhadresa R (1997). Faecal analysis and exclosure studies. In MJ Crawley (ed.) *Plant ecology*, pp. 61–71. Blackwell Scientific Publications, Oxford.

Bierzychudek P (1999). Looking backwards: assessing the projections of a transition matrix model. *Ecological Applications* **9**: 1278–1287.

Bild AH, Chang JT, Johnson WE, and Piccolo SR (2014). A field guide to genomics research. *PLoS Biology* **12**(1): e1001744.

Bingham IJ, Glass ADM, Kronzucker HJ, Robinson D, and Scrimgeour CM (2000). Isotope techniques. In AL Smit, AG Bengough, C Engels, et al. (eds) *Root methods: a handbook*, pp. 365–399. Springer-Verlag, Berlin.

Binkley D and Vitousek P (1989). Soil nutrient availability. In RW Pearcy, JR Ehleringer, HA Mooney, *et al.* (eds) *Plant physiological ecology*, pp. 75–96. Chapman and Hall, London.

Birch LC (1948). The intrinsic rate of increase of an insect population. *Journal of Animal Ecology* **17**: 15–26.

Bischoff A and Müller-Schärer H (2010). Testing population differentiation in plant species—how important are environmental maternal effects. *Oikos* **119**: 445–454.

Biswell HH (1989). *Prescribed burning in California wildlands vegetation management*. University of California Press, Berkeley, CA.

Bittelli M (2011). Measuring soil water content: a review. *Hort Technology* **21**: 293–300.

Bittner RT and Gibson DJ (1998). Microhabitat relations of the rare reed bent grass, *Calamagrostis porteri* subsp. *insperata* (Poaceae), with implications for its conservation. *Annals of the Missouri Botanical Garden* **85**: 69–80.

Bivand RS, Pedesma EJ, and Gómez-Rubio V (2008). *Applied spatial data analysis with R*. Springer, New York.

Blackman VH (1919). The compound interest law and plant growth. *Annals of Botany* **33**: 353–360.

Blomberg SP, Garland T Jr., and Ives AR (2003). Testing for phylogenetic signal in comparative data: behavioral traits are more labile. *Evolution* **57**: 717–745.

Bobblink R, Hornung M, and Roelofs JGM (1998). The effects of air-borne nitrogen pollutants on species diversity in natural and semi-natural European vegetation. *Journal of Ecology* **86**: 717–738.

Bode RF and Kessler A (2012). Herbivore pressure on goldenrod (*Solidago altissima* L., Asteraceae): its effects on herbivore resistance and vegetative reproduction. *Journal of Ecology* **100**: 795–801.

Boettiger C and Temple Lang D (2012). Treebase: an R package for discovery, access and manipulation of online phylogenies. *Methods in Ecology and Evolution* **3**: 1060–1066.

Boffey TB and Veevers A (1977). Balanced designs for two component competition experiments. *Euphytica* **26**: 481–484.

Böhm W (1979). *Methods of studying root systems*. Springer-Verlag, Berlin.

Boker SM, Neale MC, Maes HH, Wilde MJ, Spiegel M, Brick TR, et al. (2011). OpenMx: an open source extended structural equation modeling framework. *Psychmetrika* **76**: 306–317.

Bolker BM (2008). *Ecological models and data in R*. Princeton University Press, Princeton, NJ.

Bolker BM, Brooks ME, Clark CJ, Geange SW, Poulsen JR, Stevens MHH, et al. (2009). Generalized linear mixed models: a practical guide for ecology and evolution. *Trends in Ecology and Evolution* **24**: 127–135.

Bond G (1976). The results of the IBP survey of root nodule formation in non-leguminous angiosperms. In SP Nutman (ed.) *Symbiotic nitrogen fixation in plants*, pp. 443–474. Cambridge University Press, Cambridge.

Booth RE, Mackey JML, Rorison IH, Spencer RE, Gupta PL, and Hunt R (1993). ISP germination and rooting environments: sand, compost and solution culture. In GAF Hendry and JP Grime (eds) *Methods in comparative plant ecology*, pp. 19–24. Chapman and Hall, London.

Borer ET, Seabloom EW, Jones MB, and Schildhauer M (2009). Some simple guidelines for effective data management. *Bulletin of the Ecological Society of America* **90**: 205–214.

Bossard CC (1990). Tracing of ant-dispersed seeds: a new technique. *Ecology* **71**: 2370–2371.

Bossard CC and Hillier SH (1993). Response to defoliation. In GAF Hendry and JP Grime (eds) *Methods in comparative plant ecology*, pp. 45–48. Chapman and Hall, London.

Bossdorf O, Richards CL, and Pigliucci M (2008). Epigenetics for ecologists. *Ecology Letters* **11**: 106–115.

Boutin C and Harper JL (1991). A comparative study of the population dynamics of five species of *Veronica* in natural habitats. *Journal of Ecology* **79**: 199–202.

Box GEP and Tiao GC (1992). *Bayesian inference in statistical analysis*. John Wiley and Sons, Inc, New York.

Boyd RS and Barbour MG (1993). Replacement of *Cakile edentula* by *C. maritima* in strand habitat of California. *American Midland Naturalist* **130**: 209–228.

Boyd CS and Davies KW (2012). Differential seedling performance and environmental correlates in shrub canopy vs. interspace microsites. *Journal of Arid Environments* **87**: 50–57.

Boyd RS and Pitzer TR (1991). Effects of experimental defoliation on infructescences of flowering dogwood, *Cornus florida* L. (Cornaceae). *Castanea* **56**: 142–146.

Boyes DC, Zayed AM, Ascenzi R, McCaskill AJ, Hoffman NE, Davis KR, et al. (2001). Growth stage-based phenotypic analysis of Arabidopsis: a model for high throughput functional genomics in plants. *Plant Cell* **13**: 1499–1510.

Bradbeer JW (1988). *Seed dormancy and germination*. Blackie and Son Ltd, Glasgow.

Brandl H (2013). Detection of fungal infection in *Lolium perenne* by Fourier transform infrared spectroscopy. *Journal of Plant Ecology* **6**: 265–269.

Brandrud TE and Roelofs JGM (1995). Enhanced growth of the macrophyte *Juncus bulbosus* in S. Norwegian limed lakes. A regional survey. *Water, Air and Soil Pollution* **85**: 913–918.

Brandt AJ, de Kroon H, Reynolds HL, and Burns JH (2013). Soil heterogeneity generated by plant–soil feedbacks has implications for species recruitment and coexistence. *Journal of Ecology* **101**: 277–286.

Bräutigam K, Vining KJ, Lafon-Placette C, Fossdal CG, Mirouze M, Marcos JG, et al. (2013). Epigenetic regulation of adaptive responses of forest tree species to the environment. *Ecology and Evolution* **3**: 399–415.

Bréda NJJ (2003). Ground-based measurements of leaf area index: a review of methods, instruments and current controversies. *Journal of Experimental Botany* **54**: 2403–2417.

Briggs JM and Gibson DJ (1992). Effect of fire on tree spatial patterns in a tallgrass prairie landscape. *Bulletin of the Torrey Botanical Club* **119**: 300–307.

Briggs D and Walters SM (1984). *Plant variation and evolution*, 2nd edn. Cambridge University Press, Cambridge.

Broman KW, Wu H, Sen Ś, and Churchill GA (2003). R/qtl: QTL mapping in experimental crosses. *Bioinformatics* **19**: 889–890.

Brook BW, Burgman MA, and Frankham R (2000). Differences and congruencies between PVA packages: the importance of sex ratio for predictions of extinction risk. *Conservation Ecology* **4**(1): 6 [online] URL: http://www.consecol.org/vol4/iss1/art6/

Brophy C, Gibson DJ, Wayne PM, and Connolly J (2007). A modelling framework for analysing the reproductive output of individual plants grown in monoculture. *Ecological Modelling* **207**: 99–108.

Brophy C, Gibson DJ, Wayne PM, and Connolly J (2008). How reproductive allocation and flowering probability of individuals in plant populations are affected by position in stand size hierarchy, plant size and CO_2 regime. *Journal of Plant Ecology* **1**: 207–215.

Brown VK and Gange AC (1989). Differential effects of above- and below-ground insect herbivory during early plant succession. *Oikos* **54**: 67–76.

Brown VK, Gange AC, Evans IM, and Storr AL (1987). The effect of insect herbivory on the growth and reproduction of two annual *Vicia* species at different stages in plant succession. *Journal of Ecology* **75**: 1173–1189.

Bucharová A, Münzbergová Z, and Tájjek P (2010). Population biology of two rare fern species: long life and long-lasting stability. *American Journal of Botany* **97**: 1260–1271.

Bucher JB, Tarjan DP, Siegwolf RTW, Saurer M, Blum H, and Hendrey GR (1998). Growth of a deciduous tree seedling community in response to elevated CO_2 and nutrient supply. *Chemosphere* **36**: 777–782.

Buggs RJA, Chamala S, Wu WEI, Gao LU, May GD, Schnable PS, et al. (2010). Characterization of duplicate gene evolution in the recent natural allopolyploid *Tragopogon miscellus* by next-generation sequencing and Sequenom iPLEX MassARRAY genotyping. *Molecular Ecology* **19**: 132–146.

Buggs RJA, Renny-Byfield S, Chester M, Jordon-Thaden IE, Viccini LF, Chamala S, et al. (2012). Next-generation sequencing and genome evolution in allopolyploids. *American Journal of Botany* **99**: 372–382.

Bullock JM and Clarke RT (2000). Long distance seed dispersal by wind: measuring and modelling the tail of the curve. *Oecologia* **124**: 506–521.

Bullock JM, Shae K, and Skarpaas O (2006). Measuring plant dispersal: an introduction to field methods and experimental design. *Plant Ecology* **186**: 217–234.

Burdon JJ (1983). *Trifolium repens* L. *Journal of Ecology* **71**: 307–330.

Burdon JJ (1987). *Diseases and plant population biology*. Cambridge University Press, Cambridge.

Burkhardt F and Smith S (1986). Appendix VI. Darwin and William Kemp on the vitality of seeds. In F Burkhardt and S Smith (eds) *The correspondence of Charles Darwin. Volume 2: 1837–1843*, pp. 450–453. Cambridge University Press, Cambridge.

Buswell JM, Moles AT, and Hartley S (2010). Is rapid evolution common in introduced plant species? *Journal of Ecology* **99**: 214–224.

Butnor JR, Barton C, Day FP, Johnsen KH, and Mucciardi AN (2012). Using ground-penetrating radar to detect tree roots and estimate biomass. In S Mancuso (ed.) *Measuring roots: an updated approach*, pp. 213–245. Springer, Heidelberg.

Buwalda JG (1980). Growth of a clover–ryegrass association with vesicular arbuscular mycorrhizas. *New Zealand Journal of Agricultural Research* **23**: 379–383.

Cahill JF Jr (2002). Interactions between root and shoot competition vary among species. *Oikos* **99**: 101–112.

Cahill JF Jr, Kembel SW, and Gustafson DJ (2005). Differential genetic influences on competitive effect and response in *Arabidopsis thaliana*. *Journal of Ecology* **93**: 958–967.

Cain SA and Castro GM (1959). *Manual of vegetation analysis*. Harper and Brothers, New York.

Cain ML, Subler S, Evans JP, and Fortin MJ (1999). Sampling spatial and temporal variation in soil nitrogen availability. *Oecologia* **118**: 397–404.

Caldwell MM (ed.) (1971). Solar UV irradiation and the growth and development of higher plants. In AC Geise (ed.) *Phytophysiology*, Vol. 6, pp 131–177. Academic Press, New York.

Caldwell MM and Flint SD (1997). Uses of biological spectral weighting functions and the need of scaling for the ozone reduction problem. *Plant Ecology* **128**: 66–76.

Caldwell MM and Virginia RA (1989). Root systems. In RW Pearcy, J Ehleringer, HA Mooney, et al. (eds) *Plant physiological ecology*, pp. 367–398. Chapman and Hall, London.

Callaway RM (2003). Experimental designs for the study of allelopathy. *Plant and Soil* **256**: 1–11.

Campbell JE and Gibson DJ (2001). The effect of seeds of exotic species transported via horse dung on vegetation along trail corridors. *Plant Ecology* **157**: 23–35.

Campbell BD, Grime JP, and Mackey JML (1993). Foraging for light. In GAF Hendry and JP Grime (eds) *Methods in comparative plant ecology*, pp. 83–87. Chapman and Hall, London.

Capers RS and Chazdon RL (2004). Rapid assessment of understory light availability in a wet tropical forest. *Agricultural and Forest Meteorology* **123**: 177–185.

Carey PD and Watkinson AR (1993). The dispersal and fates of seeds of the winter annual grass *Vulpia ciliata*. *Journal of Ecology* **81**: 759–767.

Carlton GC and Bazzaz FA (1998). Regeneration of three sympatric birch species on experimental hurricane blowdown microsites. *Ecological Monographs* **68**: 99–120.

Carpenter SR, Frost TM, Heisey D, and Kratz TK (1989). Randomized intervention analysis and the interpretation of whole-ecosystem experiments. *Ecology* **70**: 1142–1152.

Carter MR and Gregorich EG (eds) (2006). *Soil sampling and methods of analysis*, 2nd edn. CRC Press, Boca Raton, FL.

Castillo JM, Fernández-Baco L, Castellanos EM, Luque CJ, Figueroa ME, and Davy AJ (2000). Lower limits of *Spartina densiflora* and *S. maritima* in a Mediterranean salt marsh determined by different ecophysiological tolerances. *Journal of Ecology* **88**: 801–812.

Caswell H (2001). *Matrix population models: construction, analysis, and interpretation*, 2nd edn. Sinauer Associates, Sunderland, MA.

Caswell H (2010). Life table response experiment analysis of the stochastic growth rate. *Journal of Ecology* **98**: 324–333.

Catalán P, Müller J, Hasterok R, Jenkins G, Mur LAJ, Langdon T, et al. (2012). Evolution and taxonomic split of the model grass *Brachypodium distachyon*. *Annals of Botany* **109**: 385–405.

Champely S (2009). *Package 'pwr'*. Available at: <http://cran.r-project.org/web/packages/pwr/pwr.pdf>.

Chandler C, Cheney P, Thomas P, Trabaud L, and Williams D (1983). *Fire in forestry. Vol. II. Forest fire management and organization*. John Wiley and Sons, New York.

Chanway CP, Holl FB, and Turkington R (1989). Effect of *Rhizobium leguminosarum* biovar *trifolii* genotype on specificity between *Trifolium repens* and *Lolium perenne*. *Journal of Ecology* **77**: 1150–1160.

Chanway CP, Turkington R, and Holl FB (1991). Ecological implications of specificity between plants and rhizosphere micro-organisms. *Advances in Ecological Research* **21**: 121–169.

Chapin FS III (1980). Mineral nutrition of wild plants. *Annual Review of Ecology and Systematics* **11**: 233–260.

Chapin FS III and Van Cleve K (1989). Approaches to studying nutrient uptake, use and loss in plants. In RW Pearcy, J Ehleringer, HA Mooney, *et al.* (eds) *Plant physiological ecology*, pp. 185–207. Chapman and Hall, London.

Chapin FS III, Shaver GR, Kedrowski RA (1986). Environmental controls over carbon, nitrogen and phosphorus fractions in *Eriophorum vaginatum* Alaskan tussock tundra. *Journal of Ecology* **74**: 167–195.

Chapin FS III, Autumn K, Pugnaire F (1993). Evolutionary responses to environmental stress. *American Naturalist* **142**(Suppl.): S78–S92.

Chapman SB (1986). Production ecology and nutrient budgets. In PD Moore and SB Chapman (eds) *Methods in plant ecology*, pp. 1–59. Blackwell Scientific Publications, Oxford.

Charron D and Gagnon D (1991). The demography of northern populations of *Panax quinquefolium* (American ginseng). *Journal of Ecology* **79**: 431–436.

Chazon RL and Field CB (1987). Photographic estimation of photosynthetically active radiation: evaluation of a computerized technique. *Oecologia* **73**: 525–532.

Chen M-H and Deely JJ (1996). Bayesian analysis for a constrained linear multiple regression problem for predicting the new crop of apples. *Journal of Agricultural, Biological, and Environmental Statistics* **1**: 467–489.

Chen XM, Alm DM, and Hesketh JD (1995). Effects of atmospheric CO_2 concentration on photosynthetic performance of C_3 and C_4 plants. *Biotronics* **24**: 65–72.

Cheng HH (1995). Characterization of the mechanisms of allelopathy. In Inderjit, KMM Dakshini, and FA Einhellig (eds) *Allelopathy: organisms, processes, and applications*, pp. 132–141. American Chemical Society, Washington, DC.

Cheplick GP (1992). Sibling competition in plants. *Journal of Ecology* **80**: 567–575.

Cheplick GP (1997). Effects of endophytic fungi on the phenotypic plasticity of *Lolium perenne* (Poaceae). *American Journal of Botany* **84**: 34–40.

Cheplick GP (ed.) (1998). *Population biology of grasses*. Cambridge University Press, Cambridge.

Cheplick GP and Quinn JE (1988). Subterranean seed production and population responses to fire in *Amphicarpon purshii* (Gramineae). *Journal of Ecology* **76**: 263–273.

Christianini AV and Oliveira PS (2010). Birds and ants provide complimentary seed dispersal in a Neotropical savanna. *Journal of Ecology* **98**: 573–582.

Cipriotti PA, Aguiar MR, Wiegand T, and Paruelo JM (2014). A complex network of interactions controls coexistence and relative abundance in Patagonian grass-shrub steppes. *Journal of Ecology* **102**: 776–788.

Clark JS (2005). Why environmental scientists are becoming Bayesians. *Ecology Letters* **8**: 2–14.

Clark JS and Lavine M (2001). Bayesian statistics: estimating plant demographic parameters. In SM Scheiner and J Gurevitch (eds) *Design and analysis of ecological experiments*, pp. 327–346. Oxford University Press, New York.

Clark CJ, Poulsen JR, Levey DJ, and Osenberg CW (2007). Are plant populations seed limited? A critique and meta-analysis of seed addition experiments. *American Naturalist* **170**: 128–142.

Clark RT, Famoso AN, Zhao K, Shaff JE, Craft EJ, Bustamante CD, et al. (2013). High-throughput two-dimensional root system phenotyping platform facilitates genetic analysis of root growth and development. *Plant, Cell and Environment* **36**: 454–466.

Clausen J, Keck DD, and Heisey WM (1948). *Experimental studies on the nature of species III. Environmental responses of climatic races of* Achillea. Carnegie Institution of Washington Publication 581. Carnegie Institute, Washington, DC.

Clay K (1990). Comparative demography of three graminoids infected by systemic, clavicipitaceous fungi. *Ecology* **71**: 558–570.

Clay K (1998). Fungal endophyte infection and the population dynamics of grasses. In GP Cheplick (ed.) *Population biology of grasses*, pp. 255–285. Cambridge University Press, Cambridge.

Clay K and Levin DA (1986). Environment-dependent intraspecific competition in *Phlox drummondii*. *Ecology* **67**: 37–45.

Clements FE and Goldsmith GW (1924). *The phytometer method in ecology; the plant and community as instruments*. Carnegie Institution of Washington, Washington, DC.

Clements FE, Weaver JE, and Hanson HC (1929). *Competition in cultivated crops*. Carnegie Institute of Washington Publication 398, pp. 202–233. Carnegie Institution of Washington, Washington, DC.

Cliquet JB, Murray PJ, and Boucaud J (1997). Effect of the arbuscular mycorrhizal fungus *Glomus fasciculatum* on the uptake of amino nitrogen by *Lolium perenne*. *New Phytologist* **137**: 345–349.

Cohen J (1988). *Statistical power analysis for the behavioral sciences*, 2nd edn. Academic Press, New York.

Coker K (2011). *Vegetation description and analysis: a practical approach*, 2nd edn. Wiley-Blackwell, Chichester.

Collins A, Hart EM, and Molofsky J (2010). Differential response to frequency-dependent interactions: an experimental test using genotypes of an invasive grass. *Oecologia* **164**: 959–969.

Conlisk E, Syphard AD, Franklin J, Flint L, Flint A, and Regan H (2013). Uncertainty in assessing the impacts of global change with coupled dynamic species distribution and population models. *Global Change Biology* **19**: 858–869.

Connell JH (1971). On the role of natural enemies in preventing competitive exclusion in some marine animals and in rain forest trees. In PJ dem Boer and GR Gradwell (eds) *Dynamics of populations*, pp. 298–312. Centre for Agricultural Publishing and Documentation, Wageningen.

Connell JH (1983). On the prevalence and relative importance of interspecific competition: evidence from field experiments. *American Naturalist* **122**: 661–696.

Conner JK and Zangori LA (1997). A garden study of the effects of ultraviolet-B radiation on pollination success and lifetime fitness in *Brassica*. *Oecologia* **111**: 388–395.

Connolly J (1986). On difficulties with replacement-series methodology in mixture experiments. *Journal of Applied Ecology* **23**: 125–137.

Connolly J (1987). On the use of response models in mixture experiments. *Oecologia* **72**: 95–103.

Connolly J and Wayne P (1996). Asymmetric competition between plant species. *Oecologia* **108**: 311–320.

Connolly J, Wayne P, and Murray R (1990). Time course of plant–plant interactions in experimental mixtures of annuals: density, frequency, and nutrient effects. *Oecologia* **82**: 513–526.

Connolly J, Wayne P, and Bazzaz FA (2001). Interspecific competition in plants: how well do current methods answer fundamental questions? *American Naturalist* **157**: 107–125.

Conover WJ (1999). *Practical nonparametric statistics*, 3rd edn. John Wiley and Sons, New York.

Converse RH and Martin RR (1990). ELISA methods for plant viruses. In R Hampton, E Ball, and S DeBoer (eds) *Serological methods for detection and identification of viral and bacterial plant pathogens*, pp. 179–196. American Phytopathological Society, St Paul, MN.

Coomes DA and Grubb PJ (2000). Impacts of root competition in forests and woodlands: a theoretical framework and review of experiments. *Ecological Monographs* **70**: 171–207.

Coq S, Weigel J, Bonal D, and Hättenschwiler S (2012). Litter mixture effects on tropical tree seedling growth—a greenhouse experiment. *Plant Biology* **14**: 630–640.

Corbit M, Marks PL, and Gardescu S (1999). Hedgerows as habitat corridors for forest herbs in central New York, USA. *Journal of Ecology* **87**: 220–232.

Cornell JA (1990). *Experiments with mixtures: designs, models, and the analysis of mixture data*, 2nd edn. John Wiley and Sons, New York.

Coulson T (2012). Integral projections models, their construction and use in posing hypotheses in ecology. *Oikos* **121**: 1337–1350.

Coulson T, Mace GM, Hudson E, and Possingham HP (2001). The use and abuse of population viability analysis. *Trends in Ecology and Evolution* **16**: 219–221.

Coupe MD, Stacey JN, and Cahill JF (2009). Limited effects of above and belowground insects on community structure and function in a species-rich grassland. *Journal of Vegetation Science* **20**: 121–129.

Cousens R (1996). Design and interpretation of interference studies: are some methods totally unacceptable? *New Zealand Journal of Forestry Science* **26**: 5–18.

Cox GW (1990). *Laboratory manual of general ecology*, 6th edn. Wm C Brown, Dubuque, IA.

Crawley MJ (1985). Reduction of oak fecundity by low density herbivore populations. *Nature* **314**: 163–164.

Crawley MJ (1988). Herbivores and plant population dynamics. In AJ Davy, MJ Hutchings, and AR Watkinson (eds) *Plant population ecology*, pp. 367–392. Blackwell Scientific Publications, Oxford.

Crawley MJ (1997a). Life history and environment. In MJ Crawley (ed.) *Plant ecology*, pp. 73–131. Blackwell Scientific Publications, Oxford.

Crawley MJ (1997b). Plant–herbivore dynamics. In MJ Crawley (ed.) *Plant ecology*, pp. 401–474. Blackwell Scientific Publications, Oxford.

Crawley MJ (2012). *The R book*, 2nd edn. Wiley, Oxford.

Crone EE, Ellis MM, Morris WF, Stanley A, Bell T, Bierzychudek P, et al. (2013). Ability of matrix models to explain the past and predict the future of plant populations. *Conservation Biology* **27**: 968–978.

Crowder MJ and Hand DJ (1990). *Analysis of repeated measures*. Chapman and Hall, London.

Croy CD and Dix RL (1984). Note on sample size requirements in morphological plant ecology. *Ecology* **65**: 662–666.

Cruzan MB (1998). Genetic markers in plant evolutionary ecology. *Ecology* **79**: 400–412.

Csilléry K, François O, and Blum MGB (2012). abc: an R package for approximate Bayesian computation (ABC). *Methods in Ecology and Evolution* **3**: 475–479.

Cunningham SA (1997). The effect of light environment, leaf area, and stored carbohydrates on inflorescence production by a rain forest understory palm. *Oecologia* **111**: 36–44.

Czárán T and Bartha S (1992). Spatiotemporal dynamic models of plant populations and communities. *Trends in Ecology and Evolution* **7**: 38–42.

Dafni A (1993). *Pollination ecology: a practical approach*. Oxford University Press, New York.

D'Agostino RB, Belanger A, and D'Agostino B (1990). A suggestion for using powerful and informative tests of normality. *American Statistician* **44**: 316–321.

Dahlman RC (1993). CO_2 and plants: revisited. *Vegetatio* **104/105**: 339–355.

Dai T and Wiegert RG (1996). Ramet population dynamics and net aerial primary productivity of *Spartina alterniflora*. *Ecology* **77**: 276–288.

Dale MRT (1999). *Spatial pattern analysis in plant ecology*. Cambridge University Press, Cambridge.

Dale H and Press MC (1998). Elevated atmospheric CO_2 influences the interaction between the parasitic angiosperm *Orobanche minor* and its host *Trifolium repens*. *New Phytologist* **140**: 65–73.

Dalgleish HJ, Koons DN, and Adler PB (2010). Can life-history traits predict the response of forb populations to changes in climate variability? *Journal of Ecology* **98**: 209–217.

Dalton FN (1995). In-situ root extent measurements by electrical capacitance methods. *Plant and Soil* **173**: 157–165.

Damgaard C (1998). Plant competition experiments: testing hypotheses and estimating the probability of coexistence. *Ecology* **79**: 1760–1767.

Dane JH and Topp GC (eds) (2002). *Methods of soil analysis. Part 4, Physical methods*. Soil Science Society of America, Madison, WI.

Danielson HR and Toole VK (1976). Action of temperature and light on the control of seed germination in Alta tall fescue (*Festuca arundinacea* Schreb.). *Crop Science* **16**: 320–347.

D'Antonio CM (1993). Mechanisms controlling invasion of coastal plant communities by the alien succulent *Carpobrotus edulis*. *Ecology* **74**: 83–95.

D'Arcy CJ and Burnett PA (1995). *Barley yellow dwarf: 40 years of progress*. American Phytopathological Society Press, St Paul, MN.

Darwin C (1855a). Vitality of seeds. In PH Barrett (ed.) *The collected papers of Charles Darwin, volume 1*, pp. 260–261. University of Chicago Press, Chicago, IL.

Darwin C (1855b). Effect of salt-water on the germination of seeds. In PH Barrett (ed.) *The collected papers of Charles Darwin, volume 1*, pp. 261–262. University of Chicago Press, Chicago, IL.

Darwin C (1859). *The origin of species by means of natural selection*. Murray, London [reprint 1910, Ward, Lock & Co. Ltd, London].

Davey JW, Cezard T, Fuentes-Utrilla P, Eland C, Gharbi K, and Blaxter ML (2013). Special features of RAD sequencing data: implications for genotyping. *Molecular Ecology* **22**: 3151–3164.

Davies SJ (1998). Photosynthesis of nine pioneer *Macaranga* species from Borneo in relation to life history. *Ecology* **79**: 2292–2308.

Davy AJ, Hutchings MJ, and Watkinson AR (eds) (1988). *Plant population ecology*. Blackwell Scientific Publications, Oxford.

Day RW and Quinn GP (1989). Comparisons of treatments after an analysis of variance in ecology. *Ecological Monographs* **59**: 433–463.

Dayton PK (1973). Two cases of resource partitioning in an intertidal community: making the right prediction for the wrong reason. *American Naturalist* **107**: 662–670.

Dean TA, Stekoll MS, Jewett SC, Smith-Richard O, and Hose JE (1998). Eelgrass (*Zostera marina* L.) in Prince William Sound, Alaska: effects of the Exxon Valdez oil spill. *Marine Pollution Bulletin* **36**: 201–210.

De Boeck HJ, De Groote T, and Nijs I (2012). Leaf temperatures in glasshouses and open-top chambers. *New Phytologist* **194**: 1155–1164.

Derner JD and Briske DD (1998). An isotopic (^{15}N) assessment of intraclonal regulation in C4 perennial grasses: ramet interdependence, independence or both? *Journal of Ecology* **86**: 305–315.

Derner JD, Briske DD, and Polley HW (2012). Tiller organization within the tussock grass *Schizachyrium scoparium*: a field assessment of competition–cooperation tradeoffs *Botany* **90**: 669–677.

Detling JK, Dyer MI, Procter-Gregg C, and Winn DT (1980). Plant–herbivore interactions: examination of potential effects of bison saliva on regrowth of *Bouteloua gracilis* (H.B.K.) Lag. *Oecologia* **45**: 26–31.

Deutschman DH, Levin SA, Devine C, and Buttel LA (1997). Scaling from trees to forests: analysis of a complex simulation model. *Science* **277**: 1688.

Dhingra OD and Sinclair JB (1995). *Basic plant pathology methods*, 2nd edn. CRC Press, Boca Raton, FL.

Diamond JM (1983). Laboratory, field and natural experiments. *Nature* **304**: 586–587.

Diamond JM (1986). Overview: laboratory experiments, field experiments, and natural experiments. In J Diamond and TJ Case (eds) *Community ecology*, pp. 3–22. Harper and Row, New York.

Dietrich RC, Bengough AG, Jones HG, and White PJ (2013). Can root electrical capacitance be used to predict root mass in soil? *Annals of Botany* **112**: 457–464.

Dietz H, Fisher M, and Schmid B (1999). Demographic and genetic invasion history of a 9-year-old roadside population of *Bunias orientalis* L. (Brassicaceae). *Oecologia* **120**: 225–234.

Diez J, Giladi I, Warren R, and Pulliam R (2014). Probabilistic and spatially variable niches inferred from demography. *Journal of Ecology* **102**: 544–554.

Diggle PJ (1983). *Statistical analysis of spatial point patterns*. Academic Press, London.

Dillenburg LR, Teramure AH, Forseth IN, and Whigham DF (1995). Photosynthetic and biomass allocation responses of *Liquidambar styraciflua* (Hamamelidaceae) to vine competition. *American Journal of Botany* **82**: 454–461.

Dirzo R and Sarukhán J (eds) (1984). *Perspectives in plant population ecology*. Sinauer Associates, Sunderland, MA.

Dixon P (1994). Testing spatial segregation using a nearest-neighbor contingency table. *Ecology* **75**: 1940–1948.

Dixon PM (2006). Nearest neighbor methods. *Encyclopedia of environmetrics*. doi: 10.1002/9780470057339.van007

Dixon P (2009). VEGAN, a package of R functions for community ecology. *Journal of Vegetation Science* **14**: 927–930.

Dixon P and Ellison AM (1996). Introduction: ecological applications of Bayesian inference. *Ecological Applications* **6**: 1034–1035.

Dixon AL, Herlihy CR, and Busch JW (2013). Demographic and population-genetic tests provide mixed support for the abundant centre hypothesis in the endemic plant *Leavenworthia stylosa*. *Molecular Ecology* **22**: 1777–1791.

Doak DF (1992). Lifetime impacts of herbivory for a perennial plant. *Ecology* **73**: 2086–2099.

Dodd M, Silvertown J, McConway K, Potts J, and Crawley M (1995). Community stability: a sixty-year record of trends and outbreaks in the occurrence of species in the Park Grass Experiment. *Journal of Ecology* **83**: 277–286.

Dolan RW (1994). Patterns of isozyme variation in relation to population size, isolation, and phytogeographic history in royal catchfly (*Silene regia*: Caryophyllaceae). *American Journal of Botany* **81**: 965–972.

Dombrowski JE, Baldwin JC, Azevedo MD, and Banowetz GM (2006). A sensitive PCR-based assay to detect *Neotyphodium* fungi in seed and plant tissue of tall fescue and ryegrass species. *Crop Science* **46**: 1064–1070.

Donnelly ED and Patterson RM (1969). Effect of irrigation and clipping on seed production and chasmogamy of *Sericea* genotypes. *Agronomy Journal* **61**: 501–502.

Donohue K (1998). Maternal determinants of seed dispersal in *Cakile edentula*: fruit, plant, and site traits. Ecology **79**: 2771–2788.

Dormann CF, Gruber B, and Fründ J (2008). Introducing the bipartite package: analysing ecological networks. *R News* **8**: 8–11.

Dougherty RL, Lauenroth WK, and Singh JS (1996). Response of a grassland cactus to frequency and size of rainfall events in a North American shortgrass steppe. *Journal of Ecology* **84**: 177–183.

Doust LL (1987). Population dynamics and local specialization in a clonal perennial *(Ranunculus repens)*. III. Responses to light and nutrient supply. *Journal of Ecology* **75**: 555–568.

Draper J, Mur LAJ, Jenkins G, Ghosh-Biswas GC, Bablak P, Hasterok R, et al. (2001). *Brachypodium distachyon*. A new model system for functional genomics in grasses. *Plant Physiology* **127**: 1539–1555.

Dubreuil P and Charcosset A (1998). Genetic diversity within and among maize populations: a comparison between isozyme and nuclear RFLP loci. *Theoretical and Applied Genetics* **96**: 577–587.

Dudley SA, Murphy GP, and File AL (2013). Kin recognition and competition in plants. *Functional Ecology* **27**: 898–906.

Durrett R and Levin SA (1994). Stochastic spatial models: a user's guide to ecological applications. *Philosophical Transactions of the Royal Society B: Biological Sciences* **343**: 329–350.

Dytham C (2011). *Choosing and using statistics: a biologist's guide*, 3rd edn. Wiley-Blackwell, Oxford.

Easterling MR, Ellner SP, and Dixon PM (2000). Size-specific sensitivity: applying a new structured population model. *Ecology* **81**: 694–708.

Ebert TA (1999). *Plant and animal populations: methods in demography*. Academic Press, New York.

Edwards AWF (1992). *Likelihood*. Johns Hopkins University Press, Baltimore, MD.

Egan AN, Schlueter J, and Spooner DM (2012). Applications of next-generation sequencing in plant biology. *American Journal of Botany* **99**: 175–185.

Egerton-Warburton LM (1995). An absence of ecotype evolution in three *Eucalyptus* species colonizing coal mine spoils with low pH and high aluminium content. *Water, Air and Soil Pollution* **83**: 335–349.

Ehleringer JR (1989). Temperature and energy budgets. In RW Pearcy, JR Ehleringer, HA Mooney, et al. (eds) *Plant physiological ecology*, pp. 117–136. Chapman and Hall, London.

Ehleringer JR and Field CB (eds) (1993). *Scaling physiological processes*. Academic Press, San Diego, CA.

Ekbolm R and Galindo J (2011). Applications of next generation sequencing in molecular ecology of non-model organisms. *Heredity* **107**: 1–15.

Ellis PD (2010). *The essential guide to effect sizes: statistical power, meta-analysis, and the interpretation of research results*. Cambridge University Press, Cambridge.

Ellis TW, Murray W, Paul K, Kavalieris L, Brophy J, Williams C, et al. (2013). Electrical capacitance as a rapid and non-invasive indicator of root length. *Tree Physiology* **33**: 3–17.

Ellison AM (1996). An introduction to Bayesian inference for ecological research and environmental decision making. *Ecological Applications* **6**: 1036–1046.

Ellison AM (2001). Exploratory data analysis and graphic display. In SM Scheiner and J Gurevitch (eds) *Design and analysis of ecological experiments*, 2nd edn, pp. 14–45. Chapman and Hall, New York.

Ellison AM and Farnsworth EJ (1996). Spatial and temporal variability in growth of *Rhizphora mangle* saplings on coral cays: links with variation in isolation, herbivory, and local sedimentation rate. *Journal of Ecology* **84**: 717–731.

Ellner SP and Rees M (2006). Integral projection models for species with complex demography. *American Naturalist* **167**: 410–428.

Elshire RJ, Glaubitz JC, Sun Q, Poland JA, Kawamoto K, Buckler ES, et al. (2011). A robust, simple genotyping-by-sequencing (GBS) approach for high diversity species. *PLoS ONE* **6**: e19379.

Elton CS (1966). *The pattern of animal communities*. Methuen and Co. Ltd, London.

von Ende CN (2001). Repeated measures analysis: growth and time dependent variables. In SM Scheiner and J Gurevitch (eds) *Design and analysis of ecological experiments*, pp. 134–157. Oxford University Press, New York.

Eriksson O (1998) Ramet behaviour and population growth in the clonal herb *Potentilla anserina*. *Journal of Ecology* **76**: 522–536.

Eriksson O and Jakobsson A (1998). Abundance, distribution and life histories of grassland plants: a comparative study of 81 species. *Journal of Ecology* **86**: 922–933.

Escudero A, Smolonos RC, Olano JM, and Rubio A (1999). Factors controlling the establishment of *Helianthemum squamatum*, an endemic gypsophile of semi-arid Spain. *Journal of Ecology* **87**: 290–302.

Esselman EJ, Jianqiang L, Crawford DJ, Windus, and Wolfe AD (1999). Clonal diversity in the rare *Calamagrostis porteri* spp. *insperata* (Poaceae): comparative results for allozymes and random amplified polymorphic DNA (RAPD) and intersimple sequence repeat (ISSR) markers. *Molecular Ecology* **8**: 443–451.

Ettema CH and Wardle DA (2002). Spatial soil ecology. *Trends in Ecology and Evolution* **17**: 177–183.

Evans GC (1972). *The quantitative analysis of plant growth*. University of California Press, Berkeley, CA.

Evans EW (1991). Experimental manipulation of herbivores in North American native tallgrass prairie: responses of aboveground arthropods. *American Midland Naturalist* **125**: 37–46.

Evans MEK, Holsinger KE, and Menges ES (2010). Fire, vital rates, and population viability: a hierarchical Bayesian analysis of the endangered Florida scrub mint. *Ecological Monographs* **80**: 627–649.

Fahy PC and Persely GJ (1983). *Plant bacterial diseases: a diagnostic guide*. Academic Press, New York.

Farrar JF (1999). Acquisition, partitioning and loss of carbon. In MC Press, JD Scholes, and MG Barker (eds) *Physiological plant ecology*, pp. 25–43. Blackwell Science, Oxford.

Farris MA Lechowicz MJ (1990). Functional interactions among traits that determine reproductive success in a native annual plant. *Ecology* **71**: 548–557.

Fay PA, Carlisle JD, Knapp AK, Blair JM, and Collins SL (2000). Altering rainfall timing and quantity in a mesic grassland ecosystem: design and performance of rainfall manipulation shelters. *Ecosystems* **3**: 308–309.

Feibleman JK (1972). *Scientific method*. Martinus Nijhoff, The Hague.

Feldman R, Tomback DF, and Koehler J (1999). Cost of mutualism: competition, tree morphology and pollen production in limber pine clusters. *Ecology* **80**: 324–329.

Felenstein J (1985). Phylogenies and the comparative method. *American Naturalist* **125**: 1–15.

Fenner M (1985). *Seed ecology*. Chapman and Hall, London.

Fenner M and Kitajima K (1999). Seed and seedling ecology. In FI Pugnaire and F Valladares (eds) *Handbook of functional plant ecology*, pp. 589–621. Marcel Dekker, Inc., New York.

Fernandes M, Saxena J, and Dick R (2013). Comparison of whole-cell fatty acid (MIDI) or phospholipid fatty acid (PLFA) extractants as biomarkers to profile soil microbial communities. *Microbial Ecology* **66**: 145–157.

Ferrière R, Sarrazin F, Legendre S, and Baron J-P (1996). Matrix population models applied to viability analysis and conservation: theory and practice with ULM software. *Acta Oecologica* **17**: 629–656.

Ferris RS (1984). Effects of microwave oven treatment on microorganisms in soil. *Phytopathology* **74**: 121–126.

Fetcher N, Haines BL, Cordero RA, Lodge DJ, Walker LR, Fernández DS, et al. (1996). Responses of tropical plants to nutrients and light on a landslide in Puerto Rico. *Journal of Ecology* **84**: 331–341.

Field CB, Ball TJ, and Berry JA (1989). Photosynthesis: principles and field techniques. In RW Pearcy, J Ehleringer, HA Mooney, et al. (eds) *Plant physiological ecology*, pp. 209–253. Chapman and Hall, London.

File AL, Murphy GP, and Dudley SA (2012). Fitness consequences of plants growing with siblings: reconciling kin selection, niche partitioning and competitive ability. *Proceedings of the Royal Society B: Biological Sciences* **279**: 209–218.

Finzel JA, Seyfried MS, Weltz MA, Kiniry JR, Johnson M-VV, and Launchbaugh KL (2012). Indirect measurement of leaf area index in sagebrush-steppe rangelands. *Rangeland Ecology and Management* **65**: 208–212.

Firbank LG, Watkinson AR, Norton LR, and Ashenden TW (1995). Plant populations and global environmental change: the effects of different temperature, carbon dioxide and nutrient regimes on density dependence in populations of *Vulpia ciliata*. *Functional Ecology* **9**: 432–441.

Fischer M and Matthies D (1998). Effects of population size on performance in the rare plant *Gentianella germanica*. *Journal of Ecology* **86**: 195–204.

Fischer M, Matthies D, and Schmid B (1997). Responses of rare calcareous grassland plants to elevated CO_2: a field experiment with *Gentianella germanica* and *Gentiana cruciata*. *Journal of Ecology* **85**: 681–691.

Fitter AH (1982). Influence of soil heterogeneity on the coexistence of grassland species. *Journal of Ecology* **70**: 139–148.

Fitter AH and Peat HJ (1994). The ecological flora database. *Journal of Ecology* **82**: 415–425.

Flint SD, Ryel RJ, and Caldwell MM (2003). Ecosystem UV-B experiments in terrestrial communities: a review of recent findings and methodologies. *Agricultural and Forest Meteorology* **120**: 177–189.

Flint SD, Ryel RJ, Hudelson TJ, and Caldwell MM (2009). Serious complications in experiments in which UV doses are effected by using different lamp heights. *Journal of Photochemistry and Photobiology B: Biology* **97**: 48–53.

Food and Agriculture Organization of the United Nations (1993). *Technical handbook on symbiotic nitrogen fixation: legume/Rhizobium*, 2nd edn. Food and Agriculture Organization of the United Nations, Rome.

Ford ED (2000). *Scientific method for ecological research*. Cambridge University Press, Cambridge.

Foster BL and Gross KL (1998). Species richness in a successional grassland: effects of nitrogen enrichment and plant litter. *Ecology* **79**: 2593–2602.

Fowler N (1982). Competition and coexistence in a North Carolina grassland. III. Mixtures of component species. *Journal of Ecology* **70**: 77–92.

Fowler N (1990). The 10 most common statistical errors. *Bulletin of the Ecological Society of America* **71**: 161–164.

Fowler N (1995). Density-dependent demography in two grasses: a five-year study. *Ecology* **76**: 2145–2164.

Fowler D, Cape JN, Deans JD, Leith ID, Murray MB, Smith RI, et al. (1989). Effects of acid mist on the frost hardiness of red spruce seedlings. *New Phytologist* **113**: 321–335.

Fowler J, Cohen L, and Jarvis P (1998). *Practical statistics for field biology*, 2nd edn. Wiley, New York.

Fox GA (2001). Failure-time analysis: emergence, flowering, survivorship, and other waiting times. In SM Scheiner and J Gurevitch (eds) *Design and analysis of ecological experiments*, 2nd edn, pp. 235–266. Oxford University Press, New York.

Fox J (2006). Structural equation modeling with the sem package in R. *Structural Equation Modeling* **13**: 465–486.

Franco M (1986). The influence of neighbours on the growth of modular organisms with an example from trees. *Philosophical Transactions of the Royal Society of London Series B, Biological Sciences* **313**: 209–226.

Franco M and Harper JL (1988). Competition and the formation of spatial pattern in spacing gradients: an example using *Kochia scoparia*. *Journal of Ecology* **76**: 959–974.

Franco M and Silvertown J (1997). Life history variation in plants: an exploration of the fast-slow continuum hypothesis. In J Silvertown, M Franco, and JL Harper (eds) *Plant life histories: ecology, phylogeny and evolution*, pp. 210–227. Cambridge University Press, Cambridge.

Frank K, Lorek H, Köster F, Sonnenschein M, Wissel C, and Grimm V (2003). *META-X—software for metapopulation viability analysis*. Springer-Verlag, Berlin.

Franklin J (2010). Spatial point pattern analysis of plants. In L Anselin and SJ Rey (eds) *Advances in spatial science*. Springer-Verlag, Berlin.

Franklin SB, Gibson DJ, Robertson PA, Pohlmann JT, and Fralish JS (1995). Parallel analysis: a method for determining significant principal components. *Journal of Vegetation Science* **6**: 99–106.

Fraser LH and Grime JP (1999). Interacting effects of herbivory and fertility on a synthesised plant community. *Journal of Ecology* **87**: 514–525.

Freckleton RP (2000). Phylogenetic tests of ecological and evolutionary hypotheses: checking for phylogenetic independence. *Functional Ecology* **14**: 129–134.

Freckleton RP and Watkinson AR (1998). Predicting the determinants of weed abundance: a model for the population dynamics of *Chenopodium album* in sugar beet. *Journal of Applied Ecology* **35**: 904–920.

Freckleton RP and Watkinson AR (2000). Designs for greenhouse studies of interactions between plants: an analytical perspective. *Journal of Ecology* **88**: 386–391.

Freeland JR, Kirk H, and Petersen S (2011). *Molecular ecology*, 2nd edn. Wiley-Blackwell, Oxford.

Freeman DC, McArthur ED, Sanderson SC, and Tiedemann AR (1993). The influence of topography on male and female fitness components of *Atriplex canescens*. *Oecologia* **93**: 538–547.

Frenne P, Schrijver A, Graae B, Gruwez R, Tack W, Vandelook F, et al. (2010). The use of open-top chambers in forests for evaluating warming effects on herbaceous understorey plants. *Ecological Research* **25**: 163–171.

Frostegård Å, Tunlid A, and Bååth E (2011). Use and misuse of PLFA measurements in soils. *Soil Biology and Biochemistry* **43**: 1621–1625.

Gallagher RV and Leishman MR (2012). A global analysis of trait variation and evolution in climbing plants. *Journal of Biogeography* **39**: 1757–1771.

Gange AC, Brown VK, Evans IM, and Storr AL (1989). Variation in the impact of insect herbivory on *Trifolium pratense* through early plant succession. *Journal of Ecology* **77**: 537–551.

Garcia D, Rodriguez J, Sanz JM, and Merino J (1998). Response of two populations of holm oak (*Quercus rotundifolia* Lam.) to sulfur dioxide. *Ecotoxicology and Environmental Safety* **40**: 42–48.

Gardener M (2011). *Statistics for ecologists using R and Excel: data collection, exploration, analysis and presentation*. Pelagic Publishing, Exeter.

Garnier E, Gobin O, and Poorter H (1995). Nitrogen productivity depends on photosynthetic nitrogen use efficiency and on nitrogen allocation within the plant. *Annals of Botany* **76**: 667–672.

Gartside DW and McNeilly T (1974). The potential for evolution of heavy metal tolerance in plants. II. Copper tolerance in normal populations of different plant species. *Heredity* **32**: 335–348.

Gates DM (1980). *Biophysical ecology*. Springer-Verlag, New York.

Gatsuk LE, Smirnova OV, Vorontzova LI, Zaugolnova LB, and Zhukova LA (1980). Age states of plants of various growth forms: a review. *Journal of Ecology* **68**: 675–696.

Gautier H, Varlet-Grancher C, and Baudry N (1998). Comparison of horizontal spread of white clover (*Trifolium repens* L.) grown under two artificial light sources differing in their content of blue light. *Annals of Botany* **82**: 41–48.

Geisler M, Gibson DJ, Lindsey K, Millar K, and Wood AJ (2012). Upregulation of photosynthesis genes, and downregulation of stress defense genes, is the response of *Arabidopsis thaliana* shoots to intraspecific competition. *Botanical Studies* **53**: 85–96.

Gelfand AE, Ghosh S, and Clark JS (2013). Scaling integral projection models for analyzing size demography. *Statistical Science* **28**: 641–658.

Gente R, Born N, Voβ N, Sannemann W, Léon J, Koch M, et al. (2013). Determination of leaf water content from terahertz time-domain spectroscopic data. *Journal of Infrared, Millimeter, and Terahertz Waves* **34**: 316–323.

Gerdol R, Brancaleoni L, Menghini M, and Marchsini R (2000). Response of dwarf shrubs to neighbour removal and nutrient addition and their influence on community structure in a subalpine heath. *Journal of Ecology* **88**: 256–266.

Gerrodette T (1987). A power analysis for detecting trends. *Ecology* **68**: 1364–1372.

Gersani M, Brown JS, O'Brien EE, Maina GM, and Abramsky Z (2001). Tragedy of the commons as a result of root competition. *Journal of Ecology* **89**: 660–669.

Gherardi LA and Sala OE (2013). Automated rainfall manipulation system: a reliable and inexpensive tool for ecologists. *Ecosphere* **4**: art18.

Ghosh M and Meeden G (1997). *Bayesian methods for finite population sampling*. Chapman and Hall, London.

Gibson DJ (1986). Spatial and temporal heterogeneity in soil nutrient supply measured using in-situ ion-exchange resin bags. *Plant and Soil* **96**: 445–450.

Gibson DJ (1988). The maintenance of plant and soil heterogeneity in dune grassland. *Journal of Ecology* **76**: 497–508.

Gibson DJ (2009). *Grasses and grassland ecology*. Oxford University Press, Oxford.

Gibson DJ (2012). Ordination analysis. In *Oxford bibliographies online: ecology*, doi: 10.1093/OBO/9780199830060-0003

Gibson DJ and Good RE (1987). The seedling habitat of *Pinus echinata* and *Melampyrum lineare* in oak-pine forest of the New Jersey Pinelands. *Oikos* **49**: 91–100.

Gibson DJ and Menges ES (1994). Population structure and spatial pattern in the dioecious shrub, *Ceratiola ericoides*. *Journal of Vegetation Science* **5**: 337–346.

Gibson DJ and Newman JA (2001). *Festuca arundinacea* Schreber (*F. elatior* subsp. *arundinacea* (Schreber) Hackel). *Journal of Ecology* **89**: 304–324.

Gibson DJ and Risser PG (1982). Evidence for the absence of ecotypic differentiation in *Andropogon virginicus* L. on metalliferous mine wastes. *New Phytologist* **92**: 589–599.

Gibson DJ and Skeel VA (1996). Effects of competition on photosynthetic rate and stomatal conductance of *Sorghastrum nutans*. *Photosynthetica* **32**: 503–512.

Gibson DJ, Colquhoun IA, and Greig-Smith P (1985). A new method for measuring nutrient supply rates in soils using ion-exchange resins. In AH Fitter (ed.) *Ecological interactions in soil*, pp. 73–79. Blackwell Scientific, Oxford.

Gibson DJ, Freeman CC, and Hulbert LC (1990). Effects of small mammal and invertebrate herbivory on plant species richness and abundance in tallgrass prairie. *Oecologia* **84**: 169–175.

Gibson DJ, Ely JS, Looney PB, and Gibson PT (1995). Effects of inundation from the storm surge of Hurricane Andrew upon primary succession on dredge spoil. *Journal of Coastal Research* **21**(Special Issue): 208–216.

Gibson DJ, Connolly J, Hartnett DC, and Weidenhammer JD (1999). Designs for greenhouse studies of interactions between plants. *Journal of Ecology* **87**: 1–16.

Gibson DJ, Millar K, Delong M, Connolly J, Kirwan L, Wood AJ, et al. (2008). The weed community affects yield and quality of soybean (*Glycine max* (L.) Merr.). *Journal of the Science of Food and Agriculture* **88**: 371–381.

Gibson DJ, Delong M, Chandy S, and Honu YAK (2009). Reproductive challenges of a rare grass, *Calamagrostis porteri* subsp. *insperata* (Swallen) C. Greene: implications for habitat restoration. *Applied Vegetation Science* **12**: 316–327.

Gibson DJ, Urban J, and Baer SG (2011). Mowing and fertilizer effects on seedling establishment in a successional old field. *Journal of Plant Ecology* **4**: 157–168.

Gibson DJ, Dewey J, Goossens H, and Dodd MM (2014). Intraspecific variation among clones of a naïve rare grass affects competition with a nonnative, invasive forb. *Ecology and Evolution* **4**: 186–199.

Gladbach D, Holzschuh A, Scherber C, Thies C, Dormann C, and Tscharntke T (2011). Crop–noncrop spillover: arable fields affect trophic interactions on wild plants in surrounding habitats. *Oecologia* **166**: 433–441.

Godt MJW and Hamrick JL (1998). Allozyme diversity in the grasses. In GP Cheplick (ed.) *Population biology of grasses*, pp. 11–29. Cambridge University Press, Cambridge.

Goldberg DE (1990). Components of resource competition in plant communities. In JB Grace and D Tilman (eds)

Perspectives on plant competition, pp. 27–50. Academic Press, San Diego, CA.

Goldberg DE and Barton DE (1992). Patterns and consequences of interspecific competition in natural communities: a review of field experiments with plants. *American Naturalist* **139**: 771–801.

Goldberg D and Novoplansky A (1997). On the relative importance of competition in unproductive environments. *Journal of Ecology* **85**: 409–418.

Goldberg DE and Scheiner SM (2001). ANOVA and ANCOVA: field competition experiments. In SM Scheiner and J Gurevitch (eds) *Design and analysis of ecological experiments*, 2nd edn, pp. 77–98. Oxford University Press, New York.

Goldberg DE and Turner RM (1986). Vegetation change and plant demography in permanent plots in the Sonoran desert. *Ecology* **67**: 695–712.

Goldsmith FB, Harrison CM, and Morton AJ (1986). Description and analysis of vegetation. In PD Moore and SB Chapman (eds) *Methods in plant ecology*, pp. 437–524. Blackwell Scientific Publications, Oxford.

González L (2001). Determination of water potential in leaves. In MJ Reigosa (ed.) *Handbook of plant ecophysiology techniques*, pp. 193–205. Kluwer Academic Publishers, Dordrecht.

González L and González-Vilar M (2001). Determination of relative water content. In MJ Reigosa (ed.) *Handbook of plant ecophysiology techniques*, pp. 207–212. Kluwer Academic Publishers, Dordrecht.

González L and Reigosa MJ (2001). Plant water status. In MJ Reigosa (ed.) *Handbook of plant ecophysiology techniques*, pp. 185–191. Kluwer Academic Publishers, Dordrecht.

González L, Bolaño C, and Pellissier F (2001). Use of oxygen electrode in measurements of photosynthesis and respiration. In MJ Reigosa (ed.) *Handbook of plant ecophysiology techniques*, pp. 141–153. Kluwer Academic Publishers, Dordrecht.

Gotelli NJ and Ellison AM (2013). *A primer of ecological statistics*, 2nd edn. Sinauer Associates, Sunderland, MA.

Gotelli NJ and Graves GR (1996). *Null models in ecology* [available online at: <http://www.uvm.edu/~ngotelli/nullmodelspage.html>]. Smithsonian Institution Press, Washington, DC.

Gough L, Gross K, Cleland E, Clark C, Collins S, Fargione J, et al. (2012). Incorporating clonal growth form clarifies the role of plant height in response to nitrogen addition. *Oecologia* **169**: 1053–1062.

Grabe DF (ed.) (1970). Tetrazolium testing handbook for agricultural seeds. Contribution no. 29 to the Handbook on seed testing. Association of Official Seed Analysts, North Brunswick, NJ.

Grace JB (1999). The factors controlling species density in herbaceous plant communities: an assessment. *Perspectives in Plant Ecology, Evolution and Systematics* **2**: 1–28.

Grace JB and Tilman D, eds. (1990). *Perspectives on plant competition*. Academic Press, San Diego, CA.

Grace SL and Platt WJ (1995). Effects of adult tree density and fire on the demography of pregrass stage juvenile longleaf pine (*Pinus palustris* Mill.). *Journal of Ecology* **83**: 75–86.

Grace JB, Anderson TM, Olff H, and Scheiner SM (2010). On the specification of structural equation models for ecological systems. *Ecological Monographs* **80**: 67–87.

Gram E and Bovien P (1969). *Recognition of diseases and pests of farm crops*. Blandford Press, London.

Grandillo S, Ku HM, and Tanksley SD (1999). Identifying the loci responsible for natural variation in fruit size and shape in tomato. *Theoretical and Applied Genetics* **99**: 978–987.

Gratani L, Catoni R, Pirone G, Frattaroli AR, and Varone L (2012). Physiological and morphological leaf trait variations in two Apennine plant species in response to different altitudes. *Photosynthetica* **50**: 15–23.

Gray J and Song C (2012). Mapping leaf area index using spatial, spectral, and temporal information from multiple sensors. *Remote Sensing of the Environment* **119**: 173–183.

Green RH (1979). *Sampling design and statistical methods for environmental biologists*. Wiley-Interscience, New York.

Green DS (1980). The terminal velocity and dispersal of spinning samaras. *American Journal of Botany* **67**: 1218–1224.

Greenacre M and Primicerio R (2014). *Multivariate analysis of ecological data*. Fundación BBVA, Bilbao.

Greene DF and Johnson EA (1989). A model of wind dispersal of winged or plumed seeds. *Ecology* **70**: 339–347.

Greene DF and Johnson EA (1997). Secondary dispersal of tree seeds on snow. *Journal of Ecology* **85**: 329–340.

Greig-Smith P (1979). Pattern in vegetation: presidential address to the British Ecological Society. *Journal of Ecology* **67**: 755–759.

Greig-Smith P (1983). *Quantitative plant ecology*, 3rd. edn. Blackwell Scientific Publications, Oxford.

Griffith C (1996). Distribution of *Viola blanda* in relation to within-habitat variation in canopy openness, soil phosphorus, and magnesium. *Bulletin of the Torrey Botanical Club* **123**: 281–285.

Grillo MA, Li C, Hammond M, Wang L, and Schemske DW (2013). Genetic architecture of flowering time differentiation between locally adapted populations of *Arabidopsis thaliana*. *New Phytologist* **197**: 1321–1331.

Grime JP and Booth RE (1993). Day-length responses. In GAF Hendry and JP Grime (eds) *Methods in comparative plant ecology*, pp. 79–81. Chapman and Hall, London.

Grime JP, Mackey JML, Hillier SH, and Read DJ (1987). Floristic diversity in a model system using experimental microcosms. *Nature* **328**: 420–422.

Grime JP, Hall W, and Hunt R (1989). A new development of the temperature-gradient tunnel. *Annals of Botany* **64**: 279–287.

Grime JP, Campbell BD, and Mackey JML (1993). Foraging for mineral nutrients. In GAF Hendry and JP Grime (eds) *Methods in comparative plant ecology*, pp. 87–91. Chapman and Hall, London.

Grimm V and Railsback SF (2005). *Individual-based modeling and ecology*. Princeton University Press, Princeton, NJ.

Grimm V, Lorek H, Finke J, Koester F, Malachinski M, Sonnenschein M, et al. (2004). META-X: generic software for metapopulation viability analysis. *Biodiversity and Conservation* **13**: 165–188.

van Groenendael J and de Kroon H, eds. (1990). *Clonal growth in plants: regulation and function*. SPB Academic Publishing, The Hague.

Grubb PJ (1977). The maintenance of species richness in plant communities: the importance of the regeneration niche. *Biological Reviews* **52**: 107–145.

Grubb PJ (1992). Presidential address: a positive distrust in simplicity—lessons from plant defences and from competition among plants and among animals. *Journal of Ecology* **80**: 585–610.

Grubb PJ, Lee WG, Kollmann J, and Wilson JB (1996). Interaction of irradiance and soil nutrient supply on growth of seedlings of ten European tall-shrub species and *Fagus sylvatica*. *Journal of Ecology* **84**: 827–840.

Gunaratne AMTA, Gunatilleke CVS, Gunatilleke IAUN, Madawala Weerasinghe HMSP, and Burslem DFRP (2011). Release from root competition promotes tree seedling survival and growth following transplantation into human-induced grasslands in Sri Lanka. *Forest Ecology and Management* **262**: 229–236.

Gundel PE, Helander M, Casas C, Hamilton CE, Faeth SH, and Saikkonen K (2012). Neotyphodium fungal endophyte in tall fescue (*Schedonorus phoenix*): a comparison of three northern European wild populations and the cultivar Kentucky-31. *Fungal Diversity* **60**: 15–24.

Guo H, Weiner J, Mazer SJ, Zhao Z, Du G and Li B (2012). Reproductive allometry in *Pedicularis* species changes with elevation. *Journal of Ecology* **100**: 452–458.

Gustafson DJ, Gibson DJ, and Nickrent NL (1999). Random amplified polymorphic DNA variation among remnant big bluestem (*Andropogon gerardii* Vitman) populations from Arkansas' Grand Prairie. *Molecular Ecology* **8**: 1693–1701.

Gwinn KD, Collins-Shepard MH, and Reddick BB (1991). Tissue-print-immunoblot, an accurate method for the detection of *Acremonium coenophialum* in tall fescue. *Phytopathology* **81**: 747–748.

Gwynn-Jones D, Jones AG, Waterhouse A, Winters A, Comont D, Scullion J, et al. (2012). Enhanced UV-B and elevated CO_2 impacts sub-arctic shrub berry abundance, quality and seed germination. *Ambio* **41**: 2556–2268.

Gylle AM, Nygård CA, Svan CI, Pocock T, and Ekelund NGA (2013). Photosynthesis in relation to D1, PsaA and Rubisco in marine and brackish water ecotypes of *Fucus vesiculosus* and *Fucus radicans* (Phaeophyceae). *Hydrobiologia* **700**: 109–119.

Haase P (1995). Spatial pattern analysis in ecology based on Ripley's *K*-function: introduction and methods of edge correction. *Journal of Vegetation Science* **6**: 575–582.

Haasl RJ and Payseur BA (2012). Microsatellites as targets of natural selection. *Molecular Biology and Evolution* **30**: 285–298.

Hagan DL, Jose S, and Lin CH (2013). Allelopathic exudates of cogongrass (*Imperata cylindrica*): implications for the performance of native pine savanna plant species in the southeastern US. *Journal of Chemical Ecology* **39**: 312–322.

Hahn H, Huth W, Schöberlein W, Diepenbrock W, and Weber WE (2003). Detection of endophytic fungi in *Festuca* spp. by means of tissue print immunoassay. *Plant Breeding* **122**: 217–222.

Haig SM (1998). Molecular contributions to conservation. *Ecology* **79**: 413–425.

Hairston NG (1989). *Ecological experiments: purpose, design, and execution*. Cambridge University Press, Cambridge.

Halvorson JJ, Franz EH, Smith JL, and Black RA (1992). Nitrogenase activity, nitrogen fixation, and nitrogen inputs by lupines at Mount St Helens. *Ecology* **73**: 87–98.

Hamill DN and Wright SJ (1986). Testing the dispersion of juveniles relative to adults: a new analytic method. *Ecology* **67**: 952–957.

Hamilton JA, Lexer C, and Aitken SN (2013). Genomic and phenotypic architecture of a spruce hybrid zone (*Picea sitchensis* × *P. glauca*). *Molecular Ecology* **22**: 827–841.

Hampton R, Ball E, and De Boer S (eds) (1990). *Serological methods for detection and identification of viral and bacterial plant pathogens*. American Phytopathological Society, St Paul, MN.

Hara T (1988). Dynamics of size structure in plant populations. *Trends in Ecology and Evolution* **3**: 129–133.

Harper JL (1964). The individual in the population. *Journal of Ecology* **52**(Suppl.): 149–158.

Harper JL (1967). A Darwinian approach to plant ecology. *Journal of Ecology* **55**: 247–270.

Harper JL (1977). *Population biology of plants*. Academic Press, London.

Harper JL (1982). After description. In EI Newman (ed.) *The plant community as a working mechanism*, pp. 11–25. Blackwell Scientific Publications, Oxford.

Harper JL (1986). An apophasis of plant population biology. In AJ Davy, MJ Hutchings, and AR Watkinson (eds) *Plant population ecology*, pp. 435–452. Blackwell Scientific Publications, Oxford.

Harper JL and Ogden J (1970). The reproductive strategy of higher plants: I. The concept of strategy with special reference to *Senecio vulgaris* L. *Journal of Ecology* **58**: 681–698.

Harper AM, Atkinson TG, and Smith AD (1976). Effect of *Rhopalosiphum padi* and barley yellow dwarf virus on forage yield and quality of barley and oats. *Journal of Economic Entomology* **69**: 383–385.

Harper FA, Smith SE, and Macnair MR (1997). Where is the cost in copper tolerance in *Mimulus guttatus*? Testing the trade-off hypothesis. *Functional Ecology* **11**: 764–774.

Hart SC, Stark JM, Davidson EA, and Firestone MK (1994). Nitrogen mineralization, immobilization, and nitrification. In RW Weaver, S Angle, P Bottomley, *et al.* (eds) *Methods of soil analysis: Part 2 Microbiological and biochemical properties*, pp. 985–1018. Soil Science Society of America, Madison, WI.

Harte J and Shaw R (1995). Shifting dominance within a montane vegetation community: results of a climate-warming experiment. *Science* **267**: 876–880.

Hartley SE and Jones CG (1997). Plant chemistry and herbivory, or why the world is green. In MJ Crawley (ed.) *Plant ecology*, pp. 284–324. Blackwell Scientific Publications, Oxford.

Hartnett DC (1991). Effects of fire in tallgrass prairie on growth and reproduction of prairie coneflower (*Ratibida columnifera*: Asteraceae). *American Journal of Botany* **78**: 429–435.

Hartnett DC (1993). Regulation of clonal growth and dynamics of *Panicum virgatum* (Poaceae) in tallgrass prairie: effects of neighbor removal and nutrient addition. *American Journal of Botany* **80**: 1114–1120.

Hartnett DC, Samensus RJ, Fischer LE, and Hetrick BAD (1994). Plant demographic responses to mycorrhizal symbiosis in tallgrass prairie. *Oecologia* **99**: 21–26.

Harvey PH, Read AF, and Nee S (1995). Why ecologists need to be phylogenetically challenged. *Journal of Ecology* **83**: 535–536.

Hassell MP, Comins HN, and May RM (1991). Spatial structure and chaos in insect population dynamics. *Nature* **353**: 255–258.

Hastings A (1997). *Population biology: concepts and models*. Springer-Verlag, New York.

Hastwell GT and Facelli JM (2003). Differing effects of shade-induced facilitation on growth and survival during the establishment of a chenopod shrub. *Journal of Ecology* **91**: 941–950.

Hayek LC and Buzas MA (1997). *Surveying natural populations*. Columbia University Press, New York.

Hazard L and Ghesquière M (1995). Evidence from the use of isozyme markers of competition in swards between short-leaved and long-leaved perennial ryegrass. *Grass and Forage Science* **50**: 241–248.

He T, Krauss SL, Lamont BB, Miller BP, and Enright NJ (2004). Long-distance seed dispersal in a metapopulation of *Banksia hookeriana* inferred from a population allocation analysis of amplified fragment length polymorphism data. *Molecular Ecology* **13**: 1099–1109.

Headley AD, Rumsey FJ, and Taylor I (1998). The way forward for broomrape conservation in Britain. *Naturalist* **123**: 86–89.

Heafner JW (1996). *Modeling biological systems*. Chapman and Hall, London.

Heath D (1995). *An introduction to experimental design and statistics for biology*. UCL Press, London.

Hebeisen T, Lüscher A, Zanetti S, Fischer BU, Hartwig UA, Frehner M, et al. (1997). Growth response of *Trifolium repens* L. and *Lolium perenne* L. as monocultures and bi-species mixture to free air CO_2 enrichment and management. *Global Change Biology* **3**: 149–160.

Hegazy AK, Fahmy GM, Ali MI, and Gomaa NH (2005). Growth and phenology of eight common weed species. *Journal of Arid Environments* **61**: 171–183.

Heger T, Pahl AT, Botta-Dukát Z, Gherardi F, Hoppe C, Hoste I, et al. (2013). Conceptual frameworks and methods for advancing invasion ecology. *Ambio* **42**: 527–540.

Helliwell JR, Sturrock CJ, Grayling KM, Tracy SR, Flavel RJ, Young IM, et al. (2013). Applications of x-ray computed tomography for examining biophysical interactions and structural development in soil systems: a review. *European Journal of Soil Science* **64**: 279–297.

Hemborg ÅM (1998). Costs of reproduction in subarctic *Ranunculus acris*: a five-year field experiment. *Oikos* **83**: 273–282.

Hemborg ÅM and Karlsson PS (1998). Somatic costs of reproduction in eight subarctic plant species. *Oikos* **82**: 149–157.

Hendrick RL and Pregitzer KS (1992). The demography of fine roots in a northern hardwood forest. *Ecology* **73**: 1094–1104.

Hendriks M, Mommer L, de Caluwe H, Smit-Tiekstra AE, van der Putten WH, and de Kroon H (2013). Independent variations of plant and soil mixtures reveal soil feedback effects on plant community overyielding. *Journal of Ecology* **101**: 287–297.

Hendry GR (1992). *FACE: free air CO_2 enrichment for plant research in the field*. CRC Press, Boca Raton, FL.

Hendry GF (1993). Ultra-violet (UV-B) radiation. In GAF Hendry and JP Grime (eds) *Methods in comparative plant ecology*, pp. 72–73. Chapman and Hall, London.

Hendry GAF and Grime JP (eds) (1993). *Methods in comparative plant ecology*, pp 252. Chapman and Hall, London.

Hendry RJ, McGlade JM, and Weiner J (1996). A coupled map lattice model of the growth of plant monocultures. *Ecological Modelling* **84**: 81–90.

Henig-Sever N, Poliakov D, and Broza M (2001). A novel method for estimation of wild fire intensity based on ash pH and soil microarthropod community. *Pedobiologia* **45**: 98–106.

Henry HAL (2007). Soil freeze–thaw cycle experiments: trends, methodological weaknesses and suggested improvements. *Soil Biology and Biochemistry* **39**: 977–986.

Henry M and Francki RIB (1992). Improved ELISA for the detection of barley yellow dwarf virus in grasses. *Journal of Virological Methods* **36**: 231–238.

Hernández-Barrios JC, Anten NPR, Ackerly DD, and Martínez-Ramos M (2012). Defoliation and gender effects on fitness components in three congeneric and sympatric understorey palms. *Journal of Ecology* **100**: 1544–1556.

Herrera CM and Bazaga P (2011). Untangling individual variation in natural populations: ecological, genetic and epigenetic correlates of long-term inequality in herbivory. *Molecular Ecology* **20**: 1675–1688.

Hertel D and Leuschner C (2005). The in situ root chamber: a novel tool for the experimental analysis of root competition in forest soils. *Pedobiologia* **50**: 217–224.

Hess L and De Kroon H (2007). Effects of rooting volume and nutrient availability as an alternative explanation for root self/non-self discrimination. *Journal of Ecology* **95**: 241–251.

Hetrick BAD, Hartnett DC, Wilson GWT, and Gibson DJ (1994). Effects of mycorrhizae, phosphorus availability, and plant density on yield relationships among competing tallgrass prairie grasses. *Canadian Journal of Botany* **72**: 168–176.

Heusden AWv, Koornneef M, Voorrips RE, Brüggemann W, Pet G, Vrielink-van Ginkel R, et al. (1999). Three QTLs from *Lycopersicon peruvianum* confer a high level of resistance to *Clavibacter michiganensis* spp. *michiganensis*. *Theoretical and Applied Genetics* **99**: 1068–1074.

Hewitt EJ (1966). *Sand and water culture methods used in the study of plant nutrition*, 2nd edn. Commonwealth Agricultural Bureau of Technical Communications, Farnham Royal, Buckinghamshire.

Hiatt EE, Hill NS, Bouton JH, and Stuedemann JA (1999). Tall fescue endophyte detection: commercial immunoblot test kit compared with microscopic analysis. *Crop Science* **39**: 796–799.

Hickam E (1998). Variation of *Neotyphodium coenophialum* in *Fesctuca arundinacea* as influenced by plant age and nitrogen fertilization level. MS Thesis, Southern Illinois University, Carbondale, IL.

Hilborn R and Mangel M (1997). *The ecological detective: confronting models with data*. Princeton University Press, Princeton, NJ.

Hileman DR, Ghosh PP, Bhattacharya NC, Biswas PK, Allen LH, Peresta G, et al. (1992). A comparison of the uniformity of an elevated CO_2 environment in three different types of open-type chambers. *Critical Reviews in Plant Sciences* **11**: 195–202.

Hill AC (1967). A special purpose plant environmental chamber for air pollution studies. *Journal of the Air Pollution Control Association* **17**: 743–748.

Hillier SH, Sutton F, and Grime JP (1994). A new technique for the experimental manipulation of temperature in plant communities. *Functional Ecology* **8**: 755–762.

Hilton JL and Boyd RS (1996). Microhabitat requirements and seed/microsite limitation of the rare granite outcrop endemic *Amphianthus pusillus* (Scrophulariaceae). *Bulletin of the Torrey Botanical Club* **123**: 189–196.

Ho I and Trappe JM (1984). Effects of ozone exposure on mycorrhiza formation and growth of *Festuca arundinacea*. *Environmental and Experimental Botany* **24**: 71–74.

Hobbie SE and Chapin FS III (1998). An experimental test of limits to tree establishment in Arctic tundra. *Journal of Ecology* **86**: 449–461.

Hobbs JE, Currall JE, and Gimingham CH (1984). The use of 'thermocolor' pyrometers in the study of heath fire behaviour. *Journal of Ecology* **72**: 241–250.

Hochberg ME, Menaut JC, and Gignoux J (1994). The influences of tree biology and fire in the spatial structure of the West African savannah. *Journal of Ecology* **82**: 217–226.

Hodgson JG and Booth RE (1986). Fresh weight/dry weight and related ratios. In GAF Hendry and JP Grime (eds) *Methods in comparative plant ecology*, pp. 91–92. Chapman and Hall, London.

Hodson D (1982). Is there a scientific method. *Education in Chemistry* **19**: 112–116.

Holah JC and Alexander HM (1999). Soil pathogenic fungi have the potential to affect the co-existence of two tallgrass prairie species. *Journal of Ecology* **87**: 598–608.

Holl FB, Chanway CP, Turkington R, and Radley RA (1988). Response of crested wheatgrass (*Agropyron cristatum* L.), perennial ryegrass (*Lolium perenne* L.) and white clover (*Trifolium repens* L.) to inoculation with *Bacillus polymyxa*. *Soil Biology and Biochemistry* **20**: 19–24.

Hollister RD and Webber PJ (2000). Biotic validation of small open-top chambers in a tundra ecosystem. *Global Change Biology* **6**: 835–842.

Hollister RD, Webber PJ, and Tweedie CE (2005). The response of Alaskan arctic tundra to experimental warming: differences between short- and long-term responses. *Global Change Biology* **11**: 525–536.

Hooker JD (1862). Letter to H. W. Bates. In F Burkhardt, DM Porter, J Harvey, *et al.* (eds) *Correspondence of Charles Darwin*, Vol. **10**, pp. 127–128. Cambridge University Press, Cambridge.

Hopkins B (1954). A new method for determining the type of distribution of plant individuals. *Annals of Botany* **18**: 213–227.

Hopkins B (1965). Vegetation of the Olokemeji forest reserve, Nigeria: III. the microclimates with special reference to their seasonal changes. *Journal of Ecology* **53**: 125–138.

Hornemann G, Weiss G, and Durka W (2012). Reproductive fitness, population size and genetic variation in *Muscari tenuiflorum* (Hyacinthaceae): the role of temporal variation. *Flora—Morphology, Distribution, Functional Ecology of Plants* **207**: 736–743.

Horton JL and Neufeld HS (1998). Photosynthetic responses of *Microstegium vimineum* (Trin) A. Camus, an exotic C_4, shade-tolerant grass, to variable light environments. *Oecologia* **114**: 11–19.

Horwath WR and Paul EA (1994). Microbial biomass. In RW Weaver, S Angle, P Bottomley, *et al.* (eds) *Methods of soil analysis: Part 2 Microbiological and biochemical properties*, pp. 753–773. Soil Science Society of America, Madison, WI.

Houle G and Phillips DL (1989). Seed availability and biotic interactions in granite outcrop plant communities. *Ecology* **70**: 1307–1316.

Hu JS (1995). Comparison of dot blot, ELISA, and RT-PCR assays for detection of two cucumber mosaic virus isolates infecting banana in Hawaii. *Plant Disease* **79**: 902–906.

Huang W, Siemann E, Wheeler GS, Zou J, Carrillo J, and Ding J (2010). Resource allocation to defence and growth are driven by different responses to generalist and specialist herbivory in an invasive plant. *Journal of Ecology* **98**: 1157–1167.

Hughes L, Dunlop M, French K, Leishman MR, Rice B, Rodgerson L, et al. (1994). Predicting dispersal spectra: a minimal set of hypotheses based on plant attributes. *Journal of Ecology* **82**: 933–950.

Hughes JW, Fahey TJ, and Browne B (1987). A better seed and litter trap. *Canadian Journal of Forest Research* **17**: 1623–1624.

Hulme PE (1994). Seedling herbivory in grassland: relative impact of vertebrate and invertebrate herbivory. *Journal of Ecology* **82**: 873–880.

Humboldt A von (1849). *Cosmos: a sketch of a physical description of the universe*. Henry G. Bohn, London.

Hume L and Cavers PB (1981). A methodological problem in genecology. Seeds versus clones as source material for uniform gardens. *Canadian Journal of Botany* **59**: 763–768.

Hunt R (1984). Relative growth rates of cohorts of ramets cloned from a single genet. *Journal of Ecology* **72**: 299–305.

Hunt R (1993). The ISP aerial environment: lighting, day length, temperature and humidity. In GAF Hendry and JP Grime (eds) *Methods in comparative plant ecology*, pp. 14–18. Chapman and Hall, London.

Hurlbert SH (1984). Pseudoreplication and the design of ecological field experiments. *Ecological Monographs* **54**: 187–211.

Hurlbert SH (2009). The ancient black art and transdisciplinary extent of pseudoreplication. *Journal of Comparative Psychology* **123**: 434–443.

Husband BC and Barrett SCH (1996). A metapopulation perspective in plant population biology. *Journal of Ecology* **84**: 461–469.

Huston MA (1997). Hidden treatments in ecological experiments: re-evaluating the ecosystem function of biodiversity. *Oecologia* **110**: 449–460.

Hutchings MJ (1986). Plant population biology. In PD Moore and SB Chapman (eds) *Methods in plant ecology*, pp. 377–435. Blackwell Scientific Publications, Oxford.

Hutchings MJ (1987). The population biology of the early spider orchid, *Ophrys sphegodes* Mill. I. A demographic study from 1975 to 1984. *Journal of Ecology* **75**: 711–727.

Hutchings MJ (1997). The structure of plant populations. In MJ Crawley (ed.) *Plant ecology*, pp. 325–358. Blackwell Scientific Publications, Oxford.

Hutchings MJ (2010). The population biology of the early spider orchid *Ophrys sphegodes* Mill. III. Demography over three decades. *Journal of Ecology* **98**: 867–878.

Hutchings MJ and Barkham JP (1976). An investigation of shoot interactions in *Mercurialis perennis* L., a rhizomatous perennial herb. *Journal of Ecology* **64**: 723–743.

Hutchings MJ and de Kroon H (1994). Foraging in plants: the role of morphological plasticity in resource acquisition. *Advances in Ecological Research* **25**: 159–238.

Hutchings MJ and Price EAC (1999). *Glechoma hederacea* L. (*Nepeta glechoma* Benth., *N. hederacea* (L.) Trev.). *Journal of Ecology* **87**: 347–364.

Hutchings MJ, Turkington R, and Klein E (1997). Morphological plasticity in *Trifolium repens* L.: the effects of clone genotype, soil nutrient level, and the genotype of conspecific neighbours. *Canadian Journal of Botany* **75**: 1382–1393.

Hutchinson GE (1957). Concluding remarks. *Cold Spring Harbor Symposia on Quantitative Biology* **22**: 415–427.

Hutchinson GE (1978). *An introduction to population ecology*. Yale University Press, New Haven, CT.

Ikeda H and Okutomi K (1995). Effects of trampling and competition on plant growth and shoot morphology of *Plantago, Eragrostis* and *Eleusine* species. *Acta Botanica Neerlandica* **44**: 151–160.

Inderjit, Seastedt T, Callaway R, Pollock J, and Kaur J (2008). Allelopathy and plant invasions: traditional, congeneric, and bio-geographical approaches. *Biological Invasions* **10**: 875–890.

Inderjit and Nilsen ET (2003). Bioassays and field studies for allelopathy in terrestrial plants: progress and problems. *Critical Reviews in Plant Sciences* **22**: 221–238.

Inghe O (1989). Genet and ramet survivorship under different mortality regimes—a cellular automata model. *Journal of Theoretical Biology* **138**: 257–270.

Intergovernmental Panel on Climate Change (2013). *Climate change 2013: the physical science basis. Contribution of Working Group I to the Fifth Assessment Report of the Intergovernmental Panel on Climate Change*. Intergovernmental Panel on Climate Change, Geneva.

Iyer-Pascuzzi A, Symonova O, Mileyko Y, Hao Y, Belcher H, Harer J, et al. (2010). Imaging and analysis platform for automatic phenotyping and trait ranking of plant root systems. *Plant Physiology* **152**: 1148–1157.

Izhaki I, Walton PB, Safriel UN (1991). Seed shadows generated by frugivorous birds in an eastern mediterranean scrub. *Journal of Ecology* **79**: 575–590.

Jacoby PW, Ansley RJ, and Trevino BA (1992). An improved method for measuring temperatures during range fires. *Journal of Range Management* **45**: 216–220.

Jacquemyn H, Brys R, Davison R, Tuljapurkar S, and Jongejans E (2012). Stochastic LTRE analysis of the effects of herbivory on the population dynamics of a perennial grassland herb. *Oikos* **121**: 211–218.

Jakobsen I (1994). Carbon metabolism in mycorrhiza. In JR Norris, DJ *Read*, and AK Varma (eds) *Techniques for mycorrhizal research*, pp. 149–180. Academic Press, San Diego, CA.

Jankowska-Blaszczuk M and Daws MI (2007). Impact of red : far red ratios on germination of temperate forest herbs in relation to shade tolerance, seed mass and persistence in the soil. *Functional Ecology* **21**: 1055–1062.

Jansen PA and den Ouden J (2005). Observing seed removal: remote video monitoring of seed selection, predation and dispersal. In P-M Forget, JE Lambert, PE Hulme, *et al.* (eds) *Seed fate: predation, dispersal, and seedling establishment*, pp. 363–378. CAB International, Wallingford, Oxon.

Janzen DH (1970). Herbivores and the number of tree species in tropical forests. *American Naturalist* **104**: 501–528.

Jasienski M and Bazzaz FA (1999). The fallacy of ratios and the testability of models in biology. *Oikos* **84**: 321–326.

Jeffers JNR (1978). *An introduction to systems analysis: with ecological applications*. Edward Arnold, London.

Jeffreys H (1917). On the vegetation of four Durham coal-measure fells. *Journal of Ecology* **5**: 129–154.

Jeltsch F, Moloney KA, Schurr FM, Köchy M, and Schwager M (2008). The state of plant population modelling in light of environmental change. *Perspectives in Plant Ecology, Evolution and Systematics* **9**: 171–189.

Jenkinson DS and Powlson DS (1976). The effects of biocidal treatments on metabolism in soil. V. A method for measuring soil biomass. *Soil Biology and Biochemistry* **8**: 209–213.

Jensen AM, Götmark F, and Löf M (2012). Shrubs protect oak seedlings against ungulate browsing in temperate broadleaved forests of conservation interest: a field experiment. *Forest Ecology and Management* **266**: 187–193.

Jentsch A, Kreyling J, and Beierkuhnlein C (2007). A new generation of climate-change experiments: events, not trends. *Frontiers in Ecology and the Environment* **5**: 365–374.

Jeschke JM, Gómez Aparicio L, Haider S, Heger T, Lortie CJ, Pyšek P, et al. (2012). Support for major hypotheses in invasion biology is uneven and declining. *NeoBiota* **14**: 1–20.

Jin X, Xu X, Song X, Li Z, Wang J, and Guo W (2013). Estimation of leaf water content in winter wheat using grey relational analysis—partial least squares modeling with hyperspectral data. *Agronomy Journal* **105**: 1385–1392.

John R, Ahmad P, Gadgil K, and Sharma S (2009). Heavy metal toxicity: effect on plant growth, biochemical parameters and metal accumulation by *Brassica juncea* L. *International Journal of Plant Production* **3**: 65–75.

John E and Turkington R (1997). A 5-year study of the effects of nutrient availability and herbivory on two boreal forest herbs. *Journal of Ecology* **85**: 419–430.

Johnson RA and Wichern DW (2007). *Applied multivariate statistics*, 6th edn. Pearson, Old Tappan, NJ.

Johnson HB, Polley HW, and Mayeux HS (1993). Increasing CO_2 and plant–plant interactions: effects on natural vegetation. *Vegetatio* **104/105**: 157–170.

Johnson HB, Polley HW, and Whitis RP (2000). Elongated chambers for field studies across atmospheric CO_2 gradients. *Functional Ecology* **14**: 388–396.

Johnson JA, Walse SS, and Gerik JS (2012a). Status of alternative for methyl bromide in the United States. *Outlooks on Pest Management* **23**: 53–58.

Johnson SN, Clark KE, Hartley SE, Jones TH, McKenzie SW, and Koricheva J (2012b). Aboveground–belowground herbivore interactions: a meta-analysis. *Ecology* **93**: 2208–2215.

Jolliffe P (2000). The replacement series. *Journal of Ecology* **88**: 371–385.

Jones M (1985). Modular demography and form in silver birch. In J White (ed.) *Studies on plant demography: a festschrift for John L. Harper*, pp. 223–237. Academic Press, London.

Jones HG (1992). *Plants and microclimate*, 2nd edn. Cambridge University Press, Cambridge.

Jones TH, Thompson LJ, Lawton JH, Bezemer JH, Bardgett RD, Blackburn TM, et al. (1998). Impacts of rising atmospheric carbon dioxide on model terrestrial ecosystems. *Science* **280**: 441–443.

Jones JC, Reynolds JD, and Raffaelli D (2006). Environmental variables. In WJ Sutherland (ed.) *Ecological census techniques: a handbook*, 2nd edn, pp. 370–407. Cambridge University Press, Cambridge.

Jones MP, Webb BL, Cook DA, and Jolley VD (2012). Comparing nutrient availability in low-fertility soils using ion exchange resin capsules. *Communications in Soil Science and Plant Analysis* **43**(Issue 1–2): 368–376.

Jones MP, Webb BL, Jolley VD, Hopkins BG, and Cook DA (2013). Evaluating nutrient availability in semi-arid soils with resin capsules and conventional soil tests, I: native plant bioavailability under glasshouse conditions. *Communications in Soil Science and Plant Analysis* **44**: 971–986.

Jongen M and Jones MB (1998). Effects of elevated carbon dioxide on plant biomass production and competition in a simulated neutral grassland community. *Annals of Botany* **82**: 111–123.

Jowett D (1964). Population studies on lead-tolerant *Agrostis tenuis*. *Evolution* **18**: 70–81.

Kalamees R, Püssa K, Tamm S, and Zobel K (2012). Adaptation to boreal forest wildfire in herbs: responses to post-fire environmental cues in two *Pulsatilla* species. *Acta Oecologica* **38**: 1–7.

Kalisz S, Hanzawa FM, Tonsor SJ, Thiede DA, and Voigt S (1999). Ant-mediated seed dispersal alters pattern of relatedness in a population of *Trillium grandiflorum*. *Ecology* **80**: 2620–2634.

Kaneko K (1993). The coupled map lattice: introduction, phenomenology, Lyapunov analysis, thermodynamics and applications. In K Kaneko (ed.) *Theory and applications of coupled map lattices*, pp. 1–50. John Wiley and Sons, Chichester.

Karban R (1993). Costs and benefits of induced resistance and plant density for a native shrub, *Gossypium thurberi*. *Ecology* **74**: 9–19.

Karban R and Niiho C (1995). Induced resistance and susceptibility to herbivory: plant memory and altered plant development. *Ecology* **76**: 1220–1225.

Karp A, Seberg O, and Buiatti M (1996). Molecular techniques in the assessment of botanical diversity. *Annals of Botany* **78**: 143–149.

Kattge J, Díaz S, Lavorel S, Prentice IC, Leadley P, Bönisch G, et al. (2011). TRY—a global database of plant traits. *Global Change Biology* **17**: 2905–2935.

Kaufman PB, Labavitch J, Anderson-Prouty A, and Ghosheh NS (1975). *Laboratory experiments in plant physiology*. Macmillan Publishing Co. Ltd, New York.

Kearns CA and Inouye DW (1993). *Techniques for pollination biologists*. University Press of Colorado, Niwot, CO.

Keating KA and Cherry S (2004). Use and interpretation of logistic regression in habitat-selection studies. *Journal of Wildlife Management* **68**: 774–789.

Keddy PA (2001). *Competition*, 2nd edn. Kluwer Academic Publishers, Dordrecht.

Keddy P, Gaudet C, and Fraser LH (2000). Effects of low and high nutrients on the competitive hierarchy of 26 shoreline plants. *Journal of Ecology* **88**: 413–423.

Keddy P, Nielsen K, Weiher E, and Lawson R (2002). Relative competitive performance of 63 species of terrestrial herbaceous plants. *Journal of Vegetation Science* **13**: 5–16.

Keeley JE, Keeley MB, and Bond W (1998). Stem demography and post-fire recruitment of a resprouting serotinous conifer. *Journal of Vegetation Science* **10**: 69–76.

Keinänen M, Julkunen-Tiitto R, Mutikainen P, Walls M, Ovaska J, and Vapaavouri E (1999). Trade-offs in phenolic metabolism of silver birch: effects of fertilization, defoliation and genotype. *Ecology* **80**: 1970–1986.

Kelly D and Sork VL (2002). Mast seeding in perennial plants: why, how, where? *Annual Review of Ecology and Systematics* **33**: 427–447.

Kembel SW, Cowan PD, Helmus MR, Cornwell WK, Morlon H, Ackerly DD, et al. (2010). Picante: R tools for integrating phylogenies and ecology. *Bioinformatics* **26**: 1463–1464.

Kenaga EE, Whitney WK, Hardy JL, and Doty AE (1965). Laboratory tests with Dursban insecticide. *Journal of Economic Entomology* **58**: 1043–1051.

Kennard DK, Outcalt KW, Jones D, and O'Brien J (2005). Comparing techniques for estimating flame temperature of prescribed fires *Fire Ecology* **1**: 75–84.

Kent M, Owen NW, Dale P, Newnham RM, and Giles TM (2001). Studies of vegetation burial: a focus for biogeography and biogeomorphology? *Progress in Physical Geography* **25**: 455–482.

Keuper F, van Bodegom PM, Dorrepaal E, Weedon JT, van Hal J, van Logtestijn RSP, et al. (2012). A frozen feast: thawing permafrost increases plant-available nitrogen in subarctic peatlands. *Global Change Biology* **18**: 1998–2007.

Kigathi RN, Unsicker SB, Reichelt M, Kesselmeier J, Gershenzon J, and Weisser WW (2009). Emission of volatile organic compounds after herbivory from *Trifolium pratense* (L.) under laboratory and field conditions. *Journal of Chemical Ecology* **35**: 1335–1348.

Kikvidze Z and Brooker R (2010). Towards a more exact definition of the importance of competition—a reply to Freckleton et al. (2009). *Journal of Ecology* **98**: 719–724.

Kikvidze Z, Suzuki M, and Brooker R (2011). Importance versus intensity of ecological effects: why context matters. *Trends in Ecology and Evolution* **26**: 383–388.

Kimball BA (2005). Theory and performance of an infrared heater for ecosystem warming. *Global Change Biology* **11**: 2041–2056.

King R, Morgan BJT, Gimenez O, and Brooks SP (2010). *Bayesian analysis for population ecology*. Taylor and Francis Group, Boca Raton, FL.

Kingsolver JG and Paine RT (1991). Theses, antitheses, and syntheses: conversational biology and ecological debate. In LA Real and JH Brown (eds) *Foundations of ecology*, pp. 309–317. Chicago University Press, Chicago, IL.

Kingsolver JG and Schemske DW (1991). Path analysis of selection. *Trends in Ecology and Evolution* **6**: 276–280.

Kirwan L, Lüscher A, Sebastià MT, Finn JA, Collins RP, Porqueddu C, et al. (2007). Evenness drives consistent diversity effects in intensive grassland systems across 28 European sites. *Journal of Ecology* **95**: 530–539.

Kitajima K and Augspurger CK (1989). Seed and seedling ecology of a monocarpic tropical tree, *Trachigalia versicolor*. *Ecology* **70**: 1102–1114.

Klute A (ed.) (1986). *Methods of soil analysis. Part 1, Physical and mineralogical methods* Soil Science Society of America, Madison, WI.

Knapp SJ (1994). Mapping quantitative trait loci. In RL Phillips and IK Vasil (eds) *DNA-based markers in plants*, pp. 58–96. Kluwer Academic Publishers, Dordrecht.

Knapp EE and Rice KJ (1998). Comparison of isozymes and quantitative traits for evaluating patterns of genetic variation in purple needlegrass (*Nassella pulchra*). *Conservation Biology* **12**: 1031–1041.

Knapp AK and Smith WK (1990). Stomatal and photosynthetic responses to variable sunlight. *Physiologia Plantarum* **78**: 160–165.

Knight T (2012). Using population viability analysis to plan reintroductions. In J Maschinski and KE Haskins (eds) *Plant reintroduction in a changing climate: promises and perils, the science and practice of ecological restoration*, pp. 155–169. Island Press, Washington, DC.

Kobayashi T, Ikeda H, and Hon Y (1999). Growth analysis and reproductive allocation of Japanese forbs and grasses in relation to organ toughness under trampling. *Plant Biology* **1**: 445–452.

Koh S, Vicari M, Ball JP, Rakocevic T, Zaheer S, Hik DS, et al. (2006). Rapid detection of fungal endophytes in grasses for large-scale studies. *Functional Ecology* **20**: 736–742.

Koide RT, Robichaux RH, Morse SR, and Smith CM (1989). Plant water status, hydraulic resistance and capacitance. In RW Pearcy, J Ehleringer, HA Mooney, *et al.* (eds) *Plant physiological ecology*, pp. 161–183. Chapman and Hall, London.

Kollmann J and Goetze D (1998). Notes on seed traps in terrestrial plant communities. *Flora* **193**: 31–40.

Koncz C, Chua N-H, and Schell J (eds) (1992). *Methods in Arabidopsis research*. Ward Scientific Publishing Co, Inc., River Edge, MN.

Koptur S, Smith CL, and Lawton JH (1996). Effects of artificial defoliation on reproductive allocation in the common vetch, *Vicia sativa* (Fabaceae: Papilionoideae). *American Journal of Botany* **83**: 886–889.

Kormanik P, Schultz RC, and Bryan WC (1982). The influence of vesicular-arbuscular mycorrhizae on the growth and development of eight hardwood tree species. *Forest Science* **28**: 531–539.

Korpelainen H (1992). Patterns of resource allocation in male and female plants of *Rumex acetosa* and *R. acetosella*. Oecologia **89**: 133–139.

Kos M, Baskin CC, and Baskin JM (2012). Relationship of kinds of seed dormancy with habitat and life history in the Southern Kalahari flora. *Journal of Vegetation Science* **23**: 869–879.

Kramer PJ, Hellmers H, and Downs RJ (1970). SEPEL: new phytrons for environmental research. *BioScience* **20**: 1201–1208.

Krannitz PG, Aarssen LW, and Dow JM (1991). The effect of genetically based differences in seed size on seedling survival in *Arabidopsis thaliana* (Brassicaceae). *American Journal of Botany* **78**: 446–450.

Krause GH and Weis E (1991). Chlorophyll fluorescence and photosynthesis: the basics. *Annual Review of Plant Physiology and Plant Molecular Biology* **42**: 313–349.

Krebs CJ (1999). *Ecological methodology*, 2nd edn. Benjamin/Cummings, Menlo Park, CA.

Krieger R (ed.) (2010). *Haye's handbook of pesticide technology*, 3rd edn. Elsevier, Amsterdam.

de Kroon H and van Groenendael J (eds) (1997). *The ecology and evolution of clonal plants*. Bachuys Publishers, Leiden.

de Kroon H and Hutchings MJ (1995). Morphological plasticity in clonal plants: the foraging concept revisited. *Journal of Ecology* **83**: 143–152.

de Kroon H, Plaisier A, van Groenendael JM, and Caswell H (1986). Elasticity as a measure of the relative contribution of demographic parameters to population growth rate. *Ecology* **67**: 1427–1431.

Krupnick GA and Weis AE (1999). The effect of floral herbivory on male and female reproductive success in *Isomeris arborea*. *Ecology* **80**: 135–149.

Kruschke JK (2011). *Doing Bayesian data analysis: a tutorial with R and BUGS*. Academic Press, Burlington, MA.

Ksenzhek OS and Volkov AG (1998). *Plant energetics*. Academic Press, San Diego, CA.

Kunin WE (1993). Sex and the single mustard: population density and pollinator behavior effects on seed-set. *Ecology* **74**: 2145–2160.

Kuser JE and Ledig FT (1987). Provenance and progeny variation in pitch pine from the Atlantic Coastal Plain. *Forest Science* **33**: 558–564.

Lacy RC (1993). VORTEX: a computer simulation model for population viability analysis. *Wildlife Research* **20**: 45–65.

Lacy RC and Pollak JP (2013). *Vortex: a stochastic simulation of the extinction process*, version 10.0. Chicago Zoological Society, Brookfield, IL.

Laliberté E, Turner BL, Costes T, Pearse SJ, Wyrwoll K-H, Zemunik G, et al. (2012). Experimental assessment of nutrient limitation along a 2-million-year dune chronosequence in the south-western Australia biodiversity hotspot. *Journal of Ecology* **100**: 631–642.

Laliberté E, Turner BL, Zemunik G, Wyrwoll K-H, Pearse SJ, Lambers H (2013). Nutrient limitation along the Jurien Bay dune chronosequence: response to Uren and Parsons. *Journal of Ecology* **101**: 1088–1092.

't Lam RUE (2010). Scrutiny of variance results for outliers: Cochran's test optimized. *Analytica Chimica Acta* **659**: 68–84.

Lambers H and Poorter H (1992). Inherent variation in growth rate between higher plants: a search for physiological causes and ecological consequences. *Advances in Ecological Research* **23**: 187–261.

Lambers H, Chapin FSI, and Pons TL (2008). *Plant physiological ecology*, 2nd edn. Springer-Verlag, New York.

Landhäusser SM and Lieffers VJ (1994). Competition between *Calamagrostis canadensis* and *Epilobium angustifolium* under different soil temperature and nutrients regimes. *Canadian Journal of Forest Research* **24**: 2244–2250.

Larcher W (2003). *Physiological plant ecology*, 4th edn. Springer, Berlin.

Larson AL (1971) *Two-way thermogradient plate for seed germination research: construction plans and procedures*. US Department of Agriculture Agricultural Research Service, Hyattsville, MD. Available at: <https://archive.org/details/twowaythermograd41lars> (accessed 28 May 2014).

Larson BMH and Barrett SCH (1999). The ecology of pollen limitation in buzz-pollinated *Rhexia virginica* (Melastomaceae). *Journal of Ecology* **87**: 371–381.

Lau JA, Puliafico KP, Kopshever JA, Steltzer H, Jarvis EP, Schwarzländer M, et al. (2008). Inference of allelopathy is complicated by effects of activated carbon on plant growth. *New Phytologist* **178**: 412–423.

Laughlin DC (2013). The intrinsic dimensionality of plant traits and its relevance to community assembly. *Journal of Ecology* **102**: 186–193.

Laurenroth WK and Adler PB (2008). Demography of perennial grassland plants: survival, life expectancy and life span. *Journal of Ecology* **96**: 1023–1032.

Law W and Salick J (2005). Human-induced dwarfing of Himalayan snow lotus, *Saussurea laniceps* (Asteraceae). *Proceedings of the National Academy of Sciences of the United States of America* **102**: 10218–10220.

Lawes JB, Gilbert JH, and Masters MT (1882). Agricultural, botanical and chemical results of experiments on the mixed herbage of permanent meadow, conducted for more than twenty years in succession on the same land. Part II. The botanical results. *Philosophical Transactions of the Royal Society (A&B)* **173**: 1181–1413.

Lawlor D (1970). Absorption of polyethylene glycol by plants and their effects on plant growth. *New Phytologist* **64**: 501–513.

Lawton JH, Naeem S, Woodfin RM, Brown VK, Gange A, Godfray HJC, et al. (1993). The Ecotron: a controlled environmental facility for the investigation of population and ecosystem processes. *Philosophical Transactions of the Royal Society B: Biological Sciences* **341**: 181–194.

Leadley PW and Körner C (1996). Effects of elevated CO_2 on plant species dominance in a highly diverse calcareous grassland. In C Körner and FA Bazzaz (eds) *Carbon dioxide, populations and communities*, pp. 159–175. Academic Press, New York.

Leadley PW, Niklaus P, Stocker R, and Körner C (1997). Screen-aided CO_2 control (SACC): a middle ground between FACE and open-top chambers. *Acta Oecologica* **18**: 207–219.

Leather GR and Einhellig FA (1986). Bioassays in the study of allelopathy. In AR Putnam and C-S Tang (eds) *The science of allelopathy*, pp. 133–146. Wiley-Interscience, New York.

Leck MA, Parker VT, and Simpson RL (eds) (1989). *Ecology of soil seed banks*. Academic Press, San Diego, CA.

Lee PM (1989). *Bayesian statistics: an introduction*. John Wiley and Sons, Chichester.

Lee JA (1998). Unintentional experiments with terrestrial ecosystems: ecological effects of sulphur and nitrogen pollutants. *Journal of Ecology* **86**: 1–12.

Lee J-S (2011). Combined effect of elevated CO_2 and temperature on the growth and phenology of two annual C_3 and C_4 weedy species. *Agriculture, Ecosystems and Environment* **140**: 484–491.

Lee WG and Fenner M (1989). Mineral nutrient allocation in seeds and shoots of twelve *Chionochloa* species in relation to soil fertility. *Journal of Ecology* **77**: 704–716.

Lee J-S, Usami T, and Oikawa T (2001). High performance of CO_2–temperature gradient chamber newly built for studying the global warming effect on a plant population. *Ecological Research* **16**: 347–358.

Legendre S and Clobert J (1995). ULM, a software for conservation and evolutionary biologists. *Journal of Applied Statistics* **22**: 817–834.

Legendre P and Legendre L (2012). *Numerical ecology*, 3rd English edn. Elsevier, Amsterdam.

Legendre P and Vaudor A (1991). *The R package: multidimensional analysis, spatial analysis*. Départment de Sciences Biologiques, Université de Montréal, Montreal.

Leger EA (2013). Annual plants change in size over a century of observations. *Global Change Biology* **19**: 2229–2239.

Lehtilä K and Strauss SY (1999). Effects of foliar herbivory on male and female reproductive traits of wild radish, *Raphanus raphanistrum*. *Ecology* **80**: 116–124.

Lehtilä K and Syrjänen K (1995). Positive effects of pollination on subsequent size, reproduction, and survival of *Primula veris*. *Ecology* **76**: 1084–1098.

Lemmon PE (1956). A spherical densiometer for estimating forest overstory density. *Forest Science* **2**: 314–320.

Levine JM, McEachern AK, and Cowan C (2011). Seasonal timing of first rain storms affects rare plant population dynamics. *Ecology* **92**: 2236–2247.

Li S-L, Yu F-H, Werger MJA, Dong M, and Zuidema PA (2011). Habitat-specific demography across dune fixation stages in a semi-arid sandland: understanding the expansion, stabilization and decline of a dominant shrub. *Journal of Ecology* **99**: 610–620.

Liancourt P and Tielbörger K (2009). Competition and a short growing season lead to ecotypic differentiation at the two extremes of the ecological range. *Functional Ecology* **23**: 397–404.

Liddle MJ, Budd CSJ, and Hutchings MJ (1982). Population dynamics and neighbourhood effects in establishing swards of *Festuca rubra*. *Oikos* **38**: 52–59.

Livingston NJ and Topp GC (2006). Soil water potential. In MR Carter and EG Gregorich (eds) *Soil sampling and methods of analysis*, 2nd edn. CRC Press, Boca Raton, FL.

Loehle C (1987). Hypothesis testing in ecology: psychological aspects and the importance of theory maturation. *Quarterly Review of Biology* **62**: 397–409.

Lopezaraiza–Mikel ME, Hayes RB, Whalley MR, and Memmott J (2007). The impact of an alien plant on a native plant–pollinator network: an experimental approach. *Ecology Letters* **10**: 539–550.

López-Villavicencio M, Branca A, Giraud T, and Shykoff JA (2005). Sex-specific effect of *Microbotryum violaceum* (Uredinales) spores on healthy plants of the gynodioecious *Gypsophila repens* (Caryophyllaceae). *American Journal of Botany* **92**: 896–900.

Loranger J, Meyer ST, Shipley B, Kattge J, Loranger H, Roscher C, et al. (2013). Predicting invertebrate herbivory from plant traits: polycultures show strong non-additive effects. *Ecology* **94**: 1499–1509.

Losos E (1995). Habitat specificity of two palm species: experimental transplantation in Amazonian successional forests. *Ecology* **76**: 2595–2606.

Lötscher M and Hay MJM (1997). Genotypic differences in physiological integration, morphological plasticity and utilization of phosphorus induced by variation in phosphate supply in *Trifolium repens*. *Journal of Ecology* **85**: 341–350.

Loveless MD, Hamrick JL, and Foster RB (1998). Population structure and mating system in *Tachigali versicolor*, a monocarpic Neotropical tree. *Heredity* **81**: 134–143.

Lovett Doust L (1981). Population dynamics and local specialization in a clonal perennial (*Ranunculus repens*). I. The dynamics of ramets in contrasting habitats. *Journal of Ecology* **69**: 743–755.

Lovett Doust J and Lovett Doust L, eds. (1988). *Plant reproductive ecology: patterns and strategies*. Oxford University Press, New York.

Lowry DB (2012). Ecotypes and the controversy over stages in the formation of new species. *Biological Journal of the Linnean Society* **106**: 241–257.

Lu F, Lipka AE, Glaubitz J, Elshire R, Cherney JH, Casler MD, et al. (2013). Switchgrass genomic diversity, ploidy, and evolution: novel insights from a network-based SNP discovery protocol. *PLoS Genetics* **9**: e1003215.

Ludwig D (1996). Uncertainty and the assessment of extinction probabilities. *Ecological Applications* **6**: 1067–1076.

Ludwig JA and Reynolds JF (1988). *Statistical ecology*. Wiley Interscience, New York.

Luken JO Mattimiro DT (1991). Habitat-specific resilience of the invasive shrub Amur honeysuckle (*Lonicera maackii*) during repeated clipping. *Ecological Applications* **1**: 104–109.

Lüscher A, Hendrey GR, and Nösberger J (1998). Long-term responsiveness to free air CO_2 enrichment of functional types, species and genotypes of plants from fertile permanent grassland. *Oecologia* **113**: 37–45.

Lüttge U (1997). *Physiological ecology of tropical plants*. Springer, Berlin.

McAllister CA, Knapp AK, and Maragni LA (1998). Is leaf-level photosynthesis related to plant success in a highly productive grassland? *Oecologia* **117**: 40–46.

McCarthy BC and Hanson SL (1998). An assessment of the allelopathic potential of the invasive weed *Alliaria petiolata* (Brassicaceae). *Castanea* **63**: 68–73.

McCarthy MA, Possingham HP, Day JR, and yre AJ (2001). Testing the accuracy of population viability analysis. *Conservation Biology* **15**: 1030–1038.

McCloskey M, Firbank LG, Watkinson AR, and Webb DJ (1996). The dynamics of experimental arable weed communities under different management practices. *Journal of Vegetation Science* **7**: 799–808.

McConnaughay KDM and Bazzaz FA (1991). Is physical space a soil resource? *Ecology* **72**: 94–103.

MacDonald GM and Cwynar LC (1991). Post-glacial population growth rates of *Pinus contorta* ssp. *latifolia* in western Canada. *Journal of Ecology* **79**: 417–430.

McElroy WD and DeLuca MA (1983). Firefly and bacterial luminescence: basic science and applications (*Photinus pyralis*). *Journal of Applied Biochemistry* **5**: 197–209.

McEvoy PB, Rudd NT, Cox CS, and Huso M (1993). Disturbance, competition, and herbivory effects on ragwort *Senecio jacobaea* populations. *Ecological Monographs* **63**: 55–75.

McGraw JB and Antonovics J (1983). Experimental ecology of *Dryas octopetala* ecotypes. I. Ecotypic differentiation and life-cycle stages of selection. *Journal of Ecology* **71**: 879–897.

Machlis L and Torrey JG (1956). *Plants in action*. W. H. Freeman and Co., San Francisco.

Macías FA, Castellano D, and Molinillo JMG (2000). Search for a standard phytotoxic bioassay for allelochemicals. Selection of standard target species. *Journal of Agricultural and Food Chemistry* **48**: 2512–2521.

McIntyre S, Lavorel S, and Tremont RM (1995). Plant life-history attributes: their relationship to disturbance response in herbaceous vegetation. *Journal of Ecology* **83**: 31–44.

Mack RN (1985). Invading plants: their potential contribution to population biology. In J White (ed) *Studies on plant demography: a festschrift for John L. Harper*, pp. 127–142. Academic Press, London.

Mack RN and Harper JL (1977). Interference in dune annuals: spatial pattern and neighbourhood effects. *Journal of Ecology* **65**: 345–363.

Mack RN and Pyke DA (1983). The demography of *Bromus tectorum*: variation in time and space. *Journal of Ecology* **71**: 69–94.

Mack RN and Pyke DA (1984). The demography of *Bromus tectorum*: the role of microclimate, grazing and disease. *Journal of Ecology* **72**: 731–748.

Mack KML and Rudgers JA (2008). Balancing multiple mutualists: asymmetric interactions among plants, arbuscular mycorrhizal fungi, and fungal endophytes. *Oikos* **117**: 310–320.

McLean RC and Ivimey Cook WR (1968). *Practical field ecology*, 2nd edn. George Allen and Unwin, Ltd, London.

McLellan AJ, Fitter AH, and Law R (1995). On decaying roots, mycorrhizal colonization and the design of removal experiments. *Journal of Ecology* **83**: 225–230.

McLellan AJ, Law R, and Fitter AH (1997a). Response of calcareous grassland plant species to diffuse competition: results from a removal experiment. *Journal of Ecology* **85**: 479–490.

McLellan AJ, Pratt D, Kaltz O, and Schmid B (1997b). Structure and analysis of phenotypic and genetic variation in clonal plants. In H de Kroon and J van Groenendael (eds) *The ecology and evolution of clonal plants*, pp. 185–210. Backhuys Publishers, Leiden.

McLendon T and Redente EF (1992). Effects of nitrogen limitation on species replacement dynamics during early secondary succession on a semiarid sagebrush site. *Oecologia* **91**: 312–317.

McLeod AR (1997). Outdoor supplementation systems for studies of the effects of increased UV-B radiation. *Plant Ecology* **128**: 78–92.

McMillan C (1959). The role of ecotypic variation in the distribution of the Central Grassland of North America. *Ecological Monographs* **29**: 285–308.

McNeilly T and Roose ML (1996). Co-adaptation between neighbours? A case study with *Lolium perenne* genotypes. *Euphytica* **92**: 121–128.

McPeek MA and Kalisz S (1993). Population sampling and bootstrapping in complex designs: demographic analysis. In SM Scheiner and J Gurevitch (eds) *Design and analysis of ecological experiments*, pp. 232–252. Chapman and Hall, New York.

McPhee C and Aarssen L (2001). The separation of above- and below-ground competition in plants. A review and critique of methodology. *Plant Ecology* **152**: 119–136.

McRoberts N, Finch RP, Sinclair W, Meikle A, Marshall G, Squire G, et al. (1999). Assessing the ecological significance of molecular diversity data in natural plant populations. *Journal of Experimental Botany* **50**: 1635–1645.

Mairhofer S, Zappala S, Tracy SR, Sturrock C, Bennett M, Mooney SJ, et al. (2012). RooTrak: automated recovery of three-dimensional plant root architecture in soil from x-ray microcomputed tomography images using visual tracking. *Plant Physiology* **158**: 561–569.

Mancuso S (ed.) (2012). *Measuring roots: an updated approach*. Springer, Heidelberg.

Manly BJF (2004). *Multivariate statistical methods: a primer*, 3rd edn. Chapman and Hall, London.

Marion GM, Henry GHR, Freckman DW, Johnstone J, Jones G, Jones MH, et al. (1997). Open-top designs for manipulating field temperature in high-latitude ecosystems. *Global Change Biology* **3**(Suppl. 1): 20–32.

Marks S and Strain B (1989). Effects of drought and CO_2 enrichment and competition between two old field perennials. *New Phytologist* **111**: 181–186.

Marriott CA and Zuazua MT (1996). Tillering and partitioning of dry matter and nutrients in *Lolium perenne* growing with neighbours of different species: effects of nutrient supply and defoliation. *New Phytologist* **132**: 87–95.

Martin H (ed.) (1972). *Insecticide and fungicide handbook*, 4th edn. Blackwell Scientific Publications, Oxford.

Martin JGA, Nussey DH, Wilson AJ, and Réale D (2011). Measuring individual differences in reaction norms in field and experimental studies: a power analysis of random regression models. *Methods in Ecology and Evolution* **2**: 362–374. [Erratum (2012) *Methods in Ecology and Evolution* **3**: 1099.]

Matlack GR (1987). Diaspore size, shape, and fall behavior in wind-dispersed plant species. *American Journal of Botany* **74**: 1150–1160.

Matlack GR (1989). Secondary dispersal of seed across snow in *Betula lenta*, a gap-colonizing tree species. *Journal of Ecology* **77**: 858–869.

Matlack GR (1993). Microclimate variation within and among forest edge sites in the eastern United States. *Biological Conservation* **66**: 185–194.

Matlack GR (1997). Resource allocation among clonal shoots of the fire-tolerant shrub *Gaylussacia baccata*. *Oikos* **80**: 509–518.

Matlack GR and Harper JL (1986). Spatial distribution and the performance of individual plants in a natural population of *Silene dioica*. *Oecologia* **70**: 121–127.

Mauricio R, Bowers MD, and Bazzaz FA (1993). Pattern of leaf damage affects fitness of the annual plant *Raphanus sativus* (Brassicaceae). *Ecology* **74**: 2066–2071.

Mead R (1988). *The design of experiments: statistical principles for practical application*. Cambridge University Press, Cambridge.

Medawar PB (1969). *Induction and intuition in scientific thought*. American Philosophical Society, Philadelphia, PA.

Menges ES (2000). Population viability analyses in plants: challenges and opportunities. *Trends in Ecology and Evolution* **15**: 51–56.

Menges ES and Dolan RW (1998). Demographic viability of populations of *Silene regia* in midwestern prairies: relationships with fire management, genetic variation, geographic location, population size and isolation. *Journal of Ecology* **86**: 63–78.

Menges ES, McIntyre PJ, Finer MS, Goss E, and Yahr R (1999). Microhabitat of the narrow Florida scrub endemic *Dicerandra christmanii*, with comparisons to its congener *D. frutescens*. *Journal of the Torrey Botanical Society* **126**: 24–31.

Merow C, Dahlgren JP, Metcalf CJE, Childs DZ, Evans MEK, Jongejans E, et al. (2014). Advancing population ecology with integral population models: a practical guide. *Methods in Ecology and Evolution* **5**: 99–110.

Mertz DB and McCauley DE (1980). The domain of laboratory ecology. *Synthese* **43**: 95–110.

Metcalf CJE, McMahon SM, Salguero-Gómez R, and Jongejans E (2013). *IPMpack*: an R package for integral projection models. *Methods in Ecology and Evolution* **4**: 195–200.

Meyer AH and Schmid B (1999a). Experimental demography of rhizome populations of establishing clones of *Solidago altissima*. *Journal of Ecology* **87**: 42–54.

Meyer AH and Schmid B (1999b). Experimental demography of the old-field perennial *Solidago altissima*: the dynamics of the shoot population. *Journal of Ecology* **87**: 17–27.

Meyer AH and Schmid B (1999c). Seed dynamics and seedling establishment in the invading perennial *Solidago altissima* under different experimental treatments. *Journal of Ecology* **87**: 28–41.

Meyerowitz EM and Somerville CR (eds) (1994). *Arabidopsis monograph* 27. Cold Spring Harbor Laboratory, Plainview, NY.

Middleton BA (1990). Effect of water depth and clipping frequency on the growth and survival of four wetland plant species. *Aquatic Botany* **37**: 189–196.

Middleton BA (1995). Sampling devices for the measurement of seed rain and hydrochory in rivers. *Bulletin of the Torrey Botanical Club* **122**: 152–155.

Middleton BA, van der Valk AG, Mason DH, Williams RL, and Davis CB (1991). Vegetation dynamics and seed banks of a monsoonal wetland overgrown with *Paspalum distichum* L. in northern India. *Aquatic Botany* **40**: 239–259.

Miglietta F, Lanini M, Bindi M, and Magliulo V (1997). Free air CO_2 enrichment of potato (*Solanum tuberosum*, L.): design and performance of the CO_2-fumigation system. *Global Change Biology* **3**: 417–427.

Mihail JD, Alexander HM, and Taylor SJ (1998). Interactions between root-infecting fungi and plant density in an annual legume, *Kummerowia stipulacea*. *Journal of Ecology* **86**: 739–748.

Milchunas DG (2012). Biases and errors associated with different root production methods and their effects on field estimates of belowground net primary production. In S Mancuso (ed.) *Measuring roots: an updated approach*, pp. 303–339. Springer, Heidelberg.

Milcu A, Lukac M, Subke J-A, Manning P, Heinemeyer A, Wildman D, et al. (2012). Biotic carbon feedbacks in a materially closed soil–vegetation–atmosphere system. *Nature Climate Change* **2**: 281–284.

Millar JG and Haynes KF (1998). *Methods in chemical ecology. Vol 1. Chemical methods*. Chapman and Hall, New York.

Miller TE and Schemske DW (1990). An experimental study of competitive performance in *Brassica rapa* (Cruciferae). *American Journal of Botany* **77**: 993–998.

Miller RE, ver Hoef JM, and Fowler NL (1995). Spatial heterogeneity in eight central Texas grasslands. *Journal of Ecology* **83**: 919–928.

Miller TEX, Williams JL, Jongejans E, Brys R, and Jacquemyn H (2012). Evolutionary demography of iteroparous plants: incorporating non-lethal costs of reproduction into integral projection models. *Proceedings of the Royal Society B: Biological Sciences* **279**: 2831–2840.

Milliken GA and Johnson DE (1984). *Analysis of messy data. Volume I: Designed experiments*. Lifetime Learning, Belmont, CA.

Mills LS, Doak DF, and Wisdom M (1999). The reliability of conservation actions based upon elasticities of matrix models. *Conservation Biology* **13**: 815–829.

Minchin PR (1987). An evaluation of the relative robustness of techniques for ecological ordination. *Plant Ecology* **69**: 89–107.

Mirlohi A, Sabzalian MR, Sharifnabi B, and Nekoui MK (2006). Widespread occurrence of *Neotyphodium*-like endophyte in populations of *Bromus tomentellus* Boiss. in Iran. *FEMS Microbiology Letters* **256**: 126–131.

Mitchell RJ (1993). Path analysis: pollination. In SM Scheiner and J Gurevitch (eds) *Design and analysis of ecological experiments*, pp. 211–231. Chapman and Hall, New York.

Mitchell-Olds T (1996). Genetic constraints on life-history evolution: quantitative-trait loci influencing growth and flowering in *Arabidopsis thaliana*. *Evolution* **50**: 140–145.

Mithen R, Harper JL, and Weiner J (1984). Growth and mortality of individual plants as a function of 'available area'. *Oecologia* **62**: 57–60.

Mitschunas N, Filser J, and Wagner M (2009). On the use of fungicides in ecological seed burial studies. *Seed Science Research* **19**: 51–60.

Molau U and Prentice HC (1992). Reproductive system and population structure in three arctic *Saxifraga* species. *Journal of Ecology* **80**: 149–161.

Molken T, Caluwe H, Hordijk C, Leon-Reyes A, Snoeren TL, Dam N, et al. (2012). Virus infection decreases the attractiveness of white clover plants for a non-vectoring herbivore. *Oecologia* **170**: 433–444.

Molofsky J and Augspurger CK (1992). The effect of leaf litter on early seedling establishment in a tropical forest. *Ecology* **73**: 68–77.

Moloney K (1986). A generalized algorithm for determining category size. *Oecologia* **69**: 176–180.

Mondragón D (2011). Guidelines for collecting demographic data for population dynamics studies on vascular epiphytes. *Journal of the Torrey Botanical Society* **138**: 327–335.

Mooney HA (1976). Some contributions of physiological ecology to plant population biology. *Systematic Botany* **1**: 269–283.

Mooney EH and Niesenbaum RA (2012). Population-specific responses to light influence herbivory in the understory shrub *Lindera benzoin*. *Ecology* **93**: 2683–2692.

Mooney SJ, Pridmore TP, Helliwell J, and Bennett MJ (2012). Developing x-ray computed tomography to non-invasively image 3-D root systems architecture in soil. *Plant and Soil* **352**: 1–22.

Moore KJ, Moser LE, Vogel KP, Waller SS, Johnson BE, and Pedersen JF (1991). Describing and quantifying growth stages of perennial forage grasses. *Agronomy Journal* **83**: 1073–1077.

Morris WF Doak DF (2002). *Quantitative conservation biology: theory and practice of population viability analysis*. Sinauer Associates, Sunderland, MA.

Morris K, Raulings EJ, Melbourne WH, Mac Nally R, and Thompson RM (2011). A novel trap for quantifying the dispersal of seeds by wind. *Journal of Vegetation Science* **22**: 807–817.

Morrison JA (1996). Infection of *Juncus dichotomus* by the smut fungus *Cintractia junci*: an experimental field test of the effects of neighbouring plants, environment, and host plant genotype. *Journal of Ecology* **84**: 691–702.

Morton AG (ed.) (1986). *John Hope, 1725–1786, Scottish botanist*. Edinburgh Botanic Garden (Sibbald) Trust, Edinburgh.

Mudrák O and Frouz J (2012). Allelopathic effect of *Salix caprea* litter on late successional plants at different substrates of post-mining sites: pot experiment studies. *Botany* **90**: 311–318.

Muenchen RA (2011). *R for SAS and SPSS users*, 2nd edn. Springer, New York.

Munguía-Rosas MA, Parra-Tabla V, Ollerton J, and Cervera JC (2012). Environmental control of reproductive phenology and the effect of pollen supplementation on resource allocation in the cleistogamous weed, *Ruellia nudiflora* (Acanthaceae). *Annals of Botany* **109**: 343–350.

Munns R, James RA, Sirault XRR, Furbank RT, and Jones HG (2010). New phenotyping methods for screening wheat and barley for beneficial responses to water deficit. *Journal of Experimental Botany* **61**: 3499–3507.

Münzbergová Z and Herben T (2005). Seed, dispersal, microsite, habitat and recruitment limitation: identification of terms and concepts in studies of limitations. *Oecologia* **145**: 1–8.

Murtaugh PA (1996). The statistical evaluation of ecological indicators. *Ecological Applications* **6**: 132–139.

Musselman RC, Lester DT, and Adams MS (1975). Localized ecotypes of *Thuja occidentalis* L. in Wisconsin. *Ecology* **56**: 647–655.

Nadeem K and Lele SR (2012). Likelihood based population viability analysis in the presence of observation error. *Oikos* **121**: 1656–1664.

Nägeli C (1874). Verdrängung der Pflanzenformen durch ihre Mitbewerber. *Bayerische Akademie der Wissenschaften München. Mathematisch-Physikalische Klasse: Sitzungsberichte* **11**: 109–164.

Narum SR, Buerkle CA, Davey JW, Miller MR, and Hohenlohe PA (2013). Genotyping-by-sequencing in ecological and conservation genomics. *Molecular Ecology* **22**: 2841–2847.

Nathan R, Klein E, Robledo-Arnuncio JJ, and Revilla E (2012). Dispersal kernals: review. In J Clobert, M Baguette, JG Benton, *et al.* (eds) *Dispersal ecology and evolution*, pp. 187–210. Oxford University Press, Oxford.

Neal D (2004). *Introduction to population biology*. Cambridge University Press, Cambridge.

Nenzén HK, Swab RM, Keith DA, and Araújo MB (2012). demoniche—an R-package for simulating spatially-explicit population dynamics. *Ecography* **35**: 577–580.

Newman JA, Abner ML, Dado RG, Gibson DJ, and Hickman A (2003). Effects of elevated CO_2, nitrogen and fungal endophyte-infection on tall fescue: growth, photosynthesis, chemical composition and digestibility. *Global Change Biology* **9**: 425–437.

Newman JA, Bergelson J, and Grafen A (1997). Blocking factors and hypothesis tests in ecology: is your statistics text wrong? *Ecology* **78**: 1312–1320.

Newsham KK and Robinson SA (2009). Responses of plants in polar regions to UVB exposure: a meta-analysis. *Global Change Biology* **15**: 2574–2589.

Newsham KK, Fitter AH, and Watkinson AR (1994). Root pathogenic and arbuscular mycorrhizal fungi determine fecundity of asymptomatic plants in the field. *Journal of Ecology* **82**: 805–814.

Newsham KK, Watkinson AR, and Fitter AH (1995). Rhizosphere and root-infecting fungi and the design of ecological field experiments. *Oecologia* **102**: 230–237.

Newsham KK, Lewis GC, and McLeod ARM (1998). *Neotyphodium lolii*, a fungal leaf endophyte, reduces fertility of *Lolium perenne* exposed to elevated UV-B radiation. *Annals of Botany* **81**: 397–403.

Newton AC (2007). *Forest ecology and conservation: handbook of techniques*. Oxford University Press, Oxford.

N'Guessan FK and Quisenberry SS (1994). Screening selected rice lines for resistance to the rice water weevil (Coleoptera: Curculionidae). *Environmental Entomology* **23**: 665–675.

Nickerson DM and Brunell A (1998). Power analysis for detecting trends in the presence of concomitant variables. *Ecology* **79**: 1442–1447.

Nicotra AB, Chazdon RL, and Iriartre VB (1999). Spatial heterogeneity of light and woody seedling regeneration in tropical wet forests. *Ecology* **80**: 1908–1926.

Nield AP, Ladd PG, and Yates CJ (2009). Reproductive biology, post-fire succession dynamics and population viability analysis of the critically endangered Western Australian shrub *Calytrix breviseta* subsp. *breviseta* (Myrtaceae). *Australian Journal of Botany* **57**: 451–464.

Nilsson MC and Zackrisson O (1992). Inhibition of Scots pine seedling establishment by *Empetrum hermaphroditum*. *Journal of Chemical Ecology* **18**: 1857–1870.

Nobel P (1999). *Physicochemical and environmental plant physiology*, 2nd edn. Academic Press, San Diego, CA.

Norby RJ, Edwards NT, Riggs JS, Abner CH, Wullschleger SD, and Gunderson CA (1997). Temperature-controlled open-top chambers for global change research. *Global Change Biology* **3**: 259–267.

Norby RJ and Zak DR (2011). Ecological lessons from free-air CO_2 enrichment (FACE) experiments. *Annual Review of Ecology, Evolution, and Systematics* **42**: 181–203.

Nordgren A (1988). Apparatus for the continuous, long-term monitoring of soil respiration rate in a large number of samples. *Soil Biology and Biochemistry* **20**: 955–957.

Norris JR, Read DJ, and Varma AK (eds) (1994). *Techniques for mycorrhizal research*. Academic Press, San Diego, CA.

Novoplansky A (1991). Developmental responses of *Portulaca* seedlings to conflicting spectral signals. *Oecologia* **88**: 138–140.

Novoplansky A, Cohen D, and Sachs T (1994). Responses of an annual plant to temporal changes in light environment: an interplay between plasticity and determination. *Oikos* **69**: 437–446.

Noy-Meir I and Briske DD (1996). Fitness components of grazing-induced population reduction in a dominant annual, *Triticum dicoccoides* (wild wheat). *Journal of Ecology* **84**: 439–448.

Nyland J-E and Wallander H (1994). Egosterol analysis as a means of quantifying mycorrhizal biomass. In JR Norris, DJ Read AK Varma (eds) *Techniques for mycorrhizal research*, pp. 537–548. Academic Press, San Diego, CA.

O'Brien EE and Brown JS (2008). Games roots play: effects of soil volume and nutrients. *Journal of Ecology* **96**: 438–446.

O'Dowd DJ and Gill AM (1984). Predator satiation and site alteration following fire: mass reproduction of alpine ash (*Eucalyptus delegatensis*) in southeastern Australia. *Ecology* **65**: 1052–1066.

O'Hara RB and Kotze DJ (2010). Do not log-transform count data. *Methods in Ecology and Evolution* **1**: 118–122.

Ohtani M, Kondo T, Tani N, Ueno S, Lee LS, Ng KKS, et al. (2013). Nuclear and chloroplast DNA phylogeography reveals Pleistocene divergence and subsequent secondary contact of two genetic lineages of the tropical rainforest tree species *Shorea leprosula* (Dipterocarpaceae) in South-East Asia. *Molecular Ecology* **22**: 2264–2279.

Økland RH (1995). Population biology of the clonal moss *Hylocomium splendens* in Norwegian boreal spruce forests. I. Demography. *Journal of Ecology* **83**: 697–712.

Okubo A and Levin SA (1989). A theoretical framework for data analysis of wind dispersal of seeds and pollen. *Ecology* **70**: 329–338.

Oliver FW (1913). Some remarks on Blakeney Point, Norfolk. *Journal of Ecology* **1**: 4–15.

Olson MS, Levsen N, Soolanayakanahally RY, Guy RD, Schroeder WR, Keller SR, et al. (2013). The adaptive potential of *Populus balsamifera* L. to phenology requirements in a warmer global climate. *Molecular Ecology* **22**: 1214–1230.

Orme D, Freckleton R, Thomas G, Petzoldt T, Fritz S, Isaac N, et al. (2012). *Comparative analyses of phylogenetics and evolution in R*. Available at: <http://cran.r-project.org/web/packages/caper/caper.pdf>.

Orsini L, Andrew R, and Eizaguirre C (2013). Evolutionary ecological genomics. *Molecular Ecology* **22**: 527–531.

Ouborg NJ, Piquot Y, and Van Groenendael JM (1999). Population genetics, molecular markers and the study of dispersal in plants. *Journal of Ecology* **87**: 551–568.

Ozsolak F and Milos PM (2010). RNA sequencing: advances, challenges and opportunities. *Nature Reviews Genetics* **12**: 87–98.

Pacala SW (1997). Dynamics of plant communities. In MJ Crawley (ed.) *Plant ecology*, pp. 532–535. Blackwell Scientific Publications, Oxford.

Pacala SW and Silander JA (1990). Field tests of neighborhood population dynamic models of two annual weed species. *Ecological Monographs* **60**: 113–134.

Pacala SW, Canham CD, Saponara J, Silander JA, Kobe RK, and Ribbens E (1996). Forest models defined by field measurements: estimation, error analysis and dynamics. *Ecological Monographs* **66**: 1–43.

Pacioni G (1994). Wet-sieving and decanting techniques for the extraction of spores of vesicular-arbuscular fungi. In JR Norris, DJ Read, and AK Varma (eds) *Techniques for mycorrhizal research*, pp. 777–782. Academic Press, San Diego, CA.

Packard GC and Boardman TJ (1988). The misuse of ratios, indices, and percentages in ecophysiological research. *Physiological Zoology* **61**: 1–9.

Page BG and Thomson WT (1997). *The insecticide, herbicide, fungicide quick guide*. Thomson Publications, Fresno, CA.

Pagel M (1999). Inferring the historical patterns of biological evolution *Nature* **401**: 877–884.

Paine RT (1971). The measurement and application of the calorie to ecological problems. *Annual Review of Ecology and Systematics* **2**: 145–164.

Paine CET, Marthews TR, Vogt DR, Purves D, Rees M, Hector A, et al. (2012). How to fit nonlinear plant growth models and calculate growth rates: an update for ecologists. *Methods in Ecology and Evolution* **3**: 245–256.

Pake CE and Venable LD (1995). Is coexistence of Sonoran Desert annuals mediated by temporal variability reproductive success. *Ecology* **76**: 246–261.

Palmborg C and Nordgren A (1993). Modelling microbial activity and biomass in forest soil with substrate quality measured using near infrared reflectance spectroscopy. *Soil Biology and Biochemistry* **25**: 1713–1718.

Paradis E (2006). *Analysis of phylogenetics and evolution with R*. Springer Science + Business Media, New York.

Parker MP (1995). Plant fitness variation caused by different mutualist genotypes. *Ecology* **76**: 1525–1535.

Parker PG, Snow AA, Schug MD, Booton GC, and Fuerst PA (1998). What molecules can tell us about populations: choosing and using a molecular marker. *Ecology* **79**: 361–382.

Paterson AH (1996). Mapping genes responsible for differences in phenotype. In AH Paterson (ed.) *Genome mapping in plants*, pp. 41–54. R. G. Landes Company, Austin, TX.

Paul ND, Ayres PG, and Wyness LE (1989). On the use of fungicides for experimentation in natural vegetation. *Functional Ecology* **3**: 759–769.

Pearcy RW (1989a). Measurement of transpiration and leaf conductance. In RW Pearcy RW (1989b). Radiation and light measurements. In RW Pearcy, JR Ehleringer, HA Mooney, *et al*. (eds) *Plant physiological ecology*, pp. 97–116. Chapman and Hall, London.

Pearcy, JR Ehleringer, HA Mooney, *et al*. (eds) *Plant physiological ecology*, pp. 137–160. Chapman and Hall, London.

Pearcy RW, Ehleringer J, Mooney HA, and Rundel PW (eds) (1989). *Plant physiological ecology: field methods and instrumentation*. Chapman and Hall, London.

Pearse WD and Purvis A (2013). phyloGenerator: an automated phylogeny generation tool for ecologists. *Methods in Ecology and Evolution* **4**: 692–698.

Pearson K (1930). *The life, letters and labours of Francis Galton*. Cambridge University Press, Cambridge.

Pedigo LP and Buntin GD (1994). *Handbook of sampling methods for arthropods in agriculture*. CRC Press, Boca Raton, FL.

Pe'Er G, Matsinos YG, Johst K, Franz KW, Turlure C, Radchuk V, et al. (2013). A protocol for better design, application, and communication of population viability analyses. *Conservation Biology* **27**: 644–656.

Pejic I, Ajmone-Marsan P, Morgante M, Kozumplick V, Castiglioni P, Taramino G, et al. (1998). Comparative analysis of genetic similarity among maize inbred lines detected by RFLPs, RAPDs, SSRs, and AFLPs. *Theoretical and Applied Genetics* **97**: 1248–1255.

Pereira AMN, Lister RM, Barbara DJ, and Shaner GE (1989). Relative transmissibility of barley yellow dwarf virus from sources with differing virus contents. *Phytopathology* **79**: 1353–1358.

Pérez B and Moreno JM (1998). Methods for quantifying fire severity in shrubland-fires. *Plant Ecology* **139**: 91–101.

Pérez-Harguindeguy N, Díaz S, Garnier E, Lavorel S, Poorter H, Jaureguiberry P, et al. (2013). New handbook for standardised measurement of plant functional traits worldwide. *Australian Journal of Botany* **61**: 167–234.

Perotto S, Malavasi F, and Butcher GW (1994). Use of monoclonal antibodies to study mycorrhiza: present applications and perspectives. In JR Norris, DJ Read, and AK Varma (eds) *Techniques for mycorrhizal research*, pp. 681–708. Academic Press, San Diego, CA.

Perry GLW, Miller BP, and Enright NJ (2006). A comparison of methods for the statistical analysis of spatial point patterns in plant ecology. *Plant Ecology* **187**: 59–82.

Peters RH (1980). Useful concepts for predictive ecology. *Synthese* **43**: 257–269.

Peters RH (1991). *A critique for ecology*. Cambridge University Press, Cambridge.

Petit C and Thompson JD (1997). Variation in phenotypic response to light availability between diploid and tetraploid populations of the perennial grass *Arrhenatherum elatius* from open and woodland sites. *Journal of Ecology* **85**: 657–667.

Pewsey A, Neuhäuser M, and Ruxton GD (2013). *Circular statistics in R*. Oxford University Press, Oxford.

Pfleeger TG, Plocher M, and Bichel P (2010). Response of pioneer plant communities to elevated ozone exposure. *Agriculture, Ecosystems and Environment* **138**: 116–126.

Philippi TE (1993). Multiple regression: herbivory. In SM Scheiner and J Gurevitch (eds) *Design and analysis of ecological experiments*, pp. 183–210. Chapman and Hall, New York.

Phillips NG (2013). Untangling the belowground knot. *Tree Physiology* **33**: 1–2.

Pielou EC (1977). *Mathematical ecology*. New York, Wiley.

Pierson EA and Turner RM (1998). An 85-year study of saguaro (*Carnegiea gigantea*) demography. *Ecology* **79**: 2676–2693.

Pigliucci M and Schlichting CD (1995a). Ontogenetic reaction norms in *Lobelia siphilitica* (Lobeliaceae): response to shading. *Ecology* **76**: 2134–2144.

Pigliucci M and Schlichting CD (1995b). Reaction norms of *Arabidopsis* (Brassicaceae). III. response to nutrients in 26 populations from a worldwide collection. *American Journal of Botany* **82**: 1117–1125.

Pitet M, CamprubÃ A, Calvet C, and EstaÃ°n V (2009). A modified staining technique for arbuscular mycorrhiza compatible with molecular probes. *Mycorrhiza* **19**: 125–131.

Platt JR (1964). Strong inference. *Science* **146**: 347–353.

Poczai P, Varga I, Laos M, Cseh A, Bell N, Valkonen J, et al. (2013). Advances in plant gene-targeted and functional markers: a review. *Plant Methods* **9**: 6.

van de Pol M (2012). Quantifying individual variation in reaction norms: how study design affects the accuracy, precision and power of random regression models. *Methods in Ecology and Evolution* **3**: 268–280.

Policansky D (1987). Sex choice and reproductive costs in jack-in-the-pulpit. *BioScience* **37**: 476–481.

Pons J and Pausas J (2007). Acorn dispersal estimated by radio-tracking. *Oecologia* **153**: 903–911.

Poorter H and Bergkotte M (1992). Chemical composition of 24 wild species differing in relative growth rate. *Plant, Cell and Environment* **15**: 221–229.

Poorter H and Evans JR (1998). Photosynthetic nitrogen-use efficiency of species that differ inherently in specific leaf area. *Oecologia* **116**: 26–37.

Poorter H and Garnier E (2007). Ecological significance of inherent variation in relative growth rate and its components. In FI Pugnaire and F Valladares (eds) *Functional plant ecology*, 2nd edn, pp. 67–100. Taylor and Francis Group, Boca Raton, FL.

Poorter H, Bühler J, van Dusschoten D, Climent J, and Postma JA (2012). Pot size matters: a meta-analysis of the effects of rooting volume on plant growth. *Functional Plant Biology* **39**: 839–850.

Poorter H, Remkes C, and Lambers H (1990). Carbon and nitrogen economy of 24 wild species differing in relative growth rate. *Plant Physiology* **94**: 621–627.

Popper KR (1956). Preface, 1956: on the non-existence of scientific method. In KR Popper (ed.) *Realism and the aim of science*, pp. 5–8. Rowman and Littlefield, Totowa, NJ.

Popper KR (1968). *The logic of scientific discovery*. Hutchinson and Co., London.

Popper KR (1983). *Realism and the aim of science*. Rowman and Littlefield, Totowa, NJ.

Potvin C (2001). ANOVA: experiments in controlled environments. In SM Scheiner and J Gurevitch (eds) *Design and analysis of ecological experiments*, 2nd edn, pp. 63–76. Oxford University Press, New York.

Potvin C and Roff DA (1993). Distribution free and robust statistical methods: viable alternatives to parametric statistics? *Ecology* **74**: 1617–1628.

Powell GW and Bork EW (2006). Aspen canopy removal and root trenching effects on understory vegetation. *Forest Ecology and Management* **230**: 79–90.

Powell JR, Anderson IC, and Rillig MC (2013). A new tool of the trade: plant-trait based approaches in microbial ecology. *Plant and Soil* **365**: 35–40.

Power AG (1996). Competition between viruses in a complex plant–pathogen system. *Ecology* **77**: 1004–1010.

Power AG and Gray SM (1995). Aphid transmission of barley yellow dwarf viruses: interactions between viruses, vectors, and host plants. In CJ D'Arcy and PA Burnett (eds) *Barley yellow dwarf: 40 years of progress*, pp. 259–289. APS Press, St Paul, MN.

Prasad MNV (ed.) (1997). *Plant ecophysiology*. John Wiley and Sons, New York.

Prendeville HR, Ye X, Morris TJ, and Pilson D (2012). Virus infections in wild plant populations are both frequent and often unapparent. *American Journal of Botany* **99**: 1033–1042.

Prentice HC, Lonn M, Lager H, Rosen E, and Van der Maarel E (2000). Changes in allozyme frequencies in *Festuca ovina* populations after a 9-year nutrient/water experiment. *Journal of Ecology* **88**: 331–347.

Price EA and Hutchings MJ (1996). The effects of competition on growth and form in *Glechoma hederacea*. *Oikos* **75**: 279–290.

Price MV and Reichman OJ (1987). Distribution of seeds in Sonoran Desert soils: implications for heteromyid rodent foraging. *Ecology* **68**: 1797–1811.

Procaccini G and Mazzalla L (1998). Population genetic structure and gene flow in the seagrass *Posodonia oceanica* assessed using microsatellite analysis. *Marine Ecology Progress Series* **169**: 133–141.

Purrington CB and Schmitt J (1998). Consequences of sexually dimorphic timing of emergence and flowering in *Silene latifolia*. *Journal of Ecology* **86**: 397–404.

van der Putten WH, Bardgett RD, Bever JD, Bezemer TM, Casper BB, Fukami T, et al. (2013). Plant–soil feedback: the past, the present and future challenges. *Journal of Ecology* **101**: 265–276.

Pyke DA (1986). Demographic responses of *Bromus tectorum* and seedlings of *Agropyrum spicatum* to grazing by small mammals: occurrence and severity of grazing. *Journal of Ecology* **74**: 739–754.

Pyke DA (1987). Demographic responses of *Bromus tectorum* and seedlings of *Agropyrum spicatum* to grazing by small mammals: the influence of grazing frequency and plant age. *Journal of Ecology* **75**: 825–836.

Quinn JA (1978). Plant ecotypes: ecological or evolutionary units? *Bulletin of the Torrey Botanical Club* **105**: 58–64.

Quinn JA (1987). Complex patterns of genetic differentiation and phenotypic plasticity versus an outmoded ecotype terminology. In KM Urbanska (ed.) *Differentiation patterns in higher plants*, pp. 95–113. Academic Press, London.

Quinn GP and Keough MJ (2002). *Experimental design and data analysis for biologists*. Cambridge University Press, Cambridge.

Rajapakse S and Miller JC (1994). Methods for studying vesicular-arbuscular mycorrhizal root colonization and related root physical properties. In JR Norris, DJ Read, and AK Varma (eds) *Techniques for mycorrhizal research*, pp. 761–776. Academic Press, San Diego, CA.

Ramseier D, Connolly J, and Bazzaz FA (2005). Carbon dioxide regime, species identity and influence of species initial abundance as determinants of change in stand biomass composition in five-species communities: an investigation using a simplex design and RGRD analysis. *Journal of Ecology* **93**: 502–511.

Ramstad S and Hestmark G (2000). Effective neighbourhoods for a saxicolous lichen. *Mycological Research* **104**: 198–204.

Ramula S, Rees M, and Buckley YM (2009). Integral projection models perform better for small demographic data sets than matrix population models: a case study of two perennial herbs. *Journal of Applied Ecology* **46**: 1048–1053.

Raunkiaer C (1934). *The life forms of plants and statistical plant geography*. Clarendon Press, Oxford.

R Development Core Team (2011). *R: a language and environment for statistical computing*. R Foundation for Statistical Computing, Vienna.

Reader RJ, Wilson SD, Belcher JW, Wisheu I, Keddy PA, Tilman D, et al. (1994). Plant competition in relation to neighbor biomass: an intercontinental study with *Poa pratensis*. *Ecology* **75**: 1753–1760.

Real LA (1981). Uncertainty and pollinator–plant interactions: the foraging behavior of bees and wasps on artificial flowers. *Ecology* **62**: 20–26.

Real LA and Rathcke BJ (1991). Individual variation in nectar production and its effect on fitness in *Kalmia latifolia*. *Ecology* **72**: 149–155.

Reckhow KH (1990). Bayesian inference in non-replicated ecological studies. *Ecology* **71**: 2053–2059.

Reddy PP (2013). Biofumigation. In PP Reddy (ed.) *Recent advances in crop protection*, pp. 37–60. Springer India, New Delhi.

Redente EF and Richards JL (1997). Effects of lime and fertilizer amendments on plant growth in smelter impacted soils in Montana. *Arid Soil Research and Rehabilitation* **11**: 353–366.

Reekie EG and Bazzaz FA (1987). Reproductive effort in plants. 1. Carbon allocation to reproduction. *American Naturalist* **129**: 876–896.

Reekie E and Bazzaz F (2005). *Reproductive allocation in plants*. Elsevier Academic Press, San Diego, CA.

Reekie JYC, Hicklenton PR, and Reekie EG (1994). Effects of elevated CO_2 on time of flowering in four short-day and four long-day species. *Canadian Journal of Botany* **72**: 533–538.

Rees M and Paynter Q (1997). Biological control of Scotch broom: modelling the determinants of abundance and the potential impact of introduced herbivores. *Journal of Applied Ecology* **34**: 1203–1221.

Rees M, Mangel M, Turnbull L, Sheppard A, and Briese D (2000). The effects of heterogeneity in dispersal and colonisation in plants. In MJH Hutchings,EA John, and AJA Stewart (eds) *The ecological effects of environmental heterogeneity*, pp. 237–265. Blackwell Scientific Publications, Oxford.

Rees M, Childs DZ, and Freckleton RP (2012). Assessing the role of competition and stress: a critique of importance indices and the development of a new approach. *Journal of Ecology* **100**: 577–585.

Rees M, Childs DZ, and Elllner SP (2014). Building integral projection models: a user's guide. *Journal of Animal Ecology* **83**: 528–545.

Reever-Morghan KJ and Seastedt TR (1999). Effects of soil nitrogen reduction on non-native plants in disturbed grasslands. *Restoration Ecology* **7**: 51–55.

Reich PB, Ellsworth DS, Walters MB, Vose JM, Gresham C, Volin JC, et al. (1999). Generality of leaf trait relationships: a test across size biomes. *Ecology* **80**: 1955–1969.

Reid A (2006). *Sampling and testing for plant pathogens.* Western Australian Department of Agriculture and Food, Perth.

Reigosa MJ (ed.) (2001). *Handbook of plant ecophysiology techniques.* Kluwer Academic Publishers, Dordrecht.

Reigosa MJ and Weiss O (2001). Fluorescence techniques. In MJ Reigosa (ed.) *Handbook of plant ecophysiology techniques*, pp. 155–171. Kluwer Academic Publishers, Dordrecht.

Resetarits WJ and Bernardo J (eds) (1998). *Experimental ecology.* Oxford University Press, Oxford.

Reusch TBH, Hukriede W, Stam WT, and Olsen JL (1999). Differentiating between clonal growth and limited gene flow using spatial autocorrelation of microsatellites. *Heredity* **83**: 120–126.

Revell LJ (2012). phytools: an R package for phylogenetic comparative biology (and other things). *Methods in Ecology and Evolution* **3**: 217–223.

Revilla TA, Veen GF, Eppinga MB, and Weissing FJ (2013). Plant–soil feedbacks and the coexistence of competing plants. *Theoretical Ecology* **6**: 99–113.

Reyes-Zepeda F, González-Astorga J, and Montaña C (2012). Heterozygote excess through life history stages in *Cestrum miradorense* Francey (Solanaceae), an endemic shrub in a fragmented cloud forest habitat. *Plant Biology* **15**: 176–185.

Reynolds HL and Rajaniemi TK (2007). Plant interactions: competition. In FI Pugnaire and F Valladares (eds) *Functional plant ecology*, 2nd edn, pp. 457–480. Taylor and Francis Group, Boca Raton, FL.

Ribbens E, Silander JA, and Pacala SW (1994). Seedling recuitment in forests: calibrating models to predict patterns of tree seedling dispersion. *Ecology* **75**: 1794–1806.

Rice EL (1984). *Allelopathy.* Academic Press, New York.

Rich PM (1989). *A manual for analysis of hemispherical canopy photography.* Los Alamos National Laboratory Report LA-11733-M. Available at: <http://upload.wikimedia.org/wikipedia/commons/0/0e/Hemispherical_photography_canopy_manual_rich_1989.pdf> (accessed 29 May 2014).

Richards CL, Bossdorf O, and Verhoeven KJF (2010). Understanding natural epigenetic variation. *New Phytologist* **187**: 562–564.

Richner W, Liedgens M, Bürgi H, Soldati A, and Stamp P (2000). Root image analysis and interpretation. In AL Smit, AG Bengough, C Engels, *et al.* (eds) *Root methods: a handbook*, pp. 303–341. Springer-Verlag, Berlin.

Riis T and Sand-Jensen K (1997). Growth reconstruction and photosynthesis of aquatic mosses: influence of light, temperature and carbon dioxide at depth. *Journal of Ecology* **85**: 359–372.

Rizvi SJH, Haque H, Singh VK, and Rizvi V (1992). A discipline called allelopathy. In SJH Rizvi and V Rizvi (eds) *Allelopathy: basic and applied aspects*, pp. 1–10. Chapman and Hall, London.

Roach DA and Wulff RD (1987). Maternal effects in plants. *Annual Review of Ecology and Systematics* **18**: 209–235.

Robbirt KM, Davy AJ, Hutchings MJ, and Roberts DL (2011). Validation of biological collections as a source of phenological data for use in climate change studies: a case study with the orchid *Ophrys sphegodes*. *Journal of Ecology* **99**: 235–241.

Robertson JH (1947). Responses of range grasses to different intensities of competition with sagebrush (*Artemisia tridentata* Nutt.). *Ecology* **28**: 1–16.

Robertson GP, Coleman DC, Bledsoe CS, and Sollins P (eds) (1999). *Standard soil methods for long-term ecological research.* Oxford University Press, New York.

Rockwood LL (2006). *Introduction to population ecology.* Blackwell Publishing, Oxford.

Rockwood LL and Lobstein MB (1994). The effects of experimental defoliation on reproduction in four species of herbaceous perennials from northern Virginia. *Castanea* **59**: 41–50.

Roiloa SR and Hutchings MJ (2013). The effects of physiological integration on biomass partitioning in plant modules: an experimental study with the stoloniferous herb *Glechoma hederacea*. *Plant Ecology* **214**: 521–530.

Romeo JT and Weidenhamer JD (1998). Bioassays for allelopathy in terrestrial plants. In KF Haynes and JG Millar (eds) *Methods in chemical ecology, Vol. 2. Bioassay methods*, pp. 179–211. Kluwer Academic Publishing, Norvell, MA.

Rorison IH and Robinson D (1986). Mineral nutrition. In PD Moore and SB Chapman (eds) *Methods in plant ecology*, pp. 145–213. Blackwell Scientific Publications, Oxford.

Rorison IH, Spencer RE, and Gupta PL (1993). Chemical analysis. In GAF Hendry and JP Grime (eds) *Methods in comparative plant ecology*, pp. 156–163. Chapman and Hall, London.

Rösch H, Van Rooyen MW, and Theron GK (1997). Predicting competitive interactions between pioneer plant species by using plant traits. *Journal of Vegetation Science* **8**: 489–494.

Rosenberg MS and Anderson CD (2011). PASSaGE: pattern analysis, spatial statistics and geographic exegesis. Version 2. *Methods in Ecology and Evolution* **2**: 229–232.

Roughgarden J (1998). *Primer of ecological theory.* Printice-Hall, Upper Saddle River, NJ.

Rúa MA, McCulley RL, and Mitchell CE (2013). Fungal endophyte infection and host genetic background jointly modulate host response to an aphid-transmitted viral pathogen. *Journal of Ecology* **101**: 1007–1018.

Ruete A, Wiklund K, and Snäll T (2012). Hierarchical Bayesian estimation of the population viability of an epixylic moss. *Journal of Ecology* **100**: 499–507.

Rundel PW and Jarrell WM (1989). Water in the environment. In RW Pearcy, JR Ehleringer, HA Mooney, *et al.* (eds) *Plant physiological ecology*, pp. 29–56. Chapman and Hall, London.

Rustad LE and Campbell JL (2012). A novel ice storm manipulation experiment in a northern hardwood forest. *Canadian Journal of Forest Research* **42**: 1810–1818.

Ruther J and Hilker M (1998). A versatile method for online analysis of volatile compounds from living samples. *Journal of Chemical Ecology* **24**: 525–534.

Sage RF (1994). Acclimation of photosynthesis to increasing atmospheric CO_2: the gas exchange perspective. *Photosynthesis Research* **39**: 351–368.

Salguero-Gómez R and De Kroon H (2010). Matrix projection models meet variation in the real world. *Journal of Ecology* **98**: 250–254.

Salguero-Goméz R and Shefferson R (2013). Plants don't count... or do they? New perspectives on the universality of senescence. *Journal of Ecology* **101**: 545–554.

Salisbury EJ (1942). *The reproductive capacity of plants*. Bell, London.

Salisbury FB (1996). *Units, symbols, and terminology for plant physiology*. Oxford University Press, Oxford.

Sans FX, Escarré J, Gorse V, and Lepart J (1998). Persistence of *Picris hieracioides* populations in old fields: an example of facilitation. *Oikos* **83**: 283–292.

Santas R, Koussoulaki A, and Häder D-P (1997). In assessing biological UV-B effects, natural fluctuations of solar radiation should be taken into account. *Plant Ecology* **128**: 93–97.

Santiago LS, Wright SJ, Harms KE, Yavitt JB, Korine C, Garcia MN, et al. (2012). Tropical tree seedling growth responses to nitrogen, phosphorus and potassium addition. *Journal of Ecology* **100**: 309–316.

Sarathchandra SU, Dimenna ME, Burch G, Brown JA, Watson RN, Bell NL, et al. (1995). Effects of plant parasitic nematodes and rhizosphere micro-organisms on the growth of white clover (*Trifolium repens* L.) and perennial rye grass (*Lolium perenne* L.). *Soil Biology and Biochemistry* **27**: 9–16.

Sarathchandra SU, Watson RN, Cox NR, di Menna ME, Brown JA, Burch G, et al. (1996). Effects of chitin amendment of soil microorganisms, nematodes, and growth of white clover (*Trifolium repens* L.) and perennial ryegrass (*Lolium perenne*L.). *Biology and Fertility of Soils* **22**: 221–226.

Sarukhán J (1974). Studies on plant demography: *Ranunculus repens* L., *R. bulbosus* L. and *R. acris* L. II. Reproductive strategies and seed population dynamics. *Journal of Ecology* **62**: 151–177.

Sarukhán J and Harper JL (1973). Studies on plant demography: *Ranuculus repens* L., *R. bulbosus* L. and *R. acris* L. I. Population flux and survivorship. *Journal of Ecology* **61**: 675–716.

SAS Institute Inc. (2008). *SAS/STAT® 9.2 user's guide*. SAS Institute Inc., Cary, NC.

Savolainen O and Hedrick P (1995). Heterozygosity and fitness: no association in Scots pine. *Genetics* **140**: 755–766.

Schafer DE and Chilcote DO (1969). Factors influencing persistence and depletion in buried seed populations. I. A model for analysis of parameters of buried seed persistence and depletion. *Crop Science* **10**: 342–345.

Schansker G, Tóth SZ, Holzwarth AR, and Garab G (2014). Chlorophyll *a* fluorescence: beyond the limits of the Q_A model. *Photosynthesis Research* **120**: 43–58.

Scheiner SM (2001). Introduction: theories, hypotheses, and statistics. In SM Scheiner and J Gurevitch (eds) *Design and analysis of ecological experiments*, 2nd edn, pp. 1–13. Oxford University Press, New York.

Scheiner SM and Gurevitch J (eds) (2001). *Design and analysis of ecological experiments*, 2nd edn. Oxford University Press, New York.

Schemske DW, Husband BC, Ruckelshaus MH, Goodwillie C, Parker IM, and Bishop JG (1994). Evaluating approaches to the conservation of rare and endangered pants. *Ecology* **75**: 584–606.

Schenck NC (ed.) (1982). *Methods and principles of mycorrhizal research*. The American Phytopathological Society, St Paul, MN.

Scheneiter O and Améndola C (2012). Tiller demography in tall fescue (*Festuca arundinacea*) swards as influenced by nitrogen fertilization, sowing method and grazing management. *Grass and Forage Science* **67**: 426–436.

Schiffers K, Tielbörger K, Tietjen B, and Jeltsch F (2011). Root plasticity buffers competition among plants: theory meets experimental data. *Ecology* **92**: 610–620.

Schmid B (1984a). Life histories in clonal plants of the *Carex flava* group. *Journal of Ecology* **72**: 93–114.

Schmid B (1985a). Clonal growth in grassland perennials II. Growth form and fine-scale colonizing ability. *Journal of Ecology* **73**: 808–818.

Schmid B (1985b). Clonal growth in grassland perennials III. Genetic variation and plasticity between and within populations of *Bellis perennis* and *Prunella vulgaris*. *Journal of Ecology* **73**: 819–830.

Schmid B (1994). Effects of genetic diversity in experimental stands of *Solidago altissima*—evidence for the potential role of pathogens as selective agents in plant populations. *Journal of Ecology* **82**: 165–175.

Schmid B and Harper JL (1985). Clonal growth in grassland perennials I. Density and pattern-dependent competition between plants with different growth forms. *Journal of Ecology* **73**: 793–808.

Schmid B, Puttick GM, Burgess KH, and Bazzaz FA (1988). Correlations between genet architecture and some life-history features on three species of *Solidago*. *Oecologia* **75**: 459–464.

Schoener TW (1983). Field experiments on interspecific competition. *American Naturalist* **122**: 240–285.

Schott GW (1994). A seed trap for monitoring the seed rain in terrestrial communities. *Canadian Journal of Botany* **73**: 794–796.

Schramm J and Ehrenfeld J (2010). Leaf litter and understory canopy shade limit the establishment, growth and reproduction of *Microstegium vimineum*. *Biological Invasions* **12**: 3195–3204.

Schröeder R and Prasse R (2013). Do cultivated varieties of native plants have the ability to outperform their wild relatives. *PLoS ONE* **8**(8): e71066, doi:10.1371/journal.pone.0071066.

Schwaegerle KE (1983). A method for maintaining constant soil moisture availability for potted plants. *Soil Science Society of America Journal* **47**: 608–610.

Schwaegerle KE and Bazzaz FA (1987). Differentiation among nine populations of *Phlox*: response to environmental gradients. *Ecology* **68**: 54–64.

Seastedt TR, Todd TC, and James SW (1987). Experimental manipulations of the arthropod nematode and earthworm communities in a North American tallgrass prairie. *Pedobiologia* **30**: 9–17.

Seastedt TR, James SW, and Todd TC (1988). Interactions among soil invertebrates, microbes and plant growth in the tallgrass prairie. *Agriculture, Ecosystems and Environment* **24**: 219–228.

Seifan M, Seifan T, Arizal C, and Tielbörger K (2010). Facilitating an importance index. *Journal of Ecology* **98**: 356–361.

Seiwa K (1998). Advantages of early germination for growth and survival of seedlings of *Acer mono* under different overstorey phenologies in deciduous broad-leaved forests. *Journal of Ecology* **86**: 219–228.

Semchenko M, Hutchings MJ, and John EA (2007). Challenging the tragedy of the commons in root competition: confounding effects of neighbour presence and substrate volume. *Journal of Ecology* **95**: 252–260.

Setälä H (1995). Growth of birch and pine seedlings in relation to grazing by soil fauna on ectomycorrhizal fungi. *Ecology* **76**: 1844–1851.

Shea K and Kelly D (1998). Estimating biocontrol agent impact with matrix models: *Carduus nutans* in New Zealand. *Ecological Applications* **8**: 824–832.

Shea K, Rees M, and Wood SN (1994). Trade-offs, elasticities and the comparative method. *Journal of Ecology* **82**: 951–957.

Shaffer ML (1981). Minimum population sizes for species conservation. *BioScience* **31**: 131–134.

Sharitz RB and McCormick JF (1973). Population dynamics of two competing annual plant species. *Ecology* **54**: 723–740.

Shaw RG and Mitchell-Olds T (1993). ANOVA for unbalanced design: an overview. *Ecology* **74**: 1638–1645.

Sheail J (1987). *Seventy-five years in ecology: the British Ecological Society*. Blackwell Scientific Publications, Oxford.

Sheppard LJ, Leith ID, Murray MB, Cape JN, and Kennedy VH (1998). The response of Norway spruce seedlings to simulated acid mist. *New Phytologist* **138**: 709–723.

Shipley B and Dion J (1992). The allometry of seed production in herbaceous angiosperms. *American Naturalist* **139**: 467–483.

Shugart HH (1984). *A theory of forest dynamics: the ecological implications of forest succession models*. Springer-Verlag, New York.

Shumway DL and Koide RT (1994). Within-season variability in mycorrhizal benefit to reproduction in *Abutilon theophrasti* Medic. *Plant, Cell and Environment* **17**: 821–827.

Shupert LA, Ebbs SD, Lawrence J, Gibson DJ, and Filip P (2013). Dissolution of copper and iron from automotive brake pad wear debris enhances growth and accumulation by the invasive macrophyte *Salvinia molesta* Mitchell. *Chemosphere* **92**: 45–51.

Sibbesen E (1977). A simple ion-exchange resin procedure for extracting plant-available elements from soil. *Plant and Soil* **46**: 665–669.

Siegel MR, Latch GCM, and Johnson MC (1987). Fungal endophytes of grasses. *Annual Review of Phytopathology* **25**: 293–315.

Siegel MR, Varney DR, Johnson MC, Nesmith WC, Buckner RC, Bush LP, et al. (1984). A fungal endophyte of tall fescue: evaluation of control methods. *Phytopathology* **74**: 937–941.

Siegenthaler PA and Douet-Orhant V (1994). Relationship between the ATP content measured at three imbibition times and germination of onion seeds during storage at 3, 15 and 30 degrees C. *Journal of Experimental Botany* **45**: 1365–1371.

Sigwela AM, Kerley GIH, Mills AJ, and Cowling RM (2009). The impact of browsing-induced degradation on the reproduction of subtropical thicket canopy shrubs and trees. *South African Journal of Botany* **75**: 262–267.

Silander JA and Antonovics J (1982). Analysis of interspecific interactions in a coastal plant community—a perturbation approach. *Nature* **298**: 557–560.

Silander JA and Pacala SW (1990). The application of plant population dynamic models to understanding plant competition. In JB Grace and D Tilman (eds) *Perspectives on plant competition*, pp. 67–91. Academic Press, San Diego, CA.

Silverberg JL, Noar RD, Packer MS, Harrison MJ, Henley CL, Cohen I, et al. (2012). 3D imaging and mechanical modeling of helical buckling in *Medicago truncatula* plant roots. *Proceedings of the National Academy of Sciences of the United States of America* **109**: 16794–16799.

Silvertown J (1987). *Introduction to plant population ecology*. Longman Scientific and Technical, Harlow.

Silvertown JW and Charlesworth D (2001). *Introduction to plant population biology*, 4th edn. Blackwell Science, Oxford.

Silvertown J and Dodd M (1997). Comparing plants and connecting traits. In J Silvertown, M Franco, and JL Harper (eds) *Plant life histories: ecology, phylogeny and evolution*, pp. 3–16. Cambridge University Press, Cambridge.

Silvertown JW and Lovett Doust J (1993). *Introduction to plant population biology*, 3rd edn. Blackwell Scientific Publications, Oxford.

Silvertown J, Holtier S, Johnson J, and Dale P (1992). Cellular automaton models of interspecific competition for space—the effect of pattern on process. *Journal of Ecology* **80**: 527–534.

Silvertown J, Franco M, Pisanty I, and Mendoza A (1993). Comparative plant demography—relative importance of life-cycle components to the finite rate of increase in woody and herbaceous perennials. *Journal of Ecology* **81**: 465–476.

Silvertown J, Franco M, and Harper JL (eds) (1997). *Plant life histories: ecology, phylogeny and evolution*. Cambridge University Press, Cambridge.

Sinclair G (1824). *Hortus gramineus woburnensis*. James Ridgeway, London.

Skeel VA and Gibson DJ (1998). Photosynthetic rates and vegetative production of *Sorghastrum nutans* in response to competition at two strip mines and a railroad prairie. *Photosynthetica* **35**: 139–149.

Sletvold N, Trunschke J, and Wimmergren C, Ågren J (2012). Separating selection by diurnal and nocturnal pollinators on floral display and spur length in *Gymnadenia conopsea*. *Ecology* **93**: 1880–1891.

Smit AL, Bengough AG, Engels C, Van Noordwijk M, Pellerin S, and Van De Geijn SC (eds) (2000). *Root methods: a handbook*. Springer-Verlag, Berlin.

Smith KA (ed.) (1983). *Soil analysis: instrumental techniques and related procedures*. Marcel Dekker, Inc., New York.

Smith EP (2002). BACI design. In AH El-Shaarawi and WW Piegorsch (eds) *Encyclopedia of environments*, Vol. **1**, pp. 141–148. John Wiley and Sons, Chichester.

Smith EA and Dehme FW (1992). The biological activity of glyphosate to plants and animals: a literature review. *Veterinary and Human Toxicity* **34**: 531–543.

Smith M and Moss JS (1998). An experimental investigation, using stomatal conductance and fluorescence, of the flood sensitivity of *Boltonia decurrens* and its competitors. *Journal of Applied Ecology* **35**: 553–561.

Smith GS, Johnston CM, and Cornforth IS (1983). Comparison of nutrient solutions for growth of plants in sand culture. *New Phytologist* **94**: 537–548.

Smith SD, Monson RK, and Anderson JE (1997). *Physiological ecology of North American desert plants*. Springer, Berlin.

Snaydon RW (1970). Rapid population differentiation in a mosaic environment. I. The response of *Anthoxanthum odoratum* populations to soils. *Evolution* **24**: 257–269.

Snaydon RW and Davies TM (1972). Rapid population differentiation in a mosaic environment. II. Morphological variation in *Anthoxanthum odoratum*. *Evolution* **26**: 390–405.

Snedecor GW and Cochran WG (1989). *Statistical methods*, 8th edn. Iowa State University Press, Ames, IA.

Sokal RR and Rohlf FJ (1969). *Biometry*. W. H. Freeman and Company, San Francisco, CA.

Sokal RR and Rohlf FJ (1995). *Biometry*, 3rd edn. W. H. Freeman and Company, San Francisco, CA.

Sokal RR and Rohlf FJ (2011). Biometry, 4th edition. W. H. Freeman and Company, San Francisco, CA.

Solbrig OT (ed.) (1980a). *Demography and evolution in plant populations*. University of California Press, Berkeley, CA.

Solbrig OT (1980b). Demography and natural selection. In OT Solbrig (ed.) *Demography and evolution in plant populations*, pp. 1–20. University of California Press, Berkeley, CA.

Solbrig OT, Jain S, Johnson GB, and Raven PH (eds) (1979). *Topics in plant population biology*. Columbia University Press, New York.

Soltis DE and Soltis PS (eds) (1989). *Isozymes in plant biology*. Dioscorides Press, Portland, OR.

Somasegaran P and Hoben HJ (1994). *Handbook for rhizobia: methods in legume–rhizobium technology*. Springer-Verlag, New York.

Sousa WP (1984). The role of disturbance in natural communities. *Annual Review of Ecology and Systematics* **15**: 353–391.

Sparks DL (ed.) (1996). *Methods of soil analysis. Part 3, Chemical methods*. Soil Science Society of America, Madison, WI.

Spooner PG, Lunt ID, Briggs SV, and Freudenberger D (2004). Effects of soil disturbance from roadworks on roadside shrubs in a fragmented agricultural landscape. *Biological Conservation* **117**: 393–406.

St Hilaire LR and Leopold DJ (1995). Conifer seedling distribution in relation to microsite conditions in a central

New York forested minerotrophic peatland. *Canadian Journal of Forest Research* **25**: 261–269.

Stapleton JJ (2012). Feasibility of solar tents for inactivating weedy plant propagative material. *Journal of Pest Science* **85**: 17–21.

Steel RGD, Torrie JH, and Dickey D (1996). *Principles and procedures of statistics: a biometrical approach*, 3rd edn. McGraw-Hill, New York.

Steets JA and Ashman T-L (2010). Maternal effects of herbivory in *Impatiens capensis*. *International Journal of Plant Sciences* **171**: 509–518.

Steidl RJ and Thomas L (2001). Power analysis and experimental design. In SM Scheiner and J Gurevitch (eds) *Design and analysis of ecological experiments*, 2nd edn, pp. 14–36. Oxford University Press, New York.

Stoll P, Egli P, and Schmid B (1998). Plant foraging and rhizome growth patterns of *Solidago altissima* in response to mowing and fertilizer application. *Journal of Ecology* **86**: 341–354.

Stott I, Hodgson DJ, and Townley S (2012). popdemo: an R package for population demography using projection matrix analysis. *Methods in Ecology and Evolution* **3**: 797–802.

Stowe K and Marquis R (2011). Costs of defense: correlated responses to divergent selection for foliar glucosinolate content in *Brassica rapa*. *Evolutionary Ecology* **25**: 763–775.

Strickler GS (1959). *Use of the densiometer to estimate density of forest canopy on permanent sample plots*. USDA Forest Service Pacific Northwest Forest Range Experiment Station Research Note 180, pp. 1–5.

Strong DRJ (1980). Null hypotheses in ecology. *Synthese* **43**: 271–285.

Stubben C and Milligan B (2007). Estimating and analyzing demographic models using the popbio package in R. *Journal of Statistical Software* **22**: 1–23.

Style Manual Committee Council of Science Editors (2006). *Scientific style and format: the CSE manual for authors, editors, and publishers*, 7th edn.Cambridge University Press, Cambridge.

Suding KN and Goldberg DE (1999). Variation in the effects of vegetation and litter on recruitment across productivity gradients. *Journal of Ecology* **87**: 436–449.

Sugiyama S (1998). Differentiation in competitive ability and cold tolerance between diploid and tetraploid cultivars in *Lolium perenne*. *Euphytica* **103**: 55–59.

Sultan SE, Wilczek AM, Hann SD, and Brosi BJ (1998). Contrasting ecological breadth of co-occurring annual *Polygonum* species. *Journal of Ecology* **86**: 363–383.

Sunohara Y, Ikeda H, Tsukagoshi S, Murata Y, Sakurai N, and Noma Y (2002). Effects of trampling on morphology and ethylene production in asiatic plantain. *Weed Science* **50**: 479–484.

Suppe F (1977). *The structure of scientific theories*, 2nd edn. University of Illinois Press, Urbana, IL.

Sutherland WJ (1996). *Ecological census techniques: a handbook*. Cambridge University Press, Cambridge.

Sutherland WJ, Freckleton RP, Godfray HCJ, Beissinger SR, Benton T, Cameron DD, et al. (2013). Identification of 100 fundamental ecological questions. *Journal of Ecology* **101**: 58–67.

Suwa T and Louda SM (2011). Combined effects of plant competition and insect herbivory hinder invasiveness of an introduced thistle. *Oecologia* **169**: 467–476.

Swenson N (2014). Phylogenetics and comparative methods. In *Oxford bibliographies online: ecology*. In press.

Syde MA and Peakall R (1998). Extensive clonality in the endangered shrub *Haloragodendron lucasii* (Haloragaceae) revealed by allozymes and RAPDs. *Molecular Ecology* **7**: 87–93.

Sykes JB (ed.) (1976). *The concise Oxford dictionary of current English*, 6th edn. Oxford University Press, Oxford.

Székáacs A and Darvas B (2012). Forty years with glyphosate. In MN Hasaneen (ed.) *Herbicides: properties, synthesis and control of weeds*, pp. 247–284. InTech.

Tabacknick BG and Fidell LS (2012). *Using multivariate statistics*, 6th edn. Pearson, Harlow.

Tamayo PR, Weiss O, and Sánchez-Moreiras AM (2001). Gas exchange techniques in photosynthesis and respiration infrared gas analyser. In MJ Reigosa (ed.) *Handbook of plant ecophysiology techniques*, pp. 113–139. Kluwer Academic Publishers, Dordrecht.

Tang C-S (1986). Continuous trapping techniques for the study of allelochemicals from higher plants. In AR Putnam and C-S Tang (eds) *The science of allelopathy*, pp. 113–131. Wiley-Interscience, New York.

Tang W (1990). Reproduction in the cycad *Zamia pumila* in a fire-climax habitat: an eight-year study. *Bulletin of the Torrey Botanical Club* **117**: 368–374.

Tansley AG (1917). On competition between *Galium saxatile* L. (*G. hercynicum* Weig.) and *Galium sylvestre* Poll. (*G. asperum* Schreb.) on different types of soil. *Journal of Ecology* **5**: 173–179.

Taylor BL, Wade PR, Stehn RA, and Cochrane JF (1996). A Bayesian approach to classification criteria for spectacled eiders. *Ecological Applications* **6**: 1077–1089.

Taylor BN, Beidler KV, Cooper ER, Strand AE, and Pritchard SG (2013). Sampling volume in root studies: the pitfalls of under-sampling exposed using accumulation curves. *Ecology Letters* **16**: 862–869.

Ter Heerdt GNJ, Verweij GL, Bekker RM, and Bekker JP (1996). An improved method for seed bank analysis: seedling emergence after removing the soil by sieving. *Functional Ecology* **10**: 144–151.

The International Brachypodium Initiative (2010). Genome sequencing and analysis of the model grass *Brachypodium distachyon*. *Nature* **463**: 763–768.

The Nature Conservancy (2012). *Fire management manual*. Available at: <http://www.tncfiremanual.org/guidelines.htm> (accessed 5 March 2013).

Thioulouse J, Chessel D, Dolédec S, and Olivier JM (1997). ADE-4: a multivariate analysis and graphical display software. *Statistics and Computing* **7**: 7–75.

Thomas H (1980). Terminology and definitions in studies of grassland plants. *Grass and Forage Science* **35**: 13–23.

Thomas L (1997). Retrospective power analysis. *Conservation Biology* **11**: 276–280.

Thomas SC and Bazzaz FA (1993). The genetic component in plant size hierarchies: norms of reaction to density in a *Polygonum species*. *Ecological Monographs* **63**: 231–249.

Thomas L and Krebs CJ (1997). A review of statistical power analysis software. *Bulletin of the Ecological Society of America* **78**: 126–139.

Thomas SC and Weiner J (1989). Growth, death and size distribution change in an *Impatiens pallida* population. *Journal of Ecology* **77**: 524–536.

Thomas JR, Gibson DJ, and Middleton BA (2005). Water dispersal of vegetative bulbils of the invasive exotic: *Dioscorea oppositifolia* in southern Illinois. *Journal of the Torrey Botanical Society* **132**: 187–196.

Thompson PA (1975). Characterization of the germination responses of *Silene dioica* (L.) Clairv., populations from Europe. *Annals of Botany* **39**: 1–19.

Thompson PA (1977). A note on the germination of *Narcissus bulbocodium* L. *New Phytologist* **79**: 287–290.

Thompson K (1993). Persistence in soil. In GAF Hendry and JP Grime (eds) *Methods in comparative plant ecology*, pp. 199–202. Chapman and Hall, London.

Thompson K (2000). The functional ecology of soil seed banks. In M Fenner (ed.) *Seeds: the ecology of regeneration in plant communities*. CAB International, Wallingford, Oxon.

Thompson PA and Fox DJC (1971). A simple thermo-gradient bar designed for use in seed germination studies. *Proceedings of the International Seed Testing Association* **36**: 255–263.

Thompson K and Grime JP (1979). Seasonal variation in the seed banks of herbaceous species in ten contrasting habitats. *Journal of Ecology* **67**: 893–921.

Thompson K and Stewart AJA (1981). The measurement and meaning of reproductive effort in plants. *American Naturalist* **117**: 205–211.

Thompson K, Bakker JP, and Bekker RM (1997). *Soil seed banks of north-west Europe: methodology, density and longevity*. Cambridge University Press, Cambridge.

Thornley JHM (1998). *Grassland dynamics: an ecosystem simulation model*. CAB International, Wallingford, Oxon.

Thurber CS, Jia MH, Jia Y, and Caicedo AL (2013). Similar traits, different genes? Examining convergent evolution in related weedy rice populations. *Molecular Ecology* **22**: 685–698.

van Tienderen PH and van der Toorn J (1991). Genetic differentiation between populations of *Plantago lanceolata*. I. local adaptation in three contrasting habitats. *Journal of Ecology* **79**: 27–42.

Tilman D (1996). Biodiversity: population versus ecosystem stability. *Ecology* **77**: 350–363.

Tilman D (1998). *Plant stategies and the dynamics and structure of plant communities*. Princeton University Press, Princeton, NJ.

Toft CA and Shea PJ (1983). Detecting community-wide patterns; estimating power strengthens statistical inference. *American Naturalist* **122**: 618–625.

Tompkins SP and Williams PH (1990). Fast plants for fine science—an introduction to the biology of rapid cycling *Brassica campestris* (*rapa*). *Journal of Biological Education* **24**: 239–250.

Topp GC and Reynolds WD (1998). Time domain reflectometry: a seminal technique for measuring mass and energy in soil. *Soil and Tillage Research* **47**: 125–132.

Toräng P, Ehrlén J, and Ågren J (2010). Linking environmental and demographic data to predict future population viability of a perennial herb. *Oecologia* **163**: 99–109.

Totland Ø and Nylèhn J (1998). Assessment of the effects of environmental change on the performance and density of *Bistorta vivipara*: the use of multivariate analysis and experimental manipulation. *Journal of Ecology* **86**: 989–998.

Travers SE, Tang Z, Caragea D, Garrett KA, Hulbert SH, Leach JE, et al. (2010). Variation in gene expression of *Andropogon gerardii* in response to altered environmental conditions associated with climate change. *Journal of Ecology* **98**: 374–383.

Trinder CJ, Brooker RW, and Robinson D (2013). Plant ecology's guilty little secret: understanding the dynamics of plant competition. *Functional Ecology* **27**: 918–929.

Tripathi RS (1985). Population dynamics of a few exotic weeds in north-east India. In J White (ed.) *Studies on plant demography: a festschrift for John L. Harper*, pp. 157–170. Academic Press, London.

Tsuyuzaki S (2010). Seed survival for three decades under thick tephra. *Seed Science Research* **20**: 201–207.

Tufte ER (1983). *The visual display of quantitative information*. Graphics Press, Cheshire, CT.

Tuljapurkar S (1980). *Population dynamics in variable environments*. Springer-Verlag, New York.

Turesson G (1922). The species and variety as ecological units. *Hereditas* **3**: 100–113.

Turesson G (1930). The selective effect of climate upon the plant species. *Hereditas* **14**: 99–152.

Turkington R (1994). Effect of propagule source on competitive ability of pasture grasses: spatial dynamics of six grasses in simulated swards. *Canadian Journal of Botany* **72**: 111–121.

Turkington R (2009). Professor John L. Harper FRS CBE (1925–2009). *Journal of Ecology* **97**: 835–837.

Turkington R and Harper JL (1979). The growth, distribution and neighbour relationships of *Trifolium repens* in a permanent pasture. IV. Fine-scale biotic differentiation. *Journal of Ecology* **67**: 245–254.

Turkington R and Jolliffe PA (1996). Interference in *Trifolium repens—Lolium perenne* mixtures: short- and long-term relationships. *Journal of Ecology* **84**: 563–571.

Tyler CM (1996). Relative importance of factors contributing to postfire seedling establishment in maritime chaparral. *Ecology* **77**: 2182–2195.

Ulery AL and Drees R (eds) (2008). *Methods of soil analysis. Part 5, Mineralogical methods*. Soil Science Society of America, Madison, WI.

Underwood AJ (1990). Experiments in ecology and management: their logics, functions and interpretations. *Australian Journal of Ecology* **15**: 365–389.

Underwood AJ (1992). Beyond BACI: the detection of environmental impacts on populations in the real, but variable, world. *Journal of Experimental Marine Biology and Ecology* **161**: 145–178.

Underwood AJ (1997). *Experiments in ecology: their logical design and interpretation using analysis of variance*. Cambridge University Press, Cambridge.

Uren NC and Parsons RF (2013). Nutritional characteristics of soils on an inferred chronosequence. A comment on Laliberté et al. *Journal of Ecology* **101**: 1085–1087.

Utobo EB, Ogbodo EN, and Nwogbada AC (2011). Techniques for extraction and quantification of arbuscular mycorrhizal fungi. *Libyan Agriculture Research Center Journal International* **2**: 68–78.

Valverde T and Silvertown J (1997). A metapopulation model for *Primula vulgaris*, a temperate forest understory herb. *Journal of Ecology* **85**: 193–210.

Valverde T and Silvertown J (1998). Variation in the demography of a woodland understory herb (*Primula vulgaris*) along the forest regeneration cycle: projection matrix analysis. *Journal of Ecology* **86**: 545–562.

Vandermeer JH and Goldberg DE (2003). *Population ecology: first principles*. Princeton University Press, Princeton, NJ.

Van der Pijl L (1982). *Principles of dispersal in higher plants*, 3rd edn. Springer-Verlag, New York.

Vannette RL and Hunter MD (2011). Genetic variation in expression of defense phenotype may mediate evolutionary adaptation of *Asclepias syriaca* to elevated CO_2. *Global Change Biology* **17**: 1277–1288.

Vargas-Mendoza MDIC and Fowler NL (1998). Resource-based models of competitive interactions. I. Intraspecific competition in *Ratibida columnifera* (Asteraceae). *American Journal of Botany* **85**: 932–939.

Vavrek MC, McGraw JB, and Yank HS (1997). Within-population variation in demography of *Taraxacum officianale*: season- and size-dependent survival, growth and reproduction. *Journal of Ecology* **85**: 277–287.

Vazquez-Yanes C, Orozco-Segovia A, Rincon E, Sanchez-Coronado ME, Huante P, Toledo JR, et al. (1990). Light beneath the litter in a tropical forest: effect on seed germination. *Ecology* **71**: 1952–1958.

Veiga RSL, Howard K, and van der Heijden MGA (2012). No evidence for allelopathic effects of arbuscular mycorrhizal fungi on the non-host plant *Stellaria media*. *Plant and Soil* **360**: 319–331.

Veresgolou DS and Fitter AH (1984). Spatial and temporal patterns of growth and nutrient uptake of five co-existing grasses. *Journal of Ecology* **72**: 259–272.

Ver Hoef JM and Boveng PL (2007). Quasi-Poisson vs. negative binomial regression: how should we model overdispered count data? *Ecology* **88**: 2766–2772.

Verhoeven KJF, Jansen JJ, van Dijk PJ, and Biere A (2009). Stress-induced DNA methylation changes and their heritability in asexual dandelions. *New Phytologist* **185**: 1108–1118.

Viard-Crétat F, Baptist F, Secher-Fromell H, and Gallet C (2012). The allelopathic effects of *Festuca paniculata* depend on competition in subalpine grasslands. *Plant Ecology* **213**: 1963–1973.

Vierheilig H, Schweiger P, and Brundrett M (2005). An overview of methods for the detection and observation of arbuscular mycorrhizal fungi in roots. *Physiologia Plantarum* **125**: 393–404.

Vilà M and Lloret F (2000). Seed dynamics of the mast seeding tussock grass *Ampelodesmos mauritanica* in Mediterranean shrublands. *Journal of Ecology* **88**: 479–491.

Vila-Aiub M, Goh S, Gaines T, Han H, Busi R, Yu Q, et al. (2014). No fitness cost of glyphosate resistance endowed by massive EPSPS gene amplification in *Amaranthus palmeri*. *Planta* **239**: 793–801.

Vincent JM (1970). *A manual for the practical study of root-nodule bacteria*. IBP Handbook No. 15. Blackwell Scientific Publications, Oxford.

Vinton M and Burke IC (1995). Interactions between individual plant species and soil nutrient status in shortgrass steppe. *Ecology* **76**: 1116–1133.

Violle C, Navas M-L, Vile D, Kazakou E, Fortunel C, Hummel I, et al. (2007). Let the concept of trait be functional! *Oikos* **116**: 882–892.

Violle C, Enquist BJ, McGill BJ, Jiang L, Albert CH, Hulshof C, et al. (2012). The return of the variance: intraspecific variability in community ecology. *Trends in Ecology and Evolution* **27**: 244–252.

Wade DD (1989). *A guide for prescribed fire in southern forests*. Technical Publication R8-TP11. US Department of Agriculture Forest Service Southern Region.

Wade PR (2002). Bayesian population viability analysis. In SR Beissinger and DR McCullough (eds) *Population viability analysis*, pp. 213–238. University of Chicago Press, Chicago, IL.

Waite S (2000). *Statistical ecology in practice: a guide to analyzing environmental and ecological field data*. Prentice-Hall, Harlow.

Walker MD, Walkter DA, Welker JM, Arft AM, Bardsley T, Brooks PD, et al. (1999). Long-term experimental manipulation of winter snow regime and summer temperature in arctic and alpine tundra. *Hydrological Processes* **13**: 2315–2330.

Wallace LL (1987). Mycorrhizae in grasslands: interactions of ungulates, fungi and drought. *New Phytologist* **105**: 619–632.

Wallace LL and Svejcar T (1987). Mycorrhizal and clipping efects on *Andropogon gerardii* photosynthesis. *American Journal of Botany* **74**: 1138–1142.

Wallander H, Ekblad A, Godbold DL, Johnson D, Bahr A, Baldrian P, et al. (2013). Evaluation of methods to estimate production, biomass and turnover of ectomycorrhizal mycelium in forests soils: a review. *Soil Biology and Biochemistry* **57**: 1034–1047.

Wan S, Luo Y, and Wallace LL (2002). Changes in microclimate induced by experimental warming and clipping in tallgrass prairie. *Global Change Biology* **8**: 754–768.

Wang J, Kropff MJ, Lammert B, Christensen S, and Hansen PK (2003). Using CA models to obtain insight into the mechanism of plant population spread in a controllable system: annual weeds as an example. *Ecological Modelling* **166**: 277–286.

Ward E, Tahiri-Alaoui A, and Antoniw JF (1998). Applications of PCR in fungal-plant interactions. In PD Bridge, DK Arora, CA Reddy, *et al.* (eds) *Applications of PCR in mycology*, pp. 289–307. CAB International, Wallingford, Oxon.

Wardle DA, Nicholson KS, and Rahman A (1996). Use of a comparative approach to identify allelopathic potential and relationship between allelopathy bioassays and 'competition' experiments for ten grassland and plant species. *Journal of Chemical Ecology* **22**: 933–948.

Wardle DA, Barker GM, Bonner KI, and Nicholson KS (1998). Can comparative approaches based on plant ecophysiological traits predict the nature of biotic interactions and individual plant species effects in ecosystems? *Journal of Ecology* **86**: 405–421.

Warton DI and Hui FKC (2011). The arcsine is asinine: the analysis of proportions in ecology. *Ecology* **92**: 3–10.

Waser NM and Price MV (1985). Reciprocal transplant experiments with *Delphinium nelsonii* (Ranunculaceae): evidence for local adaptation. *American Journal of Botany* **72**: 1726–1732.

Watkinson AR (1990). The population dynamics of *Vulpia fasiculata*: a nine-year study. *Journal of Ecology* **78**: 196–209.

Watkinson AR (1997). Population dynamics. In MJ Crawley (ed.) *Plant ecology*, pp. 359–400. Blackwell Scientific Publications, Oxford.

Watkinson AR and Powell JC (1997). The life history and population structure of *Cycas armstrongii* in monsoonal northern Australia. *Oecologia* **111**: 341–349.

Watson MA and Casper BB (1984). Morphological constraints on patterns of carbon distribution in plants. *Annual Review of Ecology and Systematics* **15**: 233–258.

Way DA, Ladeau SL, McCarthy HR, Clark JS, Oren RAM, Finzi AC, et al. (2010). Greater seed production in elevated CO_2 is not accompanied by reduced seed quality in *Pinus taeda* L. *Global Change Biology* **16**: 1046–1056.

Weaver JE (1919). *The ecological relations of roots*. Carnegie Institute Publication 286. Carnegie Institute, Washington, DC.

Weaver JE (1920). *Root development in the grassland formation*. Carnegie Institute Publication 292. Carnegie Institute, Washington, DC.

Weaver JE and Clements FE (1938). *Plant ecology*, 2nd edn. McGraw-Hill, New York.

Weaver RW, Angle S, Bottomley P, Bezdicek D, Smith S, Tabatabai A, et al. (eds) (1994). *Methods of soil analysis: Part 2 Microbiological and biochemical properties*. Soil Science Society of America, Madison, WI.

Wedin D and Tilman D (1993). Competition among grasses along a nitrogen gradient: initial conditions and mechanisms of competition. *Ecological Monographs* **63**: 199–229.

Weigelt A and Jolliffe P (2003). Indices of plant competition. *Journal of Ecology* **91**: 707–720.

Weiher E, van der Werf A, Thompson K, Roderick M, Garnier E, and Eriksson O (1999). Challenging Theophrastus: a common core list of plant traits for functional ecology. *Journal of Vegetation Science* **10**: 609–620.

Weiner J (1984). Neighbourhood interference amongst *Pinus rigida* individuals. *Journal of Ecology* **72**: 183–195.

Weiner J and Freckleton RP (2010). Constant final yield. *Annual Review of Ecology and Systematics* **41**: 173–192.

Weiner J and Thomas SC (1992). Competition and allometry in three species of annual plants. *Ecology* **73**: 648–656.

Weiner J, Campbell LG, Pino J, and Echarte L (2009). The allometry of reproduction within plant populations. *Journal of Ecology* **97**: 1220–1233.

Weiss O and Reigosa MJ (2001). Modulated fluorescence. In MJ Reigosa (ed.) *Handbook of plant ecophysiology techniques*, pp. 173–183. Kluwer Academic Publishers, Dordrecht.

Weldon CW and Slauson WL (1986). The intensity of competition versus its importance: an overlooked distinction and some implications. *Quarterly Review of Biology* **61**: 23–44.

Welty RE and Barker RE (1993). Reaction of twenty cultivars of tall fescue to stem rust in controlled and field environments. *Crop Science* **33**: 963–967.

Welty RE, Milbrath GM, Faulkenberry D, Azevedo MD, Meek L, and Hall K (1986). Endophyte detection in tall fescue seed by staining and ELISA. *Seed Science and Technology* **14**: 105–116.

Werner EE (1998). Ecological experiments and a reserch program in community ecology. In WJ Resetarits and J Bernardo (eds) *Experimental ecology*, pp. 3–26. Oxford University Press, New York.

West HM, Fitter AH, and Watkinson AR (1993). The influence of three biocides on the fungal associates of the roots of *Vulpia ciliata* spp. *ambigua* under natural conditions. *Journal of Ecology* **81**: 345–350.

West NM, Gibson DJ, and Minchin PR (2009). Characterizing the microhabitats of exotics species in Illinois shale barrens. *Plant Ecology* **200**: 255–265.

West NM, Gibson DJ, and Minchin PR (2010). Microhabitat analysis of the invasive exotic liana *Lonicera japonica* Thunb. *Journal of the Torrey Botanical Society* **137**: 380–390.

Westerman RL (ed.) (1994). *Soil testing and plant analysis*. Soil Science Society of America, Madison, WI.

Wheater CP and Cook PA (2000). *Using statistics to understand the environment*. Routledge, London.

Wheelwright NT and Bruneau A (1992). Population sex ratios and spatial distribution of *Octoea tenera* (Lauraceae) trees in a tropical forest. *Journal of Ecology* **80**: 425–432.

White J (1979a). The plant as a metapopulation. *Annual Review of Ecology and Systematics* **10**: 109–145.

White PS (1979b). Pattern, process, and natural disturbance in vegetation. *Botanical Review* **45**: 229–299.

White J (1980). Demographic factors in populations of plants. In OT Solbrig (ed.) *Demography and evolution in plant populations*, pp. 21–65. University of California Press, Berkeley, CA.

White J (1985a). The census of plants in vegetation. In J White (ed.) *The population structure of vegetation*, pp. 33–88. Dr. W. Junk, Dordrecht.

White J (ed.) (1985b). *The population structure of vegetation*. Dr. W. Junk, Dordrecht.

White J (ed.) (1985c). *Studies on plant demography: a festschrift for John L. Harper*. Academic Press, London.

White JF and Harper JL (1970). Correlated changes in plant size and number in plant populations. *Journal of Ecology* **58**: 467–485.

White PS and Pickett STA (1985). Natural disturbance and patch dynamics: an introduction. In STA Pickett and PS White (eds) *The ecology of natural disturbance and patch dynamics*, pp. 3–16. Academic Press, Orlando, FL.

Whitlock MC (2011). Data archiving in ecology and evolution: best practices. *Trends in Ecology and Evolution* **26**: 61–65.

Whittaker JB (1982). The effect of grazing by a chrysomelid beetle, *Gastrophysa viridula*, on growth and survival of *Rumex crispus* on a shingle bank. *Journal of Ecology* **70**: 291–296.

Widden P (2001). The use of glycerin jelly for mounting stained roots for the observation and quantification of endomycorrhizal fungi. *Mycologia* **93**: 1026–1027.

Widén B and Widén M (1990). Pollen limitation and distance-dependent fecundity in females of the clonal gynodioecious herb *Glechoma hederacea* (Lamiaceae). *Oecologia* **83**: 191–196.

Wiegand TA and Moloney K (2004). Rings, circles, and null-models for point pattern analysis in ecology. *Oikos* **104**: 209–229.

Wierzbicka M and Panufnik D (1998). The adaptation of *Silene vulgaris* to growth on a calamine waste heap (S. Poland). *Environmental Pollution* **101**: 415–426.

Wijesinghe DK and Hutchings MJ (1997). The effects of spatial scale of environmental hetergeneity on the growth of a clonal plant: an experimental study with *Glechoma hederacea*. *Journal of Ecology* **85**: 17–28.

Wildi O (2012). Classification analysis. In *Oxford bibliographies online: ecology*, doi: 10.1093/OBO/9780199830060-0025

Williams K (1976). The failure of Pearson's goodness of fit statistic. *The Statistician* **25**: 49.

Williams PH and Hill CB (1986). Rapid-cycling populations of Brassica. *Science* **232**: 1385–1389.

Williams AC and McCarthy BC (2001). A new index of interspecific competition for replacement and additive designs. *Ecological Research* **16**: 29–40.

Williams MJ, Backman PA, Clark EM, and White JF (1984). Seed treatments for control of the tall fescue endophyte. *Plant Disease* **68**: 49–52.

Williams JL, Miller TEX, and Ellner SP (2012). Avoiding unintentional eviction from integral projection models. *Ecology* **93**: 2008–2014.

Williamson GB (1990). Allelopathy, Koch's postulates, and the neck riddle. In JB Grace and D Tilman (eds) *Perspectives on plant competition*, pp. 143–162. Academic Press, San Diego, CA.

Willis F, Moat J, and Paton A (2003). Defining a role for herbarium data in Red List assessments: a case study of *Plectranthus* from eastern and southern tropical Africa. *Biodiversity and Conservation* **12**: 1537–1552.

Willis CG, Brock MT, and Weinig C (2010). Genetic variation in tolerance of competition and neighbour suppression in *Arabidopsis thaliana*. *Journal of Evolutionary Biology* **23**: 1412–1424.

Willson MF (1992). The ecology of seed dispersal. In M Fenner (ed.) *Seeds: the ecology of regeneration in plant communities*, pp. 61–85. CAB International, Wallingford, Oxon.

Wilson JB (1988). Shoot competition and root competition. *Journal of Applied Ecology* **25**: 279–296.

Wilson SD (1993). Belowground competition in forest and prairie. *Oikos* **68**: 146–150.

Wilson GB and Bell JNB (1990). Studies on the tolerance to sulphur dioxide of grass populations in polluted ares. VI. The genetic nature of tolerance in *Lolium perenne* L. *New Phytologist* **116**: 313–317.

Wilson GWT and Hartnett DC (1997). Effects of mycorrhizae on plant productivity and species abundances in tallgrass prairie microcosms. *American Journal of Botany* **84**: 478–482.

Wilson GWT and Williamson MM (2008). Topsin-M: the new benomyl for mycorrhizal-suppression experiments. *Mycologia* **100**: 548–554.

Wilson GWT, Daniels Hetrick BA, and Gerschefske Kitt D (1988). Suppression of mycorrhizal growth response of big bluestem by non-sterile soil. *Mycologia* **80**: 338–343.

Winn AA (1989). Using radionuclide labels to determine the post-dispersal fate of seeds. *Trends in Ecology and Evolution* **4**: 1–2.

Wipf S and Rixen C (2010). A review of snow manipulation experiments in arctic and alpine tundra ecosystems. *Polar Research* **29**: 95–109.

de Wit CT (1960). On competition. *Verslagen Can Landouskundige Onderzoekingen* **66**: 1–82.

Woodward FI (1997). Life at the edge: a 14-year study of a *Verbena officinalis* population's interactions with climate. *Journal of Ecology* **85**: 899–906.

Wright S (1978). *Evolution and genetics of populations. Volume 4. Variability within and among natural populations.* University of Chicago Press, Chicago, IL.

Wróbel A and Zwolak R (2013). The choice of seed tracking method influenced fate of beech seeds dispersed by rodents. *Plant Ecology* **214**: 471–475.

Wulff A, Hänninen O, Tuomainen A, and Kärenlampi L (1992). A method for open-air exposure of plants to ozone. *Annales Botanici Fennici* **29**: 253–262.

Wurst S, Vender V, and Rillig M (2010). Testing for allelopathic effects in plant competition: does activated carbon disrupt plant symbioses? *Plant Ecology* **211**: 19–26.

Wyszomirski T and Weiner JA (2009). Variation in local density results in a positive correlation between plant neighbor sizes. *American Naturalist* **173**: 705–708.

Xiao Z, Jansen PA, and Zhang Z (2006). Using seed-tagging methods for assessing post-dispersal seed fate in rodent-dispersed trees. *Forest Ecology and Management* **223**: 18–23.

Yamasaki S and Dillenburg LR (1999). Measurements of leaf relative water content in *Araucaria angustifolia*. *Revista Brasileira de Fisiologia Vegetal* **11**: 69–75.

Yates CJ and Ladd PG (2010). Using population viability analysis to predict the effect of fire on the extinction risk of an endangered shrub *Verticordia fimbrilepis* subsp. *fimbrilepis* in a fragmented landscape. *Plant Ecology* **211**: 305–319.

Yokozawa M, Kubota Y, and Hara T (1999). Effects of competition mode on the spatial pattern dynamics of wave regeneration in subalpine tree stands. *Ecological Modelling* **118**: 73–86.

You C and Petty WH (1991). Effects of Hurricane Hugo on *Manilkara bidentata*, a primary tree species in the Luquillo Experimental Forest of Puerto Rico. *Biotropica* **23**: 400–406.

Young TR (1984). The comparative demography of semelparous *Lobelia telekii* and iteroparous *Lobelia keniensis* on Mount Kenya. *Journal of Ecology* **72**: 637–650.

Young ND (1994). Constructing a plant genetic linkage map with DNA markers. In RL Phillips and IK Vasil (eds) *DNA-based markers in plants*, pp. 39–57. Kluwer Academic Publishers, Dordrecht.

Young LJ and Young JH (1998). *Statistical ecology*. Kluwer Academic Publishers, Boston, MA.

Yu ZX, Zhang Q, Yang HS, Tang JJ, Weiner J, and Chen X (2012). The effects of salt stress and arbuscular mycorrhiza on plant neighbour effects and self-thinning. *Basic and Applied Ecology* **13**: 673–680.

Yuan ZY and Chen HYH (2012). Indirect methods produce higher estimates of fine root production and turnover rates than direct methods. *PLoS ONE* **11**: e48989. doi:10.1371/journal.pone.0048989.

Yuan ZY and Chen HYH (2013). Simplifying the decision matrix for estimating fine root production by the sequential soil coring approach. *Acta Oecologica* **48**: 54–61.

Yuan T, Maun MA, and Hopkins WG (1993). Effects of sand accretion on photosynthesis, leaf-water potential and morphology of two dune grasses. *Functional Ecology* **7**: 676–682.

Zancarini A, Mougel C, Terrat S, Salon C, and Munier-Jolain N (2013). Combining ecophysiological and microbial ecological approaches to study the relationship between *Medicago truncatula* genotypes and their associated rhizosphere bacterial communities. *Plant and Soil* **365**: 183–199.

Zane L, Bargelloni L, and Patarnello T (2002). Strategies for microsatellite isolation: a review. *Molecular Ecology* **11**: 1–16.

Zar JH (2010). *Biostatistical analysis*, 5th edn. Prentice-Hall, Englewood Cliffs, NJ.

Zhang RM, Zuo ZJ, Gao PJ, Hou P, Wen GS, and Gao Y (2012). Allelopathic effects of VOCs of *Artemisia frigida* Willd. on the regeneration of pasture grasses in Inner Mongolia. *Journal of Arid Environments* **87**: 212–218.

Zielinski WJ and Stauffer HB (1996). Monitoring mertes populations in California: survey design and power analysis. *Ecological Applications* **6**: 1254–1267.

Zipperlen SW and Press MC (1997). Photosynthetic induction and stomatal oscillations in relation to light environment of two dipterocarp rain forest species. *Journal of Ecology* **85**: 491–504.

Zuidema PA, Yamada T, During HJ, Itoh A, Yamakura T, Ohkubo T, et al. (2010). Recruitment subsidies support tree subpopulations in non-preferred tropical forest habitats. *Journal of Ecology* **98**: 636–644.

Zvereva EL and Kozlov MV (2001). Effects of pollution-induced habitat disturbance on willow response to defoliation. *Journal of Ecology* **89**: 21–30.

Zvereva EL and Kozlov MV (2012). Sources of variation in plant responses to belowground insect herbivory: a meta-analysis. *Oecologia* **169**: 441–452.

Animal Index

Plant, Bacteria, Algae, and Fungi Index

General Index